THINKING Physics IS GEDANKEN PHYSICS

THIRD EDITION

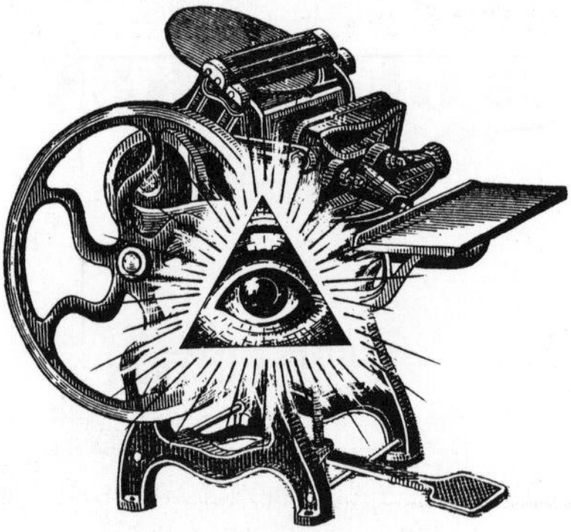

THINKING PHYSICS IS GEDANKEN PHYSICS

THIRD EDITION

Lewis Carroll Epstein

City College of San Francisco

INSIGHT PRESS

614 Vermont St., San Francisco, CA 94107

By
Lewis Carroll Epstein
Debra L. Bridges, Esq.
Insight Press

160,000 copies in print

Insight Press
614 Vermont @ 18th Street
San Francisco, California
94107-2636

DEDICATION

Most people study physics to satisfy some school requirement. A lesser number study physics to learn the tricks of Nature, so they can find out how to make things bigger or smaller or faster or stronger or more sensitive. But a tiny few, a rare tiny few, study physics because they wonder--not how things work, but why they work. They wonder what is at the bottom of things--the very bottom, if there is a bottom.

Thinking Physics is dedicated to those who wonder why. Among them was my brother, Robert Jared Epstein, Bobby.

HOW TO USE

The best way to use this book is NOT to simply read it or study it, but to read a question and STOP. Even close the book. Even put it away and THINK about the question. Only after you have formed a reasoned opinion should you read the solution. Why torture yourself thinking? Why jog? Why do push-ups?

If you are given a hammer with which to drive nails at the age of three you may think to yourself, "OK, nice." But if you are given a hard rock with which to drive nails at the age of three, and at the age of four you are given a hammer, you think to yourself, "What a marvelous invention!" You see, you can't really appreciate the solution until you first appreciate the problem.

What are the problems of physics? How to calculate things? Yes—but much more. The most important problem in physics is *perception,* how to conjure mental images, how to separate the non-essentials from the essentials and get to the heart of a problem, HOW TO ASK YOURSELF QUESTIONS. Very often these questions have little to do with calculations and have simple yes or no answers: Does a heavy object dropped at the same time and from the same height as a light object strike the earth first? Does the observed speed of a moving object depend on the observer's speed? Does a particle exist or not? Does a fringe pattern exist or not? These qualitative questions are the most vital questions in physics.

THIS BOOK

You must guard against letting the quantitative superstructure of physics obscure its qualitative foundation. It has been said by more than one wise old physicist that you really understand a problem when you can intuitively guess the answer *before* you do the calculation. How can you do that? By developing your physical intuition. How can you do THAT? The same way you develop your physical body—by exercising it.

Let this book, then, be your guide to mental pushups. Think carefully about the questions and their answers *before* you read the answers offered by the author. **You will find many answers don't turn out as you first expect. Does this mean you have no sense for physics? Not at all. Most questions were deliberately chosen to illustrate those aspects of physics which seem contrary to casual surmise. Revising ideas, even in the privacy of your own mind, is not painless work.** But in doing so you will revisit some of the problems that haunted the minds of Archimedes, Galileo, Newton, Maxwell, and Einstein.* The physics you cover here in hours took them centuries to master. Your hours of thinking will be a rewarding experience. Enjoy!

Lewis Epstein

*__Gedanken__ Physics was Einstein's expression for Thinking Physics.

CONTENTS

Physics is what physicists do late at night.
— Joe Tenn

Mechanics

Mechanics began with the energy crisis—which began with the beginning of civilization. To make a machine that would put out more work than went into it is an ancient dream. Is it an unreasonable dream? After all, a lever puts out more force at one end than is applied to the other end. But does it put out more work? Does it put out more motion? If the lever fails, might some other scheme lead to the ultimate goal, perpetual motion? It may be said the (unsuccessful) quest to make gold launched chemistry and the (unsuccessful) quest of astrology launched astronomy. The (unsuccessful) quest for perpetual motion launched mechanics.

You may have noticed that the biggest section of this book (as well as many other physics books) is the MECHANICS section. Why is mechanics so important? Because it is the goal of physics to reduce every other subject in physics to mechanics. Why? Because we understand mechanics best. Once heat was thought to be some sort of a substance; later it was found to be just mechanics. Heat could be understood as little balls called molecules bouncing about in space or connected to each other by springs and vibrating back and forth. Sound has similarly been reduced to mechanics. Much effort has been spent trying to reduce light to mechanics.

Mechanics has two parts—the easier part, **statics,** where all forces balance out to zero so nothing much happens, and the dramatic part, **dynamics,** where all the forces do not cancel each other, leaving a net force that makes things happen. How much happens depends on how long the force acts. But "long" is ambiguous. Does it mean long distance or long duration? The simple but subtle distinction between a force acting so many feet and a force acting so many seconds is the magic key to understanding dynamics.

You'll also notice that a good deal of attention is devoted to situations involving collisions (Splat, Gush, Smush, and so on). Granted collisions are interesting in their own right, but are they all that important? Many physicists believe they are. Why? Because if all the world is to be explained mechanically in terms of little balls (molecules, electrons, photons, gravitons, etc.), then the only way one ball affects another ball is if the little balls hit. If that is so, collision becomes the essence of physical interaction.

Now it may be the goal of physics to reduce every subject to mechanics and to reduce mechanics to collisions, but certainly that goal has not been and might never be reached. Nonetheless, if you are to understand physics, you must first understand mechanics. Perhaps even love mechanics.

In learning the sciences
examples are of more use than precepts.
— Sir Isaac Newton

VISUALIZE IT

Suppose you are going for a long bicycle ride. You ride one hour at five miles per hour. Then three hours at four miles per hour and then two hours at seven miles per hour. How many miles did you ride?

a) five
b) twelve
c) fourteen
d) thirty-one
e) thirty-six

ANSWER: VISUALIZE IT

The answer is: d. Remember speed multiplied by time is distance. But what is the speed? It changes during the ride. So split the trip up into segments. One hour at five mph gives five miles. Three hours at four mph gives twelve miles and two hours at seven mph gives fourteen miles. Then add the segments. Five plus twelve plus fourteen sum to thirty-one. So that is the answer, that's it.

Yes, that is the answer, but that is not it. That is just some arithmetic. Arithmetic is blind. Can you visualize what you are doing? To visualize, use geometry. Geometry has eyes.

Make a graph showing the history of the ride. For one hour it is at five mph. Then it drops to four mph and stays there for three hours. Then it jumps up to seven mph for two hours and finally it drops to zero which means the bike stops.

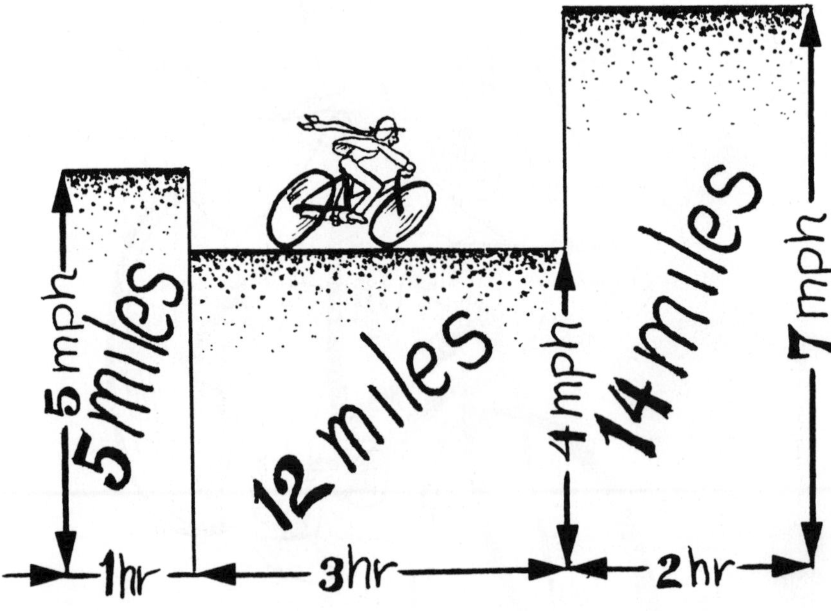

Now split the graph into three rectangles. Each rectangle represents one segment of the trip. The first rectangle is 5 mph high and 1 hour wide. What is the area of this rectangle? Multiply its height by its width—that is, multiply 5 mph by 1 hour—and you get 5 miles. The area of the rectangle is the distance covered during the first segment of the ride. Likewise the area of the second rectangle is 4 mph multiplied by 3 hours which is 12 miles. So the area of each rectangle is equal to the distance traveled on that segment of the ride.

That gives you a nice way to visualize distance traveled. Imagine a recording speedometer that gives you a graph of your speed plotted against time. The total area under the speed curve must tell how far you have traveled.

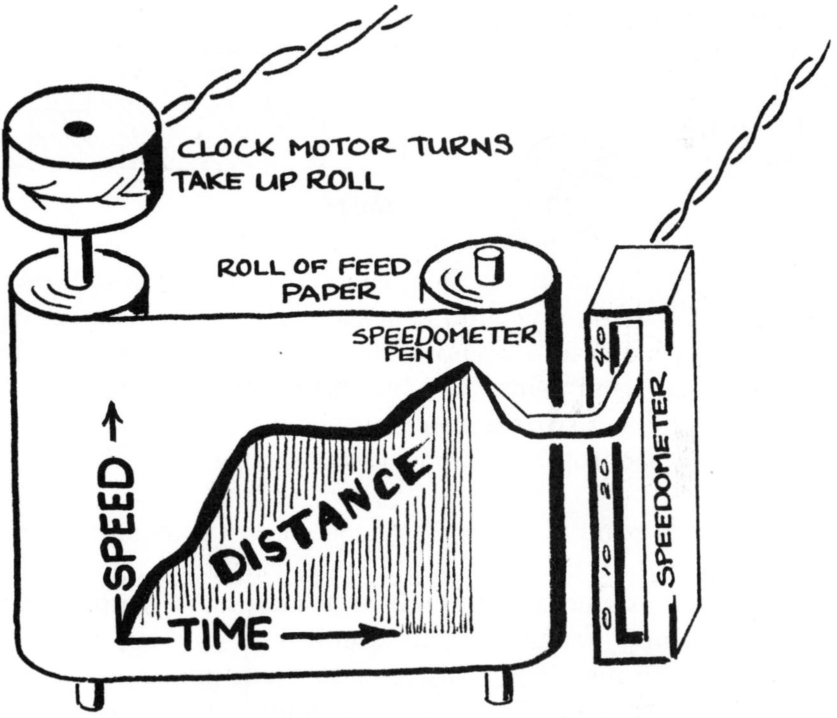

RECORDING SPEEDOMETER

INTEGRAL CALCULUS

Look at the speed graph and then answer these questions.

Two hours into the trip how fast was the thing going?

a) zero mph
b) 10 mph
c) 20 mph
d) 30 mph
e) 40 mph.

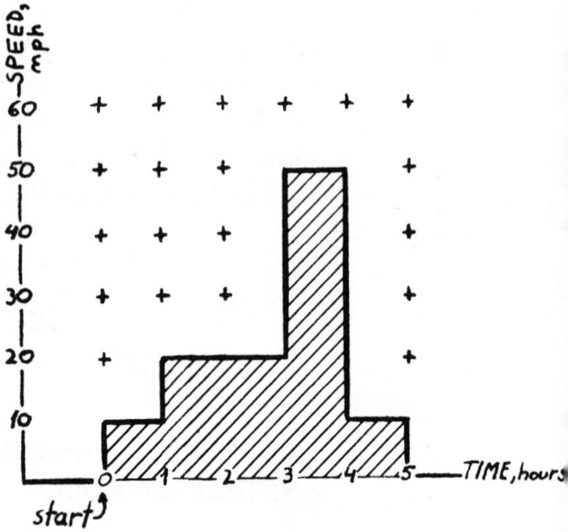

How far did the thing go during the whole trip?

a) 40 miles
b) 80 miles
c) 110 miles
d) 120 miles
e) 210 miles.

ANSWER: CALCULUS

The answer to the first question is c. Above the 2-hour mark the speed graph reads 20 mph.

The answer to the second question is also c. The area under the speed line (or curve) is divided into little squares. Each square is 1 hour wide and 10 mph high. That means the area of each little square is 10 miles. Now count the little squares under the curve. There are eleven in all. Eleven times ten miles equals 110 miles. So the total area under the graph is 110 miles and that's how far the thing went during the whole trip. How can the area of a square represent miles? Must it not represent square miles? The area of a square represents square miles if its width and height are measured in miles, but if its width is measured in hours and its height in miles per hour, and you have imagination, its area is miles.

The technique you used here for finding the distance traveled is the technique of Integral Calculus. Integral means to integrate or sum many small parts. Calculus refers to many very small parts or layers which build up to make the sum. The name comes from minerals which build up in layers such as the calculus which builds up on your teeth. As the dentist scrapes it it flakes off. Each flake is a layer.

DRAGSTER

A dragster starts from rest and accelerates to 60 mph in 10 seconds. How far does it travel during those 10 seconds?

a) 1/60 mile
b) 1/12 mile
c) 1/10 mile
d) 1/2 mile
e) 60 miles.

ANSWER: DRAGSTER

The answer is: b. First let's get everything in hours. Ten seconds is 1/6 of a minute and 1 minute is 1/60 of an hour, so 10 seconds is 1/360 of an hour.

The area under the speed line is a triangle and the area of a triangle is one-half of its height multiplied by its length. The triangle area is one-half the height multiplied by the length because the area of the triangle is half the area of the rectangle and the area of the rectangle is height mutiplied by length.

The height of the triangle is 60 mph and its length is 1/360 part of an hour so total distance moved must be $(\frac{1}{2}) \times (60 \text{ mph}) \times (1 \text{ hr}/360) = 1/12$ of a mile.

You can, if you want, think of the triangle as made up from a bunch of constant-speed steps. Each step is one layer of calculus.

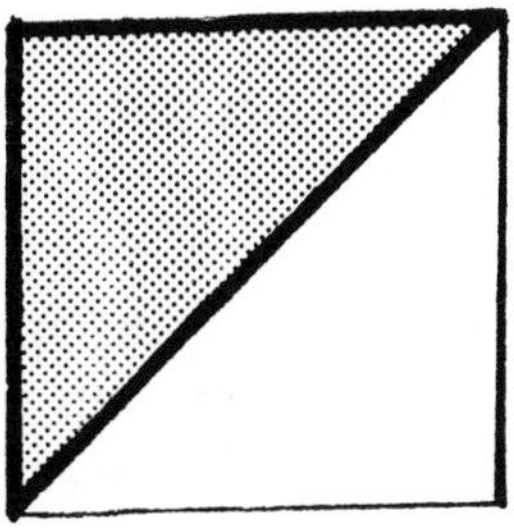

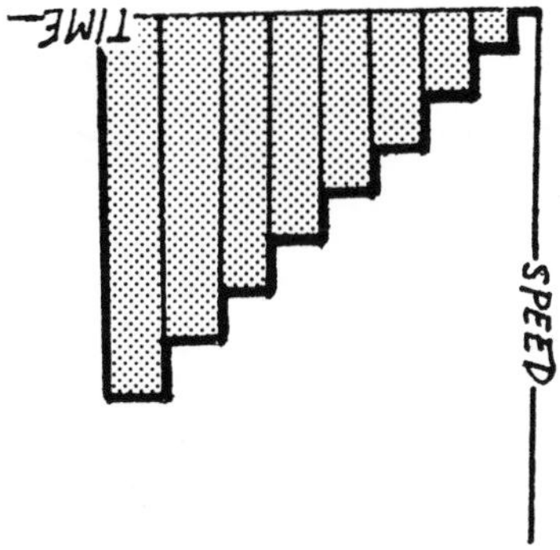

NO SPEEDOMETER

The next dragster is so stripped down that it does not even have a speedometer. At maximum acceleration from rest it goes 1/10 of a mile in 10 seconds. What speed did it get up to in those ten seconds?

a) 6 mph, b) 52 mph, c) 60 mph,
d) 62 mph, e) 72 mph.

ANSWER: NO SPEEDOMETER

The answer is: e. This is almost a rerun of DRAGSTER. The rule was: $(\frac{1}{2}) \times$ (maximum speed) $\times$ (time) = (distance).

So in this case $(\frac{1}{2}) \times (?\ \text{mph}) \times (1\ \text{hr}/360) = 1\ \text{mile}/10$.

Divide both sides of this equation by 1 hr/360. Remember 1 hr/360 divided by 1 hr/360 equals one, and 1 mile/10 divided by 1 hr/360 equals 36 mph, so: $(1/2)(?\ \text{mph}) = 36\ \text{mph}$.

Finally: $?\ \text{mph} = (2) \times (36\ \text{mph}) = 72$ miles per hour.

NOT FAR

Look at this speed graph and tell how far away from the starting point this thing ended up.

a) It is impossible to tell because the graph has no numerical scale on it.
b) It ended up at the starting point.
c) It did not end up at the starting point but where it ended can't be told.

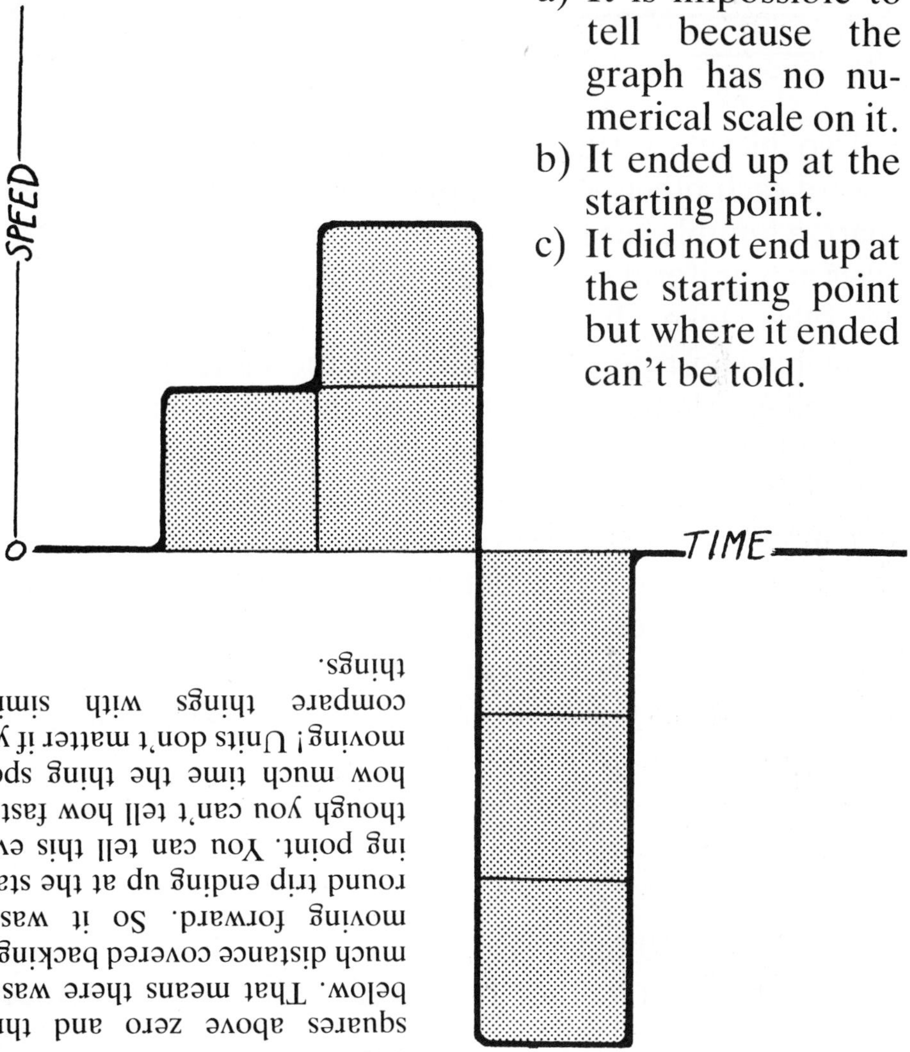

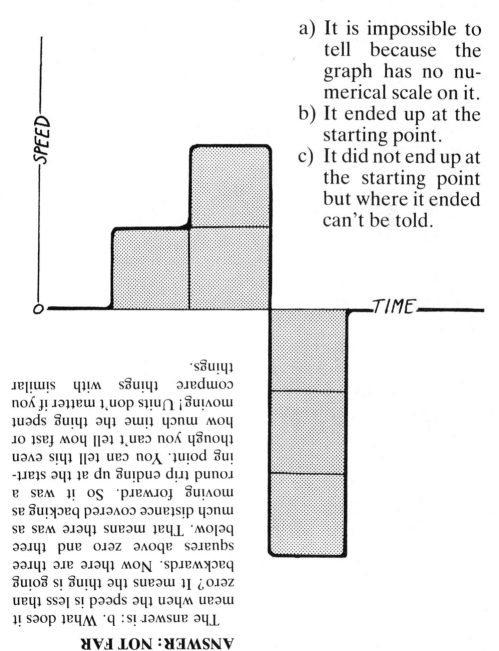

ANSWER: NOT FAR

The answer is: b. What does it mean when the speed is less than zero? It means the thing is going backwards. Now there are three squares above zero and three below. That means there was as much distance covered backing as moving forward. So it was a round trip ending up at the starting point. You can tell this even though you can't tell how fast or how much time the thing spent moving! Units don't matter if you compare things with similar things.

THE BIKES AND THE BEE

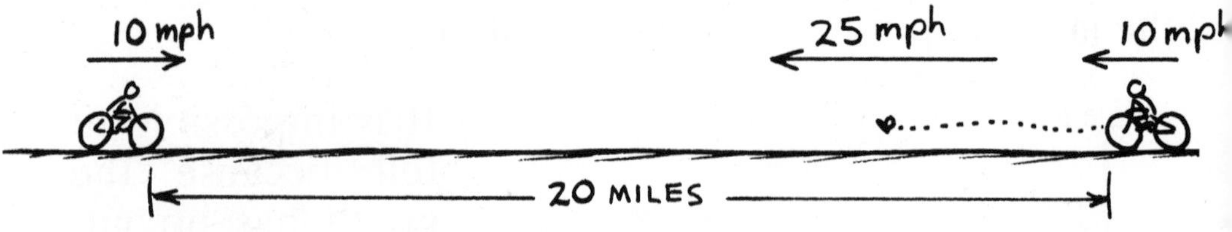

Two bicyclists travel at a uniform speed of 10 mph toward each other. At the moment when they are 20 miles apart, a bumble bee flies from the front wheel of one of the bikes at a uniform speed of 25 mph directly to the wheel of the other bike. It touches it and turns around in a negligibly short time and returns at the same speed to the first bike, whereupon it touches the wheel and instantaneously turns around and repeats the back-and-forth trip over and over again — successive trips becoming shorter and shorter until the bikes collide and squash the unfortunate bee between the front wheels. What was the total mileage of the bee in its many back-and-forth trips from the time the bikes were 20 miles apart until its hapless end? (This can be very simple or very difficult, depending on your approach.)

a) 20 miles
b) 25 miles
c) 50 miles
d) More than 50 miles
e) This problem cannot be solved with the information given

ANSWER: THE BIKES AND THE BEE

The answer is: b. The total mileage of the bee was 25 miles. The simplest approach to this solution is to consider the time involved. It will take the bicyclists an hour to meet, since each travels 10 miles at a speed of 10 mph — so the bee makes its many back-and-forth trips in an hour also. Since its speed is 25 mph, it travels a total distance of 25 miles. Again, time is an important consideration in velocity problems!

SAM

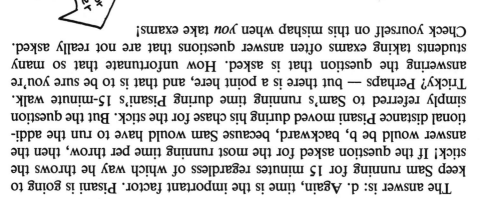

Dr. Pisani exercises his dog Sam on a 15-minute walk by throwing a stick that Sam chases and retrieves. To keep Sam running for the longest time as Dr. Pisani walks, which way should he throw the stick?

a) In front of him
b) In back of him
c) Sideways
d) In any direction, as all are equivalent

ANSWER: SAM

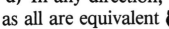

The answer is: d. Again, time is the important factor. Pisani is going to keep Sam running for 15 minutes regardless of which way he throws the stick! If the question asked for the most running time per throw, then the answer would be b, backward, because Sam would have to run the additional distance Pisani moved during his chase for the stick. But the question simply referred to Sam's running time during Pisani's 15-minute walk. Tricky? Perhaps — but there is a point here, and that is to be sure you're answering the question that is asked. How unfortunate that so many students taking exams often answer questions that are not really asked. Check yourself on this mishap when *you* take exams!

SPEED ON SPEED

The St. Charles Street streetcar is approaching Canal Street at 144 inches per second. A person in the streetcar is facing forward and walking forward in the car with a speed of 36 inches per second relative to the seats and things in the car. The person is also eating a Poor Boy which is entering his mouth at 2 inches per second (he eats fast). An ant on the Poor Boy is running to the end of the Poor Boy, away from the person's mouth. The distance between the ant and the end of the Poor Boy towards which it is running is closing at 1 inch per second. Now the question is: How fast is the ant approaching Canal Street?

a) Zero in/s
b) 100 in/s
c) 170 in/s
d) 179 in/s
e) 180 in/s

Could you convert your above answer from inches per second to miles per hour? (You don't need to do the actual arithmetical calculation.)

a) Yes b) No

If you answered "Yes" to the above, ask yourself how it is possible to say *anything* about how far the ant goes in an *hour* since the poor ant is eaten and dead in a matter of seconds.

ANSWER: SPEED ON SPEED

The answer to the first question is: d. You can picture it like this. Add the speed of the street car to the speed of the person — both going toward Canal Street. Subtract the speed of the Poor Boy sandwich (which is going the other way). Then add the speed of the ant (which is also going toward Canal Street).

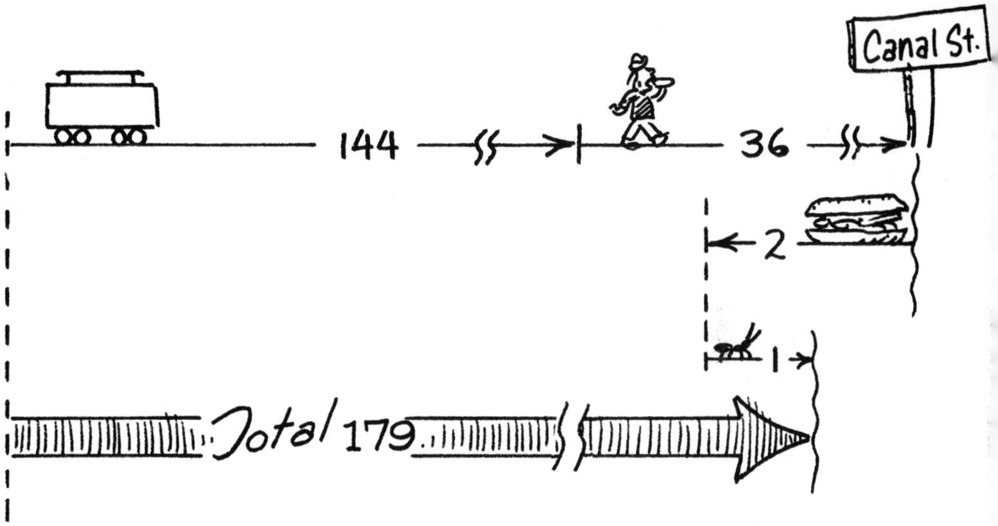

This same idea can be used to add speeds which are not back and forth in the same direction. For example, if the person walks at an angle inside the street car (forget the Poor Boy & ant).

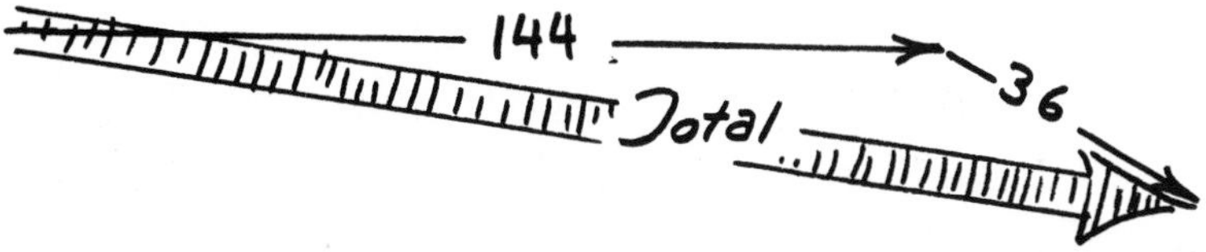

The answer to the second question is: a. You must remember, however, that when you speak of miles per hour you are making a CONDITIONAL statement. You are not saying how far a thing *will* go, but how far it *would* go IF it could go for one hour.

SPEED AIN'T ACCELERATION

As the ball rolls down this hill

a) its speed increases and acceleration decreases
b) its speed decreases and acceleration increases
c) both increase
d) both remain constant
e) both decrease

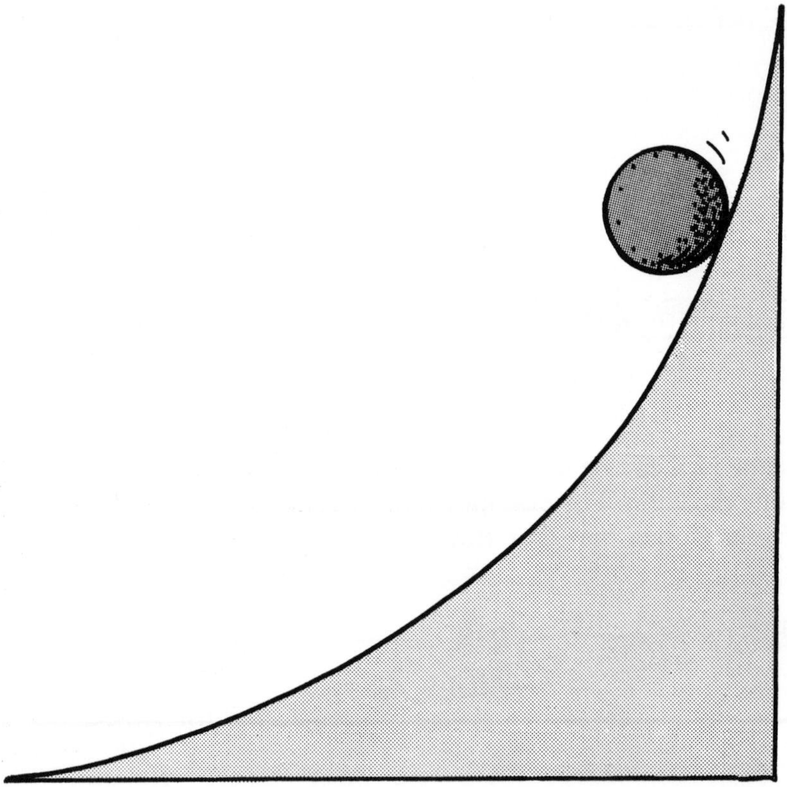

ANSWER: SPEED AIN'T ACCELERATION

The answer is: a. The speed of the ball increases as it goes down the hill, but its acceleration depends on how steep the hill is. At the top of the hill the acceleration is greatest because the hill is steepest, and as the ball goes down, the hill is less steep so the ball's acceleration decreases. So acceleration can decrease at the same time speed is increasing. Store this example in your memory and recall it if you forget the difference between speed and acceleration.

In the sketch we show the acceleration of the ball *a* parallel to the surface as a component of *g*, the acceleration of free fall — or the acceleration it would experience on a vertical "slope." The steeper the slope, the more that *a* approaches *g*. Or we could say, the less steep the slope, the more *a* approaches zero — which is the acceleration the ball would experience on a level surface. We will return to vector components later.

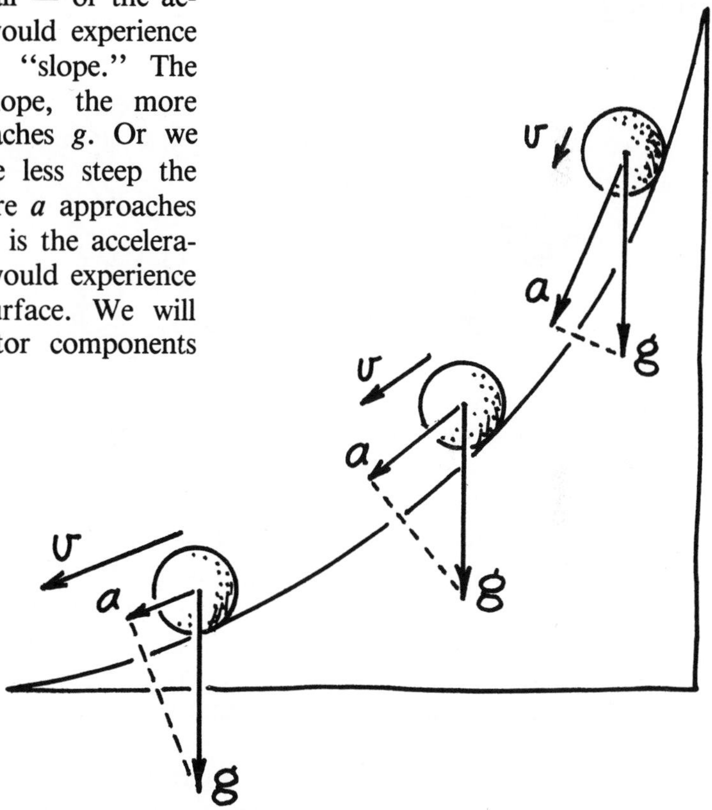

Strictly speaking, our description of the acceleration is not complete. Since the ball is moving in a curved path rather than a straight-line path, the curved motion involves another effect that we will treat later.

ACCELERATION AT THE TOP

A stone is thrown straight upward and at the tippity top of its path its velocity is momentarily zero. What is its acceleration at this point?

a) Zero
b) 32 ft/s^2
c) Greater than zero, but less than 32 ft/s^2

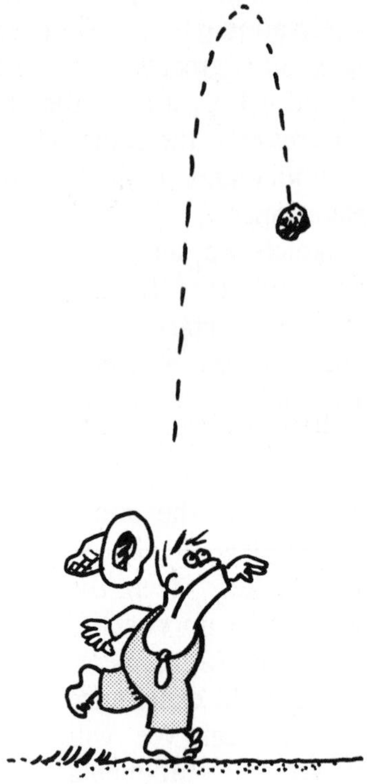

ANSWER: ACCELERATION AT THE TOP

The answer is: b. Although its velocity is instantaneously zero, it is still undergoing a RATE of *change* of velocity. This is evident if you consider its motion a moment before or after, in which case the stone is moving. For example, a second before or after reaching the top, its velocity is 32 ft/s. So it is undergoing a rate of change as it passes through the zero value of velocity just as it is undergoing the same rate of change passing through any other value of velocity. If air resistance is negligible, this rate of change of velocity is 32 ft/s^2.

Or from another point of view — Newton's 2nd Law states that an impressed net force on any object will accelerate that object. Gravitation acts on the stone at all points in its path, producing a constant acceleration at all points in its path — which includes the tippity top! After all, if the ball were momentarily stationary at the top and NOT accelerating, then it would remain stationary, which means it would not fall — ever.

TIME REVERSAL

A motion picture film is made of a falling object which shows the object accelerating downwards. Now if the film is run backwards, it will show the object accelerating

a) upward
b) still downward

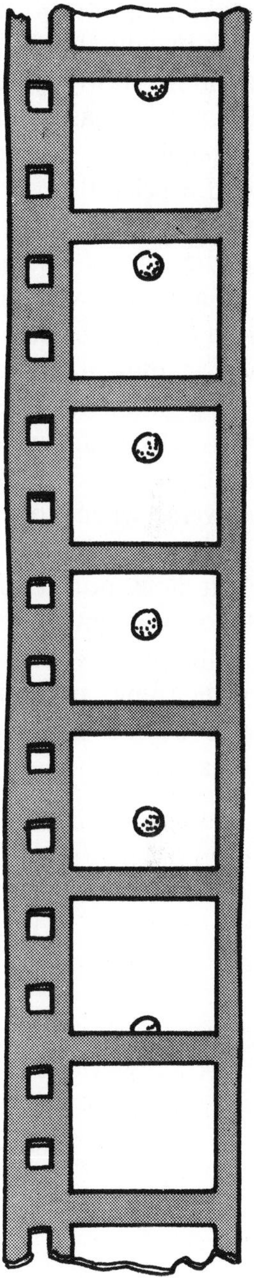

ANSWER: TIME REVERSAL

Surprise! The answer is: b. If the film is run backwards it will show the object moving up, but its acceleration will still be downward. Run the film backwards in your head. The ball is seen moving upwards initially fast, and then more slowly, just as if you threw it upward. Clearly the upward motion is not increasing so it could not be accelerating upwards. But the speed is changing so it is accelerating, and a decrease of upward speed is a downward acceleration.

What this goes to show is something about rates of change. If time is reversed the rate of change of anything will reverse, that is the rate of change of anything will decrease if it was previously increasing. But if time is reversed the rate of change of the rate of anything will not reverse. Acceleration is the rate of change of velocity and velocity is the rate of change of position so acceleration is a rate of change of a rate of change and that is why it did not reverse.

What about a rate of change of a rate of change of a rate of change — does it reverse when time reverses? Yes, it does reverse. What about the rate of change of the rate of change of the rate of change of the rate of change? It does not reverse when time reverses. Incidentally, there are symbols to represent these things which save a lot of words.

If X is the position of a thing then $\dot{X}$ is the rate of change of position or speed of the thing, and $\ddot{X}$ is the acceleration of the thing, and $\dddot{X}$ is the rate of change of the acceleration which is called the jerk, and $\ddddot{X}$ is the rate of change of the jerk which perhaps you can think up a good name for.

SCALAR

The battery output voltage, the bottle volume, the clock time, and the measure of weight all have something in common. It is that they are represented by

a) one number
b) more than one number.

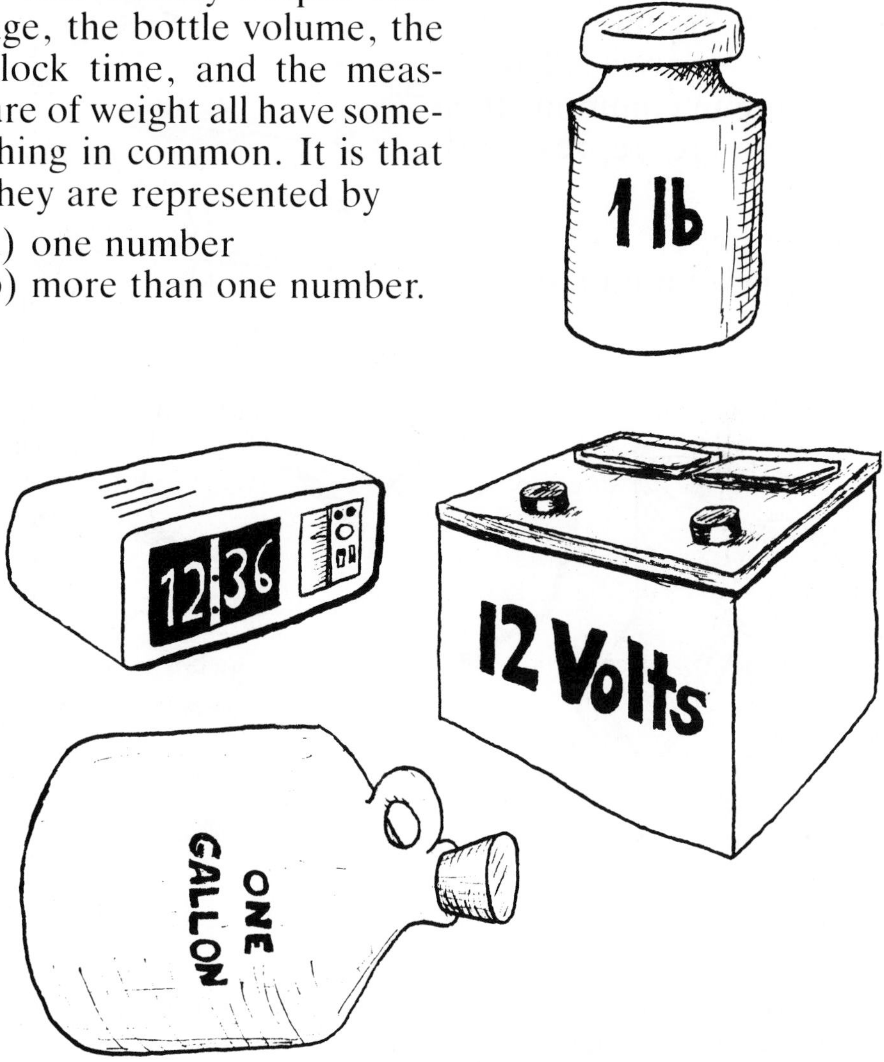

ANSWER: SCALAR

The answer is: a. Each is represented by only one number—the battery by 12 volts, the bottle by one gallon, the time by 12:36 and the weight by one pound. Things described by one number are called scalars. For example: on a scale of one to ten, how do you rate this teacher?

VECTOR

The pair of Levis, the bolt, the street intersection, and Ms. America all have something in common. It is that they are represented by

a) one number
b) a string of numbers.

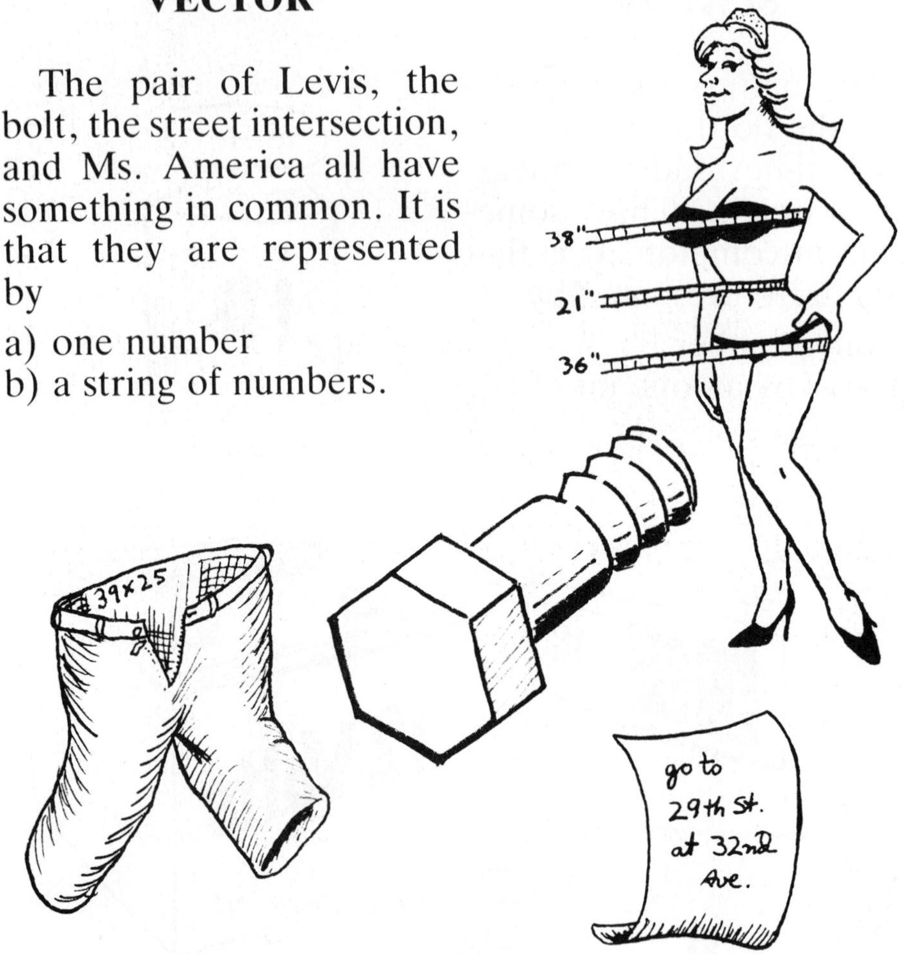

ANSWER: VECTOR

The answer is b. The Levis are designated by a length and a waist measurement; the bolt by length, diameter, and threads per inch; the street intersection by two streets; and Ms. America by at least three numbers, like 38, 21, 36. Things that must be described by more than one number are called vectors.

The name vector means carrier. In medicine a mosquito is a malaria vector. In physics the vector measures where a thing is carried. Suppose a thing is carried three feet forward, five feet to the right, and six feet up. Its displacement vector would be (3, 5, 6). Vectors can be pictured as arrows connecting the place a thing is carried from to the place it is carried to.

Forces and velocities are also vectors since they can be described as arrows. The direction of the arrow shows the direction of the force or velocity and its length shows the strength of the force or speed of the motion. But note that speed is not a vector. Why? Because it is described by one number, like seven miles per hour. Speed is a scalar. Speed does not give direction of motion. Velocity does.

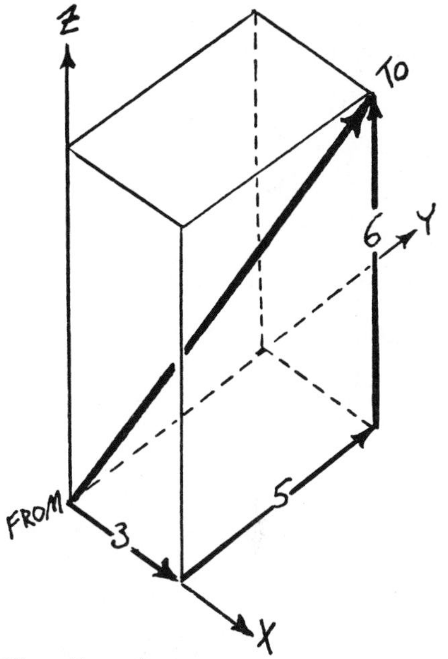

The directions forward, right, and up are sometimes called X, Y, and Z. The vector illustrated here is (X,Y,Z) = (3,5,6)

Suppose you are not going to be a physicist. Why should you be concerned about the difference between scalars and vectors? Because many people who are not physicists, for example bureaucrats and businesspeople, are asked to classify things, categorize things, or set up measuring schemes. These people frequently try to put things on a one-to-ten or A-to-F scalar scale without first stopping to think about what they are classifying. Sometimes doing this really messes up what they are trying to do.

For example, the popular measure of intelligence is related as one number called IQ. That implies that intelligence is a scalar. But is intelligence really a scalar? Some people have good memories but can't reason. Some people learn quickly and forget quickly (crammers!). Intelligence depends on many things like ability to learn, ability to remember, ability to reason, etc. So intelligence is a vector, not a scalar. That is a vital difference and the failure to recognize it has hurt thousands of people. So physicist or not, you had better get the idea of vector and scalar straight in your head.

TENSOR

Suppose you have a little rubber square. If the square was simply moved from place A to B that change could be easily expressed as a vector. But suppose the little square is not moved. Suppose it is left at A. At A the little rubber square is pulled into the shape of a parallelogram. That change from square to parallelogram can be easily represented as a

a) scalar
b) vector
c) neither

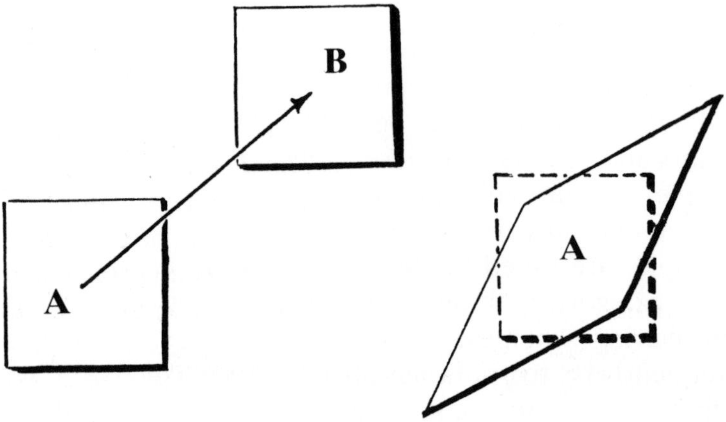

ANSWER: TENSOR

The answer is: c. There are things which cannot easily be represented as scalars or vectors—for example, the parallelogram. But parallelograms can be represented as things called TENSORS. The name comes from the tension that deformed the square into the parallelogram shape. A tensor is a super vector. A regular vector is made up from a string of scalars like (3, 5, 6,). The three, the five, and the six are by themselves each just scalars. A super vector is also made up from a string of things, but the things in the string are not scalars; each thing in the string is itself a vector. That is why I call a tensor a super vector. So a tensor must look like this: (→, ↗, →), a string of vectors.

How many vectors does it take to represent a parallelogram? Well, how many sides does a parallelogram have? Four. But you only need to know two because opposite sides of a parallelogram are parallel—that's why a parallelogram is called a parallelogram. So a parallelogram can be represented by a tensor, which is made of two vectors. For example a square (which is a parallelogram) is represented by this tensor: (↑ , →). When the square is pulled out of the square shape it is represented by this tensor (↗ , ↗).

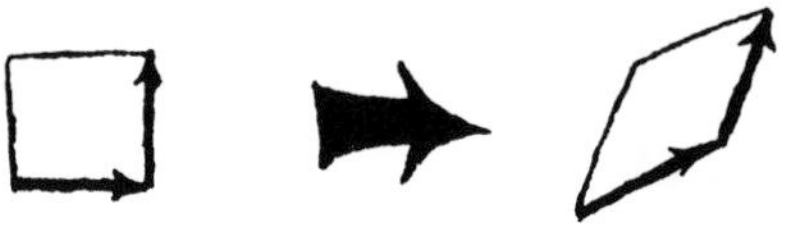

Now each vector can be represented by a string of scalar numbers and that string can be written in a column rather than a row

$$\begin{pmatrix}3\\5\\6\end{pmatrix} \text{ in place of } (3, 5, 6)$$

So a tensor can be written like this $\left[\begin{pmatrix}3\\5\\6\end{pmatrix}, \begin{pmatrix}a\\b\\c\end{pmatrix}, \begin{pmatrix}g\\h\\i\end{pmatrix}\right]$

where $\begin{pmatrix}a\\b\\c\end{pmatrix}$

and $\begin{pmatrix}g\\h\\i\end{pmatrix}$

are other vectors.

This 3×3 tensor must represent some sort of strained cube. The cube has three edges, each edge represented by a three-dimensional vector.

As you might expect, tensors are very useful to structural engineers. They represent practical things like shear strain, rotation, expansion, and deviation. You often see tensor transformations in the clouds. This happens because the wind speed aloft exceeds the wind speed near the ground, so a cubical mass of atmosphere is sheared and any cloud embedded in it is also sheared.

tensor transform of a cloud by wind shear

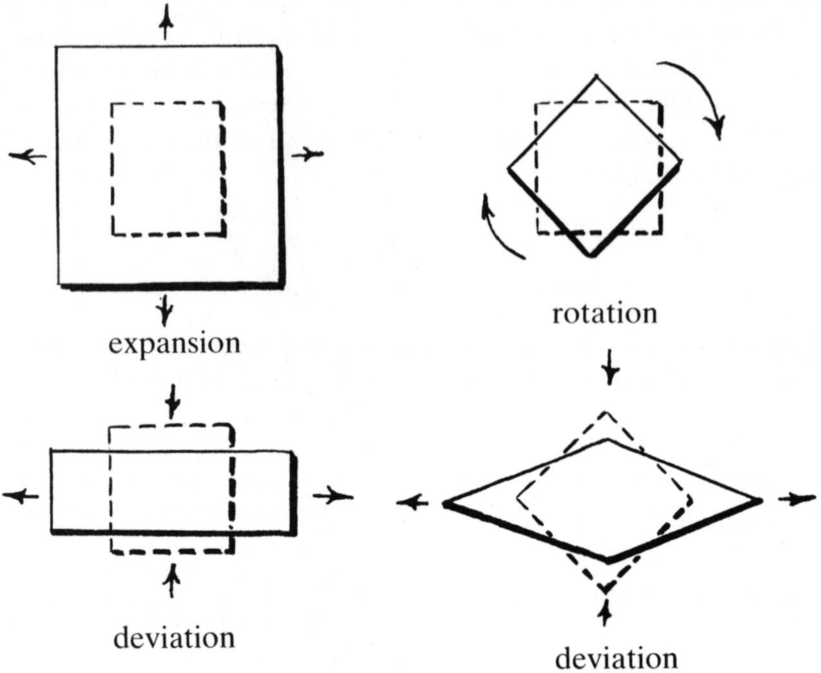

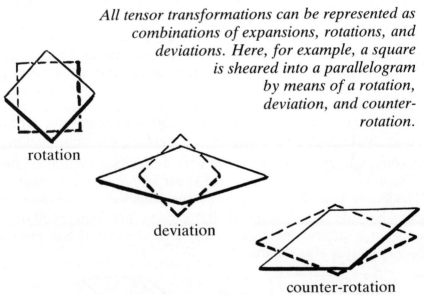

All tensor transformations can be represented as combinations of expansions, rotations, and deviations. Here, for example, a square is sheared into a parallelogram by means of a rotation, deviation, and counter-rotation.

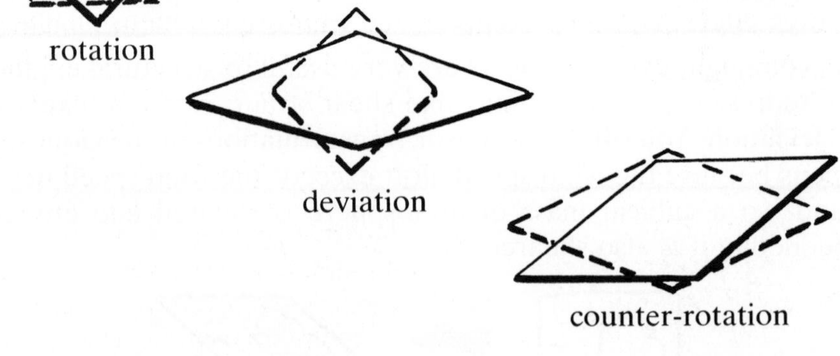

SUPER TENSORS

Do super tensors exist? a) Yes. b) No.

ANSWER: SUPER TENSORS

The answer is: a. The super tensor is a vector that is made of a string of ordinary tensors. Do super super tensors exist? Yes, they are vectors made of a string of super tensors, and on and on you can go.

Can you give an example of something that can be represented by a super tensor? Yes, the transformation of a human skull into the skull of a baboon or dog can be represented as a super tensor. Each square in the skull transforms into a parallelogram having a different shape. Since each parallelogram is different, this transformation can't be described by an ordinary tensor, which simply transforms a square into one parallelogram. More than one simple tensor is needed—a whole bunch of different tensors are needed, and a super tensor is a string of ordinary tensors. Ordinary tensors can change the direction and length of straight lines; super tensors can change straight lines into curved lines, as well as change their direction and length. Ordinary tensors cannot curve straight lines.

The equations in Einstein's General Theory of Relativity are written in terms of super tensors. Gravity, which is described by the theory, curves the paths of objects, paths that in the absence of gravity would be straight lines. And some attempts to make Unified Field Theories, which combine gravity, electricity, and other forces into one thing, are based on super super tensors.

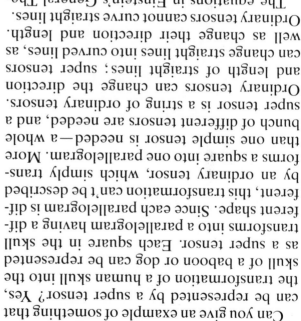

Textbooks refer to these higher tensors as third- and fourth-order tensors. These books sometimes call scalars zeroth-order tensors.

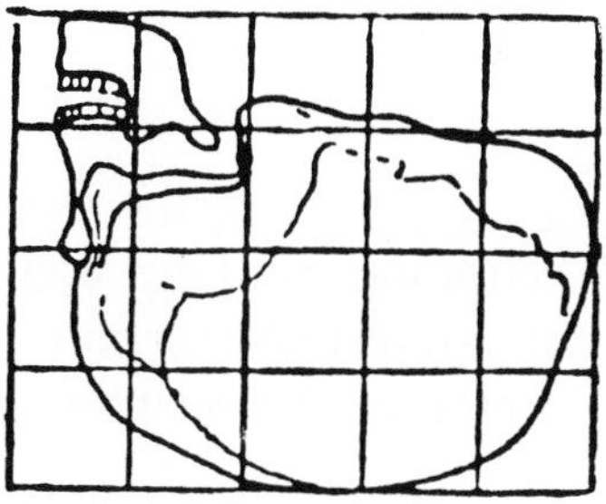

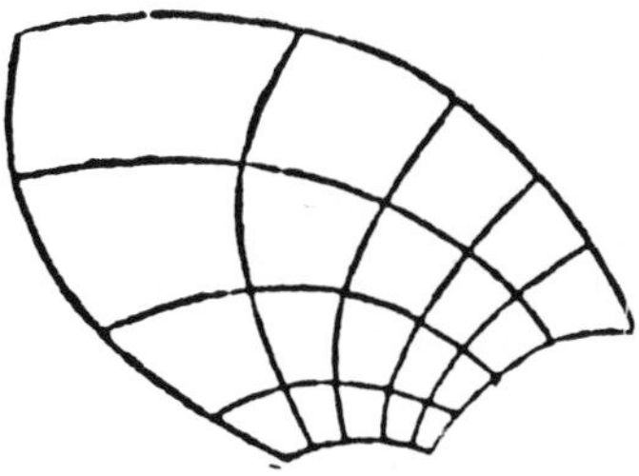

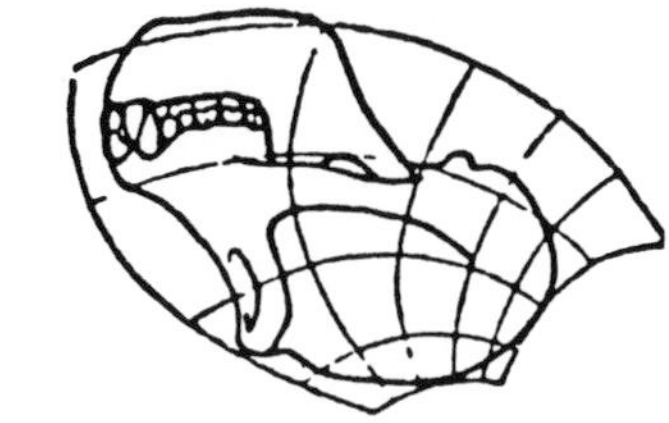

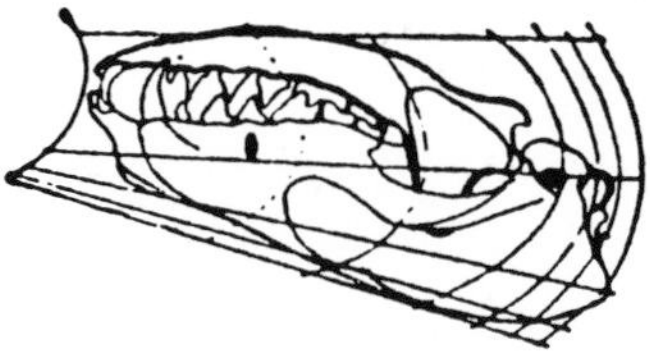

KINK

Water is shooting out of the end of a pipe. The end of the pipe is bent into a figure 6.

a) The water shoots out in a curved arc
b) The water shoots out in a straight line

Forget about the effect of gravity.

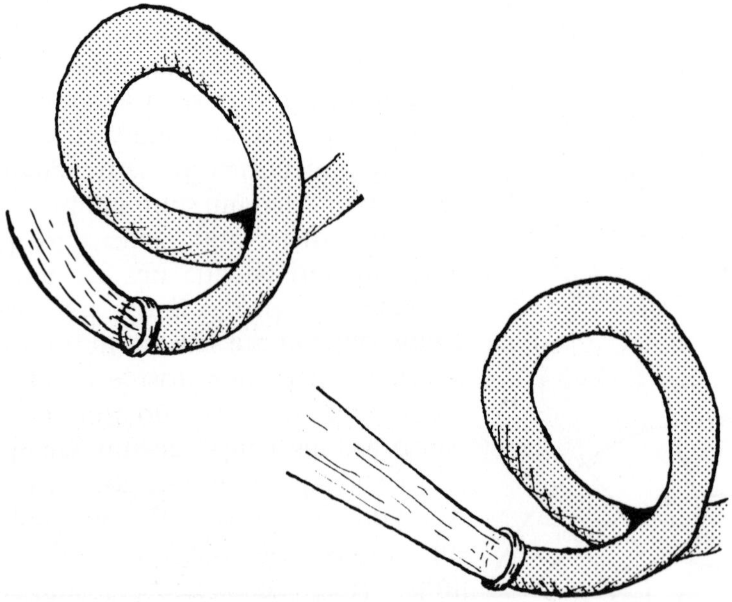

ANSWER: KINK

The answer is: b. Once out of the pipe the water goes in a straight line. Force is required to bend the water's path. If the water is inside the pipe then the pipe can force it to flow in a kinked path. But when the water gets outside it is free. Free things move in straight lines.

FALLING ROCKS

A boulder is many times heavier than a pebble — that is, the gravitational force that acts on a boulder is many times that which acts on the pebble. Yet if you drop a boulder and a pebble at the same time, they will fall together with equal accelerations (neglecting air resistance). The principal reason the heavier boulder doesn't accelerate more than the pebble has to do with

a) energy
b) weight
c) inertia
d) surface area
e) none of these

ANSWER: FALLING ROCKS

The answer is: c. It is inertia pure and simple.

If acceleration were proportional only to force, then the larger force of gravity acting on the boulder would result in it out-accelerating the lighter pebble. But the acceleration of an object also has to do with its mass — with its inertia — with its tendency to resist a change in motion. Mass resists acceleration — the greater the mass for a given force, the less the resulting acceleration. This is Newton's 2nd Law: Acceleration is directly proportional to net force and inversely proportional to mass; i.e. $a = \frac{F}{m}$. For a freely falling object, the only force acting on it is that of gravity — its weight. And weight is proportional to mass (two kilograms of sugar have twice the weight of one kilogram). A boulder that weighs 100 times more than a pebble also has 100 times the mass. It may be pulled by gravity with a force 100 times greater than that for the pebble, but it has 100 times the inertia, or 100 times the reluctance to change its state of motion.

So we see there is a reason why the ratio of force/mass and therefore the acceleration is the same (32 ft/s^2) for all freely falling objects! When air resistance is not negligible (next question), then the acceleration of fall is less than 32 ft/s^2. If air resistance builds up to the weight of the falling object, then the net force is zero and there is no acceleration.

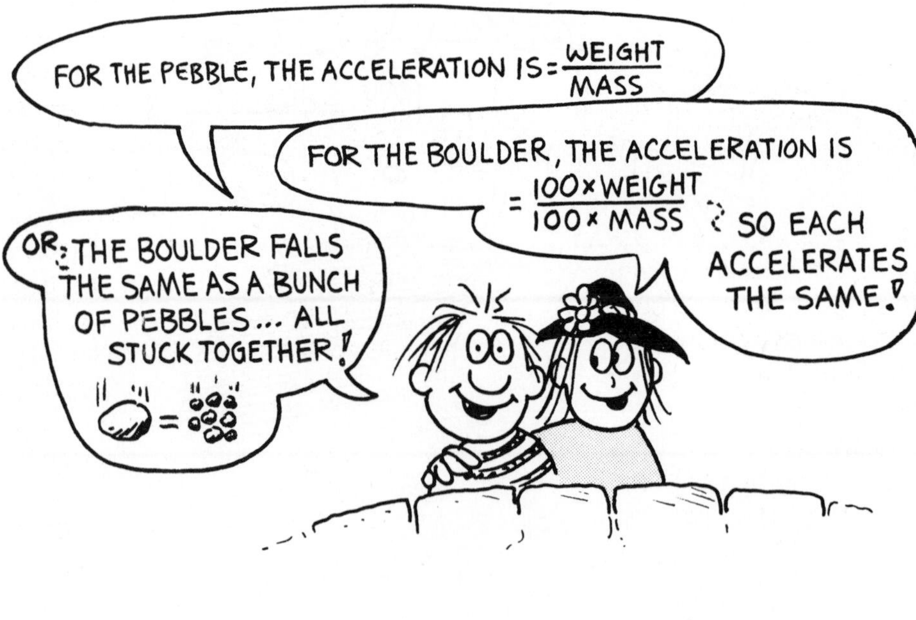

THE ELEPHANT AND THE FEATHER

Suppose that both an elephant and a feather fall from a high tree. Which encounters the greatest force of air resistance in falling to the ground?

a) The elephant
b) The feather
c) Both the same

ANSWER: THE ELEPHANT AND THE FEATHER

The answer is: a. Note that although the EFFECTS of air resistance against the feather are more pronounced, the actual FORCE of air resistance against the falling elephant is many times greater than the force against the feather. This is because the larger elephant plows through considerably more air than does the smaller feather. Also, the heavier elephant falls faster through the air, increasing air resistance even more. The feather is very light, perhaps a fraction of an ounce, so doesn't fall very fast before the air resistance it encounters builds up to the same fraction of an ounce. When this happens the feather reaches its terminal velocity — acceleration terminates and the velocity and air resistance remain unchanged throughout the remainder of the fall. An elephant falling from a high tree, on the other hand, encounters air resistance that builds up to say 20 pounds — Much greater than that of the feather, but practically negligible compared to the 2-or-so-ton unfortunate elephant who accelerates all the way to the ground.

If you answered this question incorrectly it is probably because you did not really answer the question that was asked. Be careful about distinguishing between that which is asked for and the EFFECT of that which is asked for. We'll have more such questions, so stay on your toes!

JAR OF FLIES

A bunch of flies are in a capped jar. You place the jar on a scale. The scale will register the most weight when the flies are

a) sitting on the bottom of the jar
b) flying around inside the jar
c) . . . weight of the jar is the same in both cases

ANSWER: JAR OF FLIES

The answer is: c. When the flies take off or land there might be a slight change in the weight of the jar, but if they just fly around inside a capped jar the weight of the jar is identical to the weight it would have if they sat on the bottom. The weight depends on the mass in the jar and that does not change. But how is a fly's weight transmitted to the bottom of the jar? By air currents, specifically, the downdraft generated by the fly's wings. But that downdraft of air must also come up again. Does the air current not exert the same force on the top of the capped jar as on the bottom? No. The air exerts more force on the bottom because it is going faster when it hits the bottom. What slows the air down before it hits the top? Friction. Without air friction the fly could not fly.

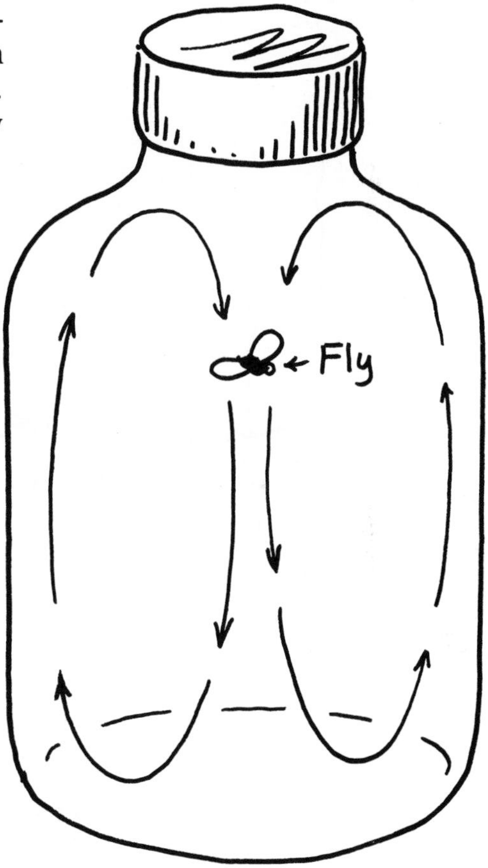

WITH THE WIND

Everyone loves a sailboat, especially on a windy day. Suppose you are sailing directly downwind with your sails full in a 20 mph wind. Then the maximum speed you could hope to attain would be

a) nearly 20 mph
b) between about 20 and 40 mph
c) . . . there would be no theoretical speed limit in this case

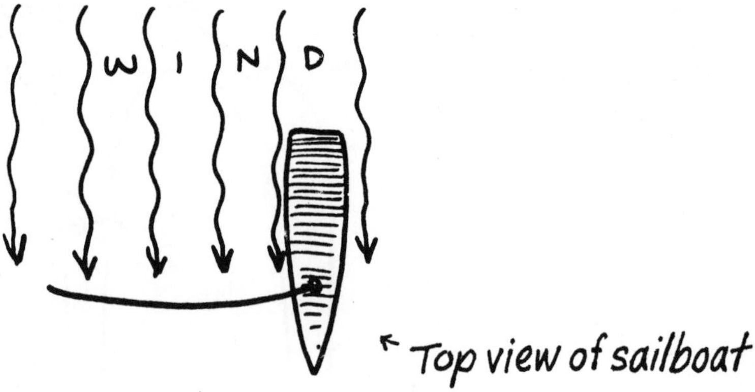

ANSWER: WITH THE WIND

The answer is: a. You could only attain the speed of the wind if the forces of water friction on the boat were zero — and even in this case you could not sail faster than the wind. Why? Because if the boat were travelling as fast as the wind there would no longer be any impact of air against the sail. The sail would sag as it would on a windless day. At the speed of the wind, there would be no wind relative to the sail.

AGAIN WITH THE WIND

You are again sailing downwind and you pull your sail in so that it no longer makes a 90° angle with the keel of the boat. This tactic will

a) decrease the speed of the boat
b) increase the speed of the boat
c) not affect the speed of the boat

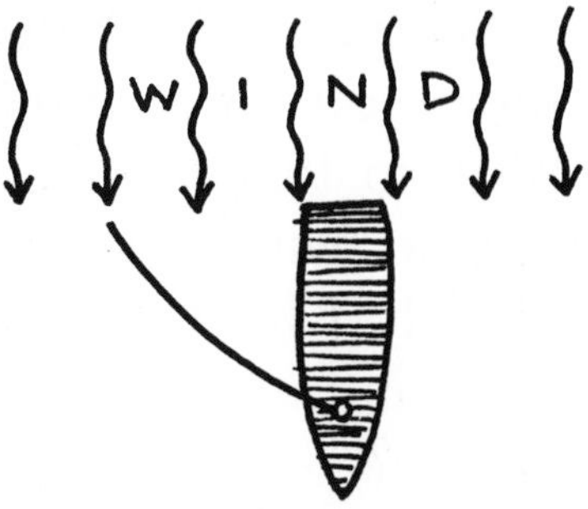

ANSWER:
AGAIN WITH THE WIND

The answer is: a, for two reasons. First, the impact against the sail is less because the sail catches less wind in the angular position. Second, the direction of the wind impact force is not in the direction of the boat's motion. We find that whenever any fluid, be it gas or liquid, interacts with a smooth surface, the force of interaction is perpendicular to the smooth surface. So the vector arrow representing this force juts 90° from the surface of the sail, as shown. Not only is this vector smaller than for the case where the sail catches maximum wind as in the last question, but only a fraction of this vector is directed along the direction of the boat's motion. It is this component that drives the boat forward. (The component that is sideways to the boat's motion tends only to tip the boat and does not contribute to forward motion.) So the boat is pushed forward by the wind, but with not as much force as before. As the sail is pulled further in, the force vector decreases in magnitude and a smaller driving component results. When the sail is pulled all the way in, so that it is parallel to the keel, it catches no wind at all and the driving force is zero.

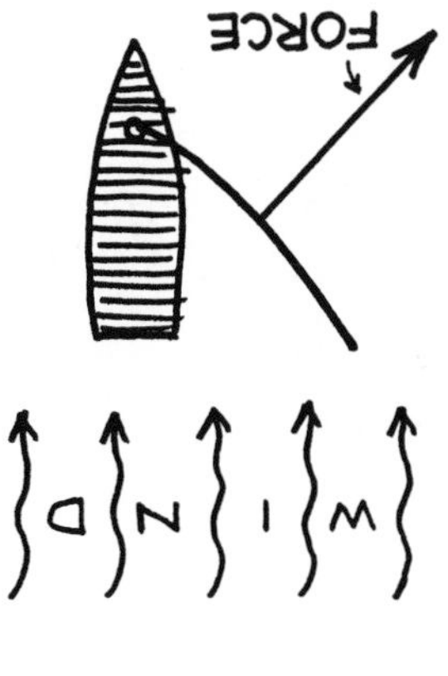

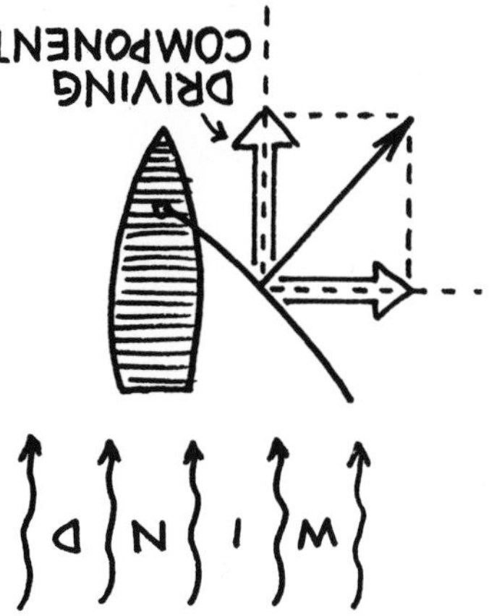

CROSSWIND

Keeping the angle of sail relative to the boat the same as in the last question, suppose you now direct your boat so that it sails directly across wind, rather than directly with the wind. Will you sail faster or slower than before?

a) Faster
b) Slower
c) The same

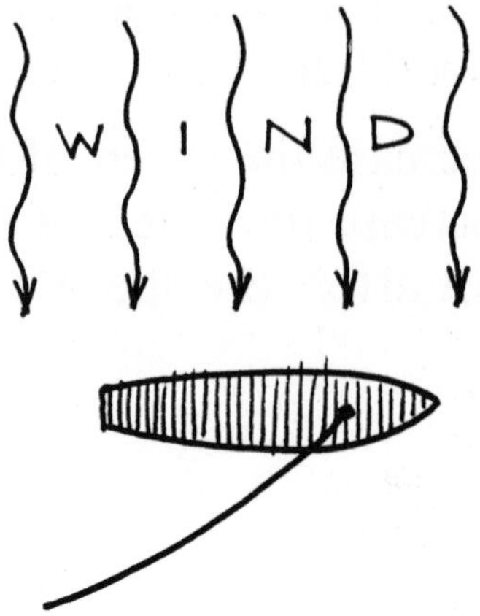

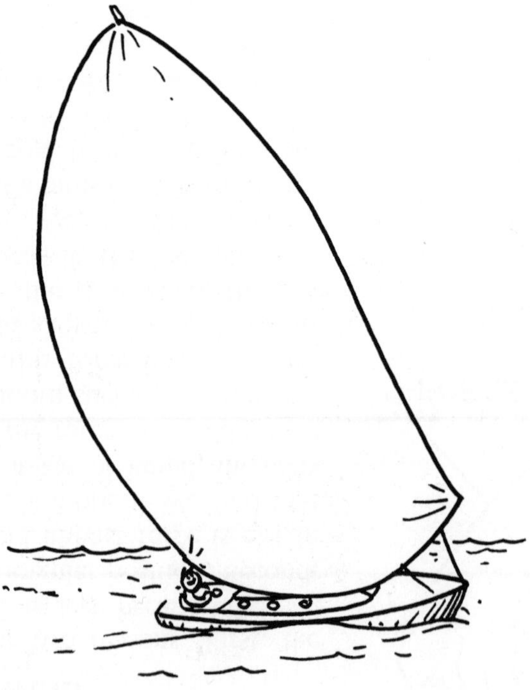

ANSWER: CROSSWIND

The answer is a. As before, the force vector perpendicular to the surface of the sail can be broken into components — one along the direction in which the boat can move, which drives the boat, and the other which is perpendicular to the boat's motion and is useless. Now, if the principal force vector in this case (wind impact against the sail) were no greater than before, the speed of the boat would be the same. But the force vector IS greater. Why? Because the sail doesn't catch up with wind speed so it will not eventually sag like before. Even when the boat is travelling as fast as the wind, there is an impact of wind against the sail. This drives the boat even faster, so it can sail faster than the wind in this position. It reaches its terminal speed when the "relative wind" (the resultant of the "natural" wind and the "artifical" wind due to the boat's motion) blows along the sail without making impact.

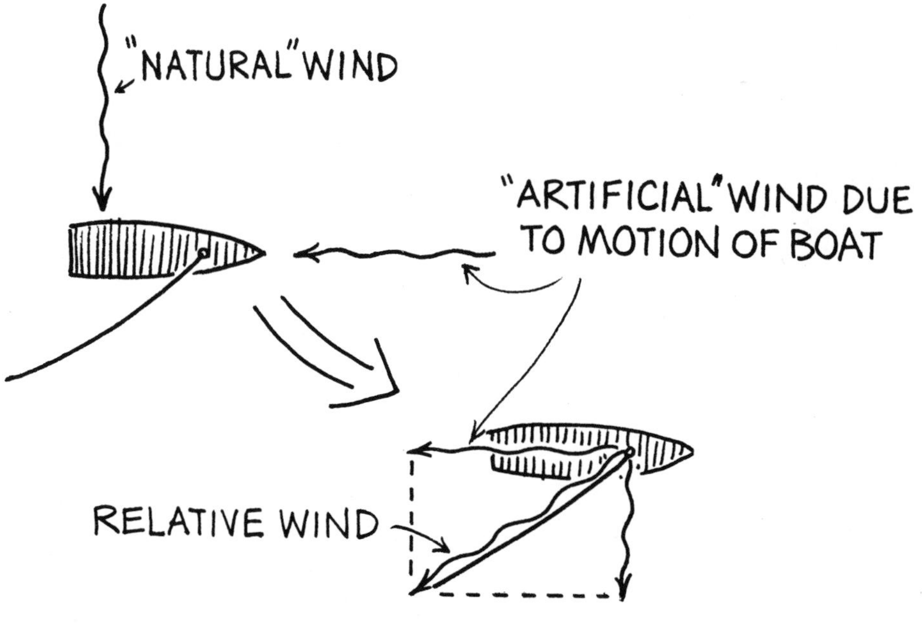

WHEN ANGLE OF RELATIVE WIND IS THE SAME AS THE SAIL ANGLE, WIND IMPACT IS NO MORE -

AGAINST THE WIND

This question is to employ your understanding of the 3 previous questions. For the four boats shown note the orientations of the sails with respect to both wind and keel directions. Which of these boats moves in a forward direction with the greatest speed?

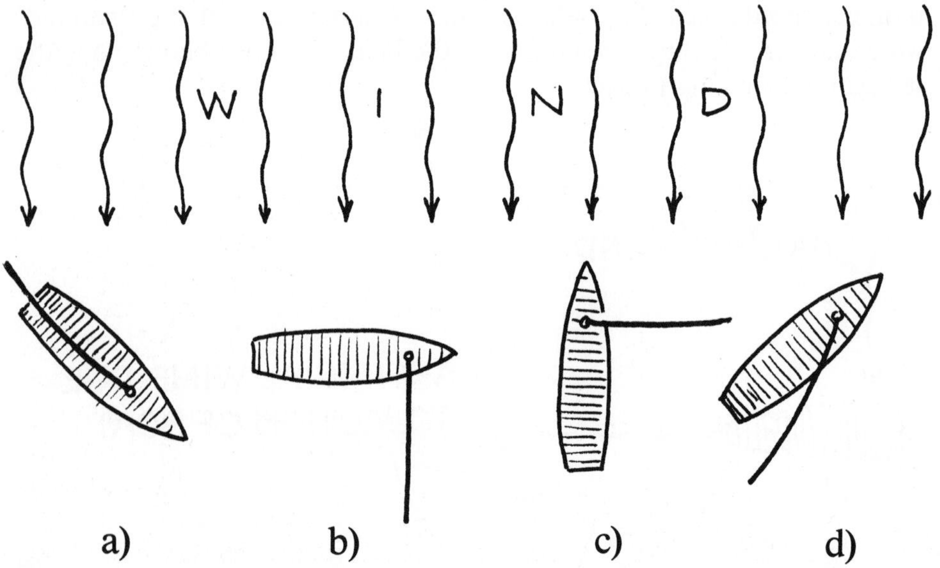

ANSWER: AGAINST THE WIND

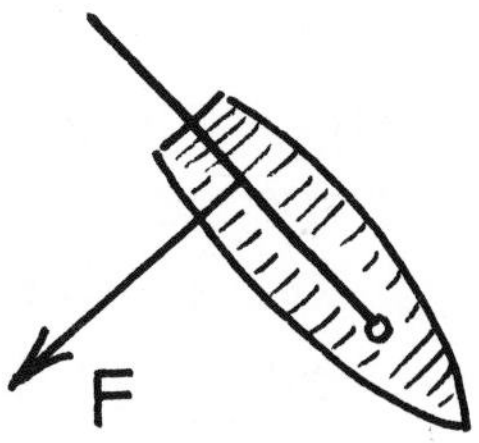

The answer is: d. It is the only boat that moves in a forward direction. The orientation of the sail for boat "a" is such that the force vector is perpendicular to the direction in which the boat can freely move (as dictated by the fin-like keel beneath). This completely sideways force has no component in a forward (or backward) direction. It is as useless in propelling the boat forward as the downward force of gravity is in pushing a bowling ball across a level surface. Sailboat "b" misses wind impact altogether, as the wind simply blows by the sail rather than into it. The sail of sailboat "c" receives full wind impact, but blows it backward rather than forward. The situation of "d" is not unlike that of the CROSSWIND question.

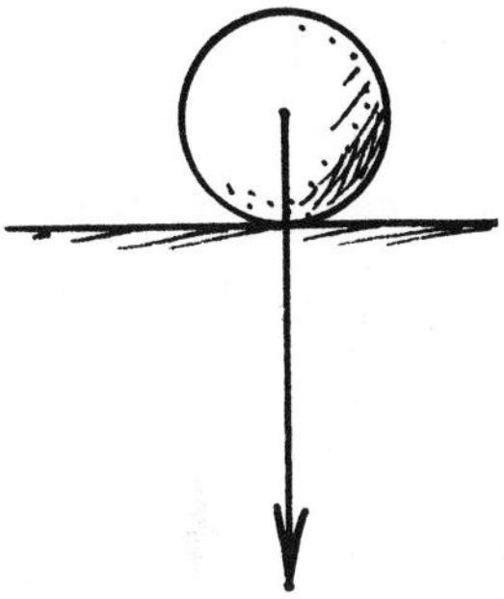

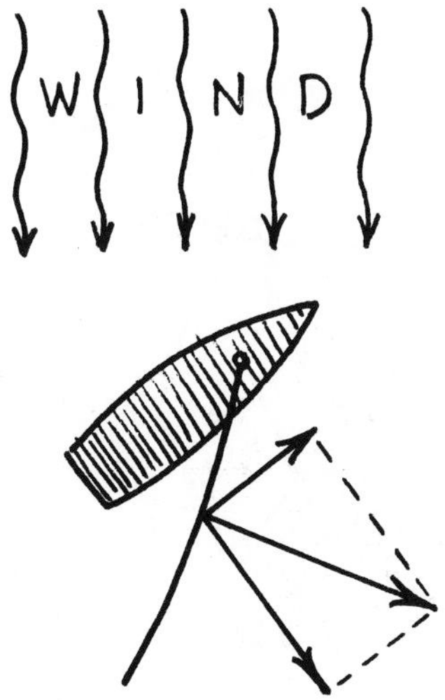

Note in the sketch that the force vector has a component along the forward direction which propels the boat at the indicated angle into and against the wind. This boat can actually sail faster than the crosswind sailboat, for the faster it travels, the GREATER is the force of the wind impact. So a sailboat's maximum speed is usually at an angle upwind! It cannot sail directly into the wind, so to reach a destination straight upwind it zigzags back and forth. This is called "tacking."

STRONGMAN

When the strongman suspends the 10-lb telephone book with the rope held vertically the tension in each strand of rope is 5 lbs. If the strongman could suspend the book from the strands pulled horizontally as shown, the tension in each strand would be

a) about 5 lbs
b) about 10 lbs
c) about 20 lbs
d) more than a million lbs

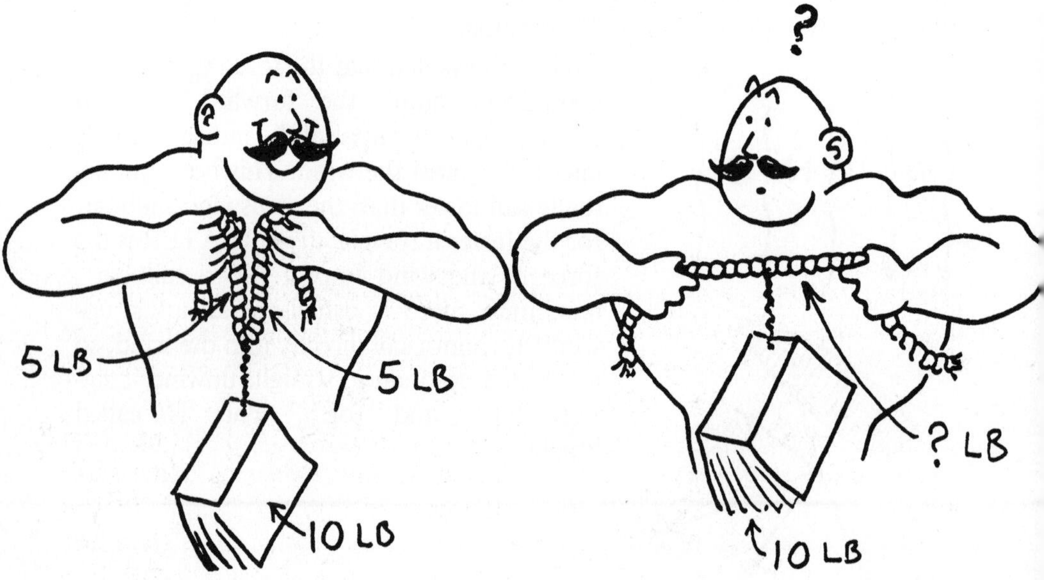

ANSWER: STRONGMAN

The answer is: d. To see why, consider the book suspended by the rope at the angle shown. We represent all the forces acting on the book with vector arrows. The 10-lb vector represents the weight of the book and acts straight downward toward the center of the earth. The length of this vector sets our scale of 10 lbs. Now, how long are the other vectors to keep the book in equilibrium? The size of these vector arrows compared to our 10-lb arrow will tell us what the tensions are in the rope.

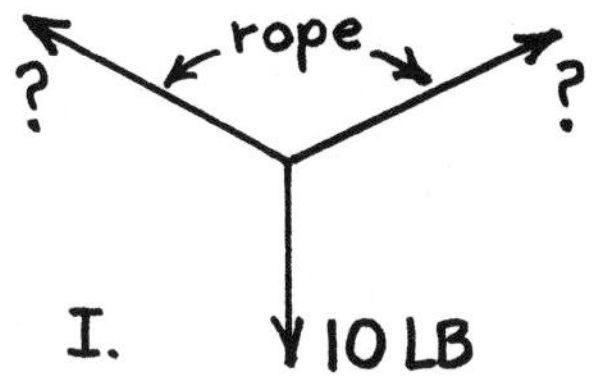

Whenever two forces act together, the total effect of the two forces can be found using the following procedure: First, draw the two forces as thin arrows (sketch II). Then complete the parallelogram with dotted lines. Now, draw in the diagonal with a fat arrow. The fat arrow is the net force.

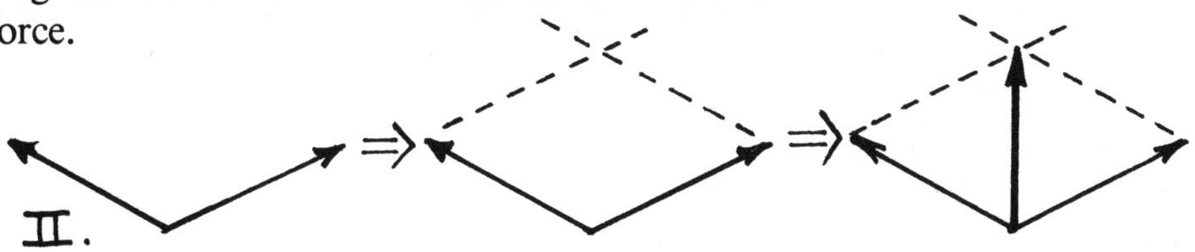

Incidentally, a physics teacher named Dave Wall says that some people think that if you pull on something with a longer rope, you should draw a longer force arrow. But that is the wrong way to think. The length of the force arrow depends only on how strong the force is — and that depends on how hard you pull — and has nothing to do with the rope length.

The total effect of the suspended ropes is to hold the 10-lb book up. That means the total effect of the ropes is an upward force of 10 lbs. How much force must the ropes exert to produce an upward force of 10 lbs? To find this force, first draw two lines extending out in the direction of the ropes (sketch III). Draw the 10-lb upward force you want the ropes to produce (fat arrow). Then draw two dotted lines from the tip of the 10-lb fat arrow so as to form a parallelogram with the ropes. Where the dotted lines hit the ropes, draw arrowheads. The arrows you have just drawn running along the ropes represents the force exerted by the ropes. As you can see, each of the rope arrows is longer and therefore represents more force than the 10-lb arrow. You can estimate the force by comparing the length of the arrows to

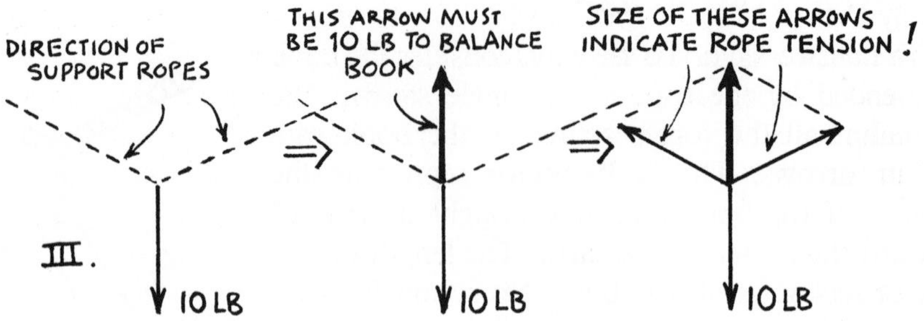

THIS ARROW MUST BE 10 LB TO BALANCE BOOK
SIZE OF THESE ARROWS INDICATE ROPE TENSION!

the length of the 10-lb arrow, for if you draw your diagram carefully, it will be to scale.

Now, our strongman did not support the book with the angle we have been discussing. We can see by the sketches that the more horizontal the supporting ropes, the greater is the tension, for our parallelogram must be longer as the base angles are flattened out. In fact, as the ropes approach the horizontal the tension required to produce a vertical 10-lb resultant approaches infinity. It is impossible that the ropes supporting the book could be completely horizontal . . . there must be a kink. So our answer of "more than a million lbs" is rather conservative.

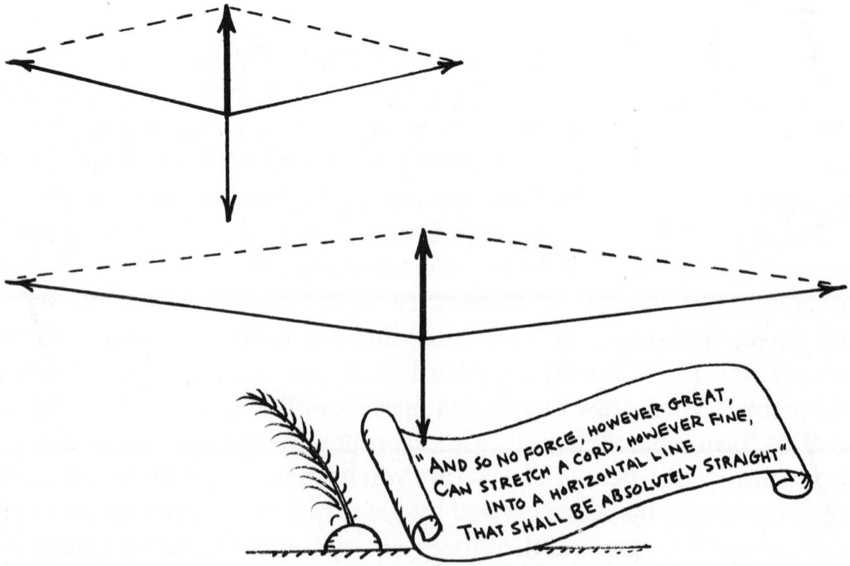

"AND SO NO FORCE, HOWEVER GREAT, CAN STRETCH A CORD, HOWEVER FINE, INTO A HORIZONTAL LINE THAT SHALL BE ABSOLUTELY STRAIGHT"

Printed in Whewell's *Elementary Treatise on Mechanics,* 1819.

By the way, in case you forgot how to draw a parallelogram, you start with the two sides you have. Then you put down a ruler so it both touches the end of one side and runs parallel to the other side. Next, draw a line. Finally, repeat the same thing for the remaining side . . . and you've got it all in the bag!

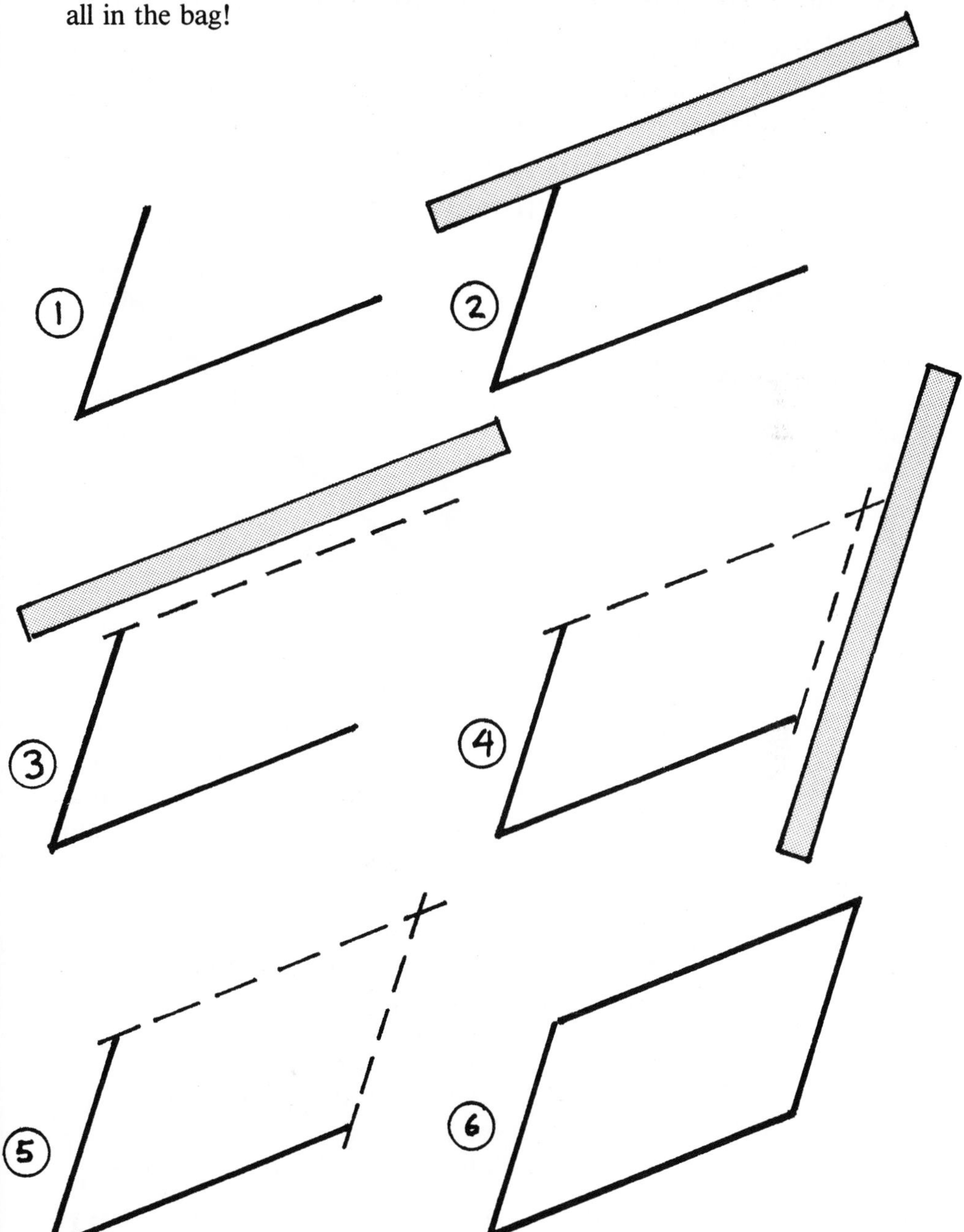

DO VECTORS ALWAYS ADD?

Can two physical things represented by vectors always be represented by one physical thing that is the sum of the two vectors? That is, if vectors A and B each represent the same kind of thing, can the two things acting together always be represented by vector C, which is the diagonal of the parallelogram formed by A and B? In short: do vectors always add?

a) Yes, vectors always add.
b) No, vectors do not always add.

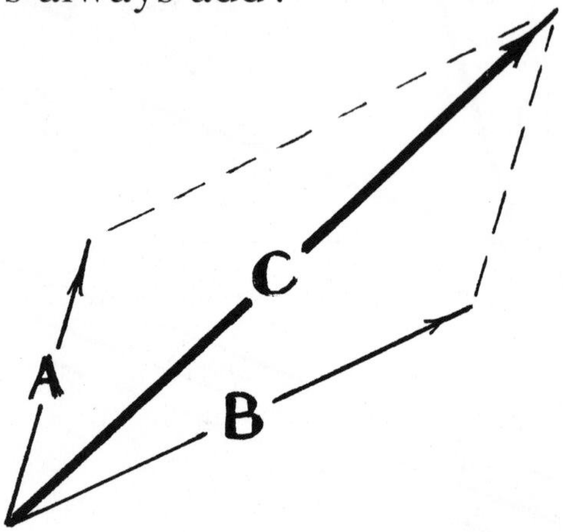

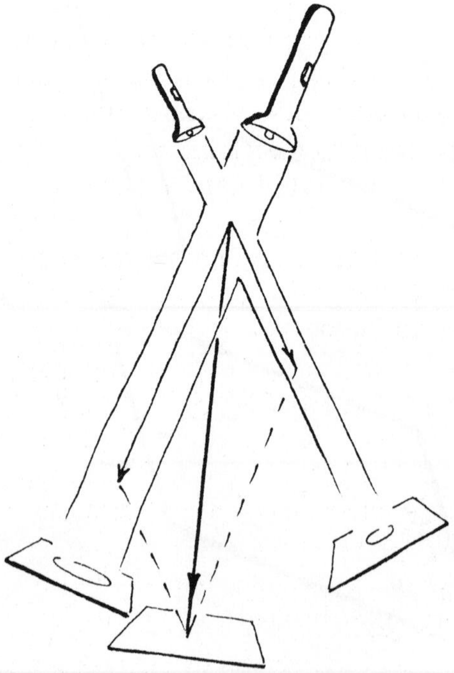

ANSWER: DO VECTORS ALWAYS ADD?

The answer is: b. Most vectors do add. Many physics books presume that anything that can be represented by vectors must add like vectors. For example, some people presume—without proof—that angular momentum vectors add. They do add, but it is not obvious that they do.

And not all vectors add. For example, the power flow in a light beam can be represented by vector. It is called a „Poynting vector." However, if two beams act together, they do not add up to a third beam

FORCE?

The strong man is pulling the spring apart. Is there a force on the spring?

a) Certainly there is.
b) There is no force on the spring.

ANSWER: FORCE

Surprise. The answer is: b. How come? Well, for one thing, a force acts in a definite direction. Which way is the force on the spring acting? To the left or to the right? For another thing, force makes things accelerate. Which way is the spring accelerating? The spring isn't accelerating. It isn't even moving.

Well, if there is no force acting on the spring, what is acting on it? A TENSION. Force and tension are very different, and not appreciating that difference causes quite a bit of confusion. Tensions result from DIFFERENT forces acting on different parts of a body. Tensions can therefore break things. If the same force acts on all parts of a body the body will accelerate in the direction of the force but it will not come apart.

When the same force acts on all parts of a body, the force is called a pure force. Pure forces *can not break things.*

When all the forces acting on a body cancel out or add to zero, as with the spring, the tension is called pure tension. Of course you can think of situations where forces and tensions are mixed.

Though force and tension can both be measured in pounds, the difference between them is very deep.* In fact they are intrinsically different kinds of mathematical quantities. A force is a vector and a tension is a tensor.

* Torque and energy also have the same units (foot-pounds) but they are certainly very different physical quantities.

PULL

Think carefully about this one: In both Case *a* and Case *b* shown below a net force of 10 lb results in the acceleration of Block A across the table toward the pulley. Disregard friction altogether.* The acceleration of Block A is

a) greater in Case *a*
b) greater in Case *b*
c) the same in both cases

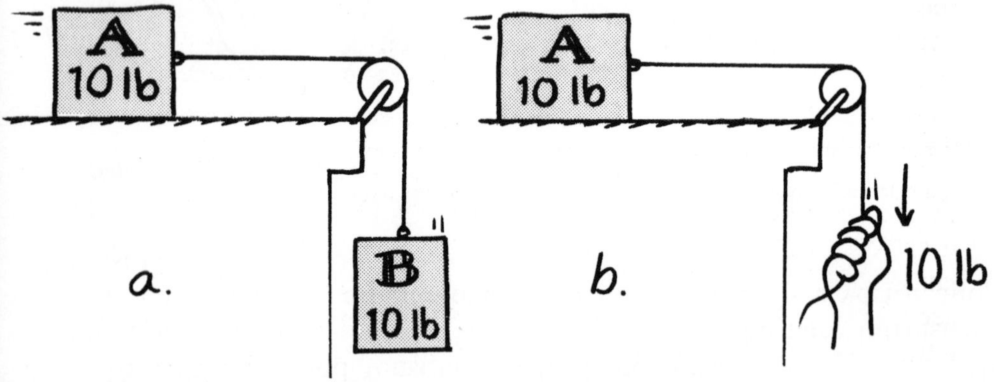

*But this case *does* involve friction — that between Block A and the surface, the friction of air on the moving blocks, and even friction at the pulley, not to mention the turning of which also tends to impede motion. So how can we simply "disregard friction altogether?" To do so may seem as if we are off in a world of hypothetical, ideal, or imaginary things which don't really exist.

That is partly true. And that is also one of the most powerful keys to understanding the real existing world. In your mind, make the situation simpler than it is by ignoring complications and details. Undress reality. That way the essential part of a situation stands alone and exposed and you can grasp it. Once you grasp the essential part, the essence can be reclothed in its details and complications.

Now how do you know that friction is a complication that can be stripped away? Because in reality you can minimise it as much as you please, if you go to pains to do so. And how do you know what other things can be minimised? Through a working sensitivity about what ultimately can and can't be done in this world. That is the art. Hopefully, this book will help you develop that art.

ANSWER: PULL

The answer is: b. This is because the string tension is not the same in both cases. The tension in the string pulled by hand in Case *b* is 10 lb, and this of course accelerates Block A. But the tension in the string when supplied by the 10-lb falling weight, Case *a*, is less than 10 lb. Why? Because if the tension were a full 10 lb the hanging weight would not accelerate downward at all, but would be in equilibrium. Like if Block A were very very massive compared to Block B and hardly moved, the tension in the string would be close to the full 10 lb, whereas if Block A were very light like a feather, the tension would be very small — the string would be slack and Block B would nearly be in a state of free fall. In our case the mass of Block A is midway between these extremes and is neither more nor less than the mass of Block B. The tension is midway between 10 and 0 lb, and is in fact 5 lb. So Block A accelerates at only half the rate when pulled by the falling weight.

We can see this another way also: Although a force of 10 lb acts in both cases, in Case *a* twice as much mass is being accelerated — the weight of one 10-lb block accelerates the mass of two 10-lb blocks. So we see the acceleration must correspondingly be half the rate as when only the mass of one 10-lb weight is involved. Another thing: Note in Case *b* that a force of 10 lb acts on the mass of a 10-lb body, which means that the acceleration must be 32 ft/s^2 — the same as in free fall (recall FALLING ROCKS).

So Block A accelerates across the table at 32 ft/s^2 when pulled by hand with a force of 10 lb, and at half this rate, 16 ft/s^2, when pulled by a hanging 10-lb weight. The two cases are not equivalent.

MAGNET CAR

Will hanging a magnet in front of an iron car, as shown, make the car go?

a) Yes, it will go
b) It will move if there is no friction
c) It will not go

ANSWER: MAGNET CAR

The answer is: c. You could just dismiss the thing by saying that no work output will result from zero work input — or perpetual motion is impossible. Or you could invoke Newton's Third Law: the force on the car is equal and opposite to the force on the magnet — so they cancel out. But these formal explanations don't illustrate why it will not work.

To see intuitively why it will not work, improve the design by putting another magnet in front of the car. Then, to streamline things, put the magnets in the car. Then comes the question: which way will it go?

POOF AND FOOP

This is a stumper. If a can of compressed air is punctured and the escaping air blows to the right, the can will move to the left in a rocket-like fashion. Now consider a vacuum can that is punctured. The air blows in the left as it enters the can. After the vacuum is filled the can will

a) be moving to the left
b) be moving to the right
c) not be moving

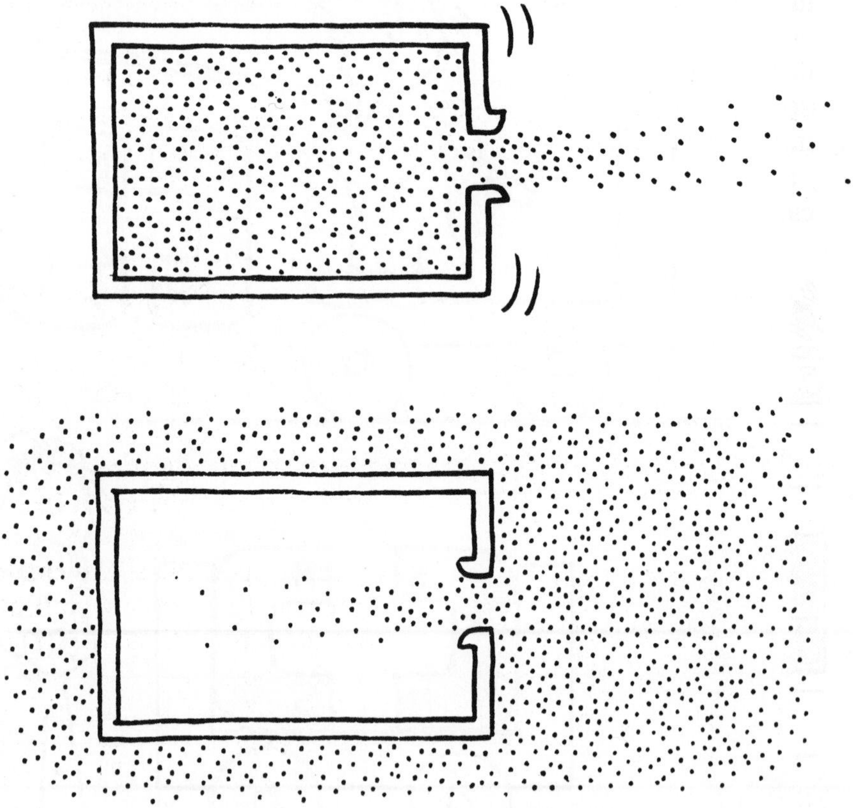

ANSWER: POOF AND FOOP

Are you reading this before you have formulated a reasoned answer in your thinking? If so, do you also exercise your body by watching others do push-ups? If the answers to both these questions is no, and if you have decided on answer c for POOF AND FOOP, you're to be congratulated!

To see why, consider the water-filled cart in Sketch I. We can see that it accelerates to the right because the force of water against its right wall is greater than the force of water against its left wall. The force against the left is less because the "force" that acts on the outlet is not exerted on the cart. Similarly with the can of compressed air. The "force" that acts on the hole is not exerted on the can, and the imbalance accelerates the can to the right.

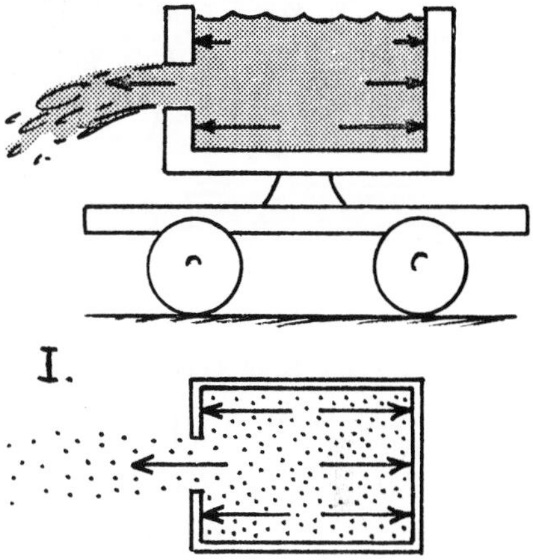

I.

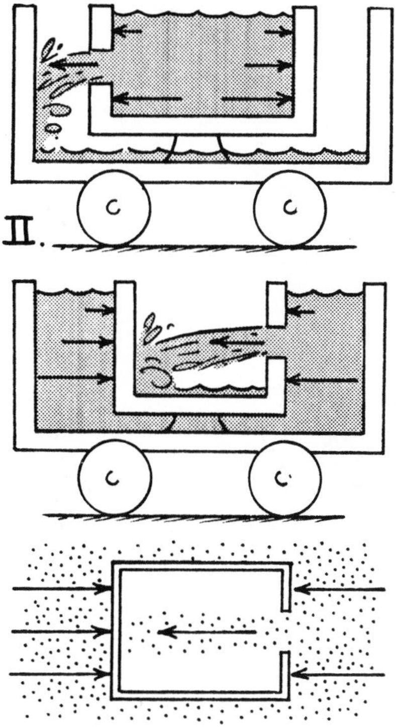

II.

Now consider Sketch II. Do these water-filled carts accelerate? No. Why not? Because the "force" of escaping water *is* exerted on the carts — on the outer wall of the top cart and on the inside wall of the bottom cart. So the water exerts no net force on the carts and no change in motion takes place (except for a momentarily slight oscillation about the center of mass). Likewise with the punctured vacuum can. The force of air that does not act at the hole nevertheless is exerted on another part of the can — on its left inner wall. So like the double-walled cart, the forces are balanced and no rocket propulsion occurs.

LEVIS

The Levi Strauss trademark shows two horses trying to pull apart a pair of pants. Suppose Levi had only one horse and attached the other side of the pants to a fencepost. Using only one horse would

a) cut the tension on the pants by one-half
b) not change the tension on the pants at all
c) double the tension on the pants

ANSWER: LEVIS

The answer is: b. Suppose one horse can exert one ton of force. Then the other horse must also exert one ton if the tug of war is to be a stand-off. Now if one horse pulls on the post with one ton of force the post must pull back with one ton, for otherwise the horse would pull the pants away. So it doesn't matter if the post pulls or if another horse pulls. Is there any force on the pants? No. Only tension.

The underlying idea developed here can be extended. Suppose two cars of identical mass traveling at 55 mph, but in opposite directions, crash head on into each other. Or suppose one of the cars drives at 55 mph into an immovable stone wall. In which case does the car suffer more damage? There is no difference. The damage is the same in each case.

The horse and the car each exert a certain force. That force can be countered by an equivalent horse or car forcing in the opposite direction, or it can be countered by an immovable object. The immovable object in effect acts as a force mirror. It puts out a force that is the mirror image, the reflection, of the force that is applied to it. The wall punches you when you punch it—and just as hard. That is Newton's third law, the reaction law.

HORSE AND BUGGY

This is perhaps the oldest and most famous "brain twister" in classical physics. Which is correct?

a) If action always equals reaction, a horse cannot pull a buggy because the action of the horse on the buggy is exactly cancelled by the reaction of the buggy on the horse. The buggy pulls backward on the horse just as hard as the horse pulls forward on the buggy, so they cannot move.
b) The horse pulls forward slightly harder on the buggy than the buggy pulls backward on the horse, so they move forward.
c) The horse pulls before the buggy has time to react, so they move forward.
d) The horse can pull the buggy forward only if the horse weighs more than the buggy.
e) The force on the buggy is as strong as the force on the horse, but the horse is joined to the earth by its flat hoofs, while the buggy is free to roll on its round wheels.

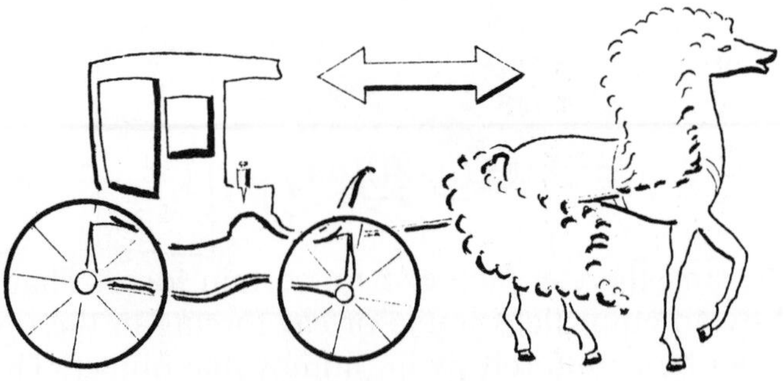

ANSWER: HORSE AND BUGGY

The answer is: e. True, the force on the horse is as strong as the force on the buggy, but you are interested in motion-acceleration, not force. A thing's acceleration depends on its mass as well as the force on it.

Well, who has the most mass—the horse or the buggy? It doesn't matter, because the horse has joined itself to the earth by its flat hoofs. So in effect one force pulls on the buggy and the equal and opposite reaction pulls on the horse AND EARTH. To pull back the horse also requires pulling back the massive earth, while the buggy, being less massive than the earth, moves much more easily. But as the buggy moves forward, the whole earth moves SLIGHTLY backward.

If the horse pulls the buggy a foot forward, how far backwards does the earth move? Suppose the buggy weighs 1000 pounds. The mass of the earth is 10,000,000,000,000,000,000,000 times greater. So the planet moves a 10,000,000,000,000,000,000,000th part of a foot backwards, which is hard to notice! Incidentally, you can save some ink by writing 10,000,000,000,000,000,000,000 (it has 22 zeros) as 10^{22}.

POPCORN NEUTRINO

In Mom's frying pan a kernel of unpopped popcorn decays into a popped popcorn which shoots off in direction "p." Therefore, during the decay it is probable that

a) a subatomic particle such as a neutrino is emitted in the opposite direction, "q"
b) a neutrino is not involved, but some invisible thing is emitted in direction "q"
c) nothing at all is emitted in direction "q"

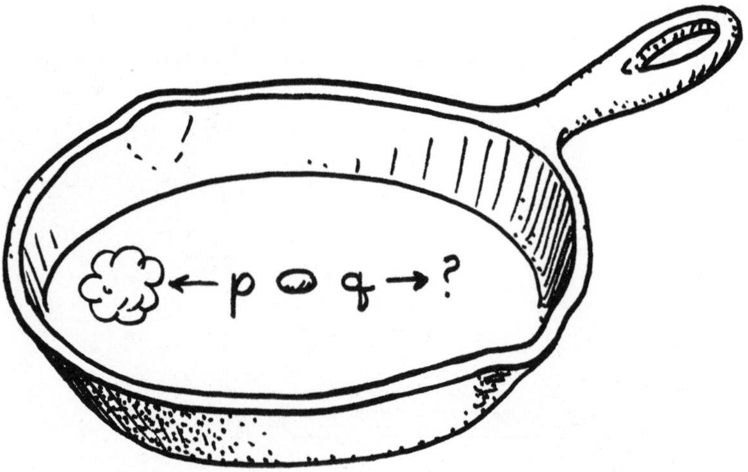

ANSWER: POPCORN NEUTRINO

The answer is: b. If neutrinos came out of popcorn they would have been discovered in the 18th Century. But some invisible thing must shoot from the popcorn. What could have propelled it in direction "p"? What is the invisible thing? Steam. There is moisture in the unpopped kernel and when heated it turns to steam — which explodes the kernel. If the steam shoots off in direction "q," the popped corn recoils in direction "p."

MOMENTUM

Momentum is inertia in motion, and is equal to the product of a body's mass and its velocity. For example, if the speed of a projected cannonball is doubled, then the momentum is doubled. Of if instead the cannonball's mass is doubled, then the momentum is likewise doubled. Suppose, however, that a cannonball's mass is somehow doubled and its velocity is also doubled. Then its momentum is

a) the same
b) doubled
c) quadrupled
d) none of these

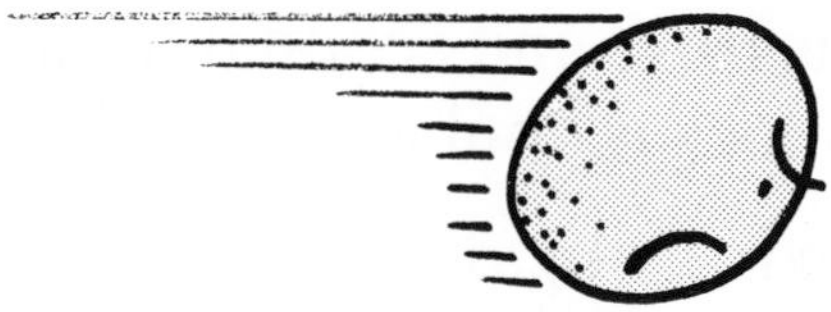

ANSWER: MOMENTUM

The answer is: c — which follows from the definition of momentum: mass x velocity. Double the mass times double the velocity equals four times the momentum. A body receives momentum by the application of an impulse — which is "FORCE multiplied by the TIME during-which-the-force-acts."

We say that: IMPULSE = change in MOMENTUM.

$$Ft = \Delta mv$$

GETTING A MOVE ON

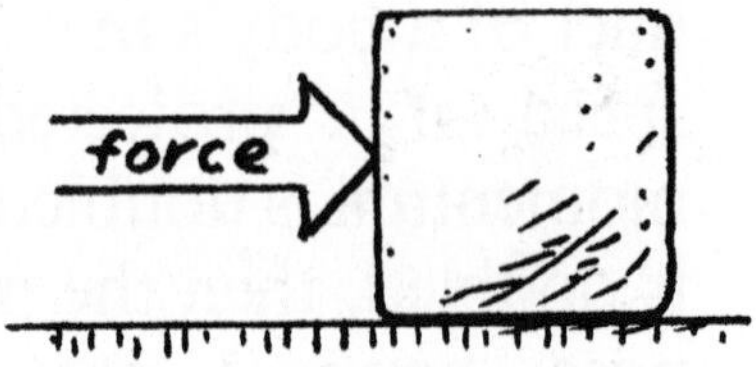

Let's develop the impulse and momentum idea further. Consider a block of ice on a friction-free frozen lake. Suppose a continuous force acts on the block. Of course, this causes the block to get a move on, to accelerate. After the force has acted for some time the speed of the block has increased a certain amount. Now if the force and mass of the block are unchanged, but the time the force acts is doubled, then the increase in speed will be

a) unchanged b) doubled c) tripled
d) four-fold e) cut in half

Next, if the force and action time are unchanged, but the mass of the block is doubled, then the speed increase will be

a) unchanged b) doubled c) cut in half
d) four-fold e) cut to one fourth

And now suppose only the force is doubled while the mass and action time are unchanged. Then the increase in speed will be

a) unchanged b) doubled c) cut in half
d) four-fold e) cut to one fourth

Finally, suppose applied force, mass, and action time are all as they were initially, but somehow the force of gravity is doubled — like if the experiment were conducted on another planet. Then the increase in speed will be

a) unchanged b) doubled c) cut in half
d) four-fold e) cut to one fourth

ANSWER: GETTING A MOVE ON

The answer to the first question is: b. The force increases the speed of the block a certain amount each second it acts. If you double the time you simply double the increase in speed.

The answer to the second question is: c. It is harder to move two blocks than one, and a block of double mass is equivalent to two original blocks. It's twice as hard to move the block of double mass, so the speed increase is cut in half. Note that this has nothing to do with gravity. Even if the block were in gravity-free space it would still take a force to change its speed. That's why spacecraft must carry engines for changing their motions even in faraway space. In space the block may have zero weight, but it still has all its mass or inertia (inertia is a synonym for mass) — that is, it still has all its resistance to a change in speed.

The answer to the third question is: b. Force makes the speed change. No force, no speed change. A little force produces a little speed change, a big force produces a big speed change. If you double the force you double the speed change — that is, you double the acceleration.

The answer to the last question is: a. Increasing gravity increases the weight of the block, but not its mass or inertia. It's the inertia of the block that counts. Now if we had to worry about friction, the weight would get into the act because weight controls friction. But the ice sliding on ice that we're concerned with is frictionless.

This whole long story can be abbreviated with one little old equation.* The equation is: Speed change equals force multiplied by time divided by mass

$$\Delta v = \frac{Ft}{m}$$

which is a rearrangement of the Impulse = change in Momentum equation ($Ft = \Delta mv$) we treated briefly in our last question, MOMENTUM.

Another idea: The change in motion needn't be a speed increase. A decrease is also a change. A speed decrease can be regarded like a speed increase if the force is reversed. Moreover, the change in motion can even be sideways if a sideways force acts on the block. Such a change might not even change how fast the block is moving. It might only change its direction of motion!

One last idea: Loosely speaking, we can use the words "speed" and "velocity" interchangeably. But strictly speaking, speed is a measure of how fast without partic-

---

*Equations are useful abbreviations of relationships and are indispensable to the physicist. But often they are misused — when they become substitutes for understanding. Never put your efforts toward memorizing equations until you understand the concepts the symbols represent. Only after a conceptual understanding are the equations truly meaningful.

ular regard for direction; and velocity is the term for speed with particular regard for direction. So velocity may be changing while speed remains constant, like a block at the end of a rope travelling in a circular path (changing direction every instant while its "speedometer" remains constant). So a change in velocity takes account of a change in speed and/or direction, whereas a change in speed is only a change in how fast. Physicists define acceleration to be the inclusive time rate of change of velocity (rather than speed) for this reason.

HURRICANE

The force exerted on a house by a 120-mph hurricane wind is

a) equally . . .
b) twice . . .
c) thrice . . .
d) four times as strong as the force exerted on the same house by a 60-mph gale wind.

ANSWER: HURRICANE

The answer is: d. If the wind speed is doubled the mass of air hitting the house per second is doubled. But the speed of that mass is also doubled. Double the mass and double the speed means four times as much momentum per second hits the house. The force on the house is proportional to how much momentum hits it per second. So if wind speed doubles force goes up four times. What if wind speed triples? Force goes up nine times.

ROCKET SLED

A little sled weighs one pound. It is set in motion over frictionless ice by a toy rocket motor. After the rocket fuel is expended, the sled is coasting over the ice at one foot per second. How much force did the rocket exert on the sled to make it go?

a) One pound
b) 4 pounds
c) 16 pounds
d) 32 pounds
e) There is no way to tell from the information given

ANSWER: ROCKET SLED

The right answer is: e. That's right, you can't tell! The rocket motor might have provided a small force for a long time or a big force for a short time, but from the information given you can't tell which. This question is very much like asking how long is a rectangle which contains 12 square inches. It might be one long and twelve high or two long and six high or three long and four high. In the case of the sled, its momentum is like the area of the rectangle and the force and time during which the force acts are like the sides of the rectangle which multiply together to make the area.

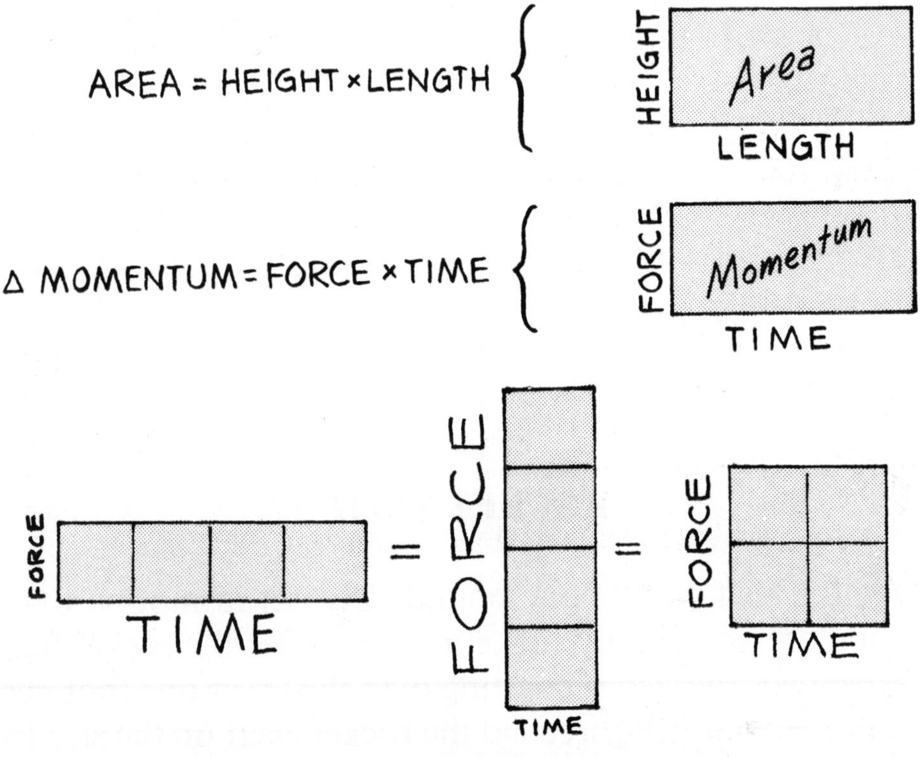

ALTHO THE FORCES AND TIMES ARE DIFFERENT IN THE THREE CASES, THE MOMENTUM EACH PRODUCES IS THE SAME.

KINETIC ENERGY

We have seen that a force multiplied by the time-during-which-the-force-acts is equal to the change in momentum of that which the force acts upon: Impulse = Δ Momentum. We now consider another central idea in physics, the work-energy principle. The Work (Force × Distance-thru-which-the-force-acts) done on a body increases the energy of the body — for example, it can increase its energy of position (gravitational potential energy = weight × height), which in turn can be converted to energy of motion which is called kinetic energy. Apply this to the following: A brick is lifted to a given height and then dropped to the ground. Next, a second identical brick is lifted twice as high as the first and also dropped to the ground. When the second brick strikes the ground it has

a) half as much kinetic energy as the first
b) as much kinetic energy as the first
c) twice as much kinetic energy as the first
d) four times as much kinetic energy as the first

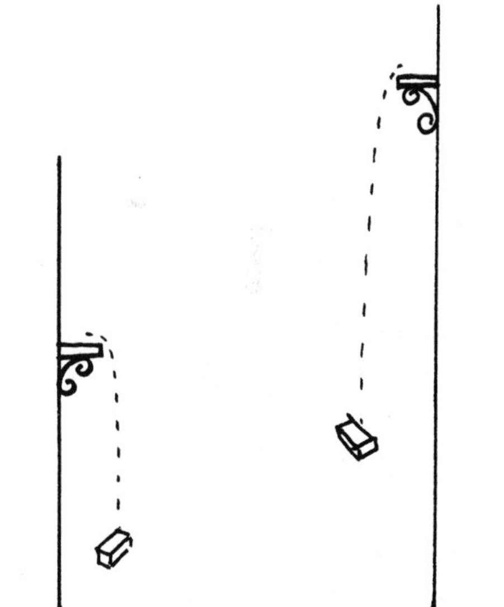

ANSWER: KINETIC ENERGY

The answer is: c. To lift a thing twice as high takes twice as much work and all the work put into lifting the brick turns into kinetic energy as it falls.

MORE KINETIC ENERGY

A brick is lifted to a given height and then dropped to the ground. Next, a second brick, which weighs twice as much as the first, is lifted just as high as the first and then it too is dropped to the ground. When the second brick strikes the ground it has

a) half as much kinetic energy
b) as much kinetic energy as the first
c) twice as much kinetic energy as the first
d) four times as much kinetic energy as the first

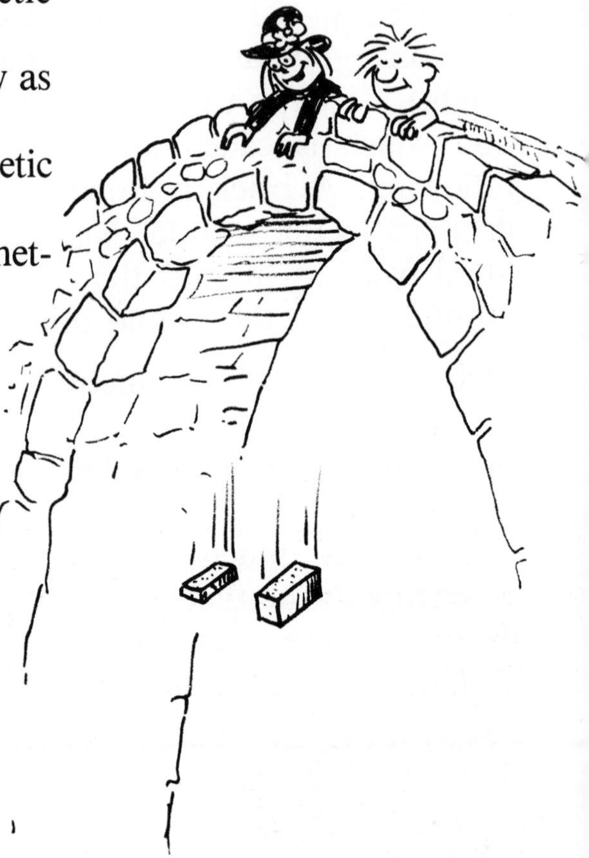

ANSWER: MORE KINETIC ENERGY

The answer is: c. Lifting the brick of double weight is like lifting the first brick twice and so takes twice as much work, which all goes into kinetic energy as it falls.

IN COURT

Preparing his case for trial, a lawyer pondered this question.* A 1-lb flower pot fell one foot from a shelf and struck his client squarely on the head. How much force did the pot exert on his client's head?

a) 1 lb
b) 4 lbs
c) 16 lbs
d) 32 lbs
e) The lawyer's question cannot be answered from the given information

ANSWER: IN COURT

The answer is: e. Why not? This sort of problem should be "duck soup" for a physics major, but no one could answer this without knowing how much "give" was in the skull (and neck and pot and hair or hat, if any). If the pot fell on something with lots of give, like a pillow, there would be very little force. If it fell on something like concrete — smash! If it struck something with no give at all the force would be infinitely large! Suppose there was one inch of give in the client, then all the energy put into the pot during the fall, one foot-pound, must be taken out as it comes to a stop in one inch. Now, an average force of 1 lb acting over one foot does as much work as an average force of 12 lbs acting over one inch. So if there was one inch of give, the average force was 12 lbs. If there was only a half-inch of give, the average force was 24 lbs. (Incidentally, the idea of a hard hat isn't so much that it's hard, but that there is lots of give in the webbing in the hat. Perhaps on another planet in another language they are called "give hats.")

---

*A question my father asked me — L. Epstein.

STEVEDORE

The stevedore is loading 100-pound drums on a truck by rolling them up a ramp. The truck bed is 3 feet above the street and the ramp is 6 feet long. How much force must she exert on the drums as she goes up the ramp?

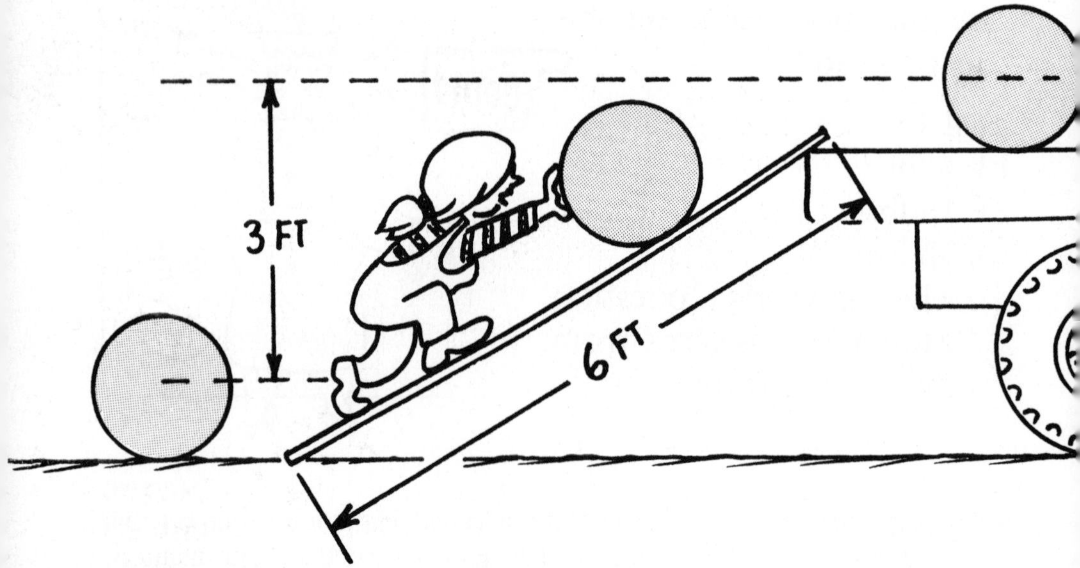

a) 200 lb
b) 100 lb
c) 50 lb
d) 10 lb
e) Can't say

ANSWER: STEVEDORE

The answer is: c. The 100-lb drums end up 3 ft higher than they were, so they end up with 3 ft × 100 lb = 300 ft lb more energy. Now what force exerted over the 6 ft length of the ramp will give 300 ft lb of work? It must be 50 lb of force because 6 ft × 50 lb = 300 ft lb. Three comments:

1) Most books solve this type of problem using a different method that involves vectors, as discussed earlier. To see how vectors are used to solve this type problem, look in other books. It's important to see things from different points of view. And it's nice to see that different viewpoints support the same conclusion.

2) If you can't add apples and oranges or feet and pounds, how come you can multiply them?

3) The idea of the inclined plane or ramp is basically the same as the idea that underlies the lever or see-saw and the block and tackle or compound pulley. In each case the force necessary to do a certain amount of work is reduced by increasing the distance over which the force acts. To lift the drum 3 ft requires only half as much force when moved twice as far (6 ft) on the inclined ramp. If lifting the heavy man on the see-saw a distance of one inch requires that the kid go down three inches, then the

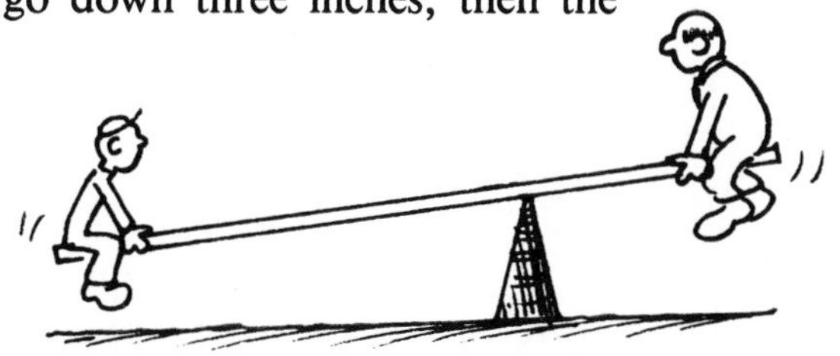

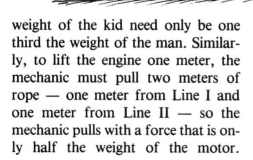

weight of the kid need only be one third the weight of the man. Similarly, to lift the engine one meter, the mechanic must pull two meters of rope — one meter from Line I and one meter from Line II — so the mechanic pulls with a force that is only half the weight of the motor.

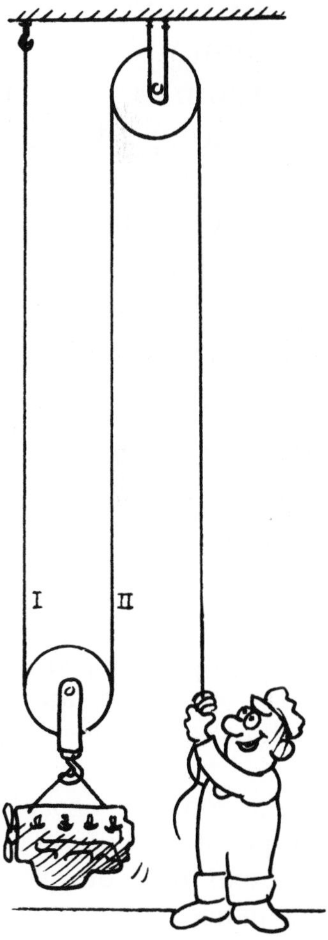

RIDING UP HILL

Suppose this block is 300 feet long, and quite steep. If I ride my bicycle up the hill along the zigzag path, which is 600 feet long, the average force I must exert is:

a) 1/4
b) 1/3
c) 1/2
d) equal to

the average force I would exert going straight up.

Again, up the same hill along the zigzag path the energy I must expend is:

a) 1/4
b) 1/3
c) 1/2
d) equal to

the energy I would spend going straight up.

ANSWER: RIDING UP HILL

The answer to the first question is: c; the answer to the second question is: d. All paths to the top require the same energy output. If some paths required more than others, I could go up those that required least and come back down those that required the most and so get back more energy than I put out in the first place—too good to be true.

Energy, in this case work, is force multiplied by distance. The energy to get to the top of the hill is the same on both paths but the distance is not the same on both paths. So if the distance gets doubled the force is cut in half.

STEAM LOCOMOTIVE

Locomotives for pulling passenger trains are different than locomotives for pulling freight. The passenger locomotive is designed for moving at high speed while the freight locomotive is designed for pulling heavy loads. Consider Locomotives I and II below: Notice the different sizes of the driving wheels and then decide which of the following statements is correct.

a) Locomotive I is for freight trains and II is for passenger trains.
b) Locomotive I is for passenger trains and II is for freight trains.
c) Both are for freight trains.
d) Both are for passenger trains.

I.

II

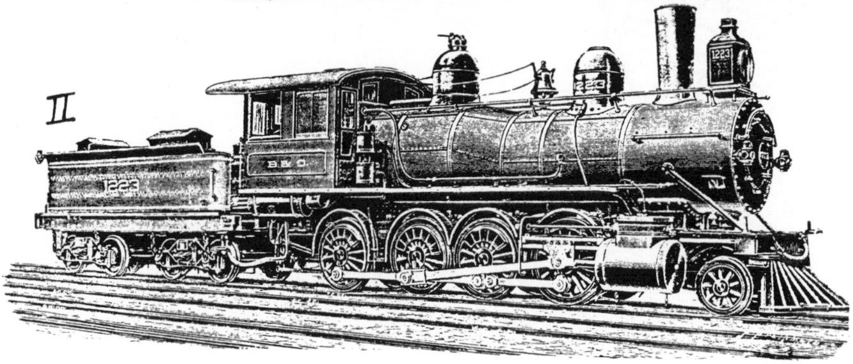

ANSWER: STEAM LOCOMOTIVE

The answer is: b. The passenger engine has the larger-diameter driving wheels. Because of the large circumference of the wheels, each piston stroke carries the passenger locomotive a longer distance. So even if both locomotives huff and puff at equal rates, the larger-wheeled passenger locomotive travels faster. The passenger locomotive makes fewer piston strokes and expends less steam per mile of travel than the smaller-wheeled freight locomotive. The freight locomotive therefore packs more steam and energy into each mile it travels. Like a heavy truck traveling in low gear, more energy is required to move the heavy freight train along a mile of track than is required to move the faster but lighter passenger train.

Incidentally, most of the steam locomotives you see these days in movies are freight locomotives, because there were many more of that type made.

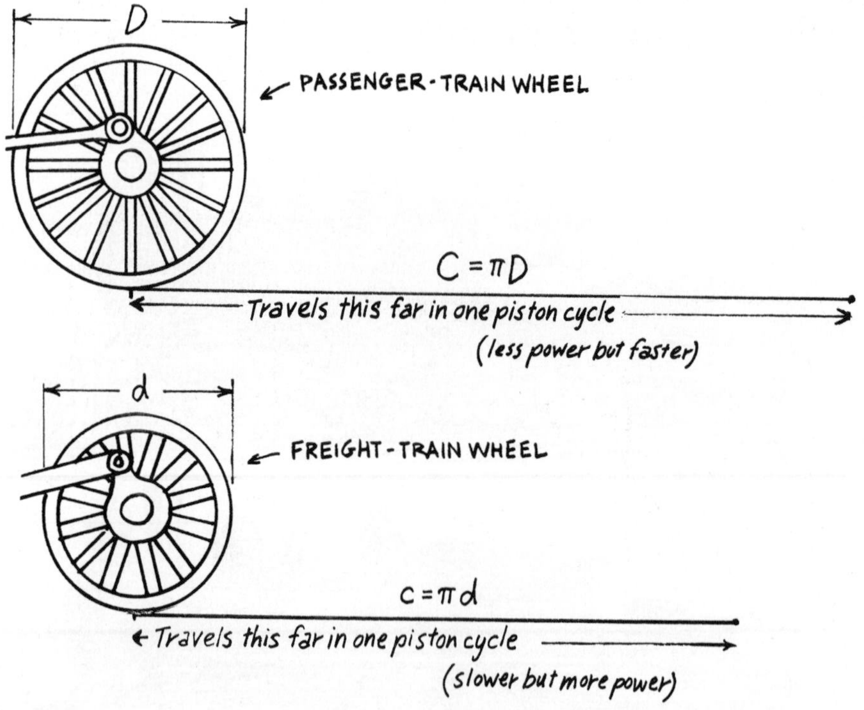

CRAZY PULLEY

The pivot point or axis of an ordinary pulley is at its center, and except for friction effects the tension in the rope that drapes over either side of the pulley is the same. But suppose that the axis were not at the center of the pulley wheel, as shown below. Then the tensions in the rope on each side of the pulley will

a) still be equal
b) be quite different

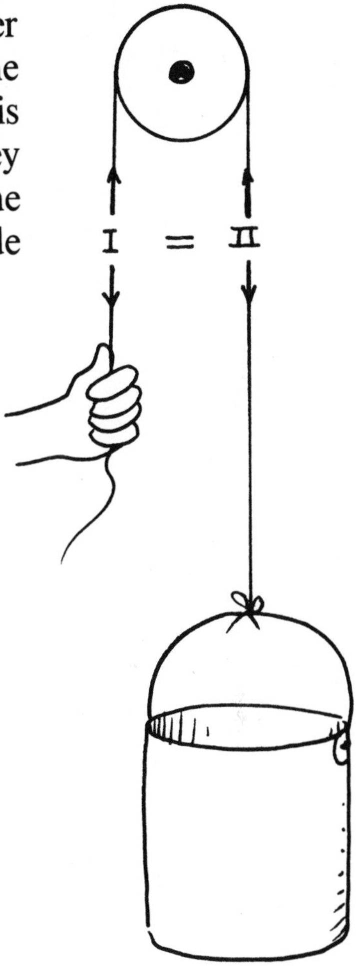

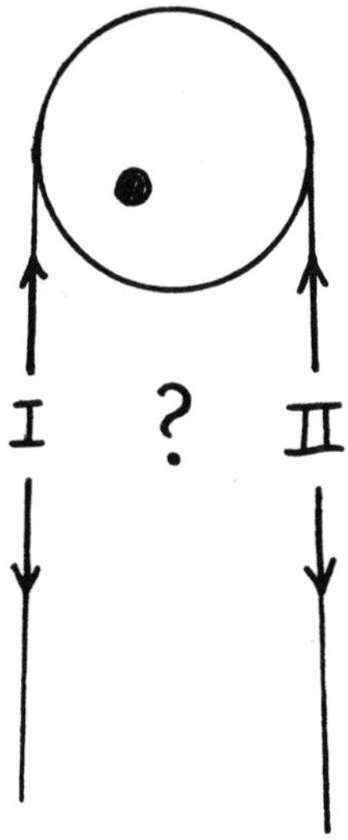

ANSWER: CRAZY PULLEY

The answer is: b. Why? The crazy pulley is a simple lever, disguised.

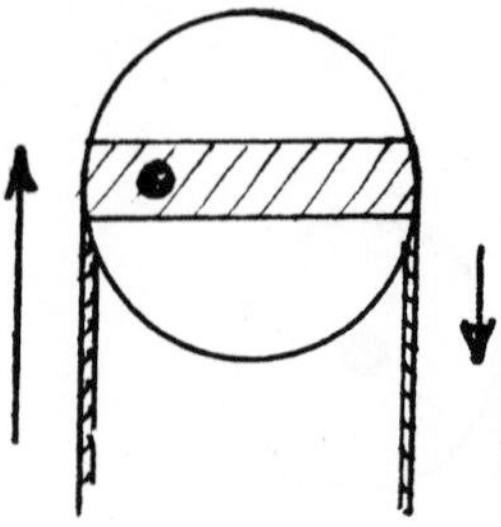

The crazy pulley is called an *eccentric* pulley. Can you see how the eccentric pulley allows the compound bow to be held in the cocked position with very little tension in the bow string? And can you see that unlike the conventional bow, the tension increases as the arrow is shot?

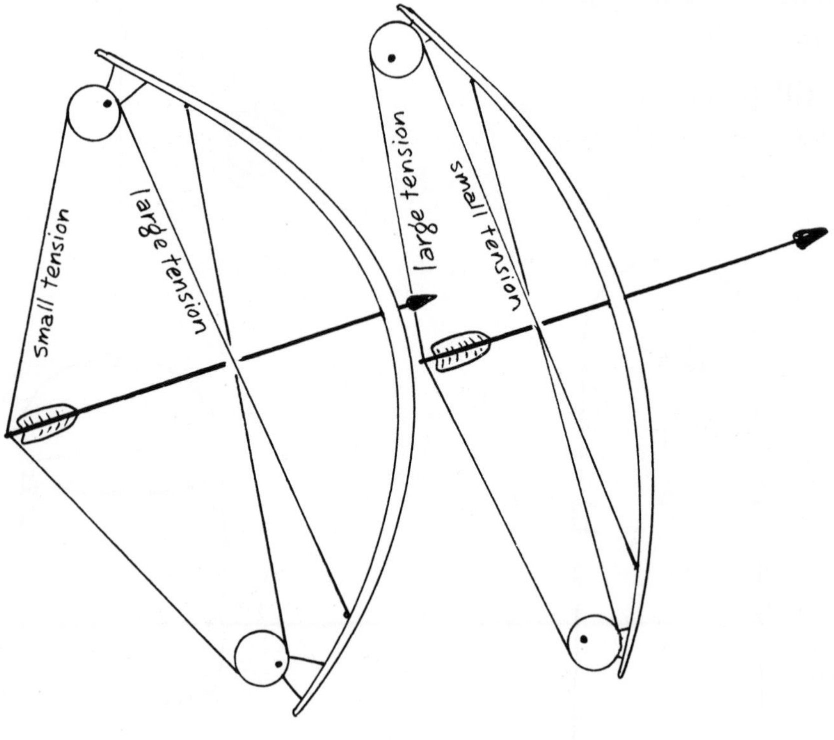

Incidently, mechanical engineers have a name for this kind of variable leverage mechanism. It is called a *toggle* action. Many electric light switches have toggle action, whence the name *toggle switch.*

RUNNER

A person starts from rest and begins to run. The runner puts a certain amount of momentum into herself and

a) more momentum into the ground
b) less momentum into the ground
c) the same amount of momentum into the ground

A person starts from rest and begins to run. The runner puts a certain amount of kinetic energy into herself and

a) more kinetic energy into the ground
b) less kinetic energy into the ground
c) the same amount of kinetic energy into the ground

ANSWER: RUNNER

The answer to the first question is: c. The answer to the second is: b. The measure of momentum is force multiplied by time. Both the force on the runner and the time during which it is exerted are equal to the force on the ground and the time during which it is exerted; of course the force on the ground is turned around so it pushes backwards. Thus the momentum put into runner and into the earth are equivalent (but opposite).

The measure of energy is force multiplied by distance. The force on runner and earth are equal (though opposite), but the distances they move while the force is exerted ARE NOT EQUAL. The runner might move several feet forward. The massive planet earth does not move a millionth of an inch backwards. So just about zero energy goes into the ground. All the runner's energy goes into the runner.

SWIMMER

A person starts from rest and begins to swim. The swimmer puts a certain amount of momentum into himself and

a) more momentum into the water
b) less momentum into the water
c) the same amount of momentum into the water

A person starts from rest and begins to swim. The swimmer puts a certain amount of kinetic energy into himself and

a) more kinetic energy into the water
b) less kinetic energy into the water
c) the same amount of kinetic energy into the water

ANSWER: SWIMMER

The answers are: c and a. The swimmer puts as much momentum into the water as (opposite) momentum into himself for the same reason the runner put equivalent (but opposite) momentum ino the earth and into herself.

But when you come to energy the story changes. The force on the swimmer and on the water are equal (though opposite). But while the swimmer pushes a handful of water a yard backwards his body moves much less than a yard forward. (This is because the mass of a handful of water is much less than the mass of the swimmer's body.) Since energy is force multiplied by distance, and since each stroke pushes the water farther than the swimmer, more energy goes into the water than into the swimmer.

So the swimmer puts most of his energy into the water while the runner puts almost all of her energy into herself. That's why a poor runner can go faster than an excellent swimmer.

AVERAGE FALLING SPEED

If a rock falls for one second, what is its average speed during that second?

a) Zero ft/s
b) 1 ft/s
c) 4 ft/s
d) 16 ft/s
e) 32 ft/s

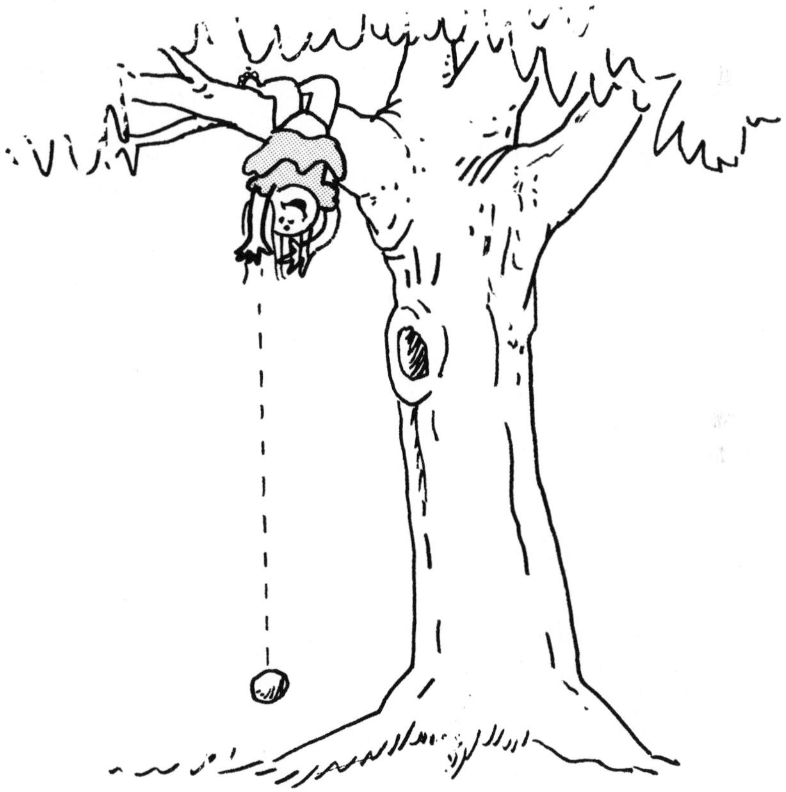

ANSWER: AVERAGE FALLING SPEED

The answer is: d. Although the speed of the rock at the end of the one-second interval is 32 ft/s, its beginning speed was zero — it was dropped from a rest position. So the average speed could not be 32 ft/s, anymore than it could be zero. Since the acceleration during the fall was constant, the average speed is simply 16 ft/s, midway between zero and 32 ft/s. We distinguish between instantaneous speed, the speed at a particular instant and the overall or average speed. Incidentally, since the stone has an average speed of 16 ft/s, it must fall a distance of 16 feet during that second.

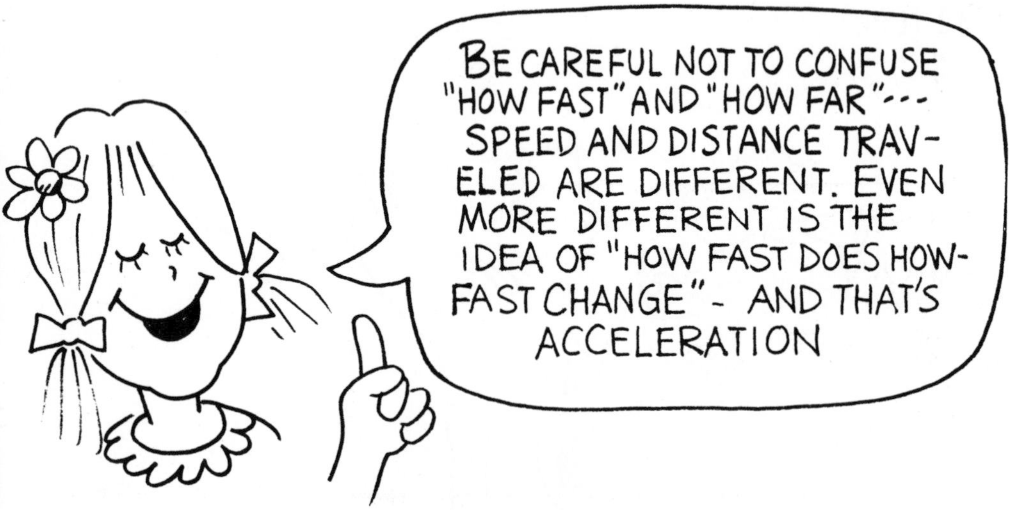
BE CAREFUL NOT TO CONFUSE "HOW FAST" AND "HOW FAR"--- SPEED AND DISTANCE TRAVELED ARE DIFFERENT. EVEN MORE DIFFERENT IS THE IDEA OF "HOW FAST DOES HOW-FAST CHANGE" - AND THAT'S ACCELERATION

MORE AVERAGE FALLING SPEED

To be sure you understand the previous answer, consider this: If a rock falls for two seconds, what would be its average speed during the two seconds?

a) 1 ft/s
b) 4 ft/s
c) 16 ft/s
d) 32 ft/s
e) 64 ft/s

ANSWER: MORE AVERAGE FALLING SPEED

The answer is: d. The speed starts from zero and goes faster and faster at a rate of 32 ft/s every second. After two seconds the speed has increased to sixty-four feet per second (64 = 2 × 32). The average between zero and sixty-four is thirty-two. So 32 ft/s is the average speed. How far did the stone fall during those two seconds? Well, total distance equals average speed multiplied by total time, so total fall distance = 32 ft/s × 2 seconds = 64 feet.

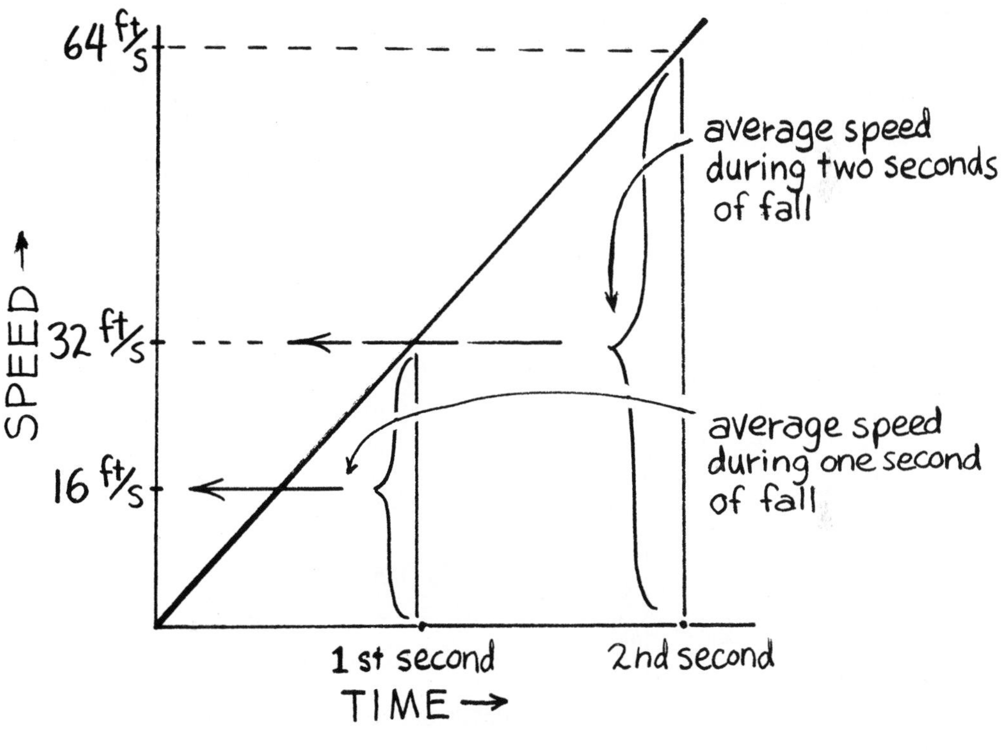

FALLING HOW FAR

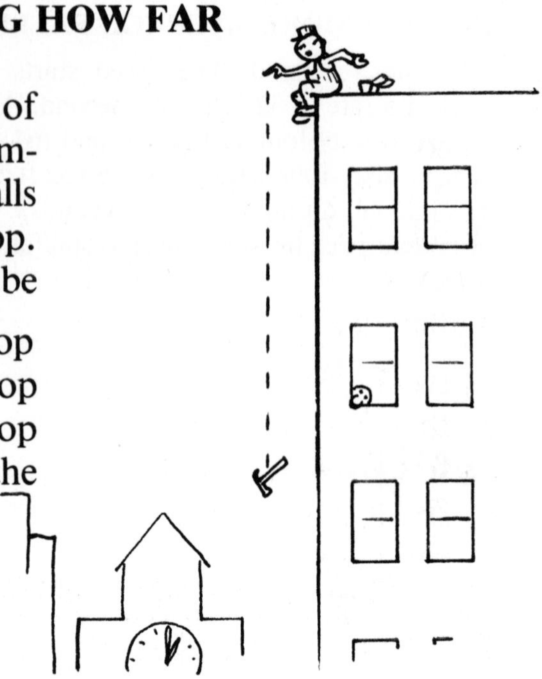

The carpenter at the top of a tall building drops his hammer. In one second it falls one story down from the top. In one more second it will be

a) two stories below the top
b) three stories below the top
c) four stories below the top
d) sixteen stories below the top
e) none of the above

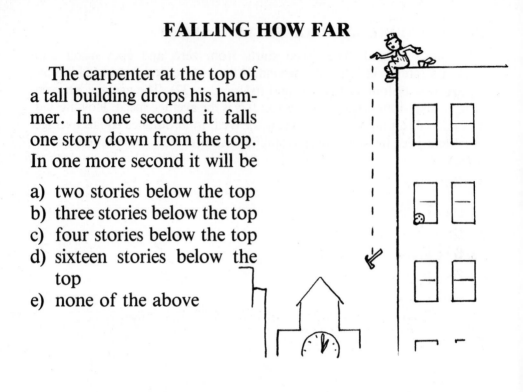

ANSWER: FALLING HOW FAR

The answer is: c. Some people might think if it went one story in one second it would have gone two stories in two seconds, and that would be true if it fell always at the same speed — but it doesn't. It goes faster and faster as it falls, so its travel during each second is not the same. Its travel increases during each second. After two seconds both the average speed and the time of travel have doubled so the total distance of travel goes up four times.

Remember: distance = average speed X time.

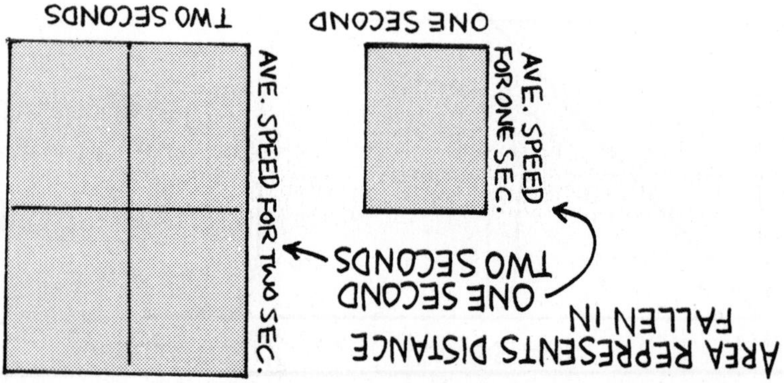

SPLAT

A bottle dropped from a balcony strikes the sidewalk with a particular speed. To double the speed of impact you would have to drop the bottle from a balcony

a) twice as high
b) three times as high
c) four times as high
d) five times as high
e) six times as high

ANSWER: SPLAT

The answer is: c. Simple street sense would seem to suggest a balcony twice as high. But to get double speed it has to fall for double time — and in double time it falls four times as far (recall FALLING HOW FAR). That means you have to lift it four times as far and that means you have to put four times as much energy into it. Now if it goes twice as fast it has twice as much momentum (see MOMENTUM) — but four times as much energy. So doubling the momentum of an object doesn't double the kinetic energy — it increases the energy four times! So we see there is a great difference between kinetic energy and momentum.

SHOOT THE CHUTE

To be sure you understand the preceding question and answer, consider this one: A roller coaster is pulled to the top of a "shoot the chute" and allowed to roll down. For a bigger thrill you might wish the car to be going twice as fast at the bottom of the run. To make your wish come true, the chute should be

a) twice as high
b) three times as high
c) four times as high
d) five times as high
e) six times as high

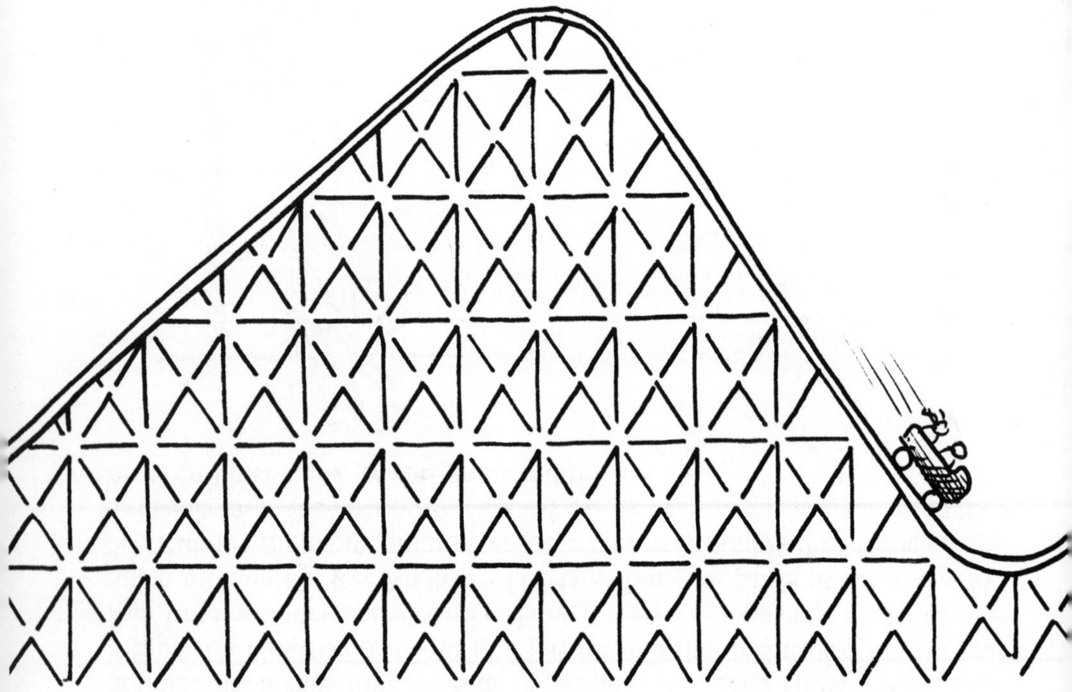

ANSWER: SHOOT THE CHUTE

The answer is: c. This is very much like SPLAT. To double speed you double momentum. To double momentum you increase kinetic energy four times. To increase kinetic energy four times you have to lift four times as far. How can the relationship between kinetic energy and momentum be conceptually visualized? As a slice of bread, the kinetic energy is the white part and the momentum is the crust. If you increase the size of a slice, which corresponds to increasing the speed of an object, you will see the amount of

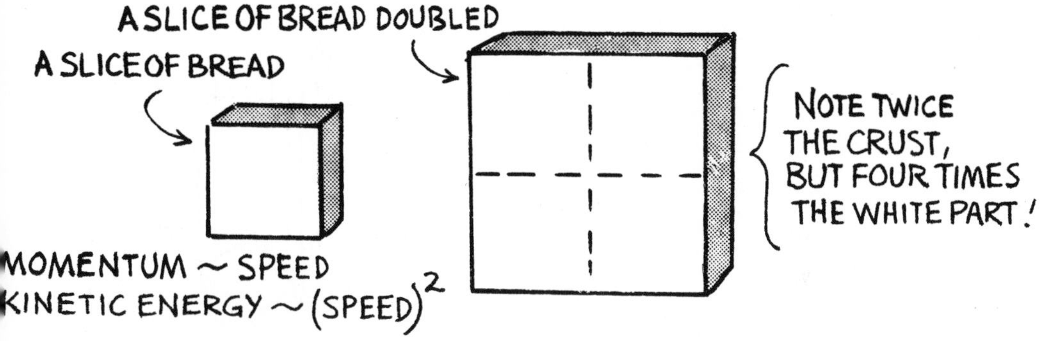

white increases four times while the amount of crust doubles. To conceptualize increasing the mass of the moving object just picture more slices or simply one thicker slice (remember, a two-pound thing is two one-pound things). The slice of bread, crust and white together corresponds to what was once called impedo in the days of Galileo before the distinction between kinetic energy and momentum was clearly recognized.

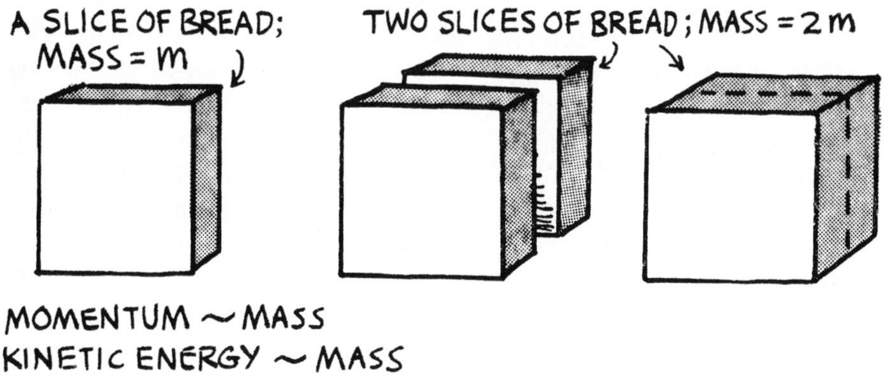

GUSH

You throw a stone into some nice soft and gushy mud. It penetrates one inch. If you wanted the stone to penetrate four inches, you would have to throw it into the mud

a) twice as fast
b) three times as fast
c) four times as fast
d) eight times as fast
e) sixteen times as fast

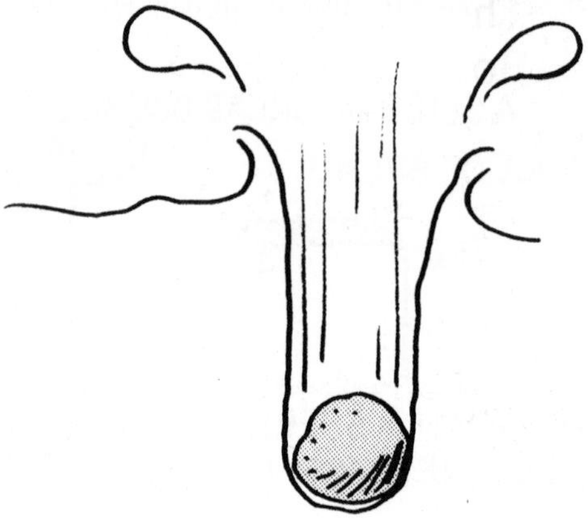

ANSWER: GUSH

The answer is: a. Again, common sense may be misleading. Four times the speed should give four times the penetrating ability, but it isn't so. Why? Push your finger into mud. To no one's surprise a certain force is required. To push four times as far the same force must be exerted over four times the distance. Therefore four times as much energy is required. To increase the kinetic energy in a stone four times, its speed need only be doubled (recall SPLAT). Consider the case of a bullet. If we double the muzzle velocity, while leaving the mass of the bullet unchanged, we will double the bullet's ability to knock things down because its momentum is doubled — but we will increase its ability to penetrate things four times. Kinetic energy and momentum are different from one another.

RUBBER BULLET

A rubber bullet and an aluminum bullet both have the same size, speed, and mass. They are fired at a block of wood. Which is most likely to knock the block over?

a) The rubber bullet
b) The aluminum bullet
c) Both the same

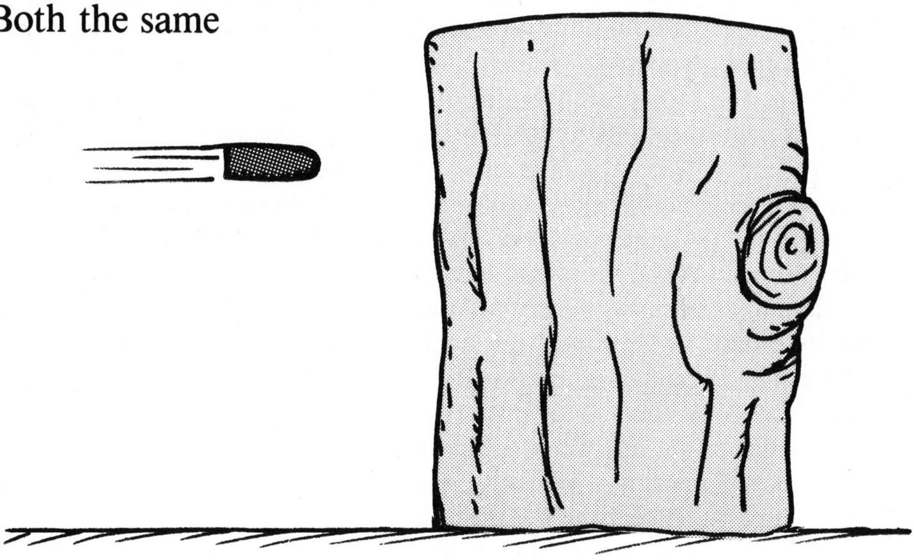

Which is most likely to damage the block?

a) The rubber bullet
b) The aluminum bullet
c) Both the same

ANSWER: RUBBER BULLET

The answer to the first question is a, and the second is b. In both cases the equal momenta of the bullets is changed upon impact with the block. But the impulses involved are different because the rubber bullet bounces and the aluminum bullet penetrates. The momentum of the aluminum bullet is completely transferred to the block which supplies the necessary impulse to stop it. But the impulse is greater for the rubber bullet, for the block not only supplies the necessary impulse to stop it, but provides additional impulse to "throw the bullet back out again." Depending on the elasticity of the rebound, this results in up to twice the impulse for the impact of the rubber bullet and therefore up to twice the momentum imparted to the block. So the rubber bullet is very much more likely to knock the block over. If you tell your sharpshooter-type acquaintances that rubber bullets are more effective for knocking things down, they may not believe you — but its true!

Now for the second half of the question. Although the rubber bullet gives the block the most momentum, it does not give it the most energy. If the bullet bounces back with a lot of speed, that means it keeps much of its kinetic energy for itself, whereas the aluminum bullet stops and therefore surrenders all its kinetic energy. The surrendered energy goes into the block. It is most important to note that this added energy the aluminum bullet puts in does not have momentum to go with it! Energy without momentum can't be kinetic energy. It must be some other kind of energy — the energy of heat, deformation and damage.

So we see that the rubber bullet puts a lot of momentum but little energy into the block, and the aluminum bullet puts much more energy but less momentum into the block.

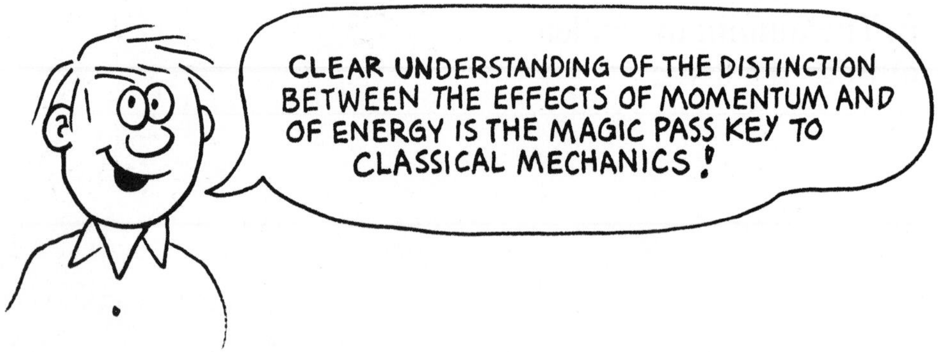
CLEAR UNDERSTANDING OF THE DISTINCTION BETWEEN THE EFFECTS OF MOMENTUM AND OF ENERGY IS THE MAGIC PASS KEY TO CLASSICAL MECHANICS!

STOPPING

A car is going 10 mph. The driver hits the brakes. The car travels 3 feet after the brakes are applied. A while later the same car is going 20 mph. The driver hits the brakes. About how far does the car go after the brakes are applied?

a) 3 ft
b) 6 ft
c) 9 ft
d) 12 ft
e) 15 ft

ANSWER: STOPPING

The answer is: d. Stopping a car is like throwing a rock straight up into the air. There is a constant stopping force acting on the car, provided by the brakes. There is a constant stopping force acting on the rock, provided by gravity. To see how a rock acts when it is thrown up into the air just make a movie of one falling down and then run it backwards in your head. In order to double the speed of a falling rock its time of fall must be doubled whereupon its distance of fall automatically increases four times. That means if you throw the rock up with double speed it goes four times as high before stopping.

So also if the car is doing 20 mph. Its stopping distance is four times longer than it was when doing 10 mph. This might be the most practical and important thing you learn from this book. If you double a car's speed, the TIME it takes to stop is doubled, but the DISTANCE it takes to stop goes up FOUR times — and the distance, not the time, determines what you will or will not hit.

SMUSH

A one-pound lump of clay travelling at one foot per second smashes into another one-pound lump of clay which is not moving. Smush! They stick and become one two-pound lump. What is the speed of the two-pound lump?

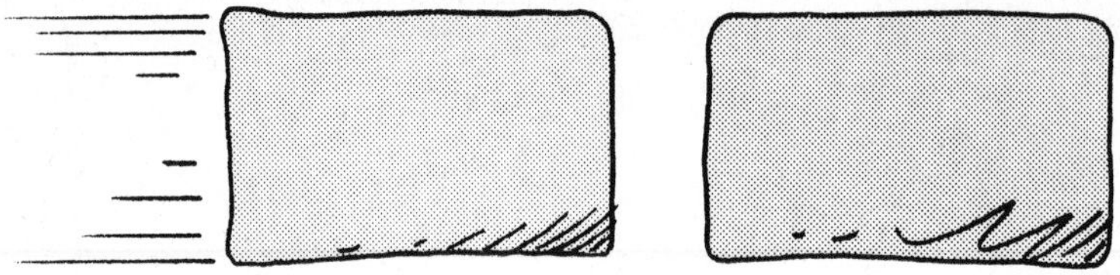

a) Zero ft/s
b) One-quarter of one ft/s
c) One-half of one ft/s
d) One ft/s
e) Two ft/s

ANSWER: SMUSH

The answer is: c. The speed lost by the moving lump goes into the one which was not moving and the exchange goes on until the speeds equalize. Another way to say this is that any momentum lost by the moving lump goes into the one which was not moving. This is because the impulse that increases the speed of the stationary lump is exactly equal and opposite to the impulse slowing the originally moving lump. Since momentum lost by one lump goes into the other, no momentum is lost. So the momentum after collision is the same as before. Momentum is mass multiplied by speed. Before collision all momentum is in the moving lump and none in the stationary lump. The effect of the collision is to double the mass of the originally moving mass without adding or deducting momentum from it. If mass doubles and momentum remains constant the speed must be cut in half. Suppose the one-pound moving lump struck a two-pound stationary lump. Then the effect of the collision is to triple the mass of the originally moving mass without adding or deducting momentum from it. If mass triples and momentum remains constant the speed must be cut to one-third of its original value.

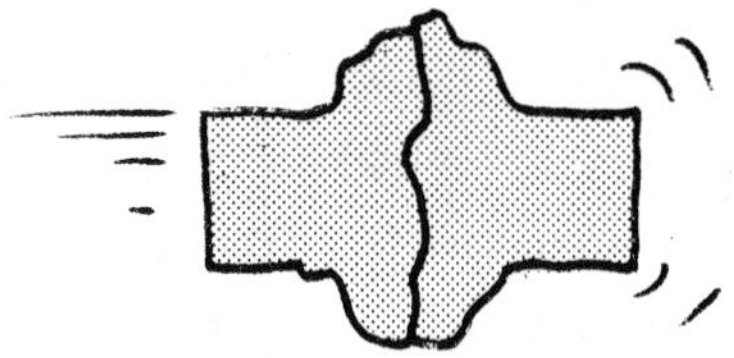

This illustrates a very important law — the Conservation of Momentum. When we make the momentum come out the same at the end as it was in the beginning we are conserving momentum. Why did we not try to think this problem out in terms of conservation of kinetic energy? Because kinetic energy is not conserved. Some of it gets turned to heat when the lumps smash and deform. The deformation takes energy; in a car accident the energy is required to "bend the tin." If a lump of clay hits a stone wall all the kinetic energy gets turned to heat. But if a perfectly elastic ball hits a stone wall it bounces off without turning any of its kinetic energy into heat and therefore doesn't lose any of its kinetic energy. There is one more thing worth saying here. In a way, heat is really hidden kinetic energy. When all the molecules are moving in the same direction it is called ordinary kinetic energy or mechanical kinetic energy. When all the molecules are bouncing about in different directions the clay as a whole does not move, the kinetic energy is mixed up and hidden, and it is called thermal kinetic energy or heat.

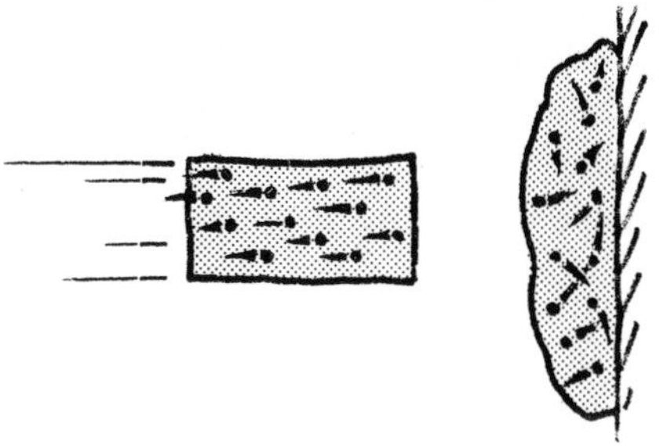

SMUSH AGAIN

A one-pound lump of clay travelling at one foot per second smashes into another one-pound lump of clay which is not moving. Smush again! They stick and become one two-pound lump of clay. What proportion of the kinetic energy in the originally moving lump was turned into heat during the collision?

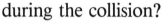

a) Zero%
b) 25%
c) 50%
d) 75%
e) 100%, that is, all the kinetic energy was turned into heat

ANSWER: SMUSH AGAIN

The answer is: c. Recall from SMUSH that the speed of the lumps when stuck together is half the speed of the single lump before smushing. Now think of the two-pound lumps as a pair of one-pound lumps. Since these lumps are moving only half as fast as the lump that did the hitting, their kinetic energies must each be one-quarter that of the single lump before smushing (recall SPLAT). Since there are two lumps in the pair, there is ¼ + ¼ = ½ as much energy as in the original single lump. So half (50%) of the kinetic energy in the originally moving lump ends up as kinetic energy in the two-pound lump. The lost half was turned into heat. What if they made a noise — like smush — when they came together? Would not that noise take some of the lost kinetic energy, so not all of it would turn into heat? But what becomes of noise? What becomes of all sound, talk included? What is a synonym for talk? Hot air. A noise turns into heat and does so very quickly. How quickly? In the time it takes an echo to die.

ROLLING IN THE RAIN

Suppose an open railroad car is rolling without friction in a vertically-falling downpour and an appreciable amount of rain falls into the car and accumulates there. Consider the effect of the accumulating rain on the speed, momentum, and kinetic energy of the car.

The *speed* of the car will

a) increase
b) decrease
c) not change

The *momentum* of the car will

a) increase
b) decrease
c) not change

And the *kinetic energy* of the car will

a) increase
b) decrease
c) not change

ANSWER: ROLLING IN THE RAIN

The answer to the first question is b; the second c; and the third, b. The rolling car has momentum only in the horizontal direction. The rain falls straight downward and has no horizontal momentum to add to the car. So the momentum of the car doesn't change. The mass of the car, however, does change — it's increased by the mass of accumulated rain. A mass increase while momentum is constant results in a decrease in velocity. So the car slows down as rain accumulates. This situation is almost a repeat of SMUSH AGAIN. Speed and kinetic energy are cut while momentum is unefected. What happens to the lost kinetic energy? It goes into heat — the water in the car is a bit warmer than the rain.

We have reasoned so far with only the conservation rules for momentum and energy. Answers to many questions are provided by these powerful rules that completely bypass often very messy forces. But let's think about force anyway to better understand our conclusions. The falling rain that hits the car ends up with the horizontal velocity of the car. So a force had to act on it, whether by interaction with the wall, floor, or surface of accumulated water. Whatever force acts on the raindrops to give them a horizontal velocity also acts on the car. It is this reaction force that slows the car.

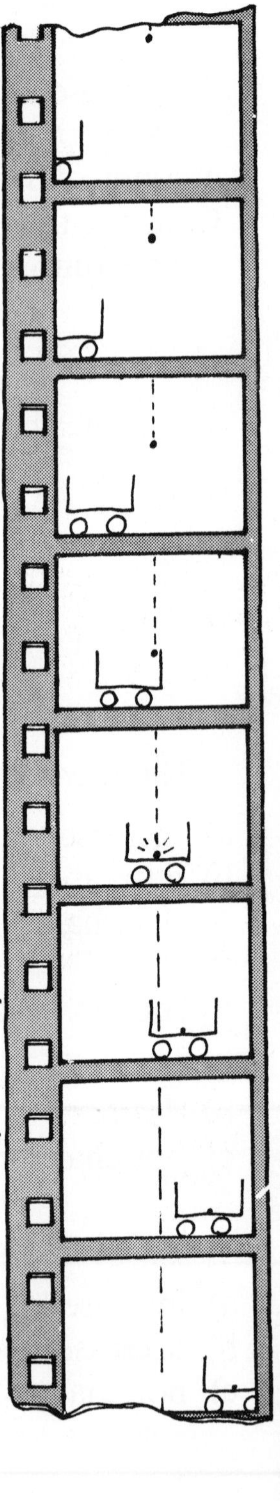

SEE HOW THE VERTICALLY-FALLING DROP IS FORCED TO THE RIGHT WHEN IT STRIKES THE CAR ? ITS THE REACTION TO THE LEFT THAT SLOWS THE CAR

ROLLING DRAIN

The rain has stopped. A drain plug is opened in the bottom of the rolling car allowing the accumulated water to run out. Consider the effects of the draining water on the speed, momentum, and kinetic energy of the rolling car.

The *speed* of the car will

a) increase
b) decrease
c) not change

The *momentum* of the car will

a) increase
b) decrease
c) not change

And the *kinetic energy* of the car will

a) increase
b) decrease
c) not change

ANSWER: ROLLING DRAIN

The answer to the first question is c; the second b; and the third, b. If you simply let go of something you are holding, it does not exert any force on you and you do not exert any force on it. As the water is released from the car it exerts no force on the car so the speed of the car is not changed. The water just drops away with the same horizontal velocity it had while in the car — like some clod dropping a beer can out of the window of a moving car. Of course the departing water takes its momentum and kinetic energy with it when it goes. So the car is left with less momentum and less kinetic energy.

HOW MUCH MORE ENERGY

This one will stump most readers. The chemical potential energy in a certain amount of gasoline is converted to kinetic energy in a car that increases its speed from 0 mph to 32 mph. To pass another car the driver accelerates to 64 mph. Compared to the energy required to go from 0 to 32 mph, the energy required to go from 32 mph to 64 mph is

a) half as much
b) as much
c) twice as much
d) three times as much
e) four times as much

ANSWER: HOW MUCH MORE ENERGY

The answer is: d. The speed increase from 0 to 32 mph was 32 mph and that from 32 mph to 64 mph was also 32 mph so you might think the energy increase was the same in both cases, but if so you are in for a new view. Pretend 0 mph, 32 mph, and 64 mph are 0 ft/s, 32 ft/s, and 64 ft/s, and recall FALLING HOW FAR. The object going 64 ft/s must have fallen for twice as much time and four times as much distance as the one going 32 ft/s. So the faster one has four times as much kinetic energy as the slow one. Let us suppose that 32 mph or 32 ft/s is equivalent to one "drop" of kinetic energy. Then 64 mph or 64 ft/s equals four "drops" of kinetic energy (say big drops for mph, and little drops for ft/s). So to get from 32 to 64 you have to add three more "drops."

SPEED AIN'T ENERGY

A truck initially at rest at the top of a hill is allowed to roll down. At the bottom its speed is 4 mph. Next, the truck is again rolled down the hill, but this time it does not start from rest. It has an initial speed of 3 mph on top, even before it starts going down the hill. How fast is it going when it gets to the bottom?

a) 3 mph
b) 4 mph
c) 5 mph
d) 6 mph
e) 7 mph

ANSWER: SPEED AIN'T ENERGY

The answer is *not* e — if you chose this answer, 7 mph, go back and *think* some more. You might review HOW MUCH MORE ENERGY!

Going down the hill adds a certain amount of kinetic energy, but *NOT a certain amount of speed.* If it is just added speed, the answer would be 3 + 4 = 7, but this is *wrong.*

But why don't the speeds add? Because to pick up 4 mph on the hill the truck must spend a certain amount of time on the hill. When it starts off at 3 mph it spends less time on the hill and so picks up less speed going down. We can, however, find something that *does* add — and that's energy. We know doubling speed increases energy four times, and tripling speed increases the energy nine times and so on. Kinetic energy is proportional to the square of the speed. So if the initial speed 3 has 9 units of energy, speed 4 (which rolling down the hill provides) has 16. So the total energy is 9 + 16 = 25 energy units. Now 25 energy units corresponds to what speed? Speed 5 — so the answer is: c.

Note that the kind of speed or energy units is not really important here. If the speed were meters per second rather than miles per hour the answer would simply have been 5 m/s rather than 5 mph. The important thing is that energies are proportional to the speeds squared, and unlike momentum, energies always combine by simple addition.

PENETRATION

A truck initially at rest rolls down Hill 1 into a very big haystack. Another identical truck also rolls from rest down Hill 2, twice as high, into an identical haystack. Compared to the truck on Hill 1, how much farther does the truck on Hill 2 penetrate into the stack?

a) Equally as far
b) Twice as far
c) Three times as far
d) Four times as far

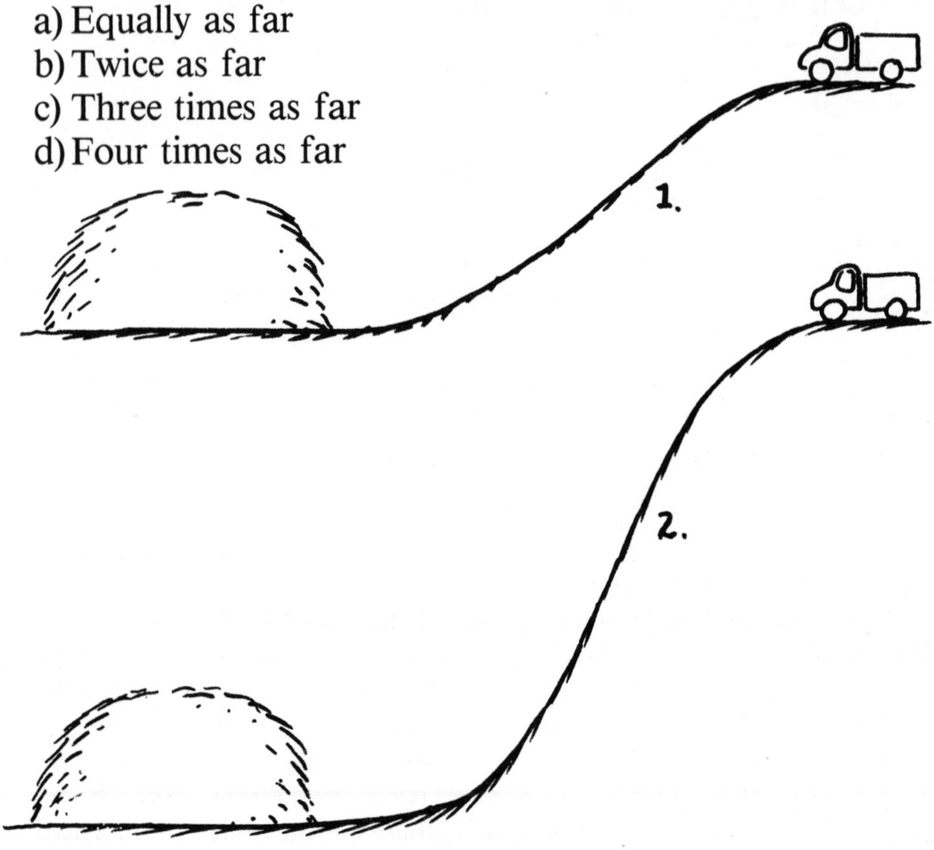

ANSWER: PENETRATION

The answer is: b. Double the distance that gravity acts on the truck rolling down the hill and you double the truck's kinetic energy, which doubles the distance the force can act as the truck penetrates the hay.

TACKLE

Mighty Mike weighs 200 lbs and is running down the football field at 8 ft/s. Speedy Gonzales weighs only 100 lbs but runs 16 ft/s, while Ponderous Poncho weighs 400 lbs and runs only 4 ft/s. In the encounter who will be more effective in stopping Mike?

a) Speedy Gonzales
b) Ponderous Poncho
c) Both the same

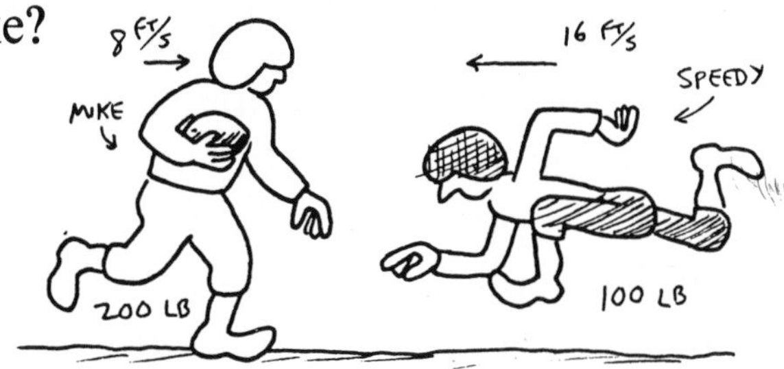

Who is more likely to break Mike's bones?

a) Speedy Gonzales
b) Ponderous Poncho
c) Both the same

ANSWER: TACKLE

The answer to the first question is: c. This is straight momentum conservation in action. Note that Mike's momentum is exactly opposed by the momentum of Speedy or Poncho ($200 \times 8 = 100 \times 16 = 400 \times 4$). So the impulse that stops Mike upon collision is the same whether by Speedy or Poncho. Both are equally effective in bringing Mike to a halt.

The answer to the second question is: a. Even though Speedy and Poncho have the same stopping effect or momentum or impulse or punch, Mike will be hurt more by the collision with Speedy (ask any football player). Why? Because Speedy has more kinetic energy than Poncho. Remember GUSH and STOPPING. When you double the speed of a thing it takes double the time to stop, but it takes *four times as much distance to stop.* It penetrates four times as deep. That means it has four times as much kinetic energy. Now Speedy is going four times faster than Poncho. Does this mean he penetrates sixteen times deeper than Poncho, using sixteen times as much kinetic energy? No. Why? Because Speedy has only one fourth as much mass as Poncho, so he has one-fourth of sixteen times as much kinetic energy — four times as much. So Speedy's four-fold energy penetrates four times farther into Mike. That's why it hurts more to be tackled by Speedy than by Poncho.

THE PILE DRIVER

Watching piles being driven into the ground is an interesting sight — the huge machine, the chug and puff and hiss of the engine and then the bang of the hammer. Suppose the hammer and the pile each weigh one ton. Then suppose the hammer drops on the pile from a height of two feet and the impact drives the pile one inch into the earth. What would be the average force of the pile on the ground as it penetrates that one inch?

a) One ton
b) Two tons
c) Twelve tons
d) Thirteen tons
e) Fourteen tons

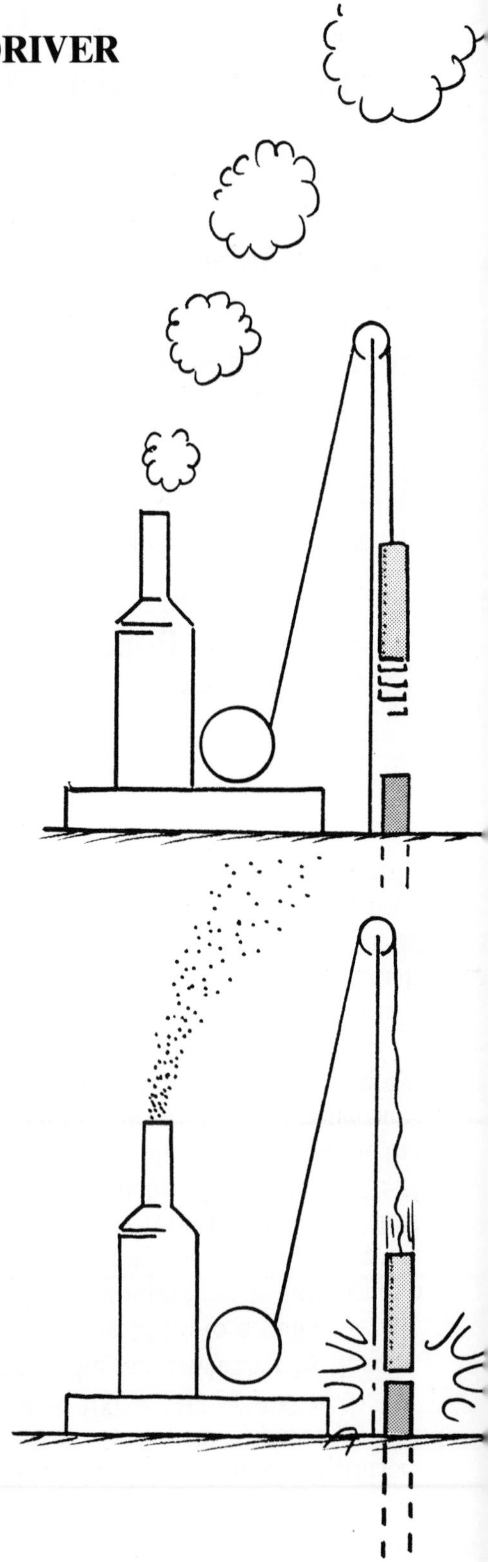

ANSWER: THE PILE DRIVER

The answer is: e. There are two foot tons of kinetic energy in the hammer when it bangs the pile. Since the hammer and the pile have equal masses, half of that energy is converted to heat by the impact (recall SMUSH AGAIN). That leaves one foot ton of kinetic energy to drive the pile one inch. Now one foot ton is equal to twelve inch tons. That is, a force of one ton pushing one foot does as much work as a force of twelve tons pushing one inch. So does this mean the pile exerts a force of twelve tons on the earth? Even more! The pile pushes on the ground with a weight of one ton even before the hammer hits it, and after the impact the hammer which also weighs one ton momentarily rests on the pile. So there are two more tons — fourteen in all. Does that mean fourteen inch tons of work were done on the earth? Yes, the two extra inch tons came from the weight of the pile and hammer as they went down that last inch. Tricky? A little. That's why engineers earn as much per hour as plumbers — or do they?

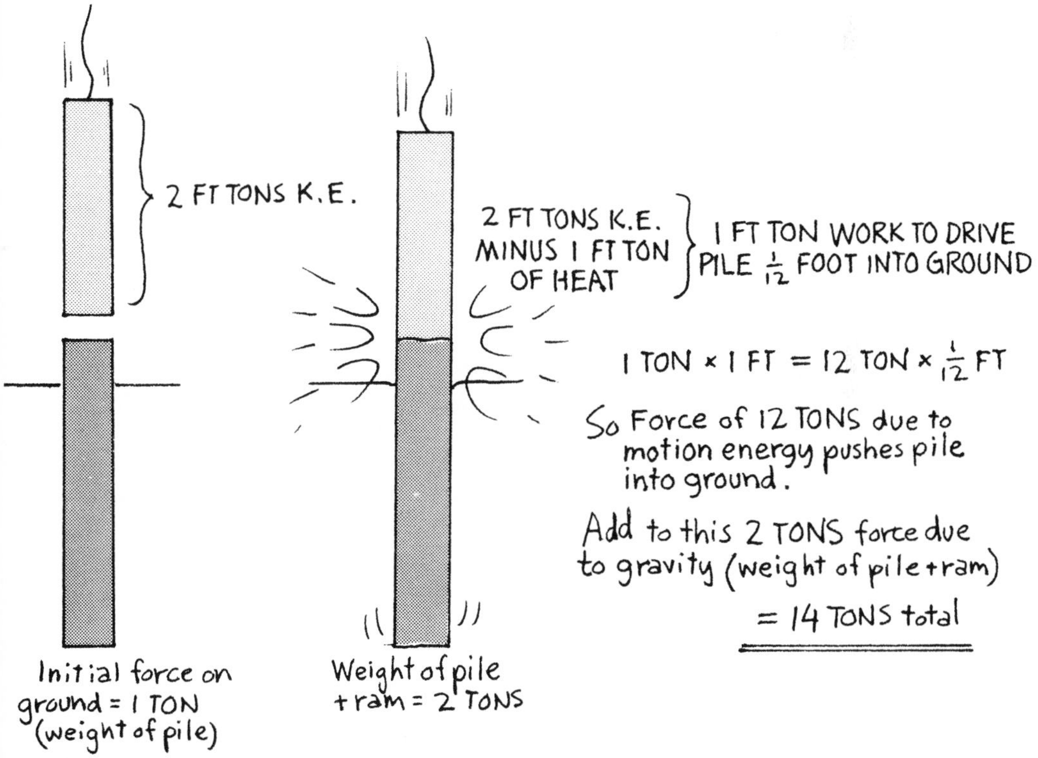

2 FT TONS K.E.
2 FT TONS K.E. MINUS 1 FT TON OF HEAT
1 FT TON WORK TO DRIVE PILE 1/12 FOOT INTO GROUND
1 TON × 1 FT = 12 TON × 1/12 FT
So Force of 12 TONS due to motion energy pushes pile into ground.
Add to this 2 TONS force due to gravity (weight of pile + ram)
= 14 TONS total
Initial force on ground = 1 TON (weight of pile)
Weight of pile + ram = 2 TONS

SLEDGE

In the classroom lecture demonstration* pictured below, the anvil shields the daring physics professor from most of the sledgehammer's

a) momentum
b) kinetic energy
c) . . . both
d) . . . actually neither

*This is a demonstration that I and my students of several semesters ago will likely never forget. I foolishly asked for a student volunteer from the class to man the sledge. In his excitement he missed the anvil and hit my hand, breaking it in two places! I now do the demonstration only with a practiced assistant. —Paul Hewitt

ANSWER: SLEDGE

The answer is: b. Every bit of momentum imparted to the anvil by the sledge is imparted to the professor (and subsequently to the earth that supports him). The anvil doesn't shield the professor from the sledge's momentum — not a bit. The shielding of kinetic energy is a different story. A significant fraction of the sledge's kinetic energy never gets to the professor — it is absorbed by the anvil in the form of heat. Have you ever noticed that a hammer head gets warm after you have been hammering vigorously? Heat is the graveyard of kinetic energy.

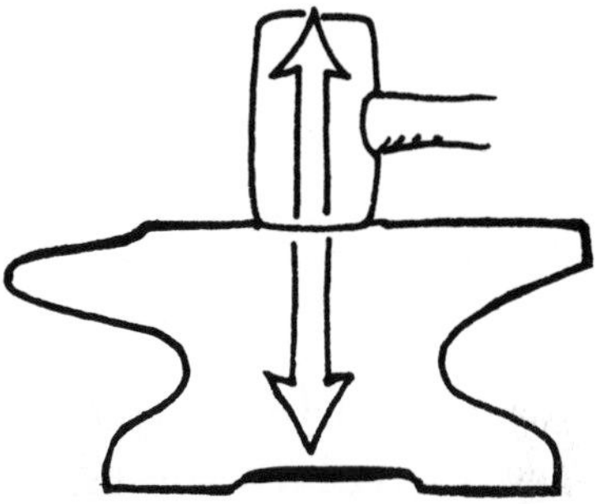

We can be a bit more insightful about this and investigate the goings on during the impact between the sledge and anvil. During impact the force on the anvil at every instant is equal and opposite to the force on the sledge at that same instant. The sledge acts on the anvil just as long and just as hard as the anvil acts (reacts) on the sledge. Therefore the impulse or punch that stops the sledge is exactly equal to the impulse or punch that goes into the anvil, and then into the professor. If the sledge comes to a dead stop, the impulse must have cancelled all the sledge's momentum, and that same impulse must put the same amount of momentum into the anvil. So we see that the anvil gets every bit of the momentum the sledge loses — the momentum is completely transferred from the sledge to the anvil. Of course, the newly acquired momentum does not make the anvil move very fast because it has much more mass than the sledge.

Now consider kinetic energy. When we analyze momentum we think about the *time* during which forces act, but when we analyze energy we think about the *distance* through which forces act. That's because the energy a body acquires is equal to force multiplied by the distance over which the force pushes the body. Consider in the sketch the relative distances that the sledge and anvil move during impact. Note that the sledge moves from I to II, while the anvil only moves from 1 to 2, which is a

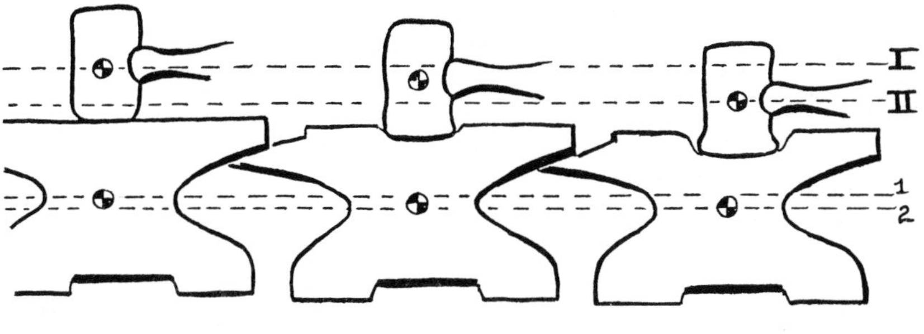

smaller distance.* Equal forces but unequal distances result in unequal changes in kinetic energy — the sledge loses more kinetic energy than the anvil gains.

So while all the momentum of the sledge is transferred to the anvil and then to the professor, all the kinetic energy is not. The professor is shielded from the kinetic energy and he will return to lecture again.

*Another way to see this is to reason that during impact, the sledge's speed drops from about 30 mph to 1 mph, while the anvil's speed increases from 0 mph to 1 mph. Although they both end up with the same speed, the sledge was moving faster than the anvil at all other instants and therefore had to move farther during the impact.

SCRATCH

(This problem is more advanced than most in this book, and involves the conservation of both energy and momentum, with a bit of vector addition to boot.) The cue ball and "8" ball are located on a pool table as illustrated. If an inexperienced player shoots the cue and successfully sinks the "8" in the corner pocket is there much danger that the cue ball will be deflected into the other corner pocket? A scratch is when the cue ball sinks in any pocket.

a) In the illustrated positions there is a great danger of scratching
b) In the illustrated positions there is little danger of scratching

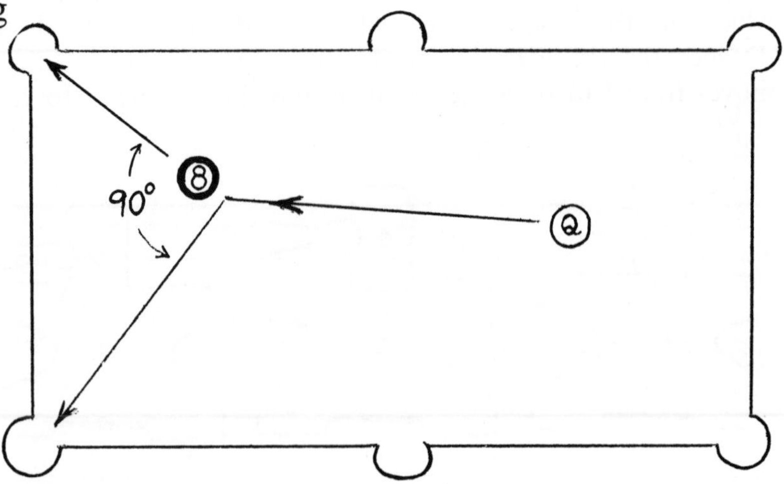

ANSWER: SCRATCH

The answer is: a. Every pool shark knows that when the “Q” ball hits the “8” ball, the balls move off in directions about 90 degrees apart, that is, they bound off at right angles to each other. From the illustrated position of the “8” ball, the corner pockets are about 90 degrees apart, so there is great danger of scratching.

But why do the balls fly off at right angles? The balls have (or should have) equal mass so their momentum is proportional to their velocity. Therefore, the vector sum of the velocity of the “Q” ball plus the “8” ball after the collision should add up to the original vector velocity of the “Q’ ball before the collision. But as the sketch shows there are many ways to pick a pair of velocities which will add up to equal the original “Q” ball velocity. Which pair should you choose?

Momentum is not the only thing to consider, for the balls are elastic and the sum of the kinetic energies of the balls after collision is about equal to the original kinetic energy of the “Q” ball. Now the kinetic energy of a ball is proportional to the square of its velocity and since the balls have equal mass, the square of the velocity of the “Q” ball after the collision plus the square of the velocity of the “8” ball after the collision should add up to equal the square of the original velocity of the “Q” ball before the collision. Now by the rules of vector addition, we know the vector velocity of the “8” and “Q” balls form the sides of a parallelogram and from the Law of Momentum Conservation the diagonal of the parallelogram is equal to the original velocity of the “Q” ball. And by conservation of kinetic energy, we know the sum of

THIS
WAY ?

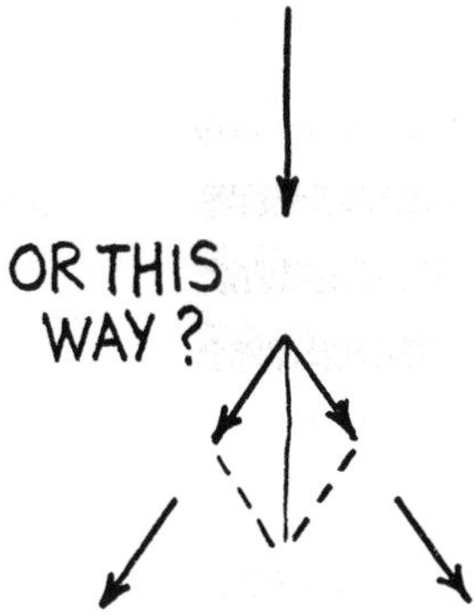
OR THIS
WAY ?

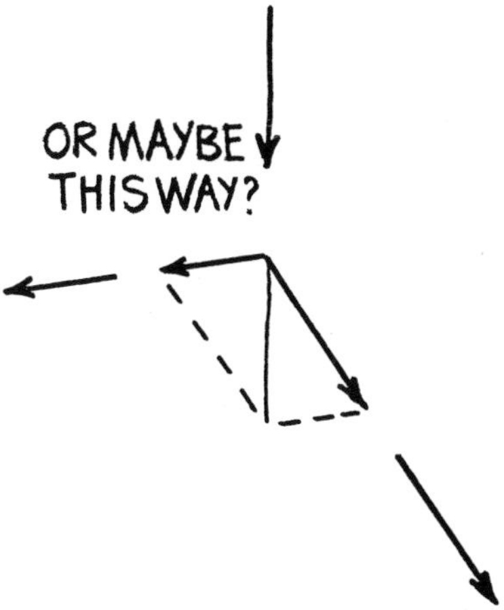
OR MAYBE
THIS WAY?

the squares of the sides of the parallelogram must equal the square of the diagonal. But that means the angle between the sides of the parallelogram must be a right angle! Remember Pythagoras.

So, the balls fly off in directions 90 degrees apart. Why did we hedge our bet and say ABOUT 90 degrees apart? Because the collision is not perfectly elastic and some of the original kinetic energy becomes heat. Also, there is some friction between the balls and the pool table. So the momentum and energy of the balls AFTER the collision does not exactly equal the momentum and energy BEFORE the collision. Also, some of the energy might go into making one of the balls spin after the collision. By taking advantage of these effects an experienced player can always find a way to shoot the "Q" ball without scratching.

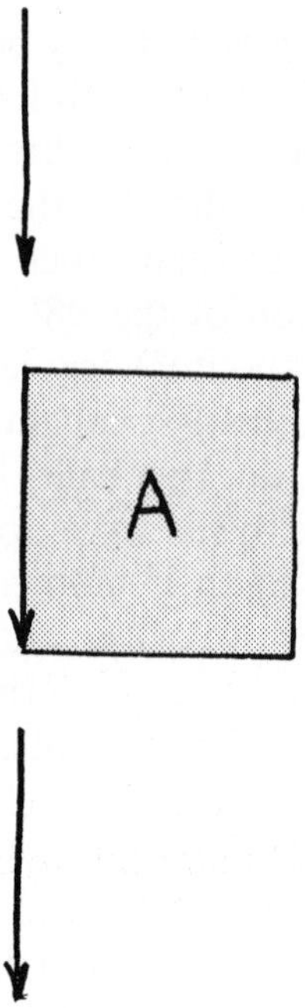

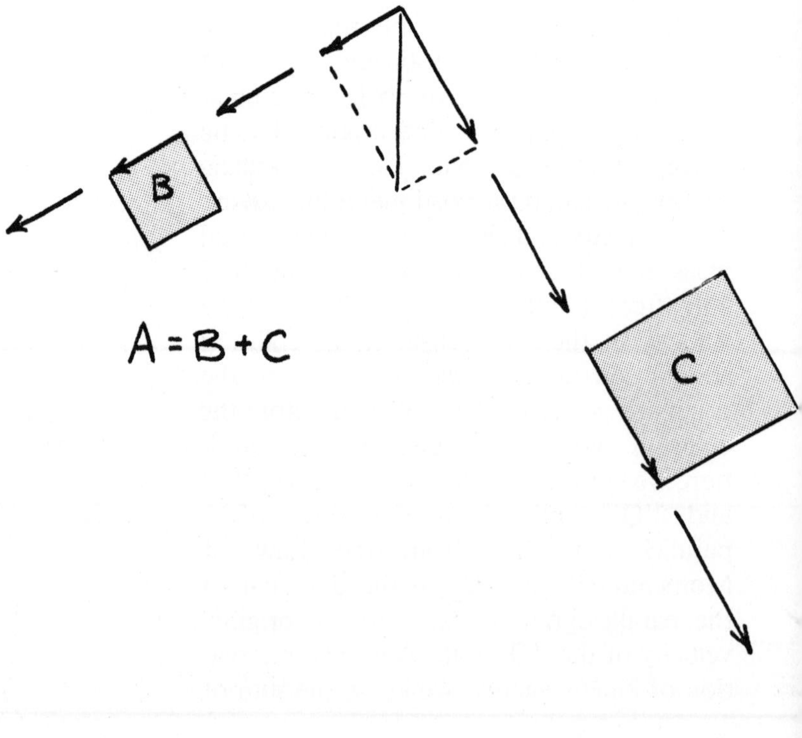

HEAVY SHADOWS

(This, too, is somewhat advanced.) Suppose that two elastic balls moving through three-dimensional space collide and rebound, conserving both kinetic energy and momentum. The balls cast shadows, and these shadows, moving over a flat two-dimensional surface come together, collide and rebound. Now pretend the shadows have mass. Suppose the shadow of a ball has a mass proportional to the mass of the ball. Then, the colliding shadows

a) conserve kinetic energy
b) conserve momentum
c) conserve both kinetic energy and momentum
d) conserve neither kinetic energy nor momentum

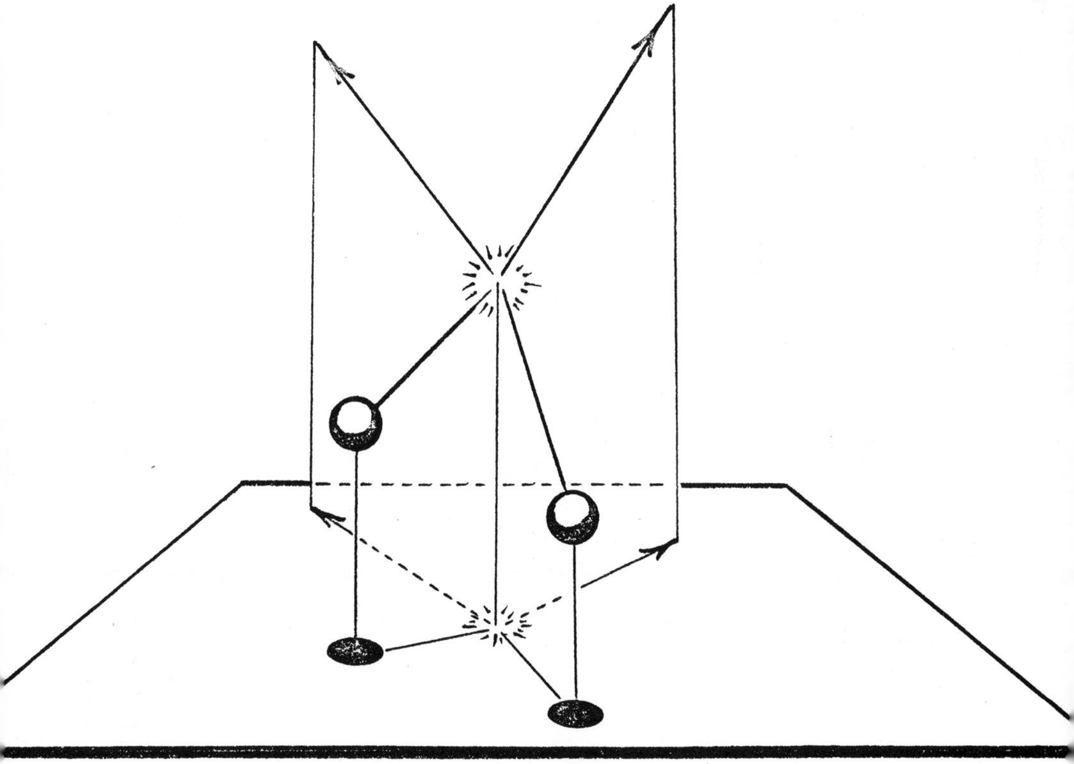

ANSWER: HEAVY SHADOWS

The answer is: b. The shadows cannot conserve kinetic energy. Suppose the balls move as illustrated below. Then the shadows are stationary after the collision, so energy cannot be conserved.

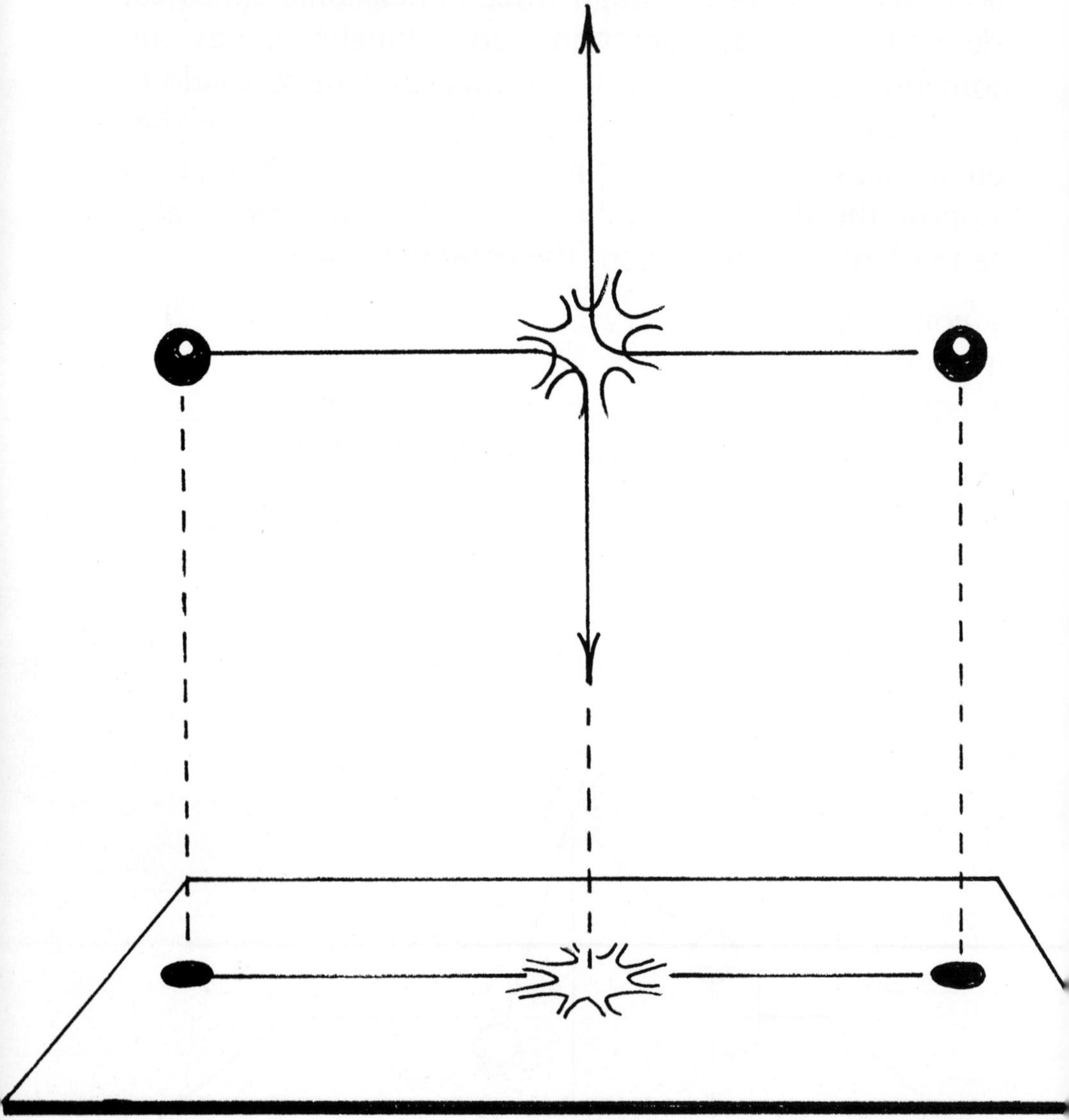

Now for momentum. The momentum of the balls before collision adds up to some total momentum vector P_1 and after collision the two momenta add up to P_2. By conservation of momentum $P_1 = P_2$. The shadow or projection of P_1 on the flat surface is p_1 and the shadow of P_2 is p_2. Then $p_1 = p_2$. Why? Because the shadows of parallel sticks (or vectors) of equal length are equal. What if the balls were not perfectly elastic? Would shadow momentum still be conserved? Yes. All collisions — elastic or inelastic — conserve momentum, but only elastic collisions conserve kinetic energy. So what does it mean if shadows conserve momentum? It means we can think of a momentum like P as made up of components, like a horizontal and vertical component or an X and Y component, and EACH of these components is itself conserved, just as if the other component did not exist.

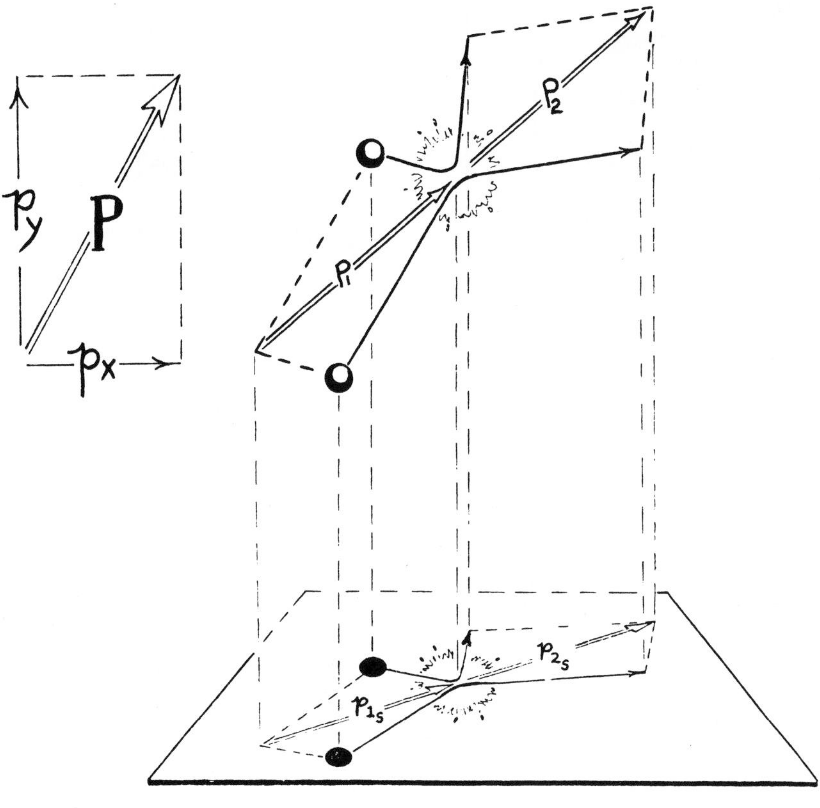

ABSOLUTE MOTION

A scientist is completely isolated inside a smoothly-moving box that travels a straight-line path through space, and another scientist is completely isolated in another box that is spinning smoothly in space. Each scientist may have all the scientific goodies she likes in her box for the purpose of detecting her motion in space. The scientist in the

a) traveling box can detect her motion.
b) spinning box can detect her motion.
c) . . . both can detect their motions.
d) . . . neither can detect their motions.

ANSWER: ABSOLUTE MOTION

The answer is: b. This question is the rotational counterpart to INERTIA. If your non-spinning box is moving smoothly in a straight-line path through space, you cannot sense its motion. For example, if you drop a coin above a cup it will fall directly into the cup whether or not your box is moving. Try it in a uniformly moving train or airplane. But if the box stops or starts or turns or jerks, you can sense motion. If the box accelerates you know that you're moving — but if it doesn't accelerate you can't tell. Do all the physics experiments that you care to in your uniformly and linearly moving box and you still can't tell. Even look outside and witness a moving background and you can't say for sure whether you or the background is moving. All you can say is you are moving relative to the background, or equivalently, vice versa. Linear motion is relative.

AMONG ALL THESE MOTIONS, WHY IS LINEAR SPECIAL IN NATURE?

But the spinning box is different. You know you're moving without consulting any background, and if the spin rate is fast enough you need only consult your stomach (have you ever become nauseous on a rotating carnival ride?). And if your box is spinning very slowly you can still tell by watching the precession of a swinging pendulum. Rotational motion is absolute.

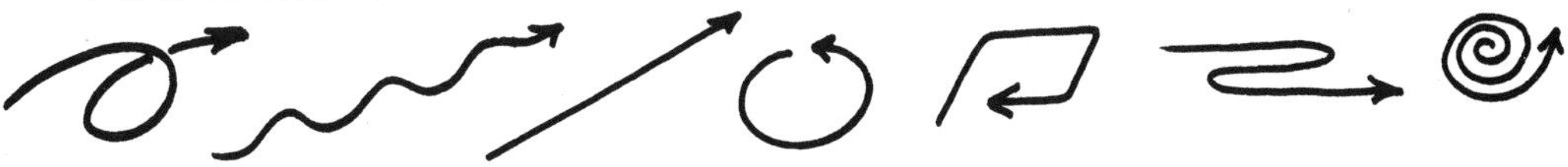

Why is one type of motion relative and the other absolute? Why are not both either relative or absolute? Or why not vice versa? After all, according to the ancient Greeks, circular motion is most preferred by the gods. These are deep and unanswered questions. All we know is that in our universe linear motion is relative and rotational motion is absolute. Were this not so the laws of motion would be very unlike those we live with now. But that does not explain why things are the way they are. That which is far away and exceedingly deep — who can find it out? Perhaps you!

TURNING

A cat runs across the floor from I to II to III without increasing or decreasing his speed. He only changes his direction of motion at II. Can we say for sure that a force was exerted on the cat at II?

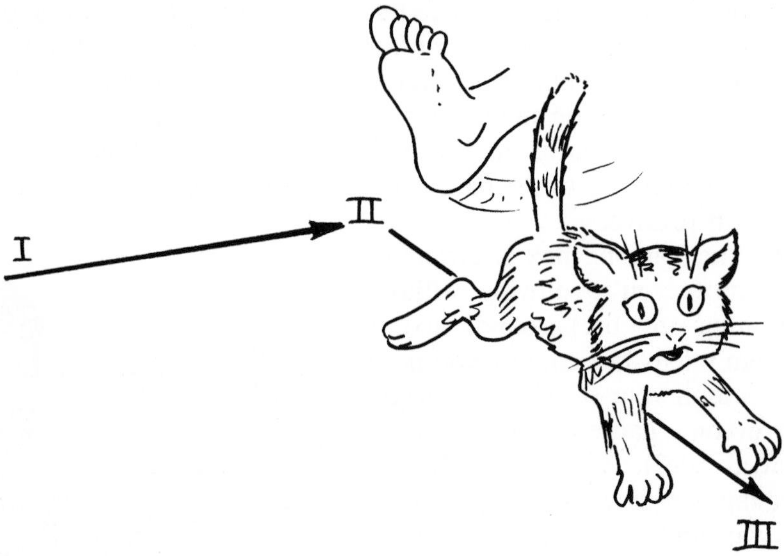

a) Yes, there had to be a force on the cat at II
b) Not necessarily, since no change in the cat's speed occurred

ANSWER: TURNING

The answer is: a. There had to be a force on the cat at II. If there were no force the cat would keep going in a straight line and go to IV instead of III. Pehaps someone kicks the unfortunate cat at II in direction V. The force of the kick turns the cat toward III. Or the cat could have turned itself by pushing on the floor with its feet. However, if the cat were on frictionless ice at II it couldn't muster a turning force. No force, therefore no turn — a "skidout."

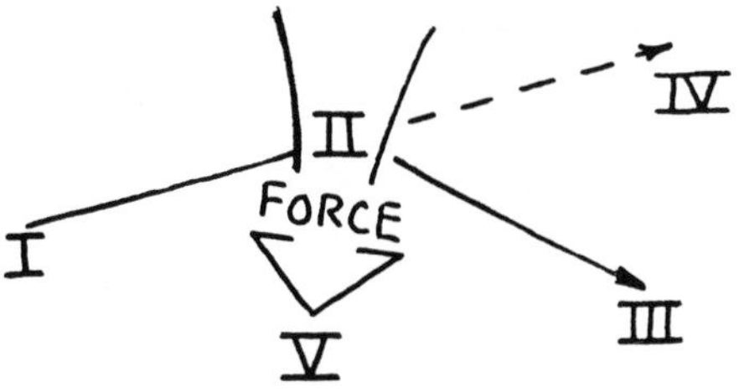

But why didn't the force change the cat's speed? Because it was a sideways force. A force pushing forward makes a thing go faster. A reverse force makes a thing slow down, stop or go backwards. But a sideways force makes a thing turn.

Physicists like to say a force always changes a thing's velocity, but does not always change its speed. What is velocity? Velocity is the "arrow" that represents a thing's motion. Physicists like to call the arrow a vector. If a thing goes faster it gets a new longer arrow, sketch A. If it goes slower, the result of deceleration or negative acceleration, it gets a new shorter arrow or vector, sketch B. And if it turns, it gets a new velocity vector which might be just as long as the original one but points in a different direction. That means same speed, but in a different direction, sketch C. So we see in this example that velocity can change while speed does not.

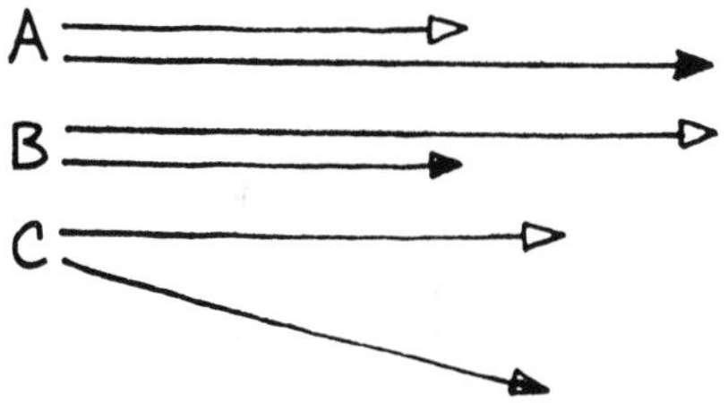

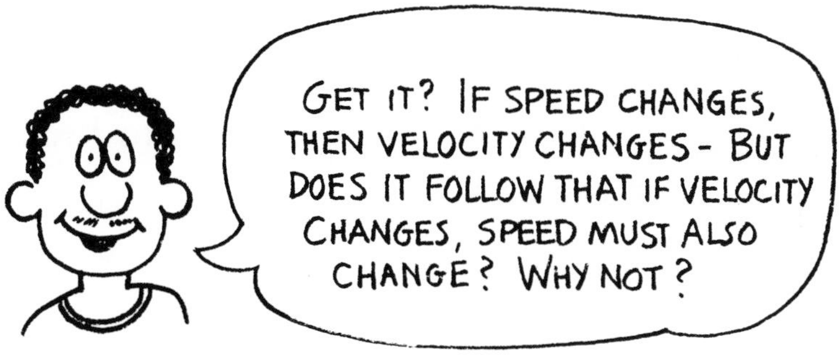

GET IT? IF SPEED CHANGES, THEN VELOCITY CHANGES - BUT DOES IT FOLLOW THAT IF VELOCITY CHANGES, SPEED MUST ALSO CHANGE? WHY NOT?

FASTER TURN

Suppose that two identical objects go around circles of identical diameter, but one object goes around the circle twice as fast as the other. The turning (centripetal)force required to keep the faster object on the circular path is

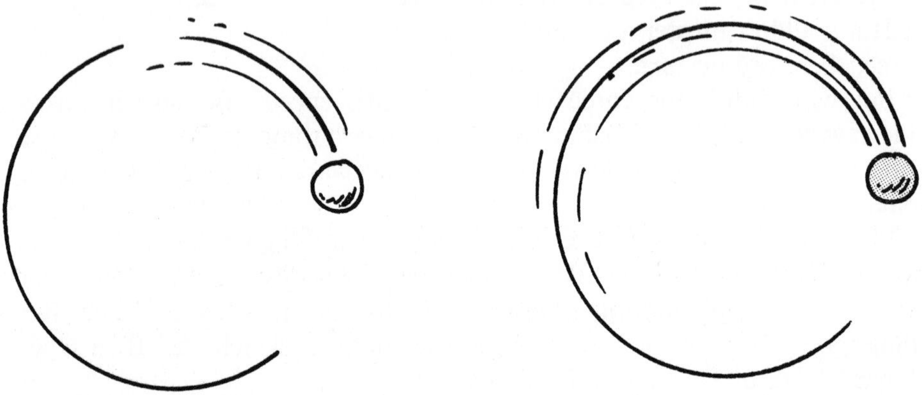

a) the same as the force required to keep the slower object on the path
b) one-fourth as much as the force required to keep the slower object on the path
c) half as much as the force required to keep the slower object on the path
d) twice as much as the force required to keep the slower object on the path
e) four times as much as the force required to keep the slower object on the path

CAN IT BE TOLD

An object with a known mass, say one kilogram, is moving from I to II with a known speed, say one meter per second. At II a force acts on the object. The force does not change the object's speed, but does change its direction of motion by 45 degrees. Is it theoretically possible to calculate how strong the force was?

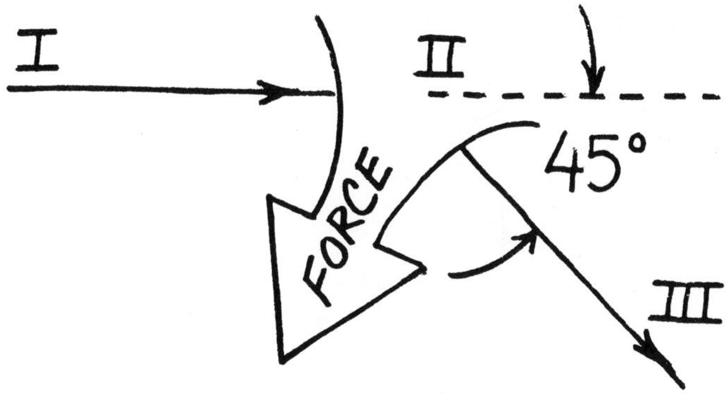

a) Yes, the strength of the force can be calculated (though I might not know how to do the calculation).
b) No, the force cannot be calculated by anyone.

ANSWER: CAN IT BE TOLD

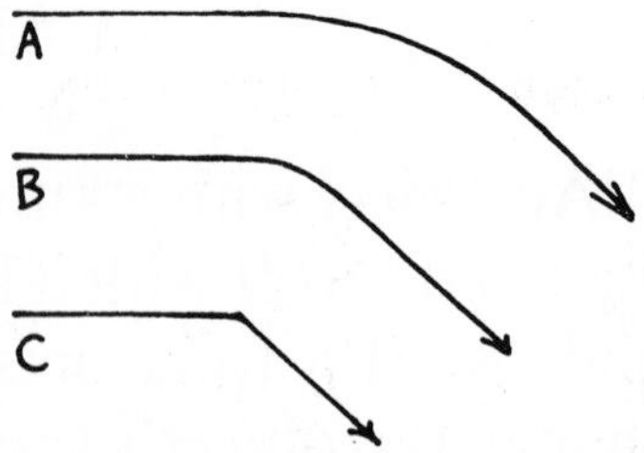

The answer is: b. You can't tell. The story here is very much like the story in ROCKET SLED. The turn could be made by a small force acting for a long time or a large force acting for a short time. If the force is small and the time long, the turn will be gradual, see sketch A, but if the force is strong and the time is short the turn will be abrupt, see sketch B. If the turn is instantaneous, as in sketch C, the force would have to be infinitely large. So, perfectly sharp turns do not exist in nature.

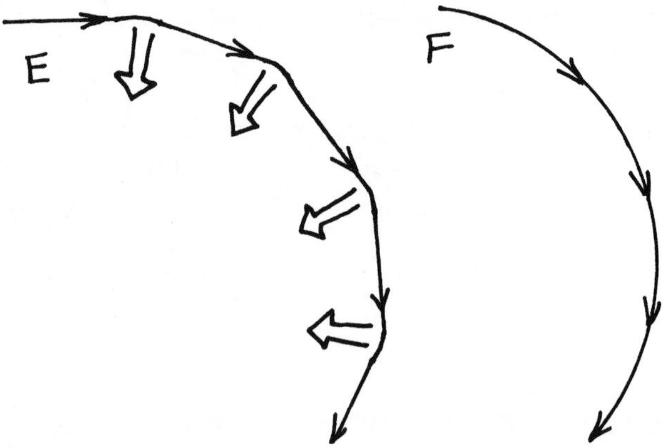

If an object goes around a bent path, like sketch E, the force is on some of the time and off at other times. The force is on in the curves, and off in the straight parts. Often it is convenient to talk about the average force. If the object goes around a smooth circular path, like F, however, then the force is exactly equal to the average force.

ANSWER: FASTER TURN

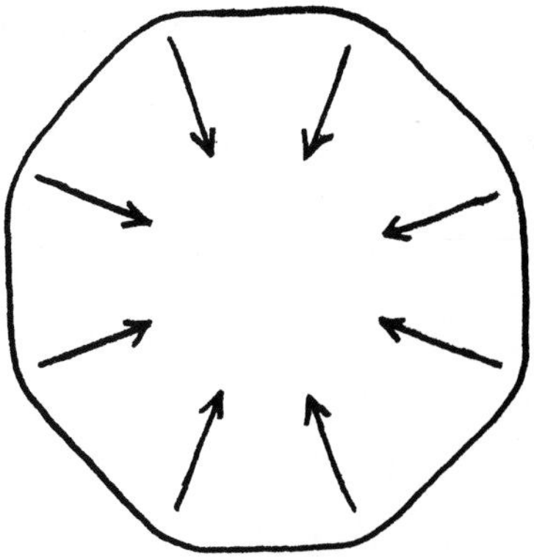

The answer is: e. Think of the circle as a bent path with many sides. As the object runs around the path it must get little kicks which bend the path. Now if the object begins running around twice as fast you have to kick it twice as hard to bend the path by the same amount. So you might think the average force on the object would be doubled. But there is more. If it runs twice as fast it comes to each bend twice as often. So you have to kick it twice as hard twice as often. That increases the average force four times. If it goes around three times as fast you must kick it three times as hard three times as often. That would increase the average force nine times.

There is a road near Lew Epstein's house with a sharp curve in it. The highway sign warns 20 miles per hour, but one day Lew decided to cheat a little and do 30 mph. What harm could an extra 10 mph do? In going from 20 to 30 he had only increased his speed 1.5 times. But increasing his speed 1.5 times increases the required centripetal force on his car $1.5 \times 1.5 = 2.25$ times. So a 50% speed increase requires a more than 100% increase in centripetal force — which didn't happen on loose gravel and Lew's car ran off the road!

SHARPER TURN

An object goes around bent path I with a speed of one mile per hour. An identical object goes around bent path II with the same speed. The diameter of path II is half the diameter of path I. The average force required to keep the object moving along path II is

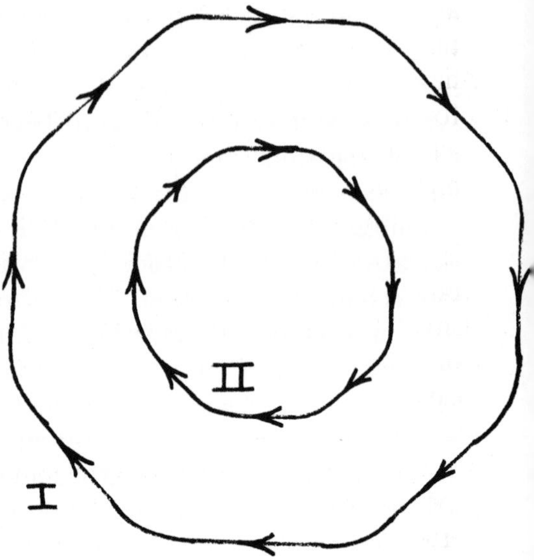

a) the same as the average force on the object in path I
b) half the average force on the object in path I
c) double the average force on the object in path I
d) four times the average force on the object in path I
e) one-fourth of the average force on the object in path I

ANSWER: SHARPER TURN

The answer is: c. The turn angles, mass, and speed of the objects moving in path II and path I are all exactly the same. So you might think the average force would be the same. But it is not. Why? Because you must average the force. On path II there is less time between bends, half as much time as on path I because path II is only half as long as path I. So there is half as much time to average over on path II and that makes the average force on path II twice as big as the average force on path I. Like a dollar averaged over twenty minutes is a nickel per minute, but averaged over ten minutes is a dime per minute.

Now if path I and II were made into perfect circles, all that has been said is still true. That is, if identical objects move with identical speeds the force on the object travelling in the smaller circle is larger. If the diameter of the small circle is one-half or one-third the diameter of the large circle, the force on the object moving in the small circle is twice or three times the force on the object moving in the large circle.

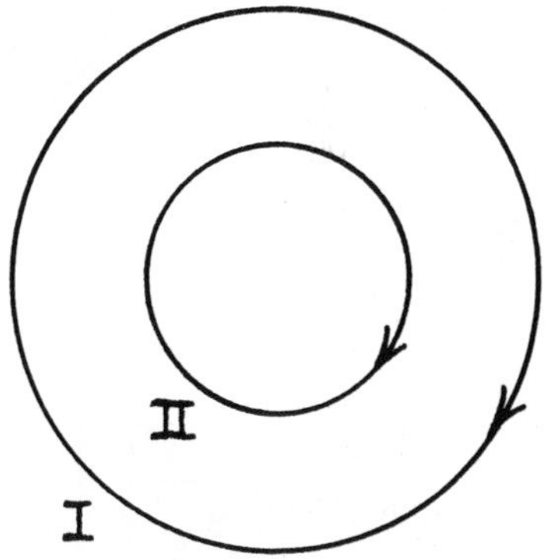

When an object moves in a circle the turning force, which always is exerted sideways, will always be directed towards the circle's center. The turning force is called centripetal force. The turning force does not keep the object going, it only keeps it turning, that is, keeps it on the circular path.

Everyone knows that as a turn gets sharper a train or motor car has a harder time going around it. But now you know why. The sharper the turn, the smaller the turning circle. The smaller the turning circle, the larger the required centripetal force. If this centripetal force is not exerted, the result is a derail or skidout.

Of course, if the objects have different masses that too must come into the story. The force needed to turn an object will increase in proportion to its mass.

In symbolic notation, the centripetal force F required to keep an object of mass m moving at speed v along a curved path of radius r is

$$F = m\frac{v^2}{r}$$

CAROUSEL

Peter and Danny are standing on a carousel which is turning as illustrated. Peter throws a ball directly towards Danny.

a) The ball gets to Danny
b) The ball goes to the right of Danny
c) The ball goes to the left of Danny

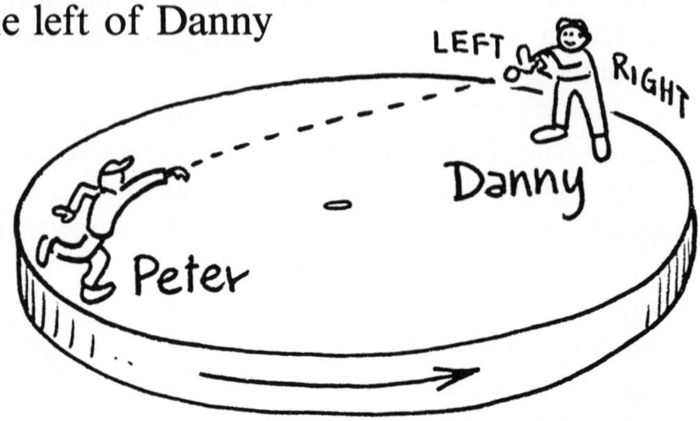

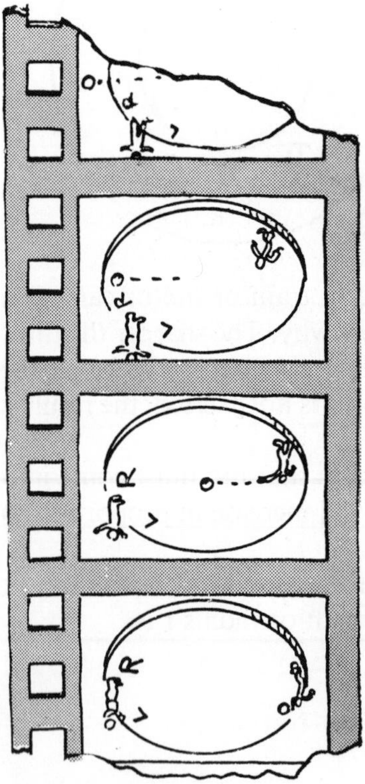

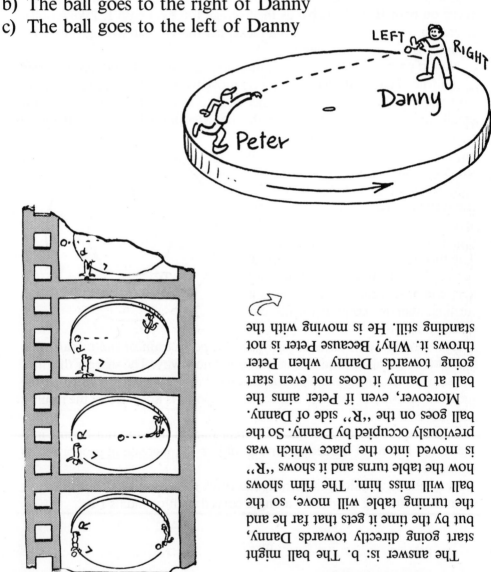

ANSWER: CAROUSEL

The answer is: b. The ball might start going directly towards Danny, but by the time it gets that far he and the turning table will move, so the ball will miss him. The film shows how the table turns and it shows "R" is moved into the place which was previously occupied by Danny. So the ball goes on the "R" side of Danny.

Moreover, even if Peter aims the ball at Danny it does not even start going towards Danny when Peter throws it. Why? Because Peter is not standing still. He is moving with the carousel. The ball carries with it Peter's velocity, that further deflects the ball towards "R."

When you live on a turning world things do not go in the direction they are aimed. In fact, they don't even SEEM to go in straight lines.

This deflection has a name. It is named after one of the first persons to study it — Coriolis. There is even a slight Coriolis effect on things moving over the earth because the earth is truly a turning world.

Suppose you forced the ball to go from Peter to Danny. How could you force it? By making it go in a pipe which started at Peter and ran to Danny (see sketch). The pipe is straight, but it turns while the ball is on its way. So even though the pipe is straight the ball moves in a curved path. It takes force to make a thing curve. The heavy-lined side of the pipe must exert the force on the ball. Do Peter and Danny think the ball curves? No. They move with the turning pipe. So they think the ball just goes straight — yet they must wonder why it keeps pushing on the side of the pipe. Of course once they realize they are turning the wonder is explained.

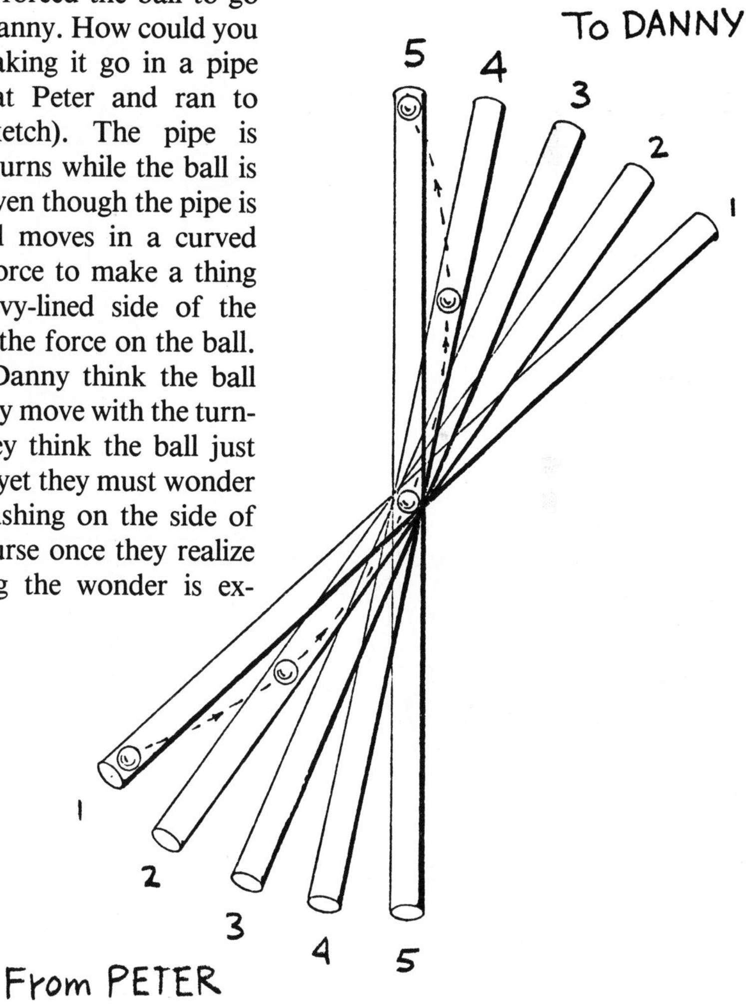

TORQUE

Harry is finding it very difficult to muster enough torque to twist the stubborn bolt with a wrench and he wishes he had a length of pipe to place over the wrench handle to increase his leverage. He has no pipe, but he does have some rope. Will torque be increased if he pulls just as hard on a length of rope tied to the wrench handle?

a) Yes
b) No

ANSWER: TORQUE

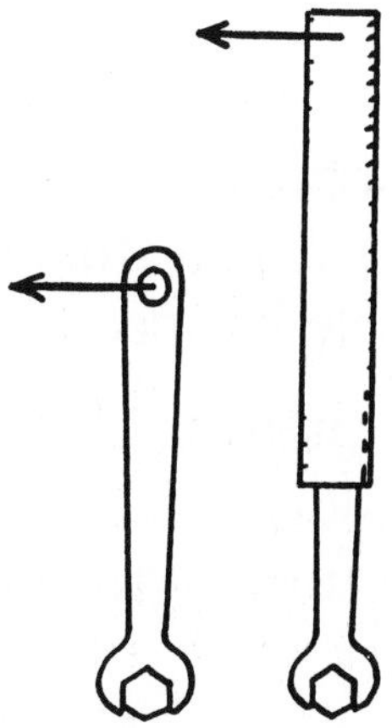

The answer is: b, no. The twisting force or torque that is applied to the stubborn bolt depends not only on the applied force but also on the length of the lever arm upon which the force acts. This relationship can be visualized by recalling your experiences with wrenches or see-saws. The greater the leverage, the greater is the torque. By attaching the rope to his wrench, Harry increases the distance from the bolt to the location of the applied force, but he doesn't increase the lever arm. That's because the lever arm is not the distance from the pivot point (bolt) to the applied force, but rather the distance to the *line of action* of the applied force. The lever arm is always at right angles to the line of action of the applied force. It is also the shortest distance between the line of action and the pivot point. When Harry uses the rope he does not change the length of the lever arm.

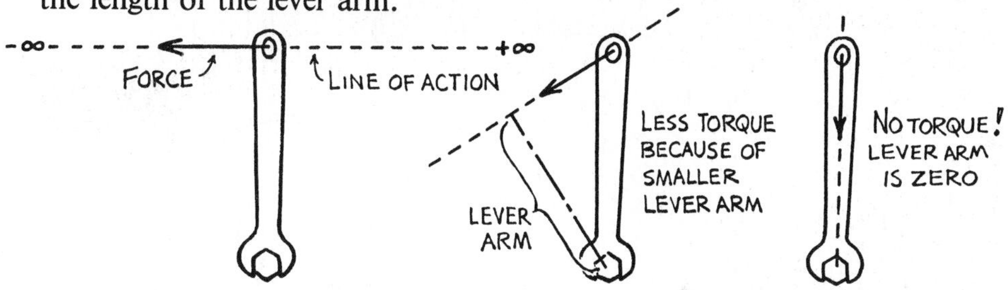

By definition, torque is equal to force multiplied by the lever arm. We can picture torque geometrically — it's twice the area of a particular triangle. Let the altitude of the triangle be the lever arm and the base of the triangle be the force vector. Remember that the area of a triangle is one-half the altitude multiplied by the base. In our case, altitude = lever arm, and base = force. So the area is equivalent to half the torque. From the sketch you can see that whether or not the force is applied directly to the wrench handle or from a rope tied to the wrench handle, the area of the triangle formed by the applied force and the pivot point is the same in both cases (same base and altitude), so the torque is the same.

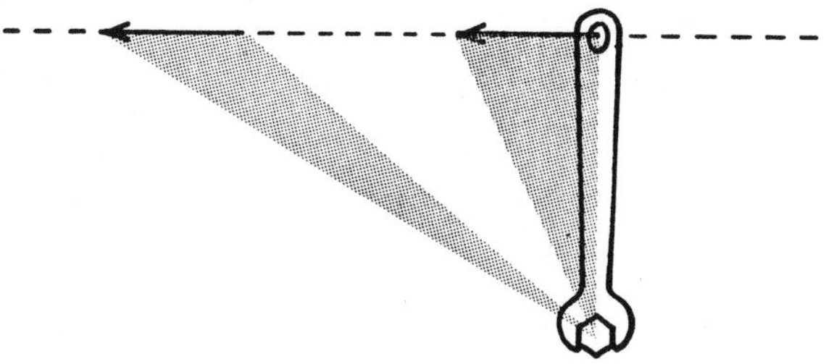

BALL ON A STRING

A ball held by a string is coasting around in a large horizontal circle. The string is then pulled in so the ball coasts in a smaller circle. When it is coasting in the smaller circle its speed is

a) greater
b) less
c) unchanged

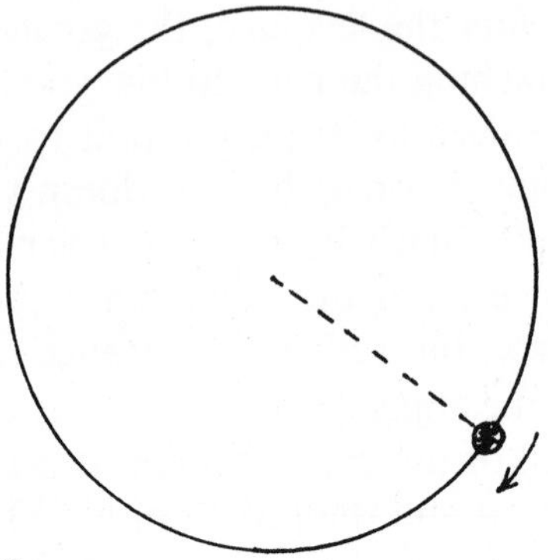

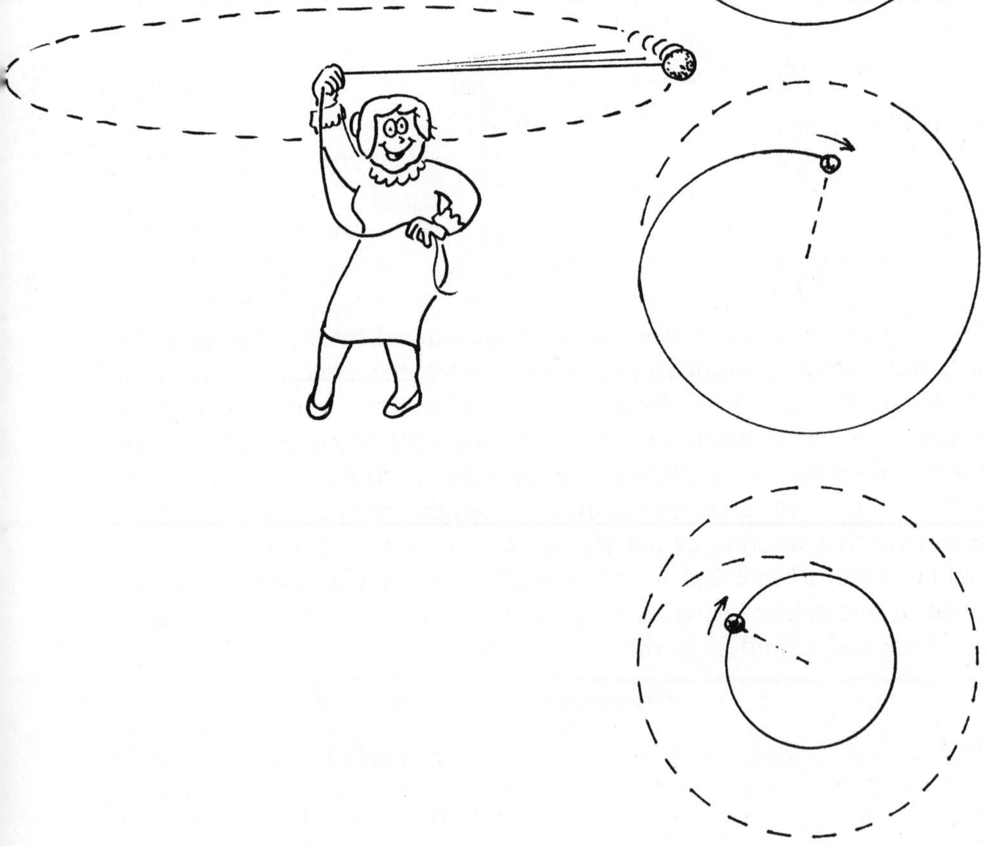

ANSWER: BALL ON A STRING

The answer is: a, greater. When the ball is moving in a circular path of constant radius the pull of the string does not speed it up — it coasts at constant speed. But when the ball is pulled into a smaller circle, the force of the string speeds it up. Why? Because in the first case where the radius remains constant the string is always pulling at right angles to the motion of the ball — in other words the pull is always sideways and serves only to keep the ball's direction of motion changing — into a circular path instead of a straight line (which would be the case if the string broke).

But when the ball is being pulled in from a large circle to a smaller circle the force of the string is not exactly sideways to the ball's motion. We can see in the sketch that the force of the string has a component of force in the direction the ball is moving. In the sketch we see that Component 1 acts to increase the speed, and Component 2 simply changes the ball's direction of motion.

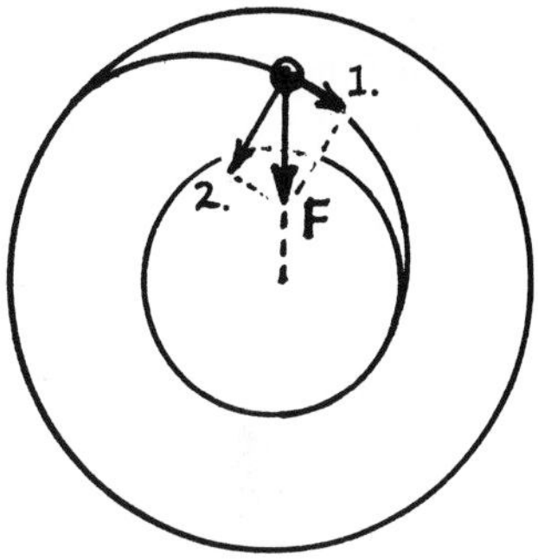

There is an ingenious way to determine this increase in speed. It depends on the idea of angular momentum, which for a body of mass m moving at speed v at a radius r is the product mvr. Just as a force is required to change the linear momentum of a body, a torque is required to change the angular momentum of a body — if no torque acts on a rotating body, its angular momentum cannot change. Does the string exert a torque on the ball? It does only if the force acts through a lever arm (recall TORQUE). Is there a lever arm? No, the line of action of the force passes straight through the pivot point. There is no torque, and therefore no change in angular momentum. The ball has the same angular momentum mvr when it arrives at the small circle that it had on the large circle. Since the product mvr doesn't change, any decrease in radius is exactly compensated by an increase in speed. Say if r decreases by half, then v increases by two; if r is pulled in to a third its initial value, the speed triples. This holds true in the absence of a torque, which would otherwise change the angular momentum.

We can represent this idea pictorially, just as we did for torque: Recall that torque is force $\times$ lever arm, and can be pictured as twice the area of the triangle formed by the force vector and lever arm, Sketch A, opposite page. So too, angular momentum $= mvr =$ (linear momentum mv) $\times$ (lever arm r) can similarly be represented by twice the area of the triangle formed by the linear-momentum vector and the lever arm, Sketch *B.* If no torque acts on any rotating system, its angular momentum will not change. In the case of our circling ball, the angular momentum is the same on the outer circle as it is on the inner circle. That means the area of the triangle formed by the linear momentum and the pivot point doesn't change. Like if the radius of the smaller circle is one third the radius of the larger circle, then the linear momentum (*mv*) of the ball on the smaller circle must be three times larger than when on the large circle. That means it goes three times as fast. Note that in this way the area of the angular momentum triangle comes out to be the same for both circles, Sketch *C.*

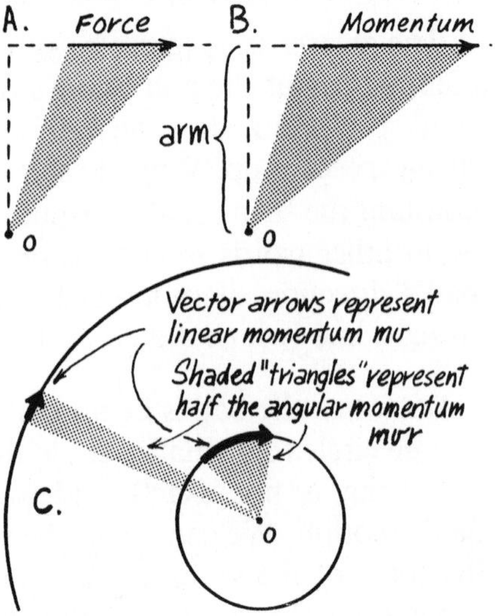

SWITCH

A streetcar is freely coasting (no friction) around the large circular track. It is then switched to a small circular track. When coasting on the smaller circle its speed is

a) greater
b) less
c) unchanged

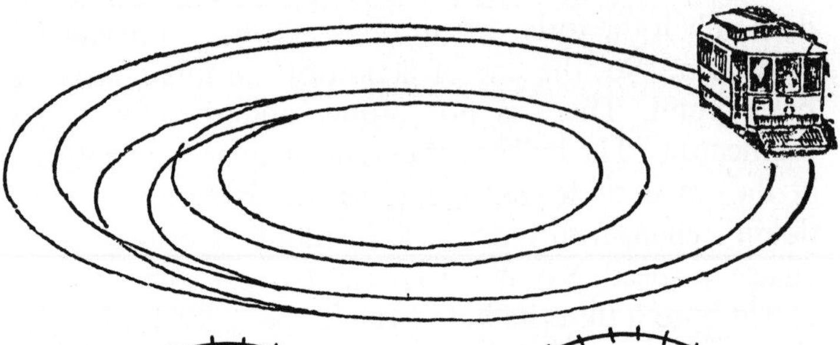

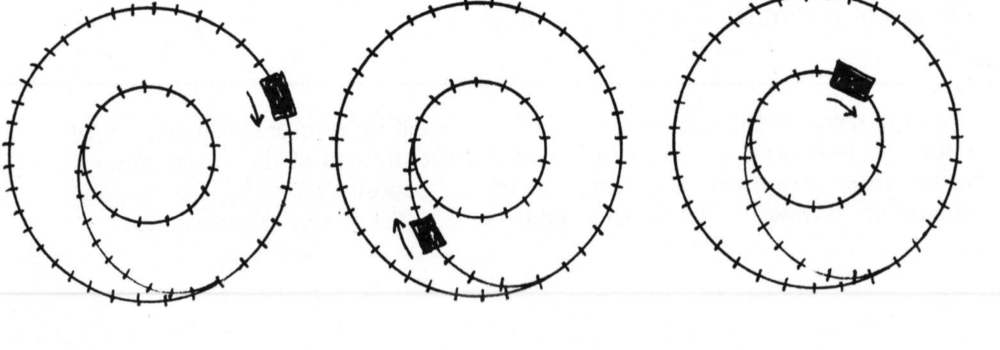

ANSWER: SWITCH

The answer is: c, unchanged. BALL ON A STRING gained speed when it was pulled in to a smaller circle because there was a force component in its direction of motion. But this is not the case for the car on the track. If the track were perfectly straight the track could exert no force to speed the car up (or slow it down if it is frictionless). A curved track, however, does exert a force that changes the state of motion of the car — a sideways force that makes the car turn. But this sideways force has no component in the coasting car's direction of motion. So the track cannot alter the speed of the freely-coasting car.

Is angular momentum conserved in this example? No. Angular momentum is conserved only in the absence of a torque. But a torque indeed exists when the car is changing circles. We can see in the sketch that the sideways track force acts about a lever arm from the center of the circles. This produces a torque that decreases the angular momentum of the car — but not the speed. The car coasts at the same speed with less angular momentum on the inner track. In this case the angular momentum is reduced because the radius is reduced.

If the car moves in the opposite direction, from the inner to the outer circle, then the torque acts to increase its angular momentum. Angular momentum increases as the radius increases. But as before, the speed of the freely-coasting car remains constant.

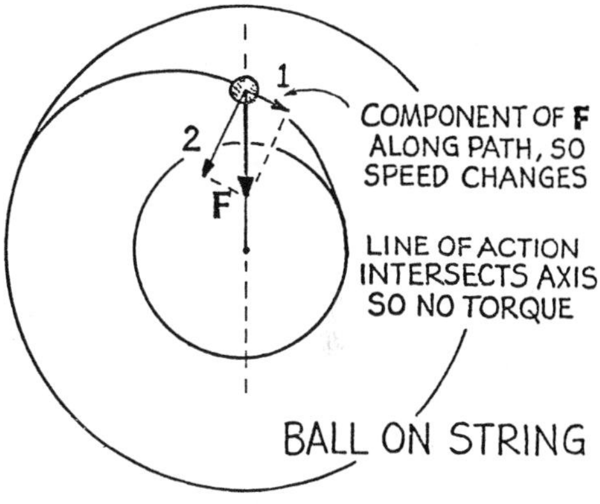

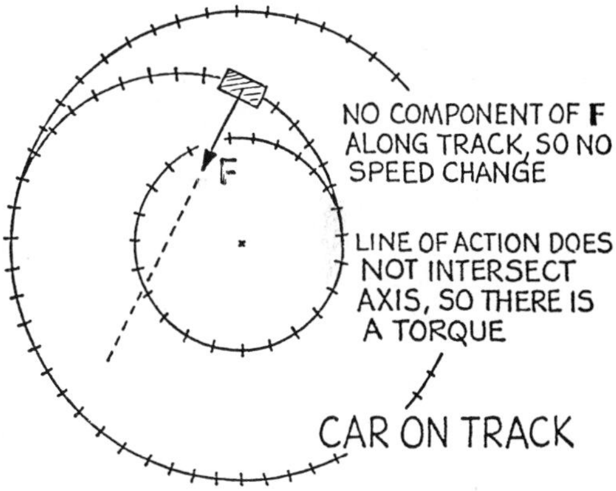

We can look at this from a work-energy point of view also: In the case of BALL ON A STRING, the Force Component 1 acted along the direction of the ball's displacement, so work (force × distance) was done on the ball which increased its kinetic energy. (It is important to see that no work is done when the ball moves in a path of constant radius, for then there exists no Component 1). In the case of the car switching tracks no Force Component 1 exists at any point, whether along the circles of constant radius or along the track joining them. The track force is everywhere perpendicular to the displacement and no work is done by the track force on the car. So no change in kinetic energy, or therefore speed, takes place.

NOSING CAR

When a car accelerates forward, it tends to rotate about its center of mass. The car will nose upward

a) when the driving force is imposed by the rear wheels (for front-wheel drive the car would nose downward).
b) whether the driving force is imposed by the rear or the front wheels

ANSWER: NOSING CAR

The answer is: b. If the car is accelerated forward, the tires push backward on the road, which in turn pushes forward on the tires. This force of the road on the tires not only accelerates the car forward, but produces a torque about the center of mass of the car. Whether this force is exerted on the rear or front or both sets of wheels, the line of action of the force is along the roadway surface and produces a rotation of the front of the car upward and the back of the car downward (which increases traction for rear-wheel drive vehicles).

We can see from the sketch that in all cases the force of friction (solid arrow) that accelerates the car tends to rotate the car (dotted arrow) counter-clockwise about its center of mass.

It's easy to see that when the brakes are applied and the force and consequently the torque is oppositely directed, the car noses downward.

This tipping effect is particularly evident in a power boat. Note in the sketch that when the boat accelerates the net force is forward and the torque tips the boat counter-clockwise, but when the boat decelerates and the force of water resistance is predominant, the net force is backward and the torque tips the boat clockwise.

Now a question for the teacher! The car noses up only while it is accelerating and goes back to level once it has achieved a constant speed. But the motor boat stays nosed up. How come? When the boat's speed is constant the friction drag on the hull bottom is equal and opposite to the force on the prop, but the prop is deeper in the drink and is farther than the hull bottom from the boat's center of mass. So the bottom friction and prop act together to produce a turning couple or torque.

MUSCLE HEAD & PIP SQUEAK

Suppose you had a car that was mostly wheels and another car that had tiny wheels. If the cars have the same total mass and their center-of-masses are equal distances from the ground and each goes from zero to 40 mph in ten seconds, which one will nose up the most?

a) Muscle Head
b) Pip Squeak
c) Both the same

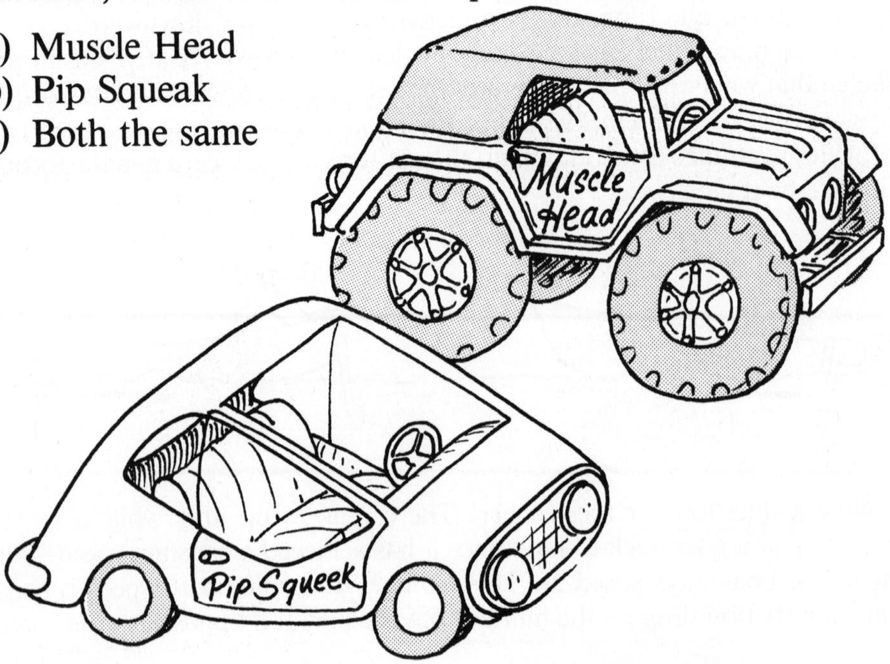

ANSWER: MUSCLE HEAD & PIP SQUEEK

The answer is: a. The wheels did not come into the NOSING CAR story at all — except that they moved the car — so you might expect that the size of the wheels makes no difference here. Even if the car had no wheels and was dragged on a sled it would still nose up when the sled accelerated!

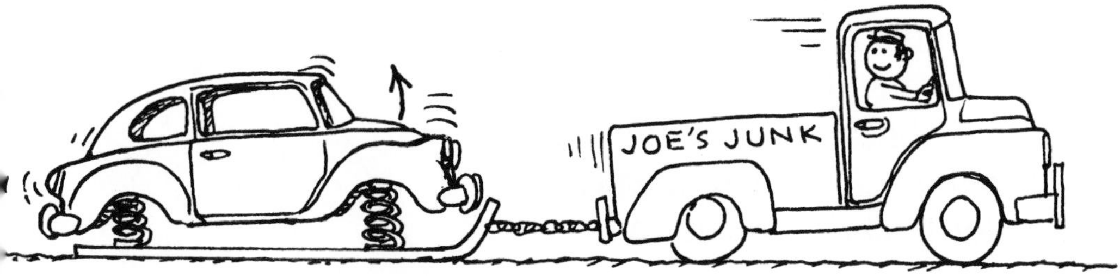

Nevertheless you might have a gut feeling that the big-wheel buggy should nose up the most — and you are right. So far, though, only half of the "nosing" story has been told. To visualize the other half, suppose the car was floating out in intersteller space and you hit the gas and made the wheels spin. What would the car do? Well, it certainly would not go anyplace. Why? Because there is no road to push on — no friction! But it would not just do nothing. If the wheels were made to spin clockwise the rest of the car body would spin anti-clockwise. Why? Because this is the rotational counterpart to action and reaction and momentum conservation.

Now back on earth that effect still operates. There are two effects which make the accelerating car nose up. One depends only on the car's acceleration and has nothing to do with the wheels. The second depends only on the spinning wheels and has nothing to do with the acceleration. If the mass of the wheels is very much less than the mass of the car only the first effect counts for much. But as the mass of the wheels becomes closer to the mass of the vehicle the second effect becomes more important.

THE SWING

Pull a swing or pendulum to one side, let it go and it will swing back and forth by itself. As it goes back and forth, the swing is conserving

a) angular and linear momentum
b) only angular momentum
c) only linear momentum
d) neither angular momentum nor linear momentum

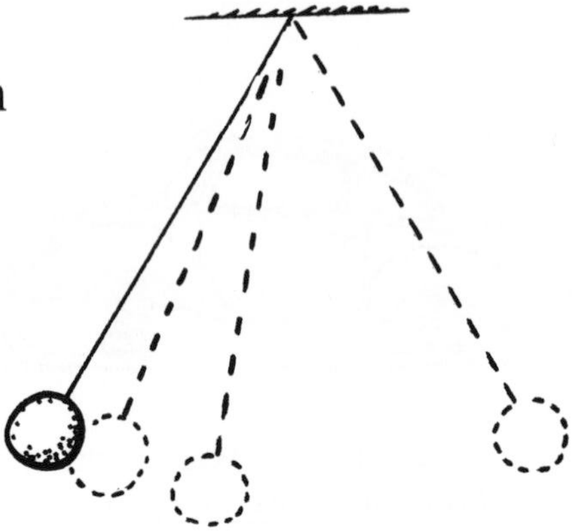

ANSWER: THE SWING

The answer is: d. First the pendulum swings to the right, next it moves back to the left. If the momentum going to the right is $+5$, then when it swings back the momentum must be -5, and when it stops to come back at the end of a swing, the momentum is zero. So the momentum changes from $+5$ to zero to -5 to zero to $+5$ to. . . . It is always changing and so it can't be conserved. What about angular momentum? Well, first it pivots clockwise about the suspension point and at the end of a swing it is momentarily not pivoting at all, and then it pivots counterclockwise, repeating the cycle over and over. So angular momentum is not covserved either. How come? Perhaps you thought momentum was always conserved? Where does the momentum go? When a ball hits a bat where does the momentum go? Into the bat. The linear momentum of the pendulum goes into the string, then into the ceiling, and then into the earth. The angular momentum of the pendulum goes directly into the earth by gravity. That means the earth gets anything the pendulum loses. As the pendulum moves to the right, the earth moves a tiny tiny tiny bit to the left. As the pendulum turns clockwise the whole earth turns a tiny tiny tiny bit anti-clockwise and of course when the pendulum swings the other way the whole earth also reverses its tiny motion. Suppose the earth's mass is a billion times larger than the mass of the pendulum and the pendulum moves one foot to the right. How far does the whole earth move? One billionth part of one foot to the left.

SWING HIGH — SWING LOW

A certain pendulum swings through an arc of one degree in one second. Next, the same pendulum is made to swing through an arc of two degrees. The time required to swing through the two-degree arc is

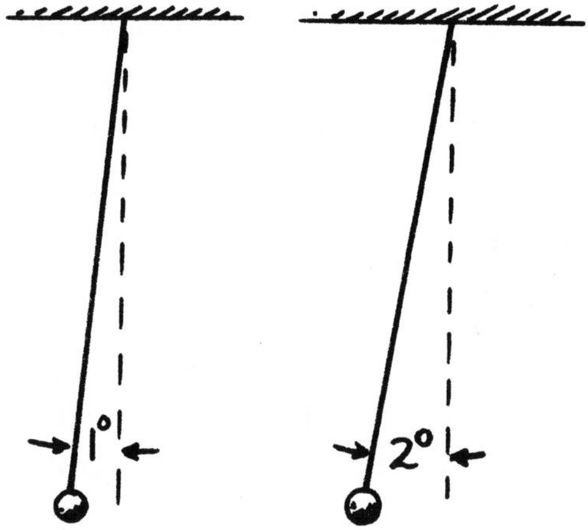

a) one-half second
b) one second
c) two seconds

ANSWER: SWING HIGH — SWING LOW

The answer is: ?. Why not try this one out like Galileo did? Use a long string and a heavy weight. Do not time single swings. Take the time for ten swings. The swings will die a little during the timing, but don't worry about that.

After you do the experiment think about this. Pull the pendulum back to B and let it swing to C. Then pull it to A and let it swing back to C. The trip from A is longer, but faster. It is length of a pendulum, not its displacement, that determines its period of vibration.

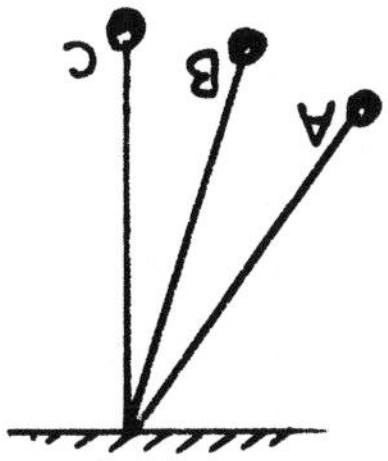

GYRO PRANK

Here's an advanced-level question: An often recounted tale tells of the physicist who hid a large spinning flywheel inside his suitcase. The hotel porter took the suitcase and proceeded to walk around a corner, as shown. What happens to the suitcase?

a) The suitcase simply pivoted around the corner in the way the porter wanted it to go, as in sketch A

b) The case tended to pivot in a way exactly opposite to the way the porter wanted it to go, as in sketch B

c) As the porter went around the corner the case pivoted top over bottom, as in sketch C

d) As the porter went around the corner the case pivoted top over bottom, as in sketch D

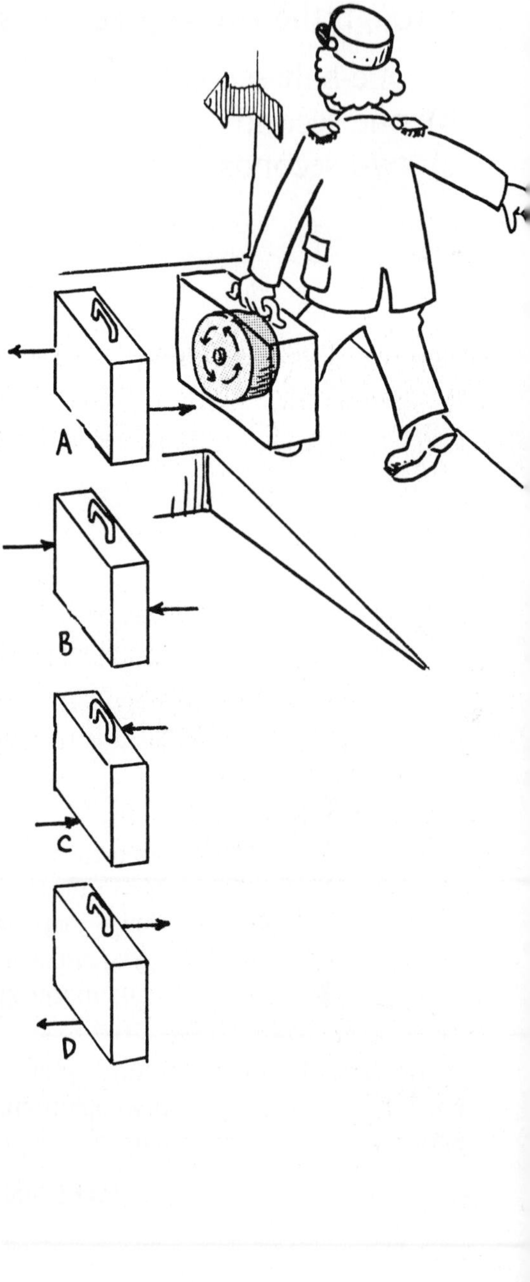

ANSWER: GYRO PRANK

The answer is: d. Gyroscopic motion is quite complicated, but we can simplify it if we picture the spinning flywheel as a round pipe or tire or doughnut in which some heavy liquid is circulating. Next, picture the ring made into a square ring. The liquid circulates by flowing up one side, then horizontally over the top, then vertically down the other side and then horizontally back on the bottom.

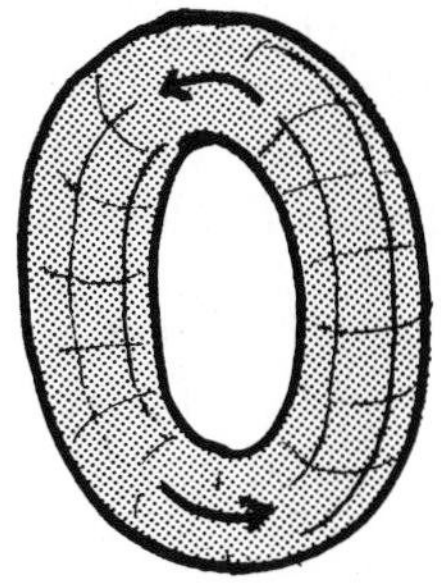

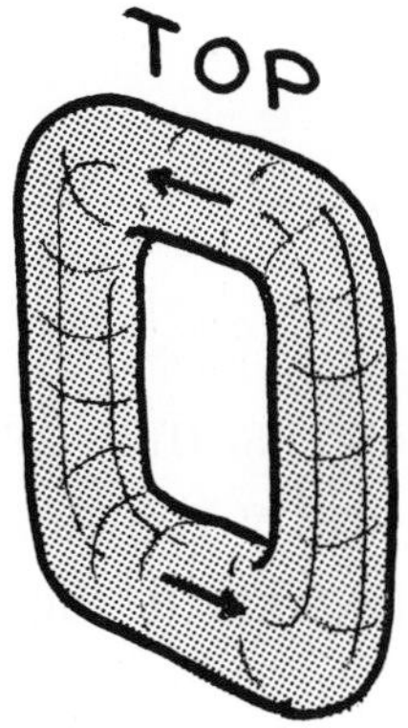

Next, picture the "square" flywheel pivoting from position I to position II as the porter goes around the corner. The part of the liquid flowing up and down in the vertical arms of the pipe does not change its direction of flow as the square flywheel or loop pivots from position I to II — the vertical sides remain vertical. But the liquid flowing in the horizontal arms does change its direction of flow. For example, the liquid in the top might begin flowing in direction 1 and finish in direction 2.

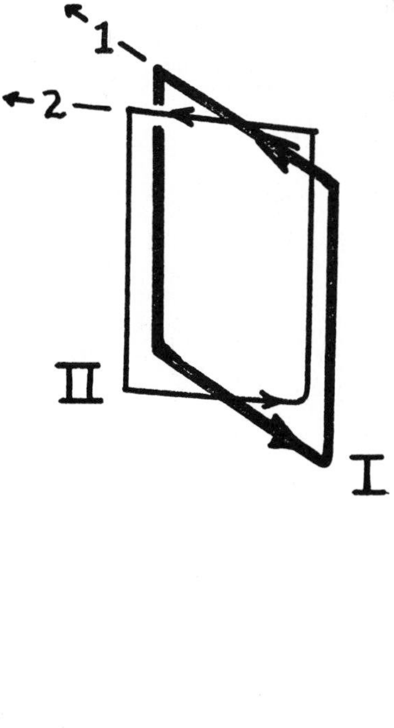

The last sketch shows how a portion of liquid flowing in the top pipe or bottom pipe is actually forced to travel in a curve as the porter turns the loop. Now, when a thing goes through a pipe which is turning it exerts a force on the side of the pipe. (Remember CAROUSEL.) Why? Because the thing tends to go straight, but is forced to turn. The arrows in the last sketch show the direction of the force on the sides of the pipe. The curves on the top and bottom are opposite and so the direction of the force on the top and bottom are opposite. The force on the pipe is the same as the force on the flywheel and that force is communicated to the suitcase. The top tilts to the right and the bottom to the left — as in sketch "D."

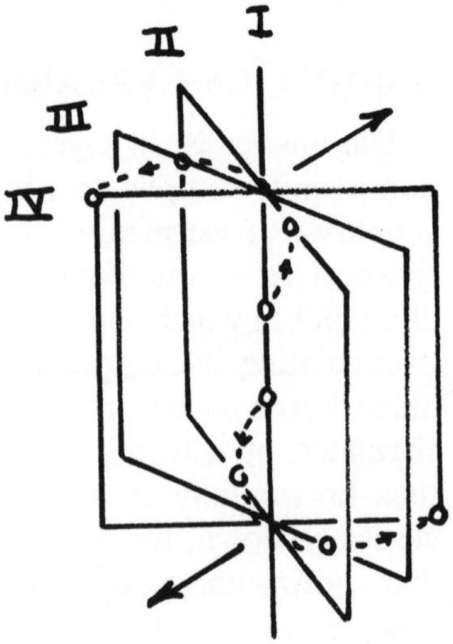

THE DROPPING BULLET

At the same time that a high-speed bullet is fired horizontally from a rifle, another bullet is simply dropped from the same height. Which bullet strikes the ground first?

a) The dropped bullet
b) The fired bullet
c) Both stike at the same time

ANSWER: THE DROPPING BULLET

The answer is: c. This is because both bullets fall the same vertical distance with the same downward acceleration. They therefore strike the ground at the same time (gravity does not take a holiday on moving objects).

This problem can be better understood by resolving the motion of the fired bullet into two parts; a horizontal component that doesn't change because no horizontal force acts on it (except for air resistance), and a vertical component that accelerates at g and is independent of its horizontal component of motion.

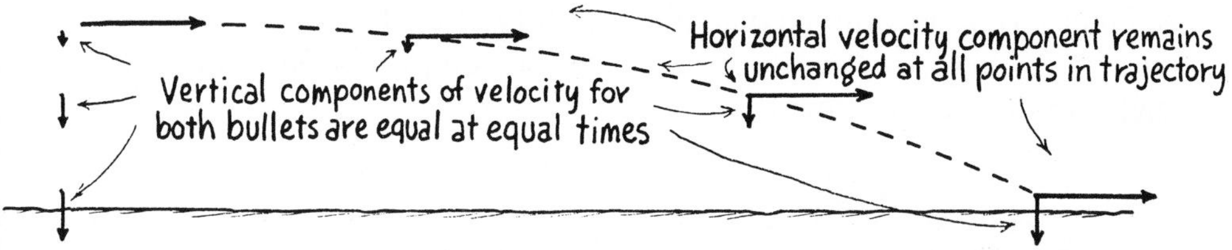

Horizontal velocity component remains unchanged at all points in trajectory
Vertical components of velocity for both bullets are equal at equal times

Or we can look at this another way. If the rifle is pointed upward above the horizontal then the dropped bullet will strike the ground first. If on the other hand, the rifle is pointed downward toward the ground, the fired bullet obviously wins. Somewhere between up and down, there will be a draw — and that's when the rifle is horizontal.

And there's another and quite different way to look at this: Suppose the ground accelerated upward, rather than the bullets downward. Can you see that the outcome would be the same?

FLAT TRAJECTORY

A bullet shot from a very high velocity rifle may travel one hundred feet or more without dropping at all.

BANG

a) True
b) False

ANSWER: FLAT TRAJECTORY

The answer is: b. There is a popular idea around, even taught in some police academies, that a sufficiently high-powered bullet will go some distance without dropping at all. But this is a real misconception. The speed of a thing cannot "turn off" gravity, not even for a moment. Even light from a laser begins to drop the moment it begins its flight. Moreover, the rate of drop is always the same, regardless of speed. If the bullet is fired horizontally in a so-called "flat trajectory," it falls 16 feet during the first second of flight. However, during that second a high speed bullet will go farther than a low speed bullet so the flight path or trajectory of the high speed bullet looks less curved than the trajectory of the low speed bullet, but both always curve. A trajectory can never be completely flat (unless it is straight up or straight down).

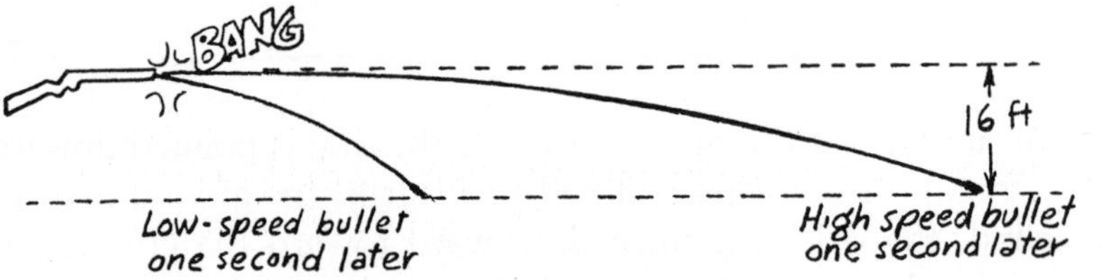

PITCHING SPEED

A boy throws a rock horizontally from an elevated position 16 feet above the ground as shown. The rock travels a horizontal distance of 40 feet. How fast did the boy throw the rock? (Again, consider time.)

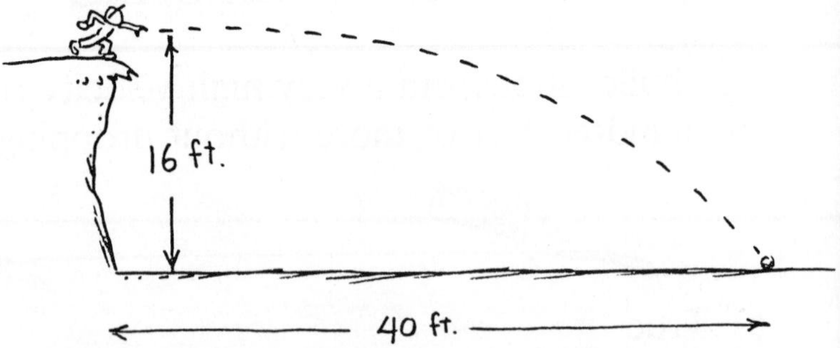

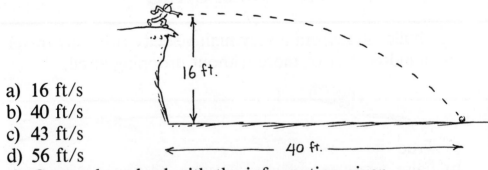

a) 16 ft/s
b) 40 ft/s
c) 43 ft/s
d) 56 ft/s
e) Cannot be solved with the information given

ANSWER: PITCHING SPEED

The answer is: b. We know that speed $= \frac{\text{distance traveled}}{\text{time taken}}$, and we are given that the rock travels 40 feet. Since the rock was thrown exactly horizontally with no initial component of velocity in the up or down direction, the time taken is the same time that would be required for the rock to fall 16 feet if it were simply dropped from rest. This time is one second, so the stone's horizontal velocity must be 40 ft/s.

WORLD WAR II

To strike the tank factory the Flying Fortress should drop its bomb load

a) before it is over the target
b) when it is directly over the target
c) after passing over the target

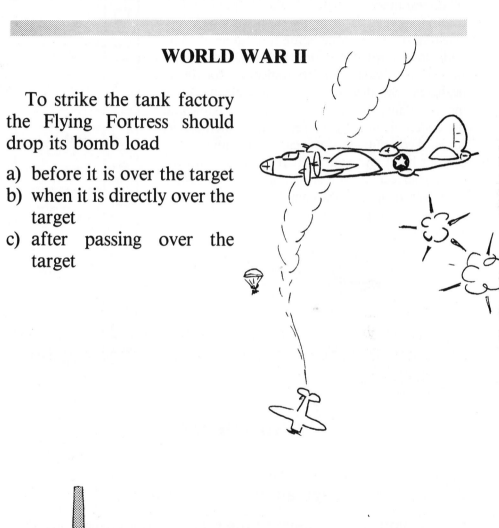

ANSWER: WORLD WAR II

The answer is: a. When the bomb is dropped it does not simply fall vertically to the ground straight below — such motion would require a zero horizontal velocity. When the bomb is released it has an initial horizontal component of velocity equal to that of the Flying Fortress (B29). If air resistance is negligible the dropping bomb continues forward with the airplane's velocity. The film shows how the bomb trails along directly under the aircraft. The bomb actually trails behind somewhat because air resistance is not really negligible. But if you drop a coin while inside the moving plane, then air resistance is negligible and the coin will fall at your feet. This is because its forward component of velocity is the same as that of your hand and your feet. It keeps up with the moving airplane. But if the plane accelerates while the coin is dropping, the coin does not land directly below its dropping position. Why?

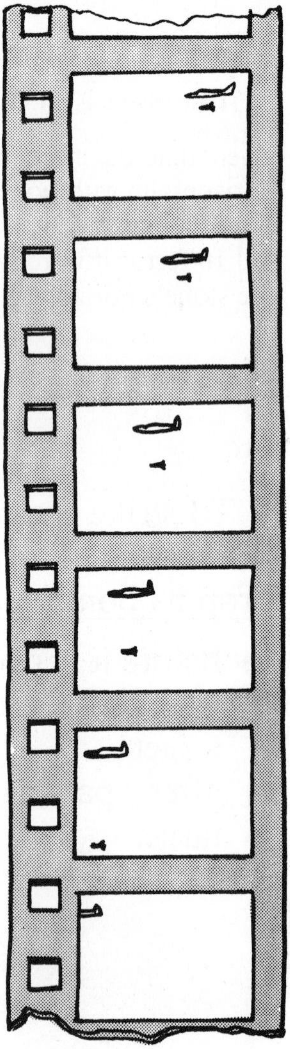

BEYOND THE REACH

Is it possible to get beyond the earth's gravitational field if you go far enough away from the earth?

a) Yes, you can get beyond the reach of the earth's gravity
b) No, you cannot get beyond the reach of the earth's gravity

ANSWER: BEYOND THE REACH

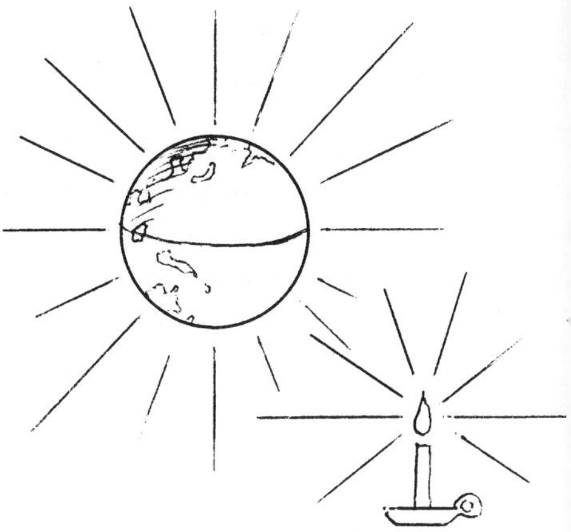

The answer is: b. Newton's vision of gravity is that "lines of flux" or "tentacles" reach out from the earth (or from any mass) into space. The lines of flux NEVER END. They go on forever. But, as they reach further and further into space they get more and more spread out so the force of gravity, which they produce, gets weaker and weaker but never disappears completely. it is like the light rays which come out from a candle, or radiation particles emanating from a piece of radioactive ore. They spread into space. The radiation gets weaker and weaker, but the rays go on forever.

How much weaker? That depends on distance. Visualize a paint spraying can. The can sprays a little square on the wall. Now suppose the wall is twice as far from the can. How much lighter or thinner is the paint spray on the more distant wall? That depends on how much more area the same spray must cover. How much larger is the area of the second square? Twice as large? No. The second square is twice as high and twice as wide, so its area goes up *four* times. So the spray is only one fourth as dense in the second square. Likewise, the intensity of gravity, or heat, sound, or light from any such point-like source is reduced to one fourth if the distance of the source is doubled. If the distance of the

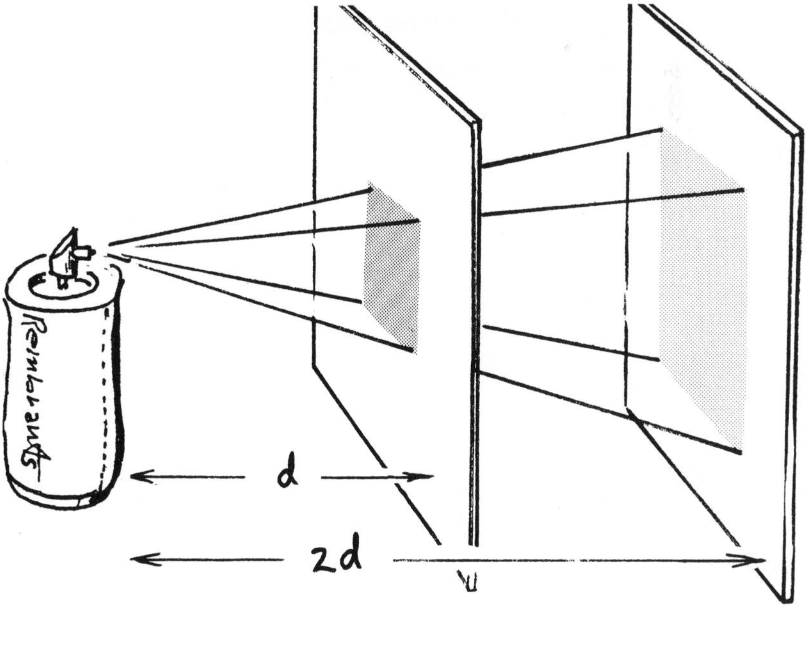

source is increased three times, the square is three times higher and wider, so spray density is reduced nine times. In general, for any stuff spreading through space from a point source: Intensity $\sim 1/(\text{distance})^2$. This results in a very sudden rise in intensity as you get close to the source. Thus, children get the notion that heat is "in" a flame and that the surrounding space is cold.

Back to gravity. Surrounding every bit of matter in the universe is a gravity field. You are matter, and you are surrounded by your own gravity field. How far does it reach? To the far ends of the universe. Your influence is everywhere. . . really!

NEWTON'S ENIGMA

This question troubled Newton for many years. A little mass, "m," is a certain distance from the center of a globular cluster of masses and there is a certain force of gravity, due to the cluster of masses, on the little mass that pulls it toward the center of the cluster. Now consider the situation whereby neither the little mass nor the center of the cluster moves, but the cluster uniformly expands. As a result of this expansion some parts of the globular cluster are closer to "m" and some are farther from "m." After the expansion, the forces of gravity of the cluster on the little mass "m" will

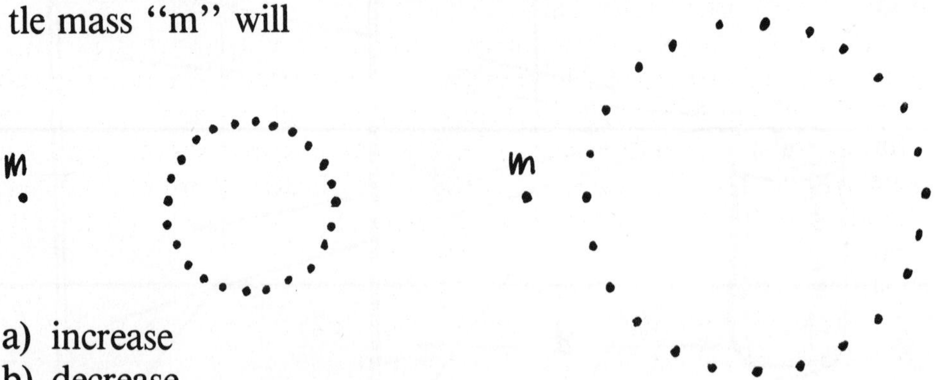

a) increase
b) decrease
c) remain unchanged

ANSWER: NEWTON'S ENIGMA

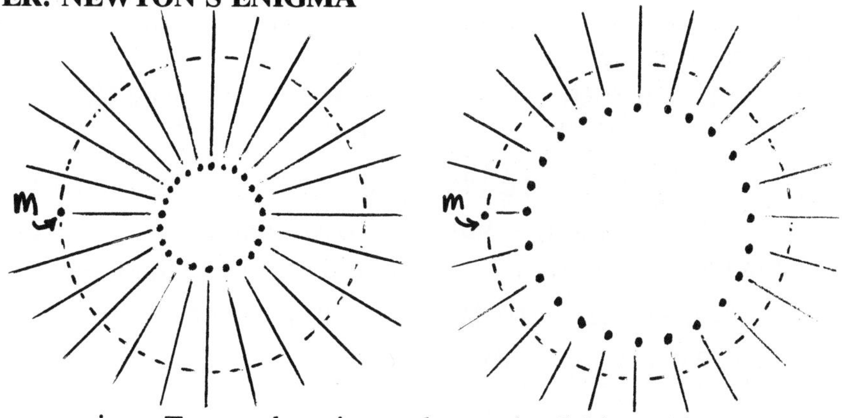

The answer is: c. To see why, picture the gravity field reaching out from the globular cluster like the tentacles of an octopus, one force tentacle from each mass in the cluster. The force of the field depends on how close together the tentacles are. Physicists call the tentacles "lines of flux" or "flux lines." Up close to the globular cluster the lines are close together and the force is strong. Further away they are spread apart and the force gets weaker.

Now picture a spherical bubble around the cluster. As long as the cluster stays inside the bubble, the number of flux lines going through the surface of the bubble will not change. So the strength of the gravity field at the bubble's surface does not change.

So the force on "m," which is at some point on the bubble, does not change. Of course, this bubble is just an imaginary thing.

If the globular cluster of mass is hollow, the sketch makes it look like there is no gravity inside the hollow, and in fact, there is no gravity in a hollow globe, at least no gravity due to the mass of the globe.

Why was Newton thinking about this? Because he pictured the earth itself as a globular cluster of little masses or atoms (but not a hollow globe). He wanted to show that the force field outside the globe was the same as the force field which would exist if all the mass in the globe was compressed to the globe's center. It makes calculations so much easier when you can think of all the mass being at one point rather than being spread all over in the cluster.

Incidentally, Newton eventually figured out the answer was "c," but he did not picture the bubble. The bubble idea came from a mathematician named Karl Friedrich Gauss, who just might have been the smartest person who ever lived. Needless to say, Gauss figured out a lot of things besides the bubble. Gauss lived during the time of Napoleon and Beethoven, which was after Newton's time.

TWO BUBBLES

This is a rather advanced problem, so either skim it or dig in!

If all space were empty except for two nearby masses, say two drops of water, the drops would, according to Newton's Law of Gravity, be attracted together. Now suppose all space were full of water except for two bubbles. How would the bubbles move?

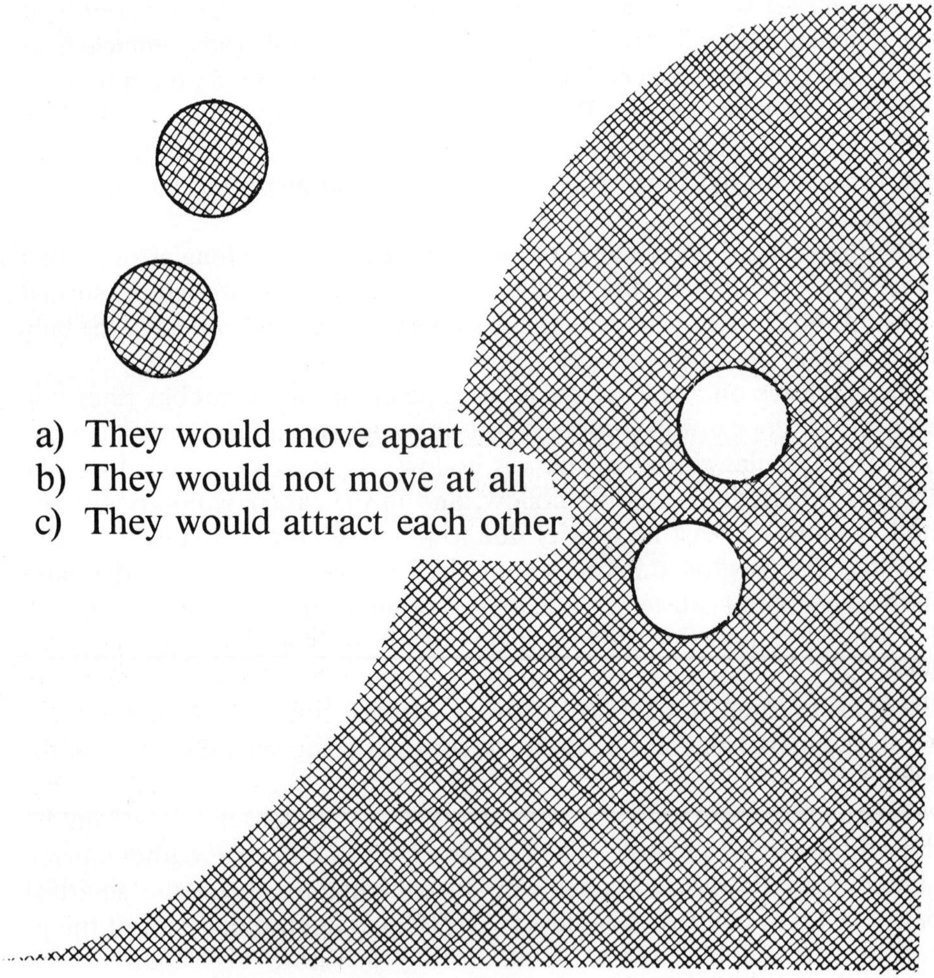

a) They would move apart
b) They would not move at all
c) They would attract each other

ANSWER: TWO BUBBLES

The answer is: c. What is the point of this question? There are at least two points. The first is to show how easily a simple situation, which we think we thoroughly understand, can be transformed (by turning it inside out) into a baffling situation. The second point will come later.

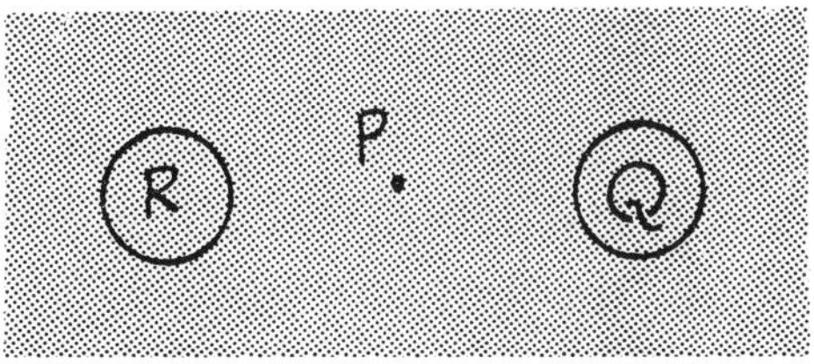

Why do the bubbles move together? If all space were completely full of water there would be no net gravity at point **P** because any attraction due to water at **Q** would be cancelled by the attraction of the water at **R**. But if the water at **R** is removed to make a bubble, the balance at **P** is upset and there is a net attraction to the water at **Q**. So now there is gravity at **P** — the attraction is towards **Q** or away from **R**. It is as if **R** repelled things.

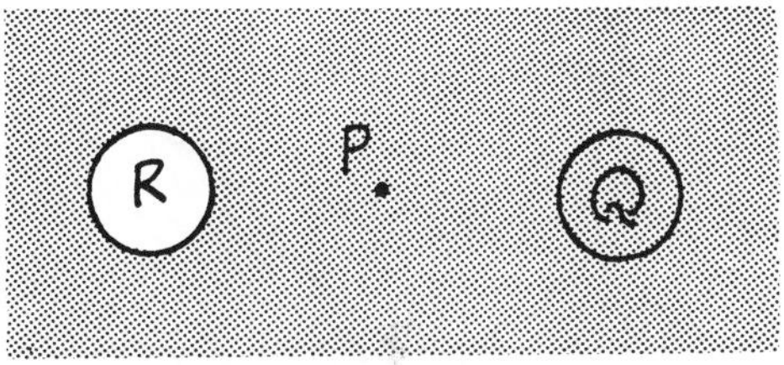

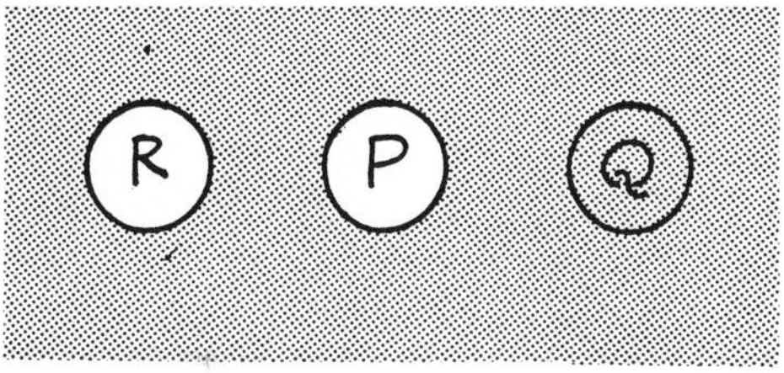

But what is meant by "things"? By things we mean particles, pebbles, and rocks and the like. A rock at **P** would move away from **R**. But what about bubbles? How does gravity affect bubbles? Gravity makes rocks and bubbles move in opposite directions — rocks down, bubbles up. So if a rock at **P** would move away from **R** then another bubble at **P** would move towards **R**. The bubbles would attract each other.

Now for the second promised point of this question. We started with the idea that mass attracts mass and from that reasoned that bubble must attract bubble or empty space attracts empty space. But we could have just as well started with the idea that empty space attracts empty space and from that reasoned that mass must attract mass. Our universe is mostly empty space with a little mass in it so we get one view of things, but if our universe were mostly mass with a little empty space in it we would get a different view of things. Or, on the other hand, we might just interchange our idea of what was mass and what was empty space.

In physics you are always close to the deep water. Take two steps off the beaten path and you may be into it. And it's not make believe.

INNER SPACE

Down in a cave below the surface of the earth there is

a) more gravity than at the earth's surface
b) less gravity than at the earth's surface
c) the same gravity as at the surface

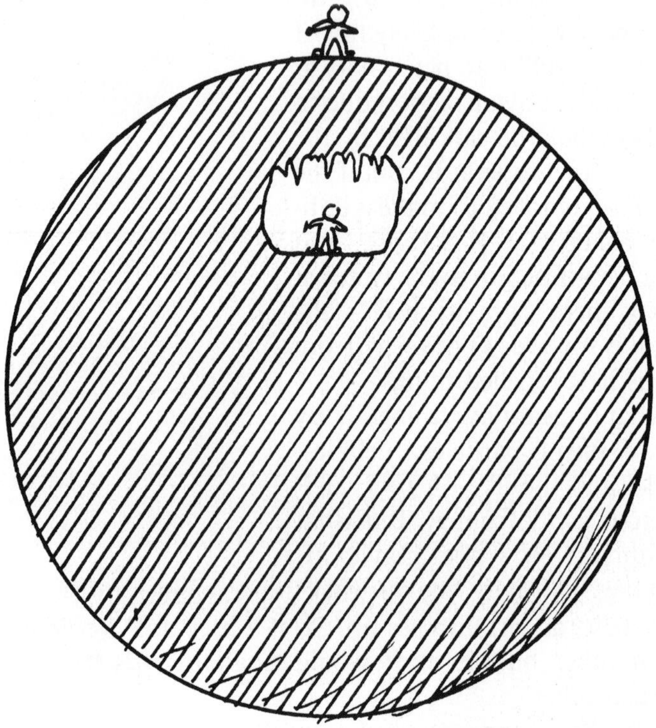

Assume that the earth's density is the same all the way through. Of course it isn't. But simplistic idealized situations must be thoroughly understood before details can be appreciated.

ANSWER: INNER SPACE

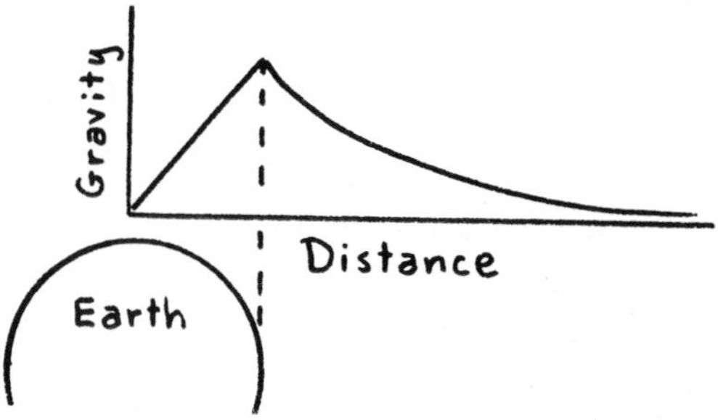

The answer is: b. There is less gravity in the cave because some of the earth's mass is above your head when you are in the cave and that mass pulls up on you and so cancels the effect of some of the mass below your feet which pulls down on you. What if the cave were at the very center of the earth? In that case there would be no gravity at all in the cave. It would be just like floating in "zero g" in a spaceship! Why? Because there would be equal amounts of earth "above" and "below" you.

Things on the surface of the earth, where we live, experience the most gravity. If you go up from the surface away from the earth into space, gravity gets weaker and if you go down inside the earth it also gets weaker.

FROM EARTH TO MOON

Of the five places listed below it would be easiest to launch a spacecraft from the earth from

a) New Mexico (south over Mexico)
b) California (north over the Pacific Ocean)
c) Florida (east over the Atlantic Ocean)
d) Moscow (east over Siberia)

The first trip to the moon was, in fact, launched from (see above)

a) b) c) d) e)

A century ago Jules Verne thought the journey to the moon would commence from (see above)

a) b) c) d) e)

ANSWER: FROM EARTH TO MOON

The answer is: c. The earth spins from west to east, which is why the sun rises in the east and sets in the west. (Stop and visualize that.) So if you launch the spacecraft to the east you can use some of the earth's spin "for free." Since the earth spins around its pole, the pole stands still, and those parts of the earth near the pole move slowly; those parts farther away move fastest. The equator is farthest away from the poles and moves fastest. Therefore the best way to launch is to the east from the equator — or as close to it as you can get.

The answer to the second and third questions is also: c. It is history now.

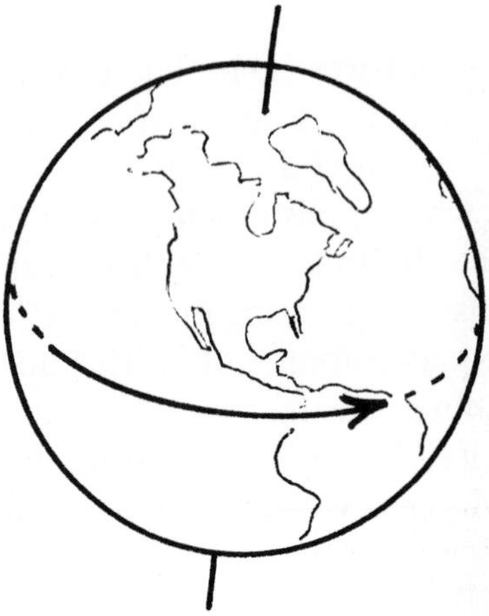

EARTH SET

Suppose you lived on the moon. Suppose the earth was directly over your head. How long would it be before you would see the earth set?

a) one day (one earth day, 24 hours)
b) one-fourth of a day (6 hours)
c) one month (the time it takes the moon to orbit the earth
d) one-fourth of a month
e) you would never see the earth set

ANSWER: EARTH SET

The answer is: e. As seen from earth you always see one side of the moon. That's why the ancient astronomers thought the moon was something stuck on the dome of the sky, rather than another world. The back of the moon was a subject of mystery until first photographed by a Russian spacecraft.

If you lived on the side of the moon that faced the earth you would never see the earth set because that side of the moon continually faces the earth.

PERMANENT NIGHT

Suppose the earth still goes around the sun every year but you keep one side of the earth always facing the sun, so one side always sees the sun and the other side never sees it. As seen from the earth, the sun is stationary in the sky. Now in this supposed situation the stars would

a) also seem stationary in the sky
b) would seem to go once around the earth every day
c) would seem to go once around the earth every year

ANSWER: PERMANENT NIGHT

The answer is: c. As seen by someone standing in the middle of the dark side of the earth, the star is overhead when the earth is at position **A**. The star is underfoot one-half year later when the earth is at **C** and back overhead again one year later when the earth gets back to **A**. So even though the sun is frozen in the sky, causing one side of the earth to have a permanent day and the other side a permanent night, the stars would still seem to go around the sky once every year.

Thank God this is not now the existing situation, though there is reason to believe that some time in the distant future (when the earth's spin is reduced by tidal friction) this will become the existing situation.

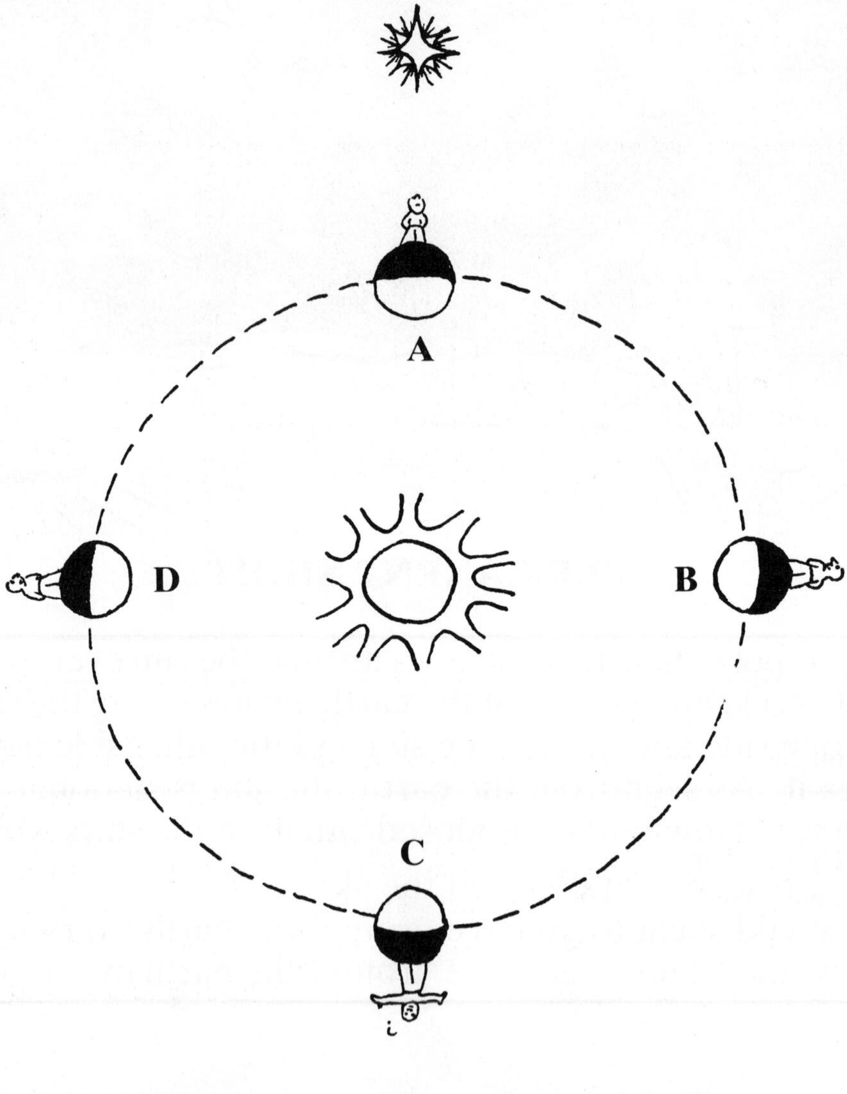

STAR SET

Using a good digital watch, get the exact time when a bright star goes behind a distant building or tower. A day later time the disappearance again. Sighting over a nail fixed in a window sash will help you return your eye to the same location for each sighting. It will be found that

a) the star disappears at the same time each night
b) the star disappears a little earlier each night
c) the star disappears a little later each night

ANSWER: STAR SET

The answer is: b. The motion of the stars in the night sky is due to the earth spinning, so the stars, like the sun, should seem to go around the earth every twenty-four hours, but not quite. Why not quite? Because you have two circular motions: one around the earth every day and one around the sun every year.

If the earth kept one side facing the sun all the time the star would go around the earth once every year. But the earth does not keep one side facing the sun. It spins around so one side does not always face the sun. It spins so fast that you see the sun go around about 365 times each year. So you also see the star go around 365 times each year. NO. You see the star go around 366 times! Why the extra time? *Because it would go around once each year even if the earth kept one side always facing the sun.* Remember, there are two circular motions, and the total circular motion is the sum of both.

How do you know the circular motions should be added together? Maybe one should be subtracted from the other. The two motions add because the earth's daily spin and yearly orbit both circle the same way. The sun also spins the same way. The earth's spin and orbital motion probably came with the material taken from the sun when the earth was created. That's why all circle the same way.

So the star must go around the sky a little faster than the sun. The sun goes around once in about 24 hours so the star goes around once in a little less than 24 hours. How much less? The sun goes around the sky in about 1,440 minutes (1,440 minutes = 24 hours × 60 minutes per hour). The star must go around the sky a bit faster so that after one year it has made it around one more time than the sun. In other words the star has to get in an extra turn in one year. If a normal turn (sun turn) takes 1,440 minutes and if you cut that time by 4 minutes (4 times 365 approximately equals 1,440) then after 365 turns (one year) you will be one turn ahead.

So stars don't set (or rise) at the same time each night. Each night they come up (and go down) four minutes earlier than they did the night before. That adds up to roughly half an hour in a week. So at the same time of night you don't always see the same stars in the sky, even though at the same time of day (say noon) you always see the same star—the sun. It is this effect that causes the constellations in the winter night to be different from those in the summer night.

In 1931 a radio engineer named Jansky who worked for the Bell Telephone Labs picked up radio noise on a particularly sensitive short wave receiver at about the same time each day. No one could tell or even guess where the radio noise came from. Then Jansky noted that the noise came four minutes earlier each day. He said that means the radio noise must be coming from the stars. For years no one would believe it, but he had in fact discovered the first extraterrestrial radio source—the center of the Milky Way.

LATITUDE AND LONGITUDE

By glancing at the night sky you can immediately estimate your

a) latitude
b) longitude
c) both
d) neither

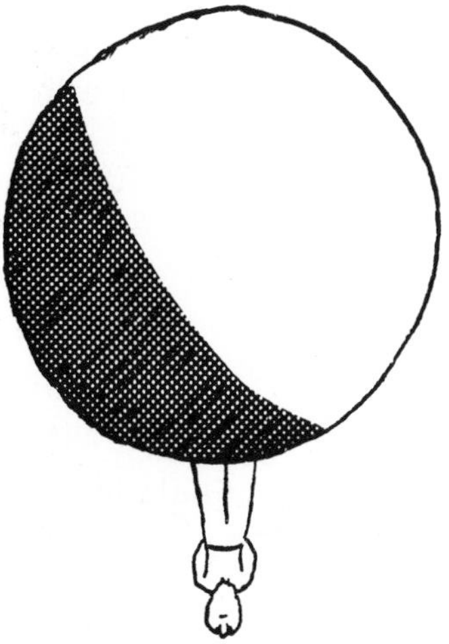

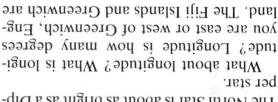

ANSWER: LATITUDE AND LONGITUDE

The answer is: a. If the North Star is directly over your head you must be at the north pole. (If it is directly under your feet you must be at the south pole.) If it is halfway between overhead and underfoot—that means on the horizon—you are at the equator. If the North Star is halfway between overhead and the horizon—that means 45° up from the horizon (or down from the zenith)—you must be at 45° north latitude. Why not 45° south? Because at 45° south you could not see the North Star.

Is the North Star the brightest star in the sky? No. But it is easy to find—see the sketch. The Big Dipper points the way. I have drawn two Big Dippers because the Dipper spins around the North Star every day, so sometimes it will be above the star and sometimes below it and sometimes to the right or left of it. The North Star is about as bright as a Dipper star.

What about longitude? What is longitude? Longitude is how many degrees you are east or west of Greenwich, England. The Fiji Islands and Greenwich are

on opposite sides of the earth so the Fiji Islands are at 180° east or west longitude (it doesn't matter when you go halfway around).

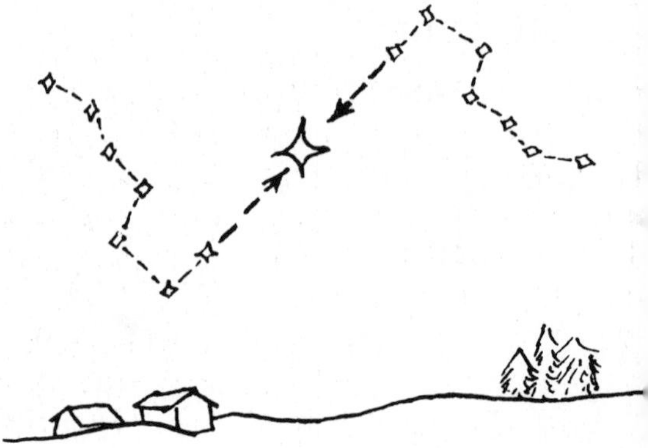

New Orleans, Louisiana, is west of England a quarter of the way around the world and so at 90° west longitude. Calcutta, India, is a quarter of the way around the world east of England and so at 90° east longitude.

How can you tell what your longitude is? By looking at the stars? That won't work because the sky over New Orleans at midnight tonight (New Orleans time) will be identical to the sky over Cairo, Egypt, at midnight tonight (Cairo time). You need more than a look at the sky. You need to know the time.

You need to know not only the local time, which you can tell by looking at a sundial, you also need to know what time it is in Greenwich, England. How does it work? Well, suppose it is noon where you are and you get on the telephone and ask Greenwich what time they have. Greenwich answers "midnight." Where are you? On the side of the earth opposite Greenwich, 180° east or west longitude. Before telephones and radios sailors had to carry

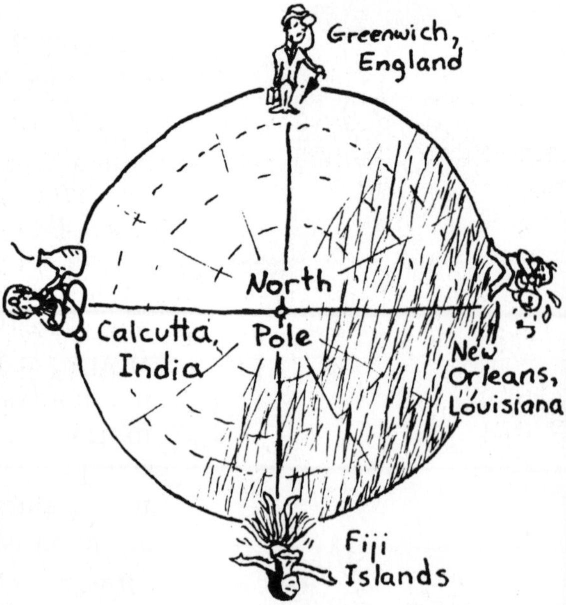

clocks reading Greenwich time. If the sailors' clock was only one minute slow or fast the ship's navigation could be 16 miles in error (16 miles = 24,000 miles around the world divided by 1,440 minutes around the day).

Why is it so much more difficult to tell longitude than latitude? Because latitude was made by God; longitude is man-made. That is, Nature designated the north pole by the way of the earth spins, but people designated Greenwich. According to some people, longitude should not even be measured from Greenwich. According to the metric system the earth should be divided into one thousand metric degrees (not 360°) starting from Paris, France. On some homestead and U.S. land grant documents longitude is designated from Washington, D.C.

A SINGULARITY

New Orleans is exactly one-quarter the distance around the globe from London (New Orleans is at 90° west longitude). When it is noon in London, what time is it in New Orleans?

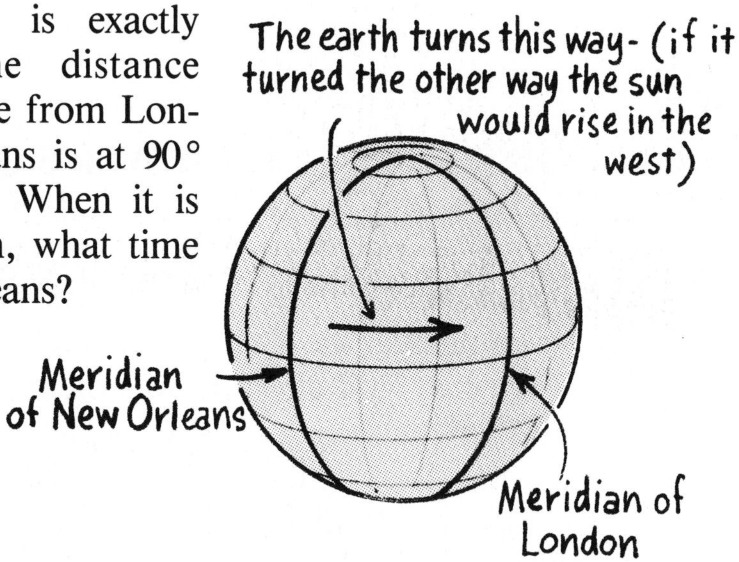
The earth turns this way- (if it turned the other way the sun would rise in the west)
Meridian of New Orleans
Meridian of London

a) Noon
b) Midnight
c) 6 a.m.
d) 6 p.m.
e) anytime

When it is noon in London, what time is it at the North Pole?

a) Noon b) Midnight c) 6 a.m. d) 6 p.m.
e) anytime

ANSWER: A SINGULARITY

The answer to the first question is: c. One-quarter of 24 hours is 6 hours. The sun is now high over London. In six hours the earth will turn New Orleans under the sun. So, it must be 6 AM in New Orleans.

The answer to the second question is: e. It is noon at any place on earth when the sun is over the meridian running through that place, but all meridians run through the North Pole, therefore the rule for telling time breaks down at the North Pole.

A place where a rule breaks down is called a singularity. Usually the rules that break down are rules for calculating something. (For example, what is $\frac{1}{x-2}$ when x equals 2? Infinity? Plus infinity or minus infinity?) There is the fantastic possibility that there are singularities in the universe where the laws of physics break down, as for example, at the center of a black hole.

PRESIDENT EISENHOWER'S QUESTION

Almost everyone in the U.S.A. was astonished and concerned when the U.S.S.R. won the first round of the space race by launching the first earth satellite, SPUTNIK, in 1957. The most vital question was the mass of the payload (spacecraft) which the U.S.S.R. could orbit. This was the question the President of the United States asked his science advisors: "All we know for certain about Sputnik is its altitude and orbital speed. From that information can you calculate Sputnik's mass?" The science advisors answered:

a) "Yes, we can."
b) "No, we cannot."

ANSWER: PRESIDENT EISENHOWER'S QUESTION

The answer is b. Just as rocks of different masses falling without air resistance fall identically (recall FALLING ROCKS), and just as projectiles of various masses thrown with equal velocities follow identical trajectories, satellites of whatever mass moving at equal velocities at equal altitudes fall around the earth in the same orbit. The force that holds a satellite in orbit is simply the weight of the object at that altitude which in turn is proportional to its mass. So if the mass of a satellite should be doubled, the force holding it in orbit is also doubled — it would have twice the weight. Sputnik would stay in the same orbit at the same speed whatever its mass.

Only by actually probing Sputnik could we determine its mass. For example, if it were probed with another satellite of known mass and velocity, and the velocity of rebound measured, then the mass of Sputnik could be determined by momentum conservation. But without interacting with it, its mass cannot be determined by means of the orbit it follows.

TREETOP ORBIT

If the earth had no air (atmosphere) or mountains to interfere, could a satellite given adequate initial velocity orbit arbitrarily close to the earth's surface — provided it did not touch?

a) Yes, it could
b) No, orbits are only possible at a sufficient distance above the earth's surface where gravitation is reduced

ANSWER: TREETOP ORBIT

The answer is: a. It is important to realize that a satellite is simply a freely falling body that has enough tangential or sideways speed to enable it to fall around the earth rather than into it. If a projectile is launched horizontally at treetop level at everyday speeds, its curved path would quickly intercept the earth and it would crash. But if it were launched at 5 miles per second (18000 mph) its curved path would match the curvature of the earth's surface. If no obstructions or air resistance were present, it would continuously fall without intercepting the earth — it would be in circular orbit. If it were projected at higher speeds it would follow elliptically-shaped orbits — at projection speeds beyond 7 miles per second (25000 mph) would escape the earth altogether. So the essence of launching satellites is to get them up above air drag and boost their tangential speeds high enough so that the curvature of their paths at least matches the curvature of the earth.

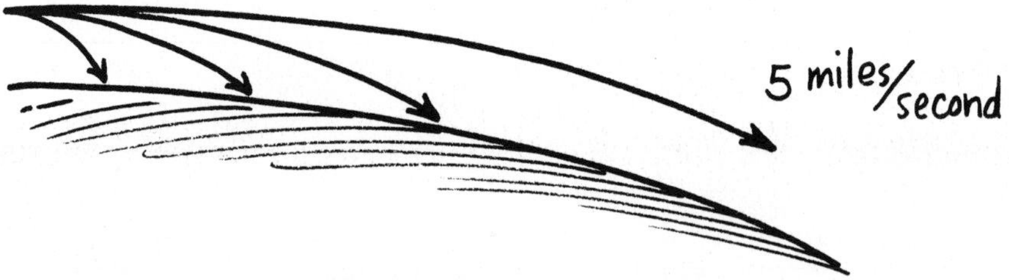

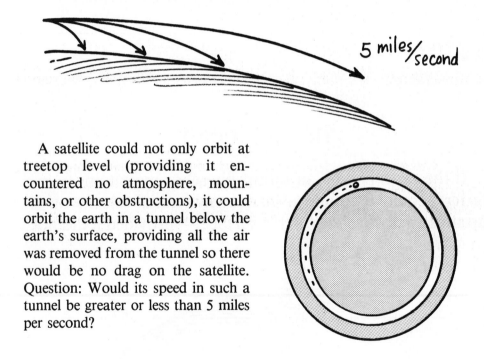

A satellite could not only orbit at treetop level (providing it encountered no atmosphere, mountains, or other obstructions), it could orbit the earth in a tunnel below the earth's surface, providing all the air was removed from the tunnel so there would be no drag on the satellite. Question: Would its speed in such a tunnel be greater or less than 5 miles per second?

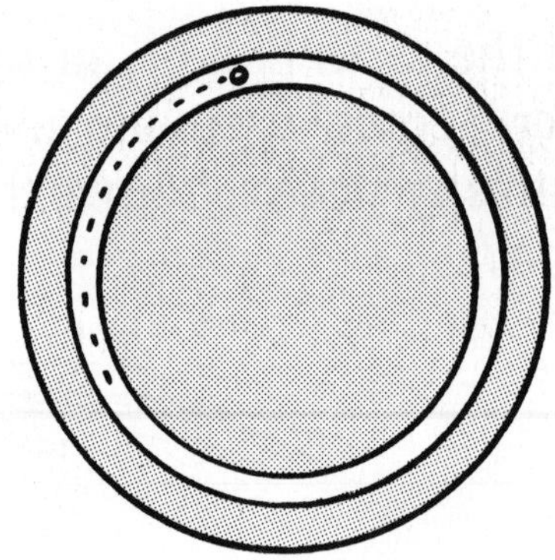

GRAVITY OFF

Kepler's Second Law states that the imaginary line between any planet and the sun sweeps equal areas in equal times as it orbits the sun. If the gravitation between the sun and the planets were somehow switched off and the planets no longer followed elliptical paths, would Kepler's equal area in equal time law still hold true?

a) Yes, the Second Law would still hold, even with gravity off

b) No, Kepler's laws relate to elliptical orbits produced by gravity. Turn off gravity and Kepler's Second Law is meaningless.

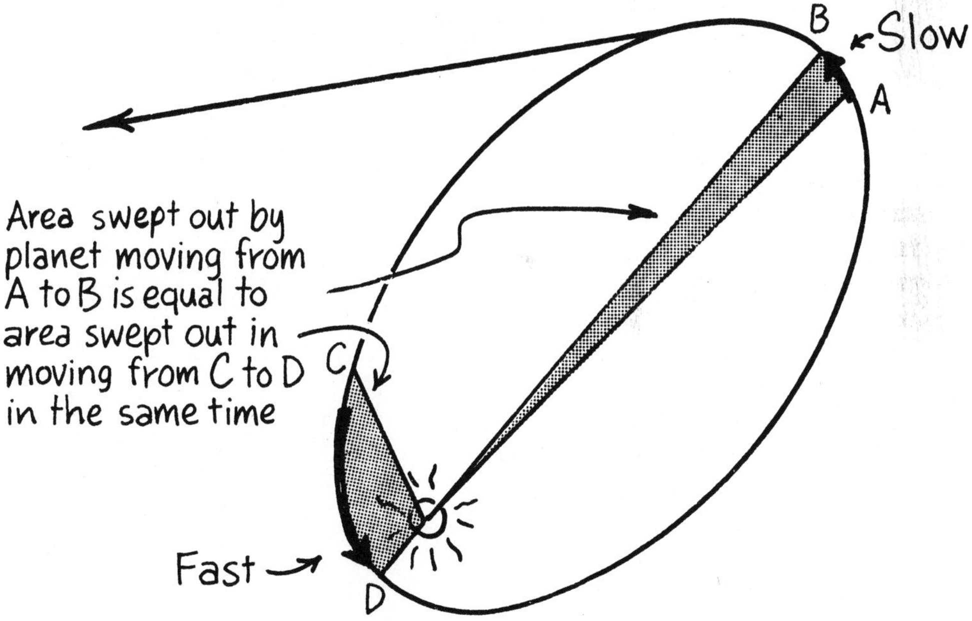
B
Slow
A
Area swept out by planet moving from A to B is equal to area swept out in moving from C to D in the same time
C
Fast
D

ANSWER: GRAVITY OFF

The answer is: a. Kepler's Second Law states that the imaginary line joining a planet to the sun sweeps out equal areas of space in equal time. This just means that the planet's angular momentum around the sun is unchanging (Remember BALL ON A STRING). Now if gravity is shut off, say when the planet is at position **e**, the planet shoots away along the line **e f g h i j** at constant speed. If the distance between **e** and **f**, **g** and **h**, **i** and **j** are all equal, then the planet must travel between them in equal times.

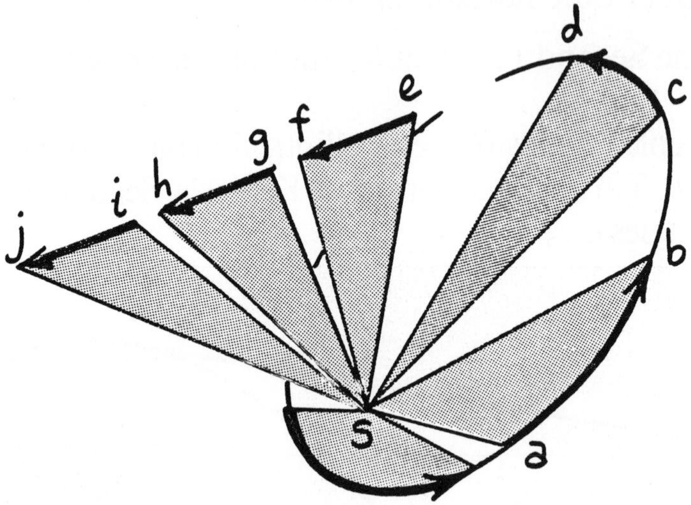

But the areas of the triangles △ **efs** and △ **ghs** and △ **ijs** are all equal. Why? Because all the triangles have equal bases and a common altitude at s.

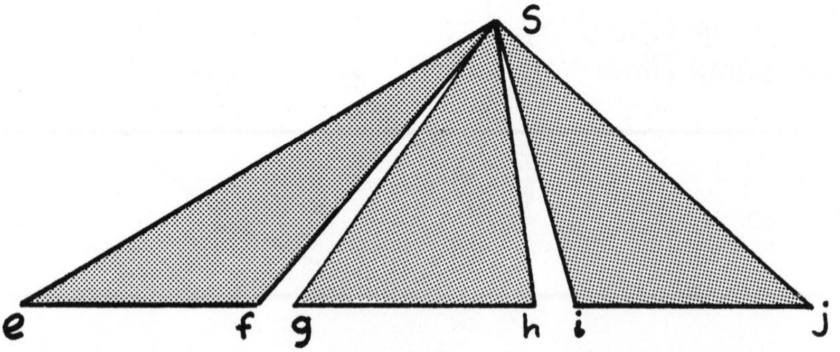

If **ef** = **gh** = **ij** then the area of △**efs** = the area of △**ghs** = the area of △ **ijs**.

SCIENCE FICTION

As you move away from the earth its gravity gets weaker. But suppose it did not? Suppose it got stronger? If that fictitious law were so, would it be possible for things, like the moon, to orbit the earth?

a) Yes, just as they presently do
b) Yes, but unlike they presently do
c) No, orbital motion could not occur

ANSWER: SCIENCE FICTION

The answer is: b. The earth and moon are held together by the invisible force of gravity, but suppose they were held together by a spring? (You must imagine some way for the spring

to pivot without getting wrapped around the earth.) Even if the moon were held by a spring it could still orbit the earth. But now the force, being the force of the spring, would get stronger rather than weaker as it was stretched further into space.

Incidentally, if the planets orbited the sun on springs they would still move in elliptical orbits, but the sun would be at the center of the ellipse, not at one focus and all the planets would require about the same time to orbit the sun regardless of the size of their orbits.

Before Newton's time a man named Robert Hooke (a big name in the development of the microscope) had the idea that gravity must work like a spring. Hooke could never understand why Newton got all the glory for the theory of gravity, but as you can see, the spring idea does not lead to the kind of elliptical orbits which are, in fact, found in nature.

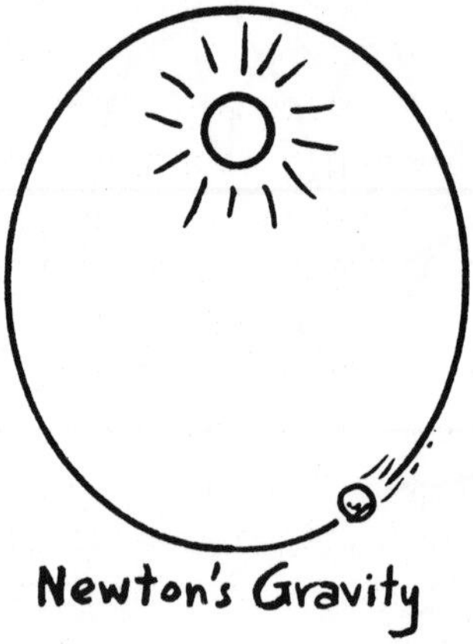

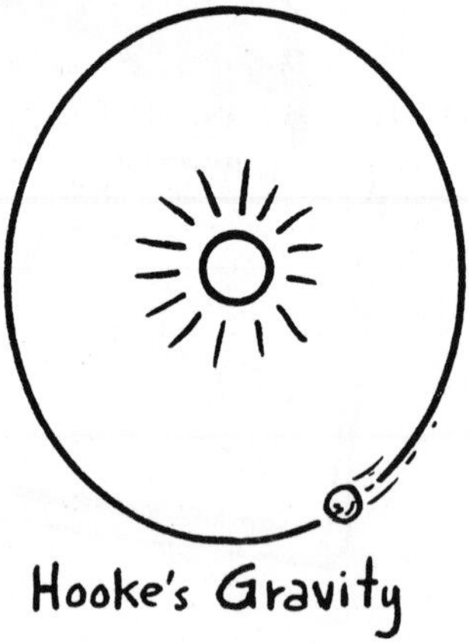

REENTRY

Sputnik I, the first artificial earth satellite, fell back to earth because friction with the outer part of the earth's atmosphere slowed it down. (Of course, this happens to all low orbiting spacecraft.) As Sputnik spiralled closer and closer to the earth its speed was observed to

a) decrease
b) remain constant
c) increase

ANSWER: REENTRY

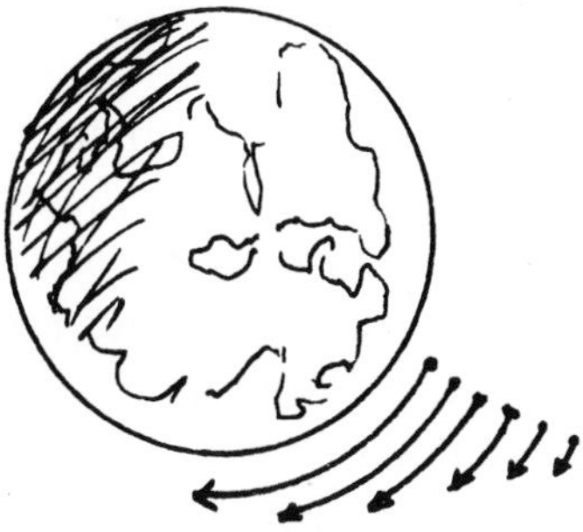

The answer is: c. This very much surprised many people back in 1958 because they were told atmospheric friction was slowing the satellite. The explanation is as follows.

At any height above the earth's surface there is a critical speed. The critical speed is maximum near the earth's surface and is progressively less at higher altitudes. If the spacecraft at a given altitude slows to less than the critical speed for that altitude, it falls closer to the earth and gains speed in doing so — but because of increasing atmospheric friction, it doesn't gain enough speed to attain the even greater speed required for orbit closer to the earth. The speed it in fact gains while approaching the surface of the earth is always too little too late.

BARYCENTER

Which is correct?

a) The moon goes around the center of the earth
b) The earth goes around the center of the moon
c) They both go around some point between the centers

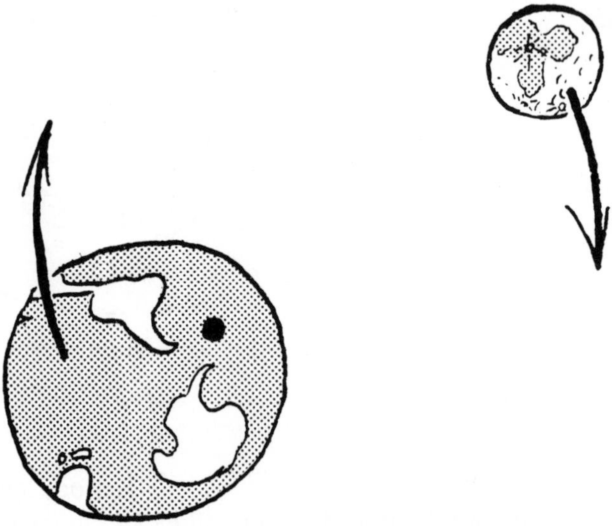

ANSWER: BARYCENTER

The answer is: c. They must go around their common center of mass, which is much closer to the earth than the moon because the earth has eighty times as much mass as the moon. In fact the common center, called the barycenter, is so close to the earth that it is inside the earth—but not at its center. The barycenter is one thousand miles below the earth's surface. The diameter of the earth is 8000 miles and the distance to the moon is about 30 earth diameters.

How long does it take the earth to orbit around the barycenter? Exactly as long as it takes the moon to go around it—one month.

OCEAN TIDE

The ocean tide produced by the moon is deepest on the side of the earth

a) under the moon
b) away from the moon
c) about equally deep on both sides

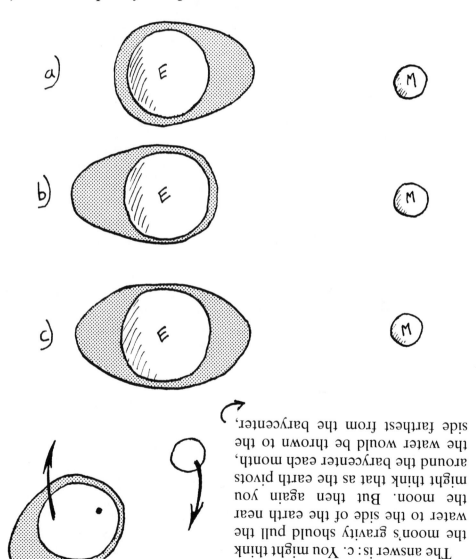

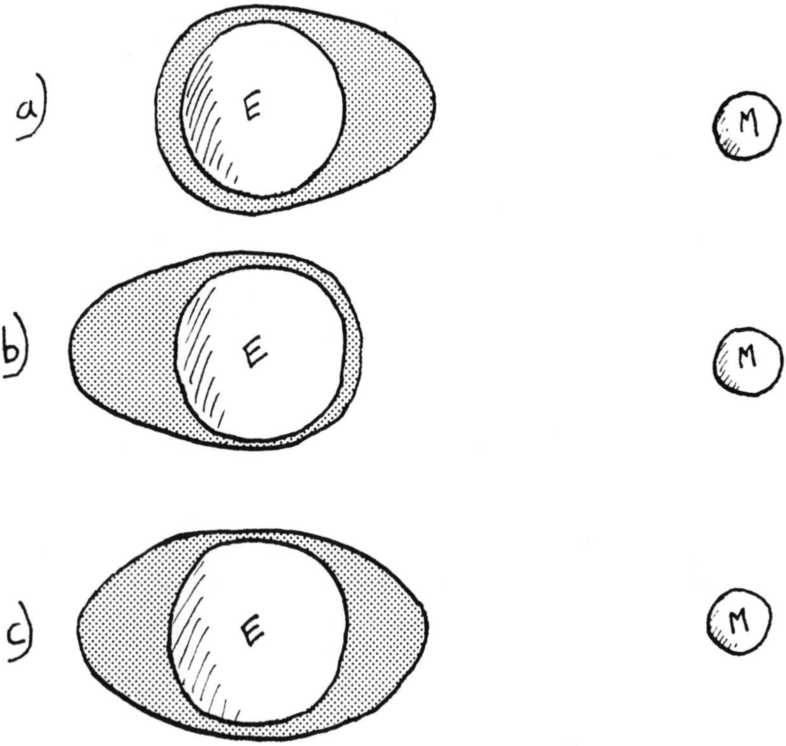

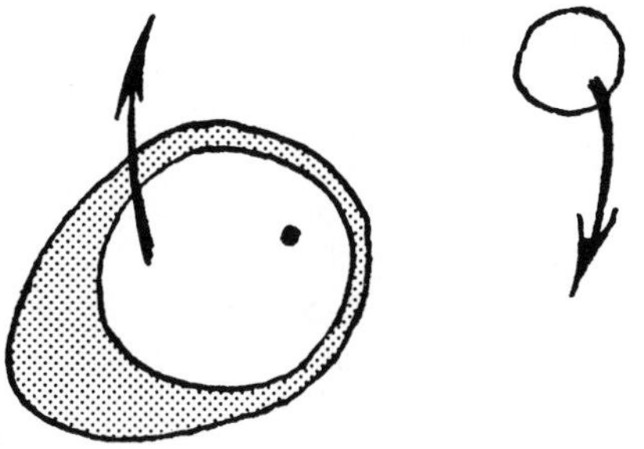

ANSWER: OCEAN TIDE

The answer is: c. You might think the moon's gravity should pull the water to the side of the earth near the moon. But then again you might think that as the earth pivots around the barycenter each month, the water would be thrown to the side farthest from the barycenter, which is the side away from the moon.

Actually both of these thoughts are correct. On the side near the moon lunar gravity is strongest so the water is pulled towards the moon. On the side farthest from the barycenter the centrifuge effect is strongest so the water is also thrown out there. The surface of the ocean ends up being shaped like a football, one end towards the moon, one end away. But don't suppose the water is actually lifted by gravity or the centrifuge effect. The water is slid off regions of the earth where the moon is on the horizon and accumulated on the face looking towards and the face looking away from the moon.

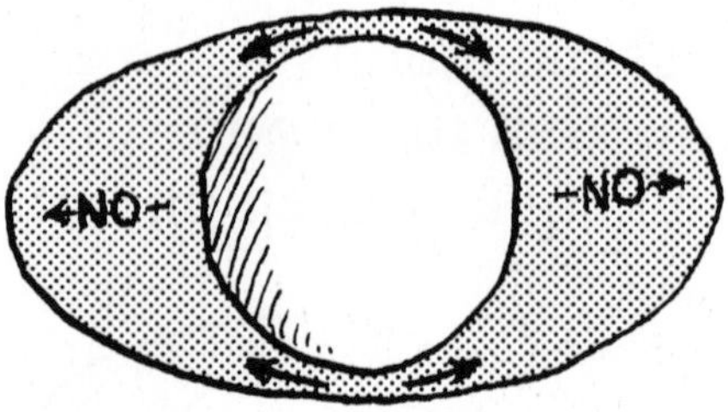

What about the force of the earth's own gravity and the centrifuge effect of the earth's daily spin around its center? These forces must be stronger than those due to the moon's gravity and the monthly spin around the barycenter. Yes, they are stronger. But they act equivalently all around the earth, so whatever they do, they do the same all around the earth. On the other hand, the effect of the moon's gravity and the barycenter spin are not the same all around the earth. That's why the water depth is not the same all around the earth.

SATURN'S RINGS

Over a century ago, J.C. Maxwell calculated that if Saturn's rings were cut from a piece of sheet metal they would not be strong enough to withstand the tidal tension or gravitational gradient tension that Saturn would put on them and would, therefore, rip apart. But suppose the rings were cut from a piece of thick-plate, rather than thin-sheet, iron. Might the thick-plate hold? The thick-plate would

a) fail as easily as the thin-sheet
b) fail more easily than the thin-sheet
c) not fail as easily as the thin-sheet

ANSWER: SATURN'S RINGS

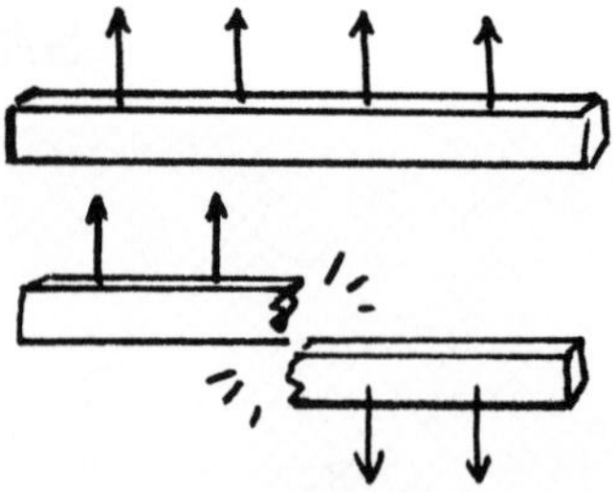

The answer is: a. First of all, let us understand why material breaks or fails. Material does not break simply because of force. A uniform force applied to a bar for example simply makes the bar accelerate in the direction of the force. Material breaks because DIFFERENT forces are applied to different parts of the material. The force DIFFERENCES produce tensions, compressions or shears and it is these stresses that cause material to fail.

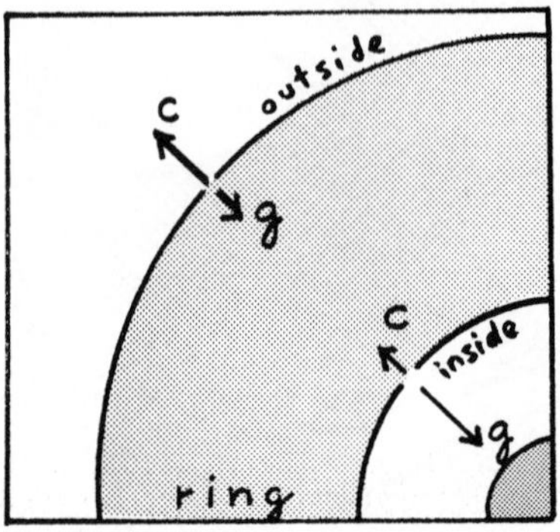

Now suppose that Saturn's rings were solid discs. Saturn will exert a greater gravitational force on the inside of the disc than on the outside and this puts a tension strain into the disc that tends to pull it apart. The gravitational stress is proportional to the mass of the ring. If the solid disc ring is made to spin this only makes the strain worse because there is more centrifugal force on the outer part of the disc than on the inner part. (This is why flywheels sometimes explode.) The centripetal stress is also proportional to the mass of the ring. So increasing the mass of the ring by going from sheet to plate-iron will not help. The rings would be stronger, but the stress on them would also be stronger. So a solid ring about Saturn, whether thick or thin, would be ripped apart.

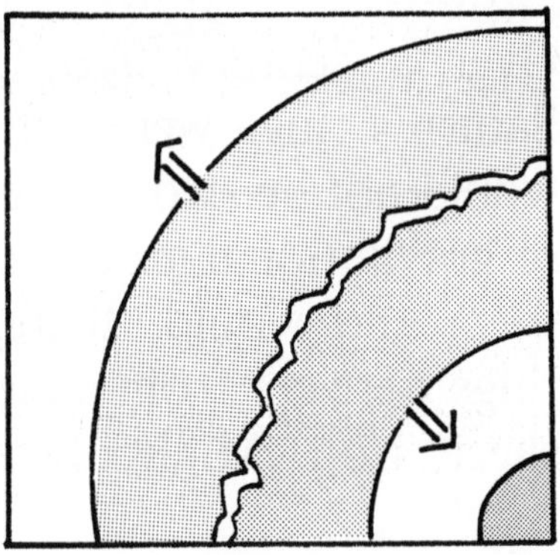

Maxwell decided the rings could not be solid discs. They had to be made of many individual particles. That way each particle could orbit Saturn with its own individual speed adjusted to the gravitational force at its particular distance from the planet. So the inner part of Saturn's rings orbit faster than the outer parts, and all stress is avoided.

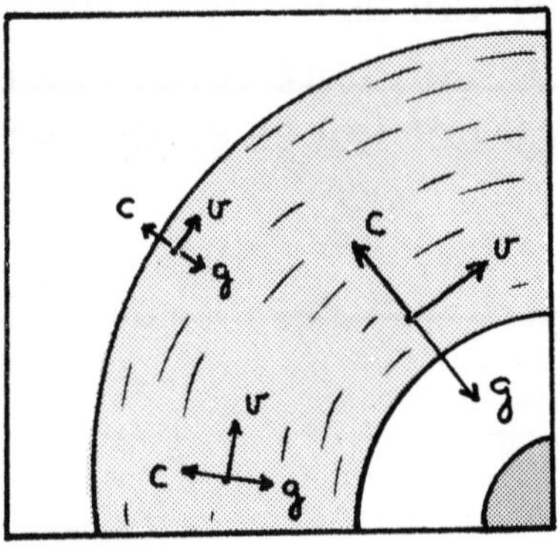

HOLE MASS

The mass of a "black hole" must be infinite, or at least nearly infinite.

a) True
b) False

ANSWER: HOLE MASS

The answer is: b. If an object is moving fast enough it can escape from a planet or star. If it is not moving fast enough it falls back. The minimum speed required to get away is called ESCAPE VELOCITY. The larger the mass of the planet or star, the larger the escape velocity — provided the planet's size does not change.

Now, a black hole is a thing for which the escape velocity is larger than the speed of light. But the speed of light is not infinite. It is a large number, but it is a finite number. The mass of the black hole is also finite and not infinite. In fact, black holes can be made from very small masses if the mass is sufficiently compressed. If you could figure out how to compress a mass and make a black hole it would be a fantastic science fair project!

MORE QUESTIONS (WITHOUT EXPLANATIONS)

You're on your own with these questions which parallel those on the preceding pages. Think Physics!

1. Suppose you wish to average 40 mph on a particular trip and find that when you are half way to your destination you have averaged 30 mph. How fast should you travel the remaining half of your trip to attain the overall average of 40 mph?
   a) 50 mph   b) 60 mph   c) 70 mph   d) 80 mph   e) none of these

2. A dragster accelerates from zero to 60 mph in a distance *D* and time *T*. Another dragster accelerates from zero to 60 mph in a distance *"?"* and time *2T*. That is, it took the second dragster twice as long to get to 60 mph. How much distance did the second dragster cover while getting up to 60 mph?
   a) ¼D   b) ½D   c) D   d) 2D   e) 4D

3. As the ball rolls down this hill
   a) its speed increases and acceleration decreases
   b) its speed decreases and acceleration increases
   c) both increase   d) both remain constant   e) both decrease

4. As an object freely falls (zero air resistance) its velocity increases and its acceleration
   a) increases   b) decreases   c) remains constant

5. As an object falls through air and *is* affected by air resistance its velocity increases and its acceleration
   a) increases   b) decreases   c) remains constant

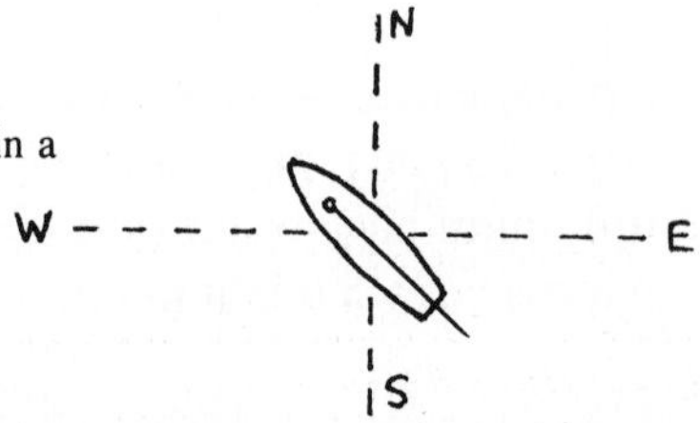

6. The sailboat shown will sail fastest when the wind blows in a direction from the
   a) north   b) east   c) south   d) west   e) none of these

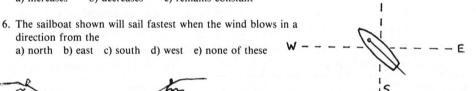

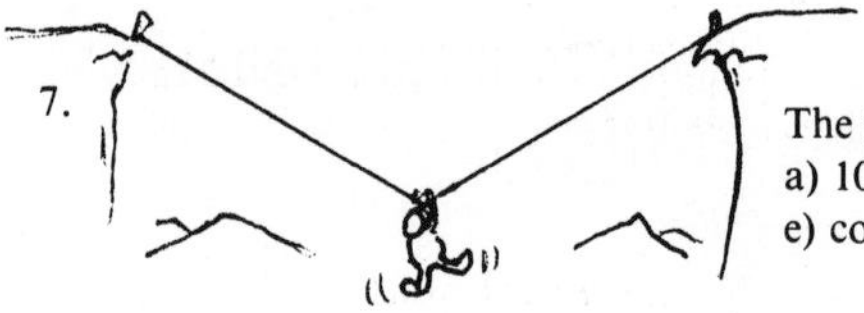

7. The tension in the rope that supports the 200 lb. man is about
   a) 100 lb   b) 200 lb   c) 400 lb   d) 800 lb
   e) considerably more than 800 lb

8. The tension in the string that joins the downward accelerating 20-lb weight and the upward accelerating 10-lb weight is
   a) less than 10 lb
   b) 10 lb
   c) more than 10 lb but less than 20 lb
   d) 20 lb
   e) more than 20 lb

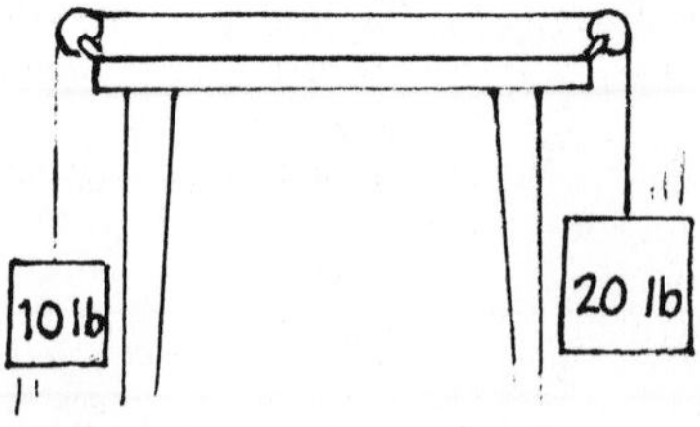

9. The cart will
   a) accelerate toward the left
   b) accelerate toward the right
   c) not accelerate

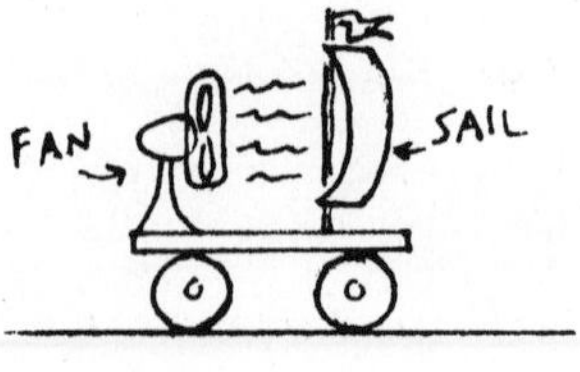

10. When air is blown into the garden sprinkler it spins clockwise. When air is sucked the sprinkler spins
a) also clockwise b) anti-clockwise c) not at all

11. Suppose you are driving along the highway and a bug splatters into your windshield. Which experiences the greatest impact force?
a) the bug b) the windshield c) both the same

12. Suppose that when you jump from an elevated position to the ground below you bend your knees upon making contact and thereby extend the time your momentum is reduced by 10 times that of a stiff-legged abrupt landing. Then the average forces your body experiences upon landing are reduced
a) less than 10 times b) 10 times c) more than 10 times

13. A Cadillac and a Volkswagen rolling down a hill at the same speed are forced to a stop. Compared to the force that halts the Volkswagen, the force to bring the Cadillac to rest
a) must be greater b) must be the same c) may be less

14. When a 10-lb object falls from rest to the ground 10 ft below, it strikes with an impact of about
a) 10 lb b) 50 lb c) 100 lb d) can't tell

15. One block of ice slides down the plane while the other drops off the end. The block that reaches ground with the greatest speed is the
a) sliding block
b) block that drops
c) both the same

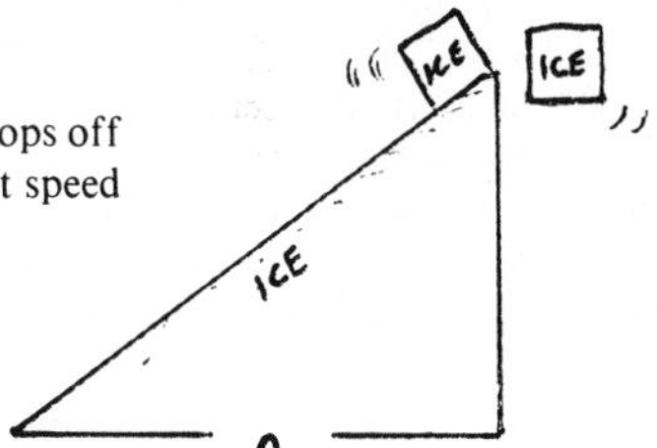

16. From the top of a building we throw three rocks each with the same speed. One is thrown straight up, one sideways, and one straight down. Which rock is moving fastest when it hits the ground?
a) the one thrown up
b) the one thrown sideways
c) the one thrown down
d) all hit with the same speed

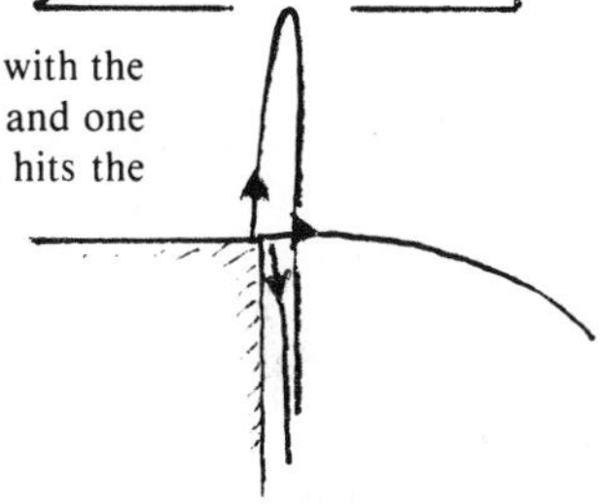

17. A rock is dropped from the top of a tall tower. Half a second later a second rock is dropped. The separation between the rocks as they fall
a) increases b) decreases c) remains constant

18. In the above situation the second rock hits the ground
a) less than half a second after the first
b) a half second after the first
c) more than a half second after the first

19. Two rocks are simultaneously dropped from locations one a foot higher than the other. As they fall the separation between the rocks
a) increases b) decreases c) remains constant

20. In the above situation there is a time lapse between impacts as the rocks hit the ground. Suppose the rocks are dropped the same way, one a foot higher than the other, but from a greater height. Then the time lapse between impacts
a) increases b) decreases c) remains the same

21. If the momentum of a falling body is doubled, its kinetic energy is
a) doubled b) quadrupled c) can't tell

22. A driver going 10 mph hits the brakes and skids 25 feet to a stop. If she were instead going 40 mph the car would skid
a) 100 ft. b) 200 ft. c) 400 ft. d) more than 400 ft.

23. Suppose a pair of pool balls moving at 10 ft/s collide, and then roll off in the *same* direction, each at 10 ft/s. This unlikely collision violates the conservation of

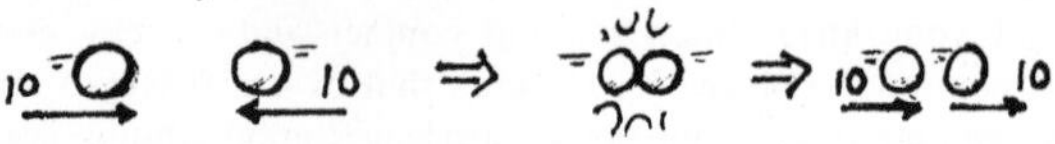

a) kinetic energy b) momentum c) both d) neither

24. Suppose in the preceding situation that after collision the two balls bounced apart each with a speed of 15 ft/s. In this unlikely collision there would be a violation of the conservation of

a) kinetic energy b) momentum c) both d) neither

25. A lump of clay slides down a hill from a height of 25 ft. A second lump of clay only half as heavy as the first slides down another hill and the two lumps come to a dead stop when they collide at the bottom of the hills. How high is the hill from which the small lump slid?
a) 37½ ft. b) 50 ft. c) 75 ft. d) 100 ft.
e) can't say

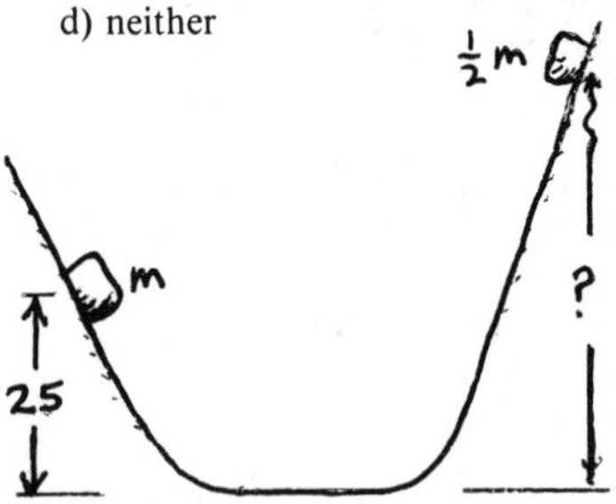

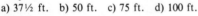

26. A one-ton pickup truck going 12 mph drives into a giant haystack. An identical one-ton pickup, loaded with one ton of hay bales and going only 6 mph, drives into the same haystack. Which truck penetrates farther into the haystack before stopping?
a) the fast empty truck b) the slow loaded truck c) both the same

27. A railroad car is rolling without friction in a vertical downpour. A drain plug is opened in the bottom of the car that allows the accumulating water to run out. If the water drains as fast as it accumulates the speed of the car
a) increases b) decreases c) remains the same

28. A gun recoils when it is fired. So the gun and the bullet each get some kinetic energy and momentum. Both the gun and bullet get equal
a) but opposite amounts of momentum
b) amounts of kinetic energy
c) both
d) neither

29. Would the anvil afford much protection if another anvil were dropped on it?
a) Yes b) No

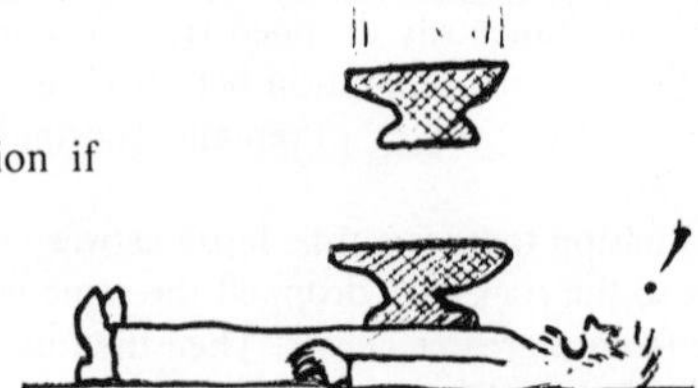

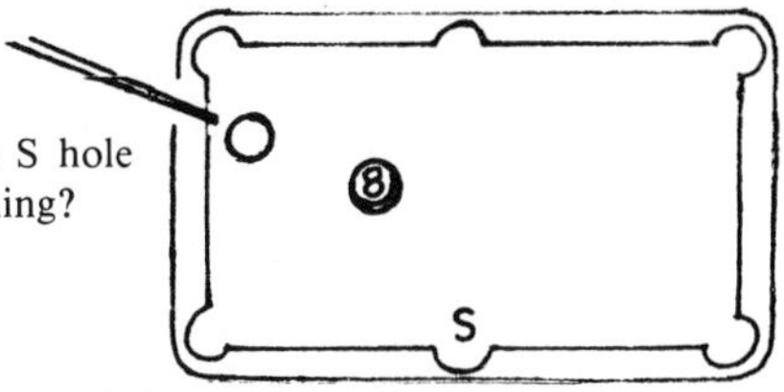

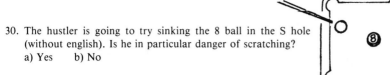

30. The hustler is going to try sinking the 8 ball in the S hole (without english). Is he in particular danger of scratching?
 a) Yes b) No

31. If the speed of a thing changes, then its velocity changes. And if the velocity of a thing changes, then its speed
 a) must change also
 b) may or may not change
 c) must not change

32. The net force that acts on a car traveling in a circle at constant speed on a flat surface is
 a) in the direction the car is traveling
 b) in the direction of the radius of the circle
 c) equal to zero

33. The driver in a car turning in a given radius and moving with a given speed experiences a certain centrifugal force. This force will be MOST increased by
 a) doubling the speed of the car
 b) doubling the radius in which it turns
 c) halving the radius in which it turns
 d) "a" and "b" produce exactly the same effect
 e) "a" and "c" produce exactly the same effect

34. An object dropped down a deep mine shaft at the earth's equator will be slightly deflected to the
 a) north b) east c) south d) west e) . . . not at all

35. An object dropped down a deep mine shaft at the north pole will be slightly deflected to the
 a) north b) east c) south d) west e) . . . not at all

36. We know if forces I and II act together, they are equivalent to force III. Now, is the torque on the nut produced by force I and force II acting together *always* equivalent to the torque produced by force III?
 a) Yes b) No

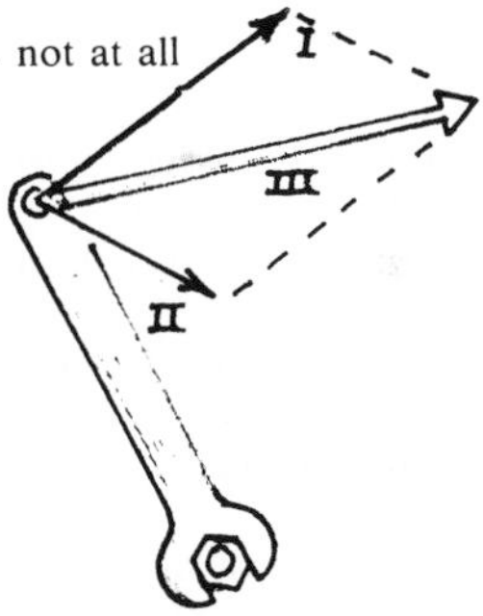

37. 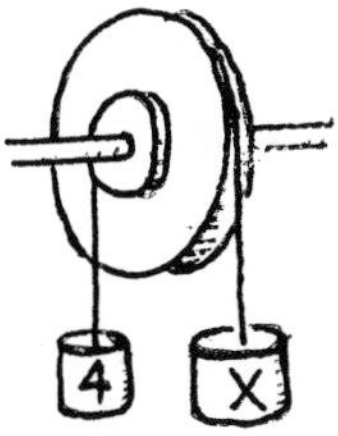

The rim of a wheel is 16 ft. in circumference and the hub is 8 ft. in circumference. A 4-lb weight is wound around and hung from the left side of the hub. What weight must be wound around and hung from the right side of the rim to balance the wheel so it will not turn?
 a) 2 lb. b) 3.14 lb. c) 6 lb. d) 8 lb.

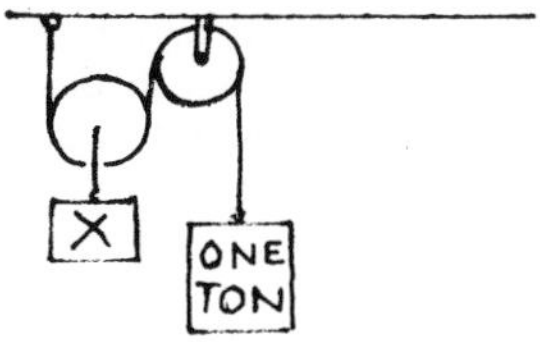

38. Two weights are suspended from a ceiling by means of a cord and two pulleys as illustrated. If the one-ton weight is held in balance by X, then X weighs
 a) 1 ton b) 2 tons c) ½ ton d) ⅓ ton e) 3 tons

39. The answer to BALL ON A STRING stated that no torque acted on the ball because the force was directed to the center C and therefore had no lever arm. But aren't there lever arms for Components 1 and 2? Why then do we say no torque exists?
a) But there are *no* lever arms about C for Components 1 and 2!
b) Components 1 and 2 are perpendicular to each other and cancel about C
c) Torques due to Components 1 and 2 cancel each other
d) Misprint: A net torque certainly does exist!

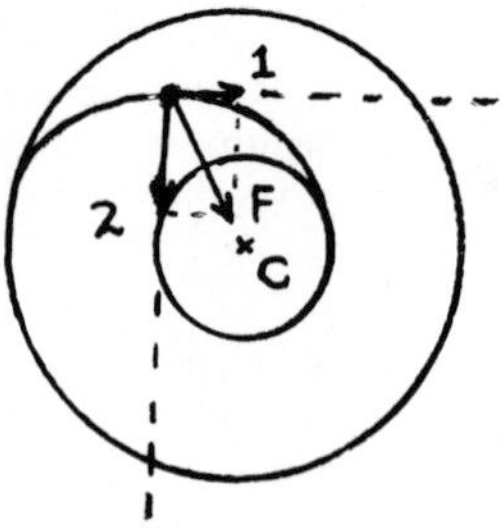

40. An object isolated in space and not interacting with any external agency can by internal interactions change its own
a) linear momentum b) linear kinetic energy c) both d) neither

41. It is possible for the same isolated object in the preceding question to change its own
a) angular momentum b) rotational kinetic energy c) both d) neither

42. The illustrated collision between the dumbbells violates conservation of
a) kinetic energy
b) momentum
c) angular momentum
d) all of these
e) none of these

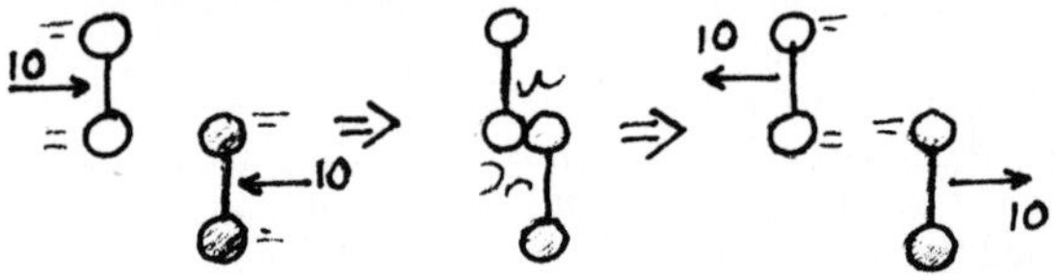

43. When a 4-wheel-drive jeep accelerates from rest, it noses
a) upward b) downward c) not at all

44. The bob of a swinging pendulum has its greatest speed at the bottom of its swing and its greatest acceleration
a) at the bottom also
b) at its highest point where it momentarily stops
c) none of these

45. A marksman fires at a distant target. If it takes one second for his bullet to get there, he should aim above the target
a) 16 ft. b) 32 ft. c) can't say unless we know the target is at eye level

46. A ball is thrown horizontally at 40 ft/s. from an elevated position and one second later strikes the *ground* at a speed of about
a) 32 ft/s. b) 40 ft/s. c) 50 ft/s.

47. Suppose a planet orbits about a giant star that begins to collapse to a black hole. After the collapse, the orbital radius for the planet is
a) smaller b) larger c) unchanged d) non-existent

48. Strictly speaking, in the ground floor lobby of a giant skyscraper you weigh
a) a bit less b) a bit more c) the same as outside the skyscraper

49. The moon does not fall to earth because
a) it is in the earth's gravitational field
b) the net force on it is zero
c) it is beyond the main pull of the earth's gravity
d) it is being pulled by the sun and planets as well as by the earth
e) all of the above
f) none of the above

50. Here is a one pound Bank of England note, British money, approximately equal to \$2.00 in 1979. Take a careful look at the elliptical orbit of the planet G, D, B, P, K with the sun at C. This diagram represents
a) Newton's theory of gravity
b) Hooke's theory of gravity

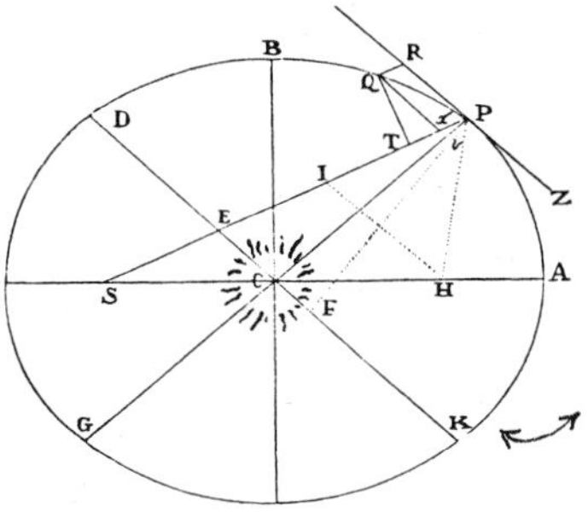

51. The earth's orbit about the sun is slightly elliptical, the earth being closest to the sun in December and farthest in June. The orbital speed of the earth is therefore
a) greater in December
b) greater in June
c) the same throughout the year

52. A planet is found on the average to orbit the sun a bit faster than it should according to Newton's laws. It is suspected an undiscovered planet is responsible. The orbit of the undiscovered planet lies
a) inside the orbit of the fast-moving planet
b) outside the orbit of the fast-moving planet

53. Suppose a US and USSR space craft are both freely coasting in circular orbits moving in the same direction. If the altitude of the USSR craft is 100 miles, and the altitude of the US craft is 110 miles, then the
a) US craft gradually pulls ahead of the USSR craft
b) USSR craft gradually pulls ahead of the US craft
c) crafts remain abreast of each other

Physics is like sex.
It has practical applications.
But that is not the main reason you do it.
— Richard Feynman

Fluids

Fluid is the substance of dreams, infinitely flexible, without a shape of its own. Free from most of the friction encountered by solids, fluid vividly exhibits the laws of motion. Fluid is a natural teacher. Fluids allow some solids the privilege of floating on them while other solids sink. But solids that sink can be hollowed out and made to float—iron can float! How much should a solid that is too heavy to float be hollowed out in order to float? That might well be the most ancient question in physics.

WATER BAG

One cubic foot of seawater weighs about 64 pounds. Suppose you pour one cubic foot of seawater into a plastic baggie, tie the baggie closed with no air bubbles inside, attach a string to it and lower it into the ocean. When the baggie of water is completely submerged, how much force do you have to exert on the string to hold it up?

a) Zero lb
b) 32 lbs
c) 64 lbs
d) 128 lbs
e) You have to push it down because it tends to come up

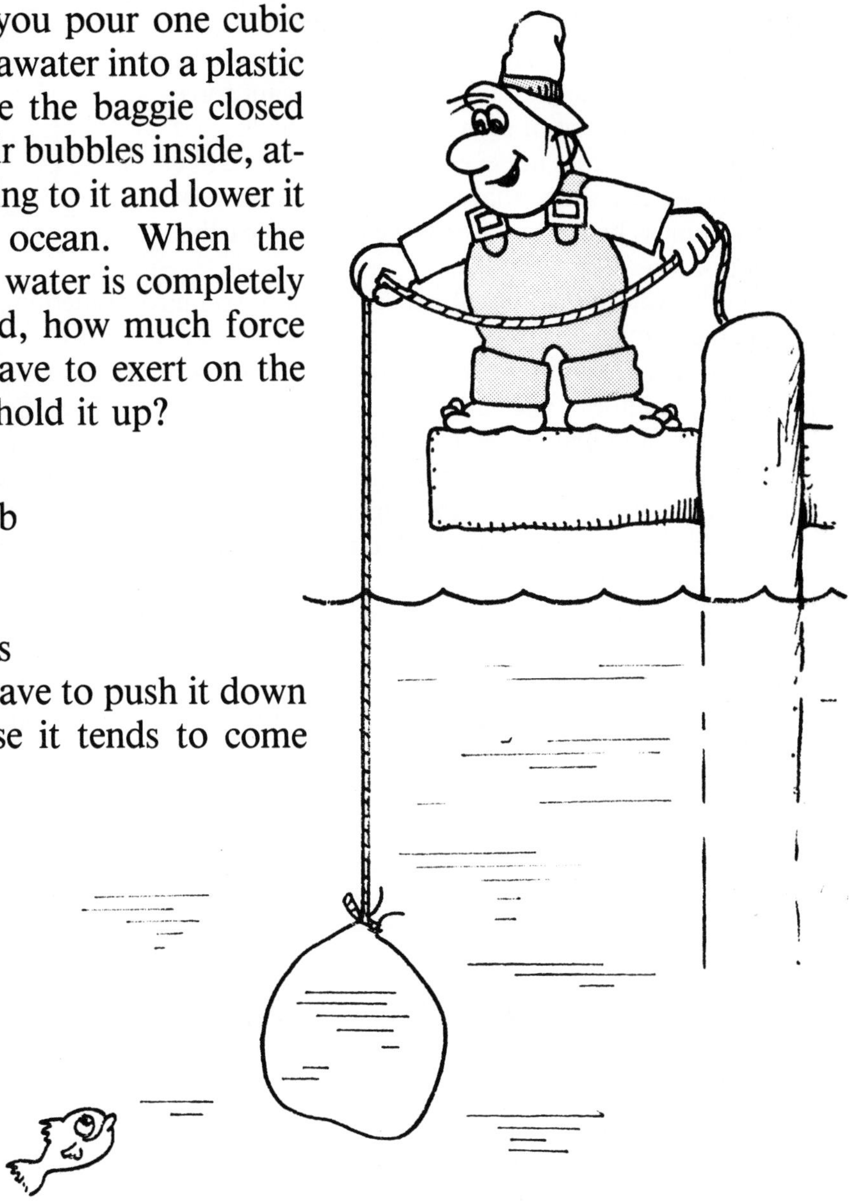

ANSWER: WATER BAG

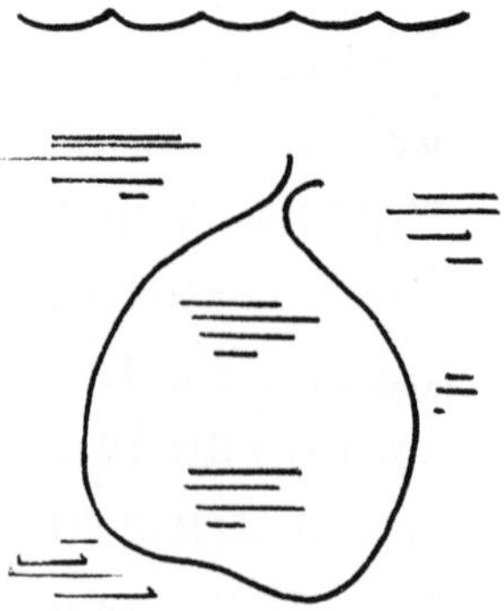

The answer is: a. If you have a parcel of water completely surrounded by other water, the parcel doesn't go up or go down. It stays put – stagnant water. The parcel can be any size or any shape. The weight of the water in the parcel is exactly equal to and supported by the buoyancy of the surrounding water. That means the buoyant force on each cubic foot of seawater must be 64 lbs and the cubic foot does not even have to be shaped like a cube — any shape that occupies one cubic foot of volume is buoyed up by the weight of one cubic foot of seawater — 64 lbs.

Let's explore this topic further. Suppose you have something that has a volume of one cubic foot but weighs 128 lbs, that is, it has twice the density of seawater. The water surrounding the thing does not care what is inside of it. If its volume is one cubic foot the buoyant force is 64 lbs. So, the water lifts 64 lbs and that leaves a surplus of 64 lbs for you to lift. Can we say that any 128 lb object has an apparent weight of 64 lbs when submerged? No. Under water a 128 lb object weights 64 lbs only if its volume is one cubic foot. If the volume of the 128 lb object were two cubic feet for example, its apparent weight would be zero. To determine the apparent weight of a submerged body, subtract the weight of the volume of water the object displaces. Consider, for example, a 128 lb object with a volume of three cubic feet. The weight is 128 minus three times 64, which is 128 minus 192, which is a negative number, minus 64! Minus 64 means the buoyant force

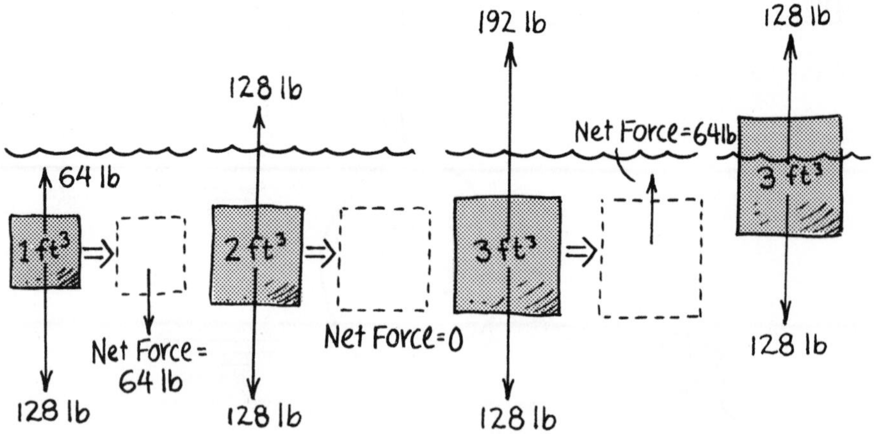

pushing up is 64 lbs greater than the object's weight and so the object goes up. It floats to the surface of the water. But it does not stop when it meets the surface. It goes up until part of it extends above the surface. How far does it extend? It goes up until one cubic foot of the object is above the water-line. That leaves two cubic feet in the water. Two cubic feet of water weigh 128 lbs and so the buoyant force on the part under water is 128 lbs, and 128 lbs is exactly enough to balance the weight of the object. So we see that the buoyant force that is exerted on a submerged body equals the weight of water displaced by the body — and in the special case where the body floats, this buoyant force is equal to the weight of the body.

This rather long-winded answer is an overview of Archimedes' Principle. More on this idea in the following questions.

GOING DOWN

You suspend a boulder weighing fifty pounds and lower it beneath the surface of the water. When the boulder is fully submerged you find you have to support less than 50 lbs. As the boulder is submerged still further the support force needed to hold the boulder is

a) less
b) the same
c) more than just beneath the surface

ANSWER: GOING DOWN

The answer is: b. The submerged boulder is buoyed up with a force equal to the weight of water displaced by the boulder, which makes the support force less than 50 lbs when the boulder is suspended beneath the surface of the water. As the boulder is lowered deeper and deeper, the volume, and hence the weight of water it displaces, doesn't change. Because water is practically incompressible, its density near the surface or very deep is the same. Thus the buoyant force does not change with depth and the support force needed to hold the boulder is the same whether the boulder is just beneath the surface or deep beneath the surface.

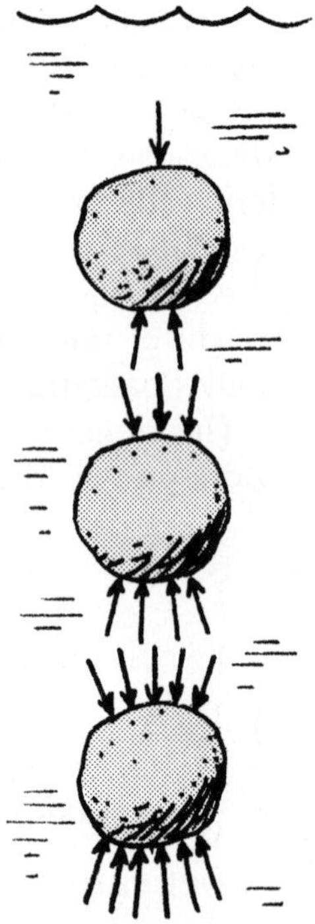

Water pressure, on the other hand, does increase with depth, which is why submerged objects are buoyed up to begin with. The bottom of a submerged body is always deeper than the top, so there is always more pressure up against the bottom than down against the top. But this doesn't mean that the buoyant force on a submerged body increases with depth, for the DIFFERENCE in upward and downward pressures is the same at all depths, as indicated in the sketch.

THE DIFFERENCES IN PRESSURE ARE 1 UNIT EVERYWHERE BENEATH THE SURFACE
GREATER PRESSURE AGAINST BOTTOM

(In the unusual circumstance where a submerged body rests against the bottom with no water film between it and the container bottom, there would be no upward buoyant force.)

HEAD

By properly shaping the head of a piston (like the pistons in the cylinders of an automobile engine), can the downward force on the piston be increased?

a) There is more downward force on the dome head than on the flat head because the dome has more surface area to push on.

b) There is more downward force on the flat head than on the dome head because all the pressure on the flat head pushes straight down.

c) If the diameter of and pressure on the cylinders are equal, the downward force on the pistons is equal.

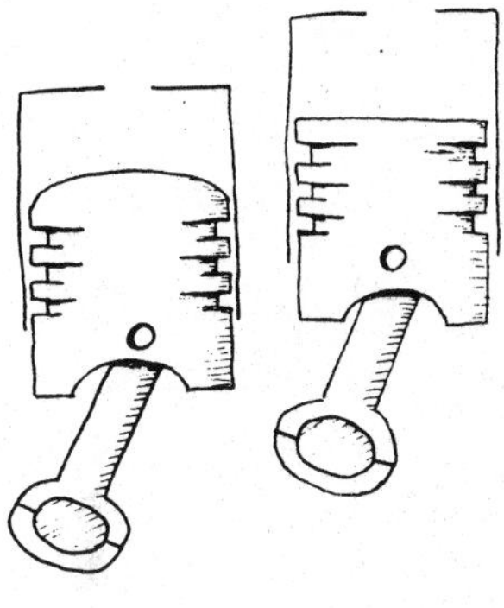

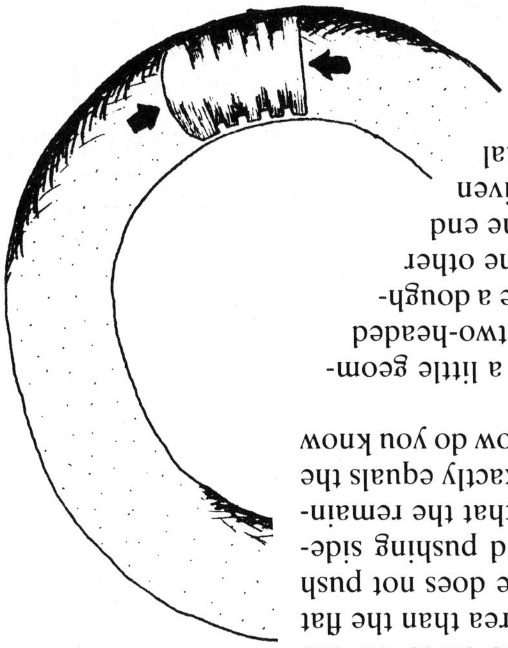

ANSWER: HEAD

The answer is: c. There is more force on the dome because it has more surface area than the flat head. But all the force on the dome does not push straight down. Some of it is wasted pushing sideways. Exactly enough is wasted so that the remaining downward force on the dome exactly equals the downward force on the flat head. How do you know the forces are exactly equal?

Well, you could work it out with a little geometry, or better yet you could put a two-headed piston into a big cylinder shaped like a doughnut. One end of the piston is flat, the other domed. If there is more force on one end than on the other, the piston is driven around the doughnut in perpetual motion with a perpetual force pushing it. That would make a great engine—too good to be true.

THE GROWING BALLOON

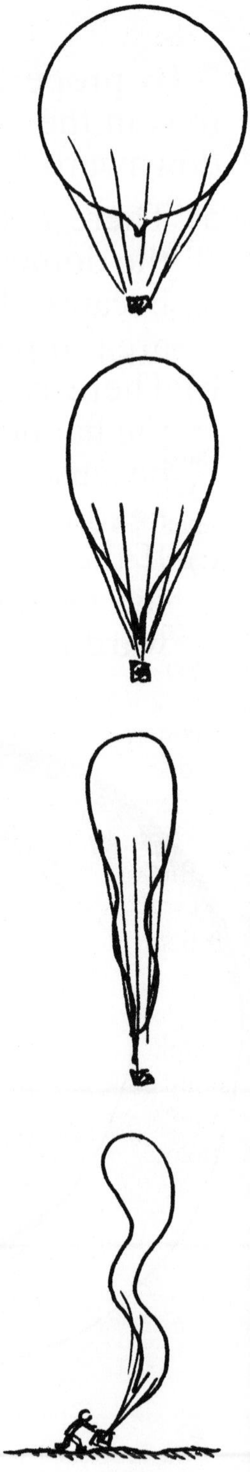

Very large air bags are used for high-altitude weather balloons. At ground level the bag is only partly filled with helium, just buoyant enough to rise. As the balloon rises, the less dense surrounding air allows the helium gas to gradually expand and the balloon grows larger. As it grows larger and larger, the buoyancy of the balloon

a) increases
b) decreases
c) stays the same

ANSWER: THE GROWING BALLOON

The answer is: c, the buoyancy stays the same. The reason the balloon expands as it rises is because the air pressure decreases with altitude. When the atmospheric pressure decreases, say to one-half its ground-level value, the balloon will have expanded to twice its original size. Half as much air pressure means that the density of the surrounding air is one-half the ground-level density. Buoyancy depends on the *weight* of the displaced air, and the weight of twice the volume at half the density is unchanged from the ground value. Since the volume of the balloon and the air density both depend on the air pressure in exactly opposite ways, the buoyancy doesn't change while the balloon is growing.

Will temperature change the story? After all, the air generally cools with increasing altitude which in turn tends to shrink the size of the balloon and the volume of air it displaces. But interestingly enough, it does not lessen the *weight* of air displaced. Why? Because the lower temperature also increases the density of the air by the same amount. A volume decrease of say 10% is also accompanied by a density increase of 10%, so the weight of displaced air and hence the buoyancy stays the same. So neither changes in air pressure or temperature affect the buoyancy of the balloon.

But something else does — the balloon fabric. Buoyancy ultimately decreases at very high altitudes when the balloon is fully inflated and the fabric begins to stretch. The stretching of fabric resists further balloon growth and lessens the volume of air that would be displaced by simple pressure and temperature factors. Now as the non-growing balloon further ascends into less dense air, buoyancy decreases until it matches the weight of the balloon. When buoyancy and weight are equal to each other the balloon has reached its maximum altitude.

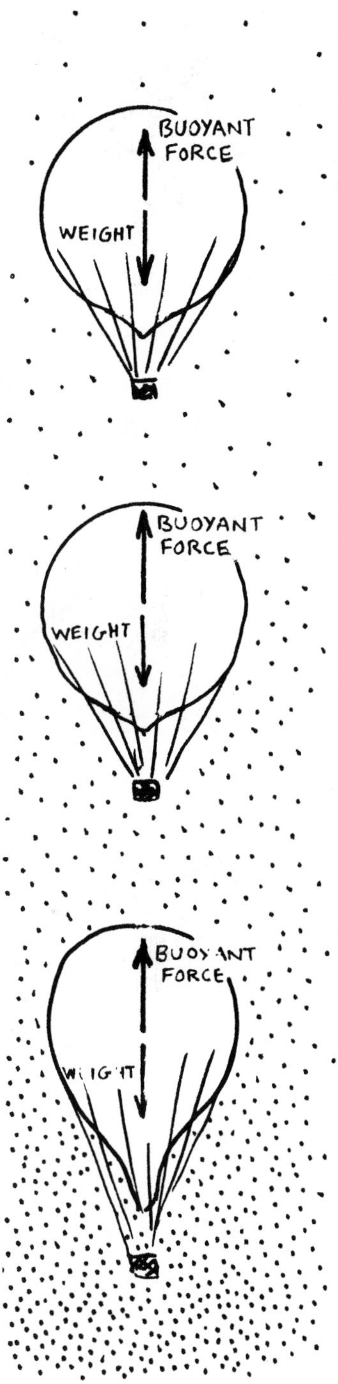

WATER SEEKS ITS OWN LEVEL

It is a common saying that water seeks its own level. This can be demonstrated by pouring water into a U-shaped container and noting that the water-surface levels in each side equalize. But *why* does water seek its own level? The reason has most to do with

a) atmospheric pressure on both surfaces
b) water pressure depending on depth
c) the density of water

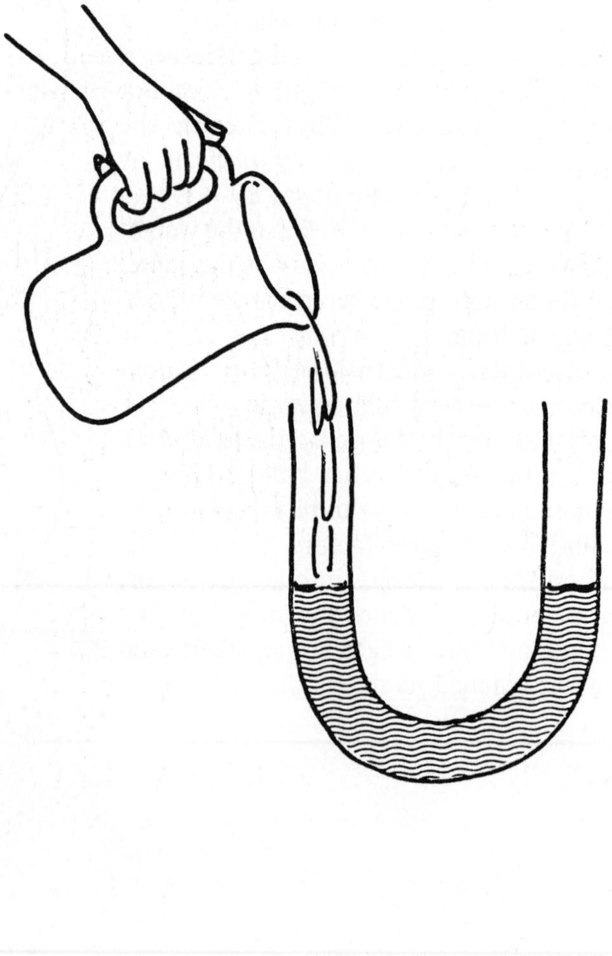

ANSWER: WATER SEEKS ITS OWN LEVEL

The answer is: b. Water would seek its own level in the open or in a vacuum, so atmospheric pressure is of little concern here. The pressure in a liquid depends on the density of the liquid and its depth (also its velocity, which doesn't apply here since any motion is momentary). Since the density of liquid in each side of the U-tube is equal, regardless of the amount of water in each side, we are left with depth as the principal factor.

Consider the two positions marked "X" in the sketch. If the water is at rest, then the pressures at these positions must be equal — otherwise a flow would occur from the region of greater pressure to that of lower pressure until the pressures equalize. But since the pressure depends on water depth, equal pressures must result from equal depths — so the weight of water above each "X" (or in each column), must be the same (which is clearly not the case in our sketch). So we see there is a reason why water seeks its own level. We'll look at this idea once more in the following question.

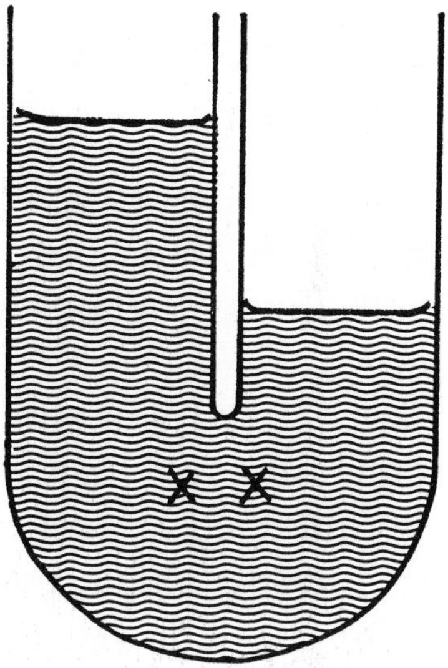

BIG DAM LITTLE DAM

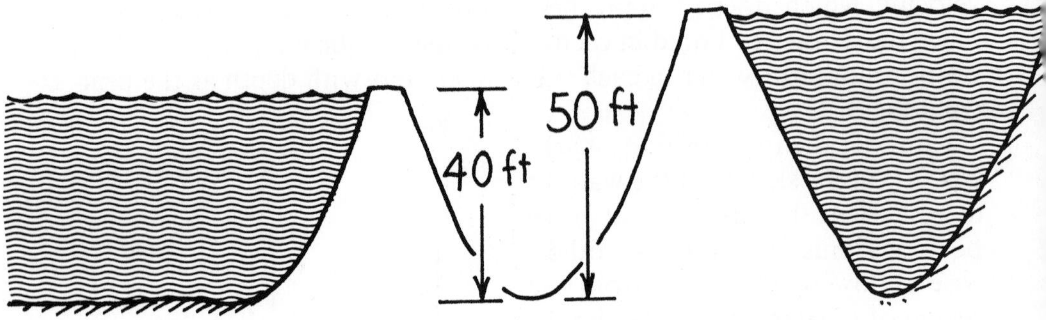

Dams are constructed thicker at the bottom than at the top because water pressure against the dam increases with depth. But how about the volume of water held back by a dam?

The Sierra Light and Power Company in California has a dam fifty feet high with a little lake behind it. Not far away the Department of Reclamation also has a dam — only forty feet high, but with a much larger lake behind it. Which dam must be strongest?

a) The Sierra Light and Power dam must be strongest
b) The Department of Reclamation dam must be strongest
c) Both must have the same strength

ANSWER: BIG DAM LITTLE DAM

The answer is: a. The strength of the dam must withstand the pressure of the water behind it — and the pressure of the water depends only on how deep the lake is and not at all on how long the lake is. So the pressure is greatest against the dam having the deepest water behind it, and not necessarily against the dam having the most water.

To follow up on WATER SEEKS ITS OWN LEVEL, imagine the two bodies of water are joined by a pipe as shown in the sketch.

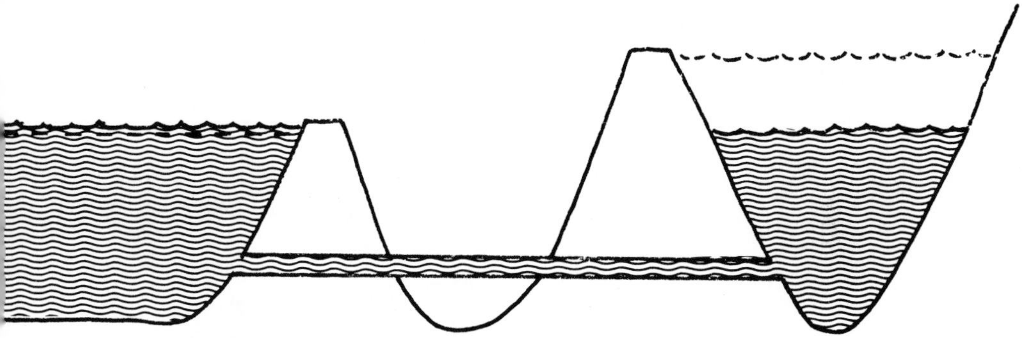

Can you see that water will flow through the pipe in a direction from greater to lower pressure and will keep flowing until the pressures are equalized? Pressures at each end of the pipe and against the dams are equal when the water levels in each are equal. Water pressure depends on depth, not volume.

BATTLESHIP FLOATING IN A BATHTUB

Can a battleship float in a bathtub?* Of course, you have to imagine a very big bathtub or a very small battleship. In either case, there is just a bit of water all around and under the ship. Specifically, suppose the ship weighs 100 tons (a very small ship) and the water in the tub weighs 100 pounds. Will it float or touch bottom?

a) It will float if there is enough water to go all around it
b) It will touch bottom because the ship's weight exceeds the water's weight

* This was my father's favorite physics question. —L. Epstein

ANSWER: BATTLESHIP FLOATING IN A BATHTUB

The answer is: a. There are a lot of ways to show why. This way was suggested by a student. Consider the ship floating in the ocean (sketch I). Next, surround the ship with a big plastic baggie — this is actually done sometimes with oil tankers — (sketch II). Next, let the ocean freeze except for the water in the baggie next to the ship (sketch III). Finally, get an ice sculptor to cut a bathtub out of the solid ice and you have it (sketch IV).

This question points out the danger of thinking in words, rather than thinking in pictures and ideas. If you just think in words you might reason: "To float, the battleship must displace its own weight in water. Its own weight is 100 tons, but there is only 100 pounds of water available — so it cannot float." But if you picture the idea you will see the displacement refers to the water that would fill the ship's hull if the inside of the ship's hull were filled to the water-line. And this displacement is 100 tons.

Don't rely on words, or equations, until you can picture the idea they represent.

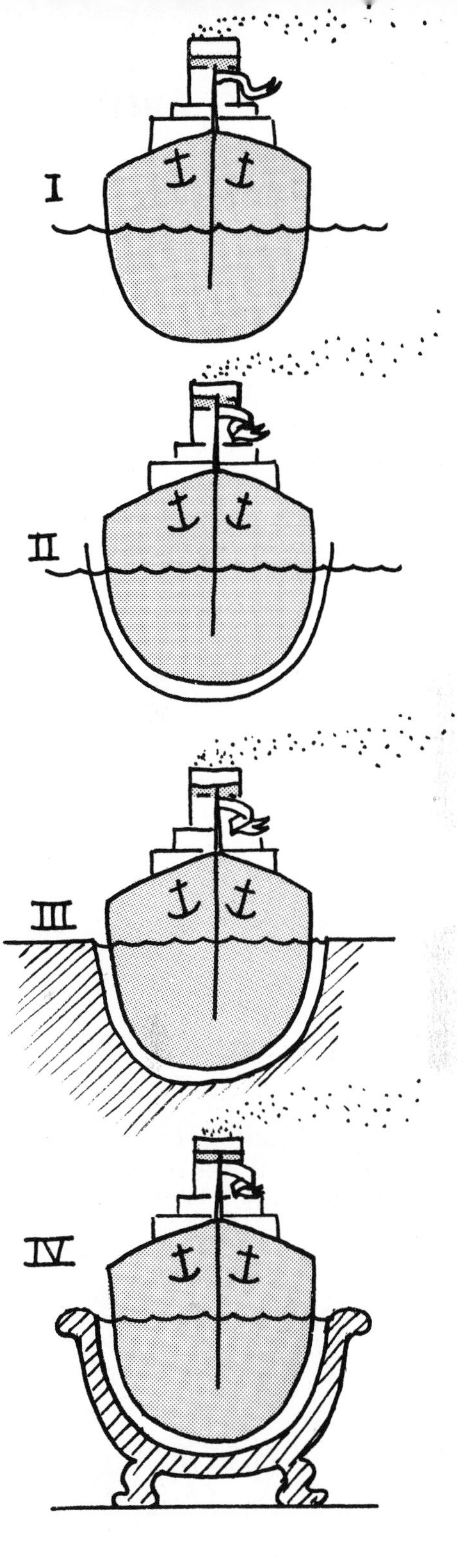

BOAT IN THE BATHTUB

Which weighs more?

a) A bathtub brim full of water
b) A bathtub brim full of water with a battleship floating in it
c) Both weigh the same amount

ANSWER: BOAT IN THE BATHTUB

COLD BATH

This is a bathtub brim full of ice-cold water with an iceberg floating in it. When the iceberg melts, the water in the tub will

a) go down a little
b) spill over
c) stay exactly brim full without spilling

ANSWER: COLD BATH

The answer is: c. The weight of the water displaced by the iceberg exactly equals the weight of the iceberg. When the iceberg melts it "shrinks" and turns back to water and fits exactly into the volume of water it displaced. Incidentally, the volume of ice above water must be exactly equal to the increase in volume of the water that froze and expanded to become ice.

THREE BERGS

This question is to bust the curve busters. (Curve busters are people who answer all the questions right.) Three icebergs float in bathtubs brim full with ice-cold water. Iceberg A has a big air bubble in it. Iceberg W has some unfrozen water in it. Iceberg S has a railroad spike frozen in it. When they melt what will happen?

a) Only the water in S will spill over
b) The water in S will get lower and the water in A and W will stay exactly brim full
c) The water in A will stay brim full, the water in W will spill over and the water in S will spill over
d) All will spill over
e) All stay exactly brim full

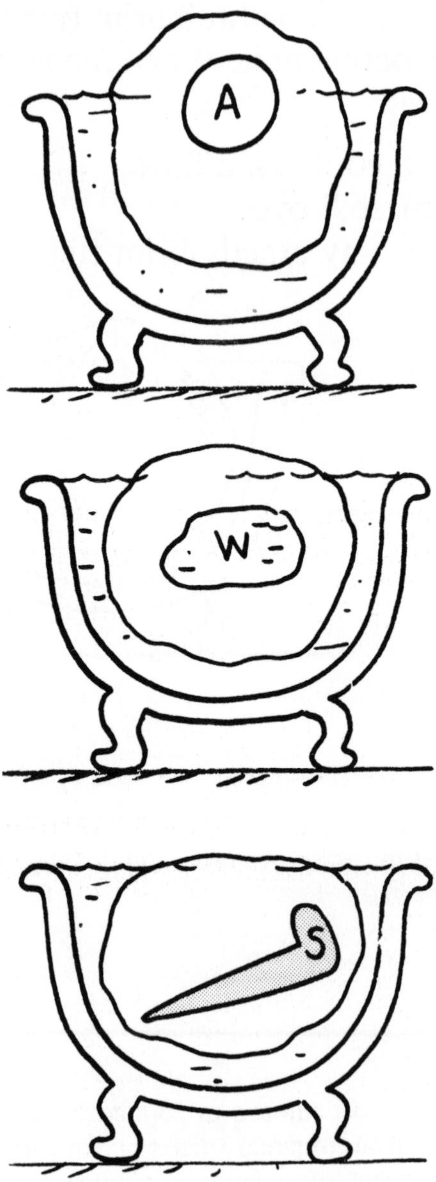

ANSWER: THREE BERGS

The answer is: b. First, remember the lesson from COLD BATH. An iceberg floating in a brim full bathtub will melt and the tub will stay exactly brim full without spilling. Now in your imagination, shift the air bubble to the upper surface of the berg. This will not affect the berg's weight so it can't affect its displacement. Now pierce the bubble. What was the bubble is now only a little cavity. No change in weight involved, but the berg is now a "regular" berg without an air bubble.

Next, take the berg with the unfrozen water in it and let your imagination shift the water hole to the lower surface of the berg. This will not affect the berg's weight so it cannot affect its displacement. Now pierce the water hole. What was the water hole is now only a little cavity. No change in weight involved, but the berg is now a "regular" berg without a water hole. So the bergs with the air bubble and water hole will melt like "regular" bergs and not raise or lower the water in the tub.

Now in your imagination, move the spike to the bottom of the berg. No change in weight or displacement. Next, melt or break the spike free from the berg. The spike goes to the bottom of the tub, but in no way increases or decreases its displacement. However, the berg is relieved of its heavy load, so like an unloaded boat it pops up on the water. As the berg goes up, the water in the tub goes down. The berg is now a "regular" berg, so when it melts the water level will be unaffected — it will be as far below brim level as it was before melting.

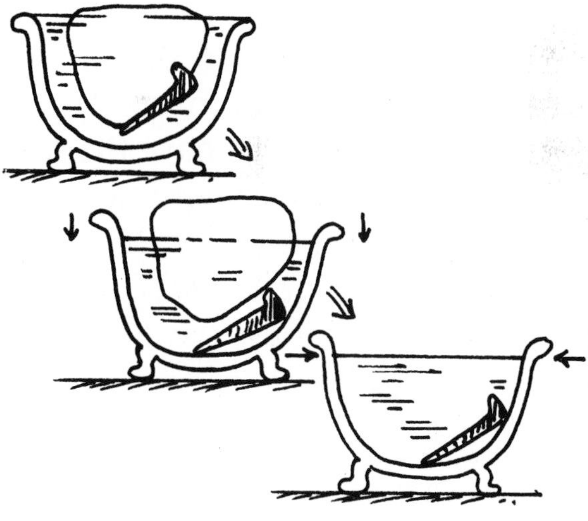

PANCAKE OR MEATBALL

A drop of liquid with a large surface tension and a drop of liquid with a small surface tension are put on clean pieces of glass. One drop looks like a little pancake; the other looks like a little meatball. Which liquid has the largest surface tension?

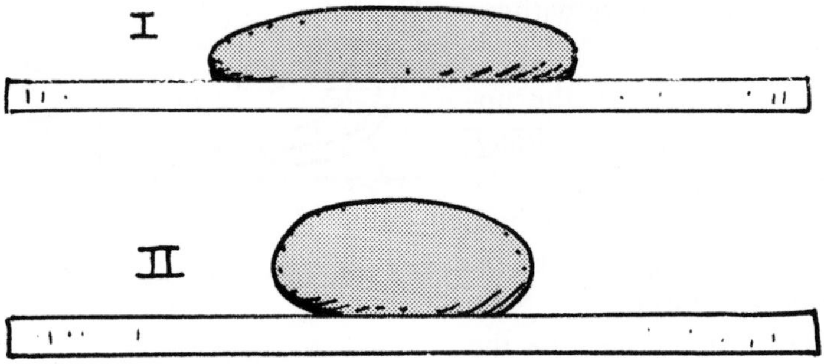

a) The pancake (drop I)
b) The meatball (drop II)
c) If both drops have the same volume they will have the same surface tension

ANSWER:
PANCAKE OR MEATBALL

The answer is: b. Surface tension, the contractive force of the surface of liquids, is caused by molecular attractions. A molecule beneath the surface of a liquid is attracted by neighboring molecules in every direction, resulting in no tendency to be pulled in a preferred direction. But a molecule at the surface is pulled only to the sides and downward — not upward. These molecular attractions therefore tend to pull the molecule from the surface into the liquid, which causes the surface to become as small as possible. So to minimize surface area, the liquid drops on the clean glass ball up, just like kittens on a cold night ball up to minimize surface exposure. The liquid with the greatest attractions between its molecules and hence the most surface tension balls up the most.

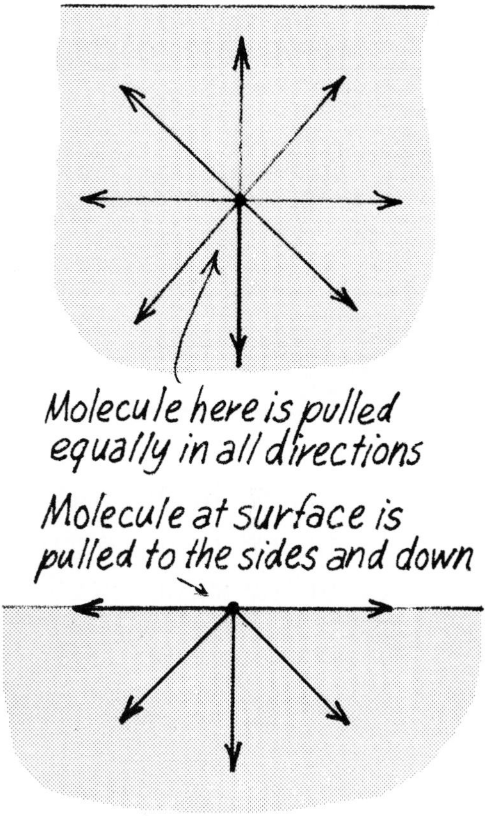
Molecule here is pulled equally in all directions
Molecule at surface is pulled to the sides and down

BOTTLE NECK

Ten gallons of water per minute is flowing through this pipe. Which is correct: The water goes

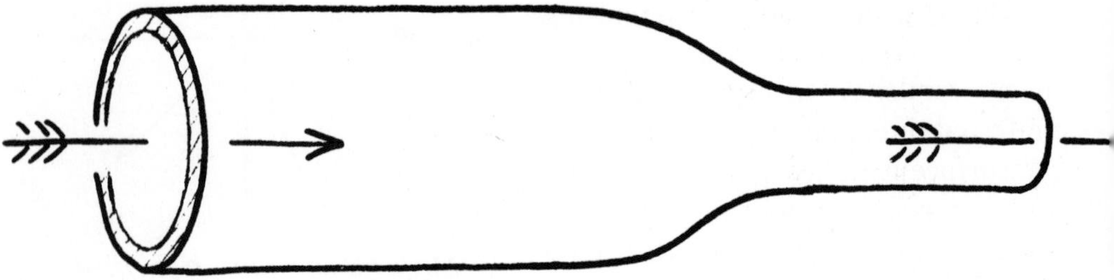

a) fastest in the wide part of the pipe
b) fastest in the narrow part
c) at the same speed in both the wide and narrow parts

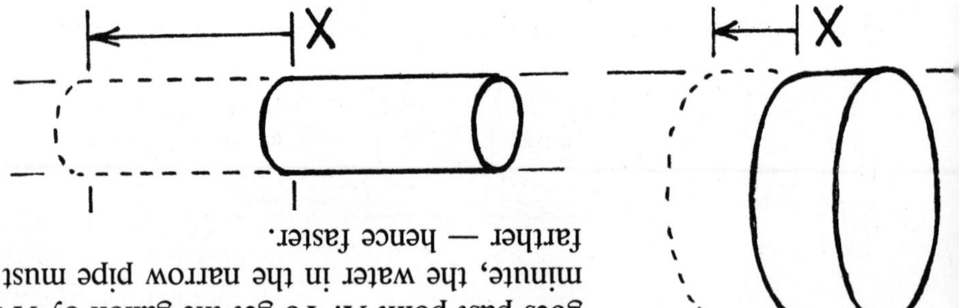

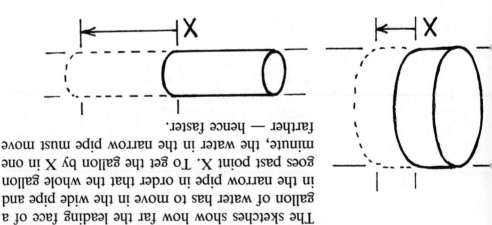

ANSWER: BOTTLENECK

The answer is: b. Water flows fast in narrow parts and slow in wide parts, just as in a creek. The sketches show how far the leading face of a gallon of water has to move in the wide pipe and in the narrow pipe in order that the whole gallon goes past point X. To get the gallon by X in one minute, the water in the narrow pipe must move farther — hence faster.

WATER FAUCET

Water coming out of a sink faucet necks down as it falls, as shown in the sketch. It necks down because

a) its speed increases as it falls
b) of surface tension
c) both of the above
d) of air resistance
e) of atmospheric compression

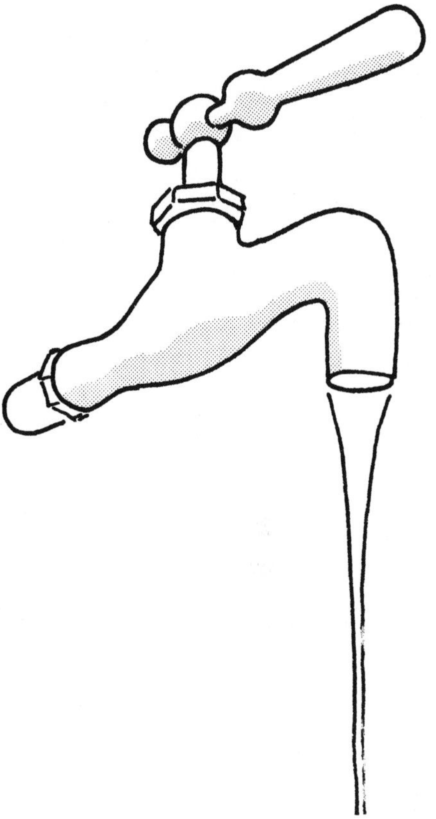

ANSWER: WATER FAUCET

The answer is: c. The same number of gallons per minute pass through the top of the neck at "T" and the bottom of the neck at "B." However, due to the acceleration of fall, the speed of the water at "B" is greater so the water stream is narrower (remember BOTTLE NECK?). But that does not completely explain things. Why doesn't the falling water neck down into a number of separate streams? Because surface tension holds the streams together, like the hairs of a wet brush. To make the water break into separate streams you can force them apart by forcing the stream through something like a screen.

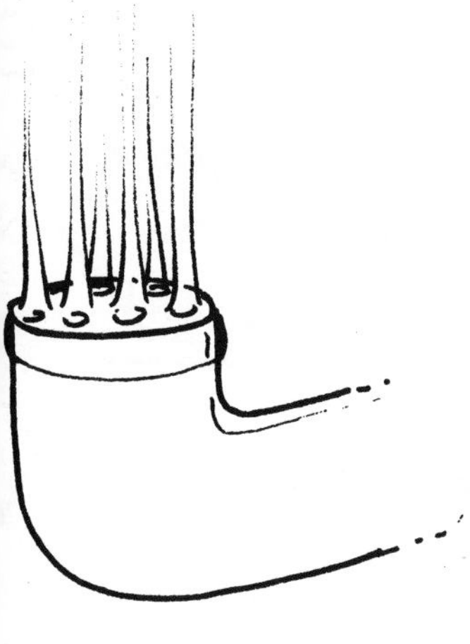

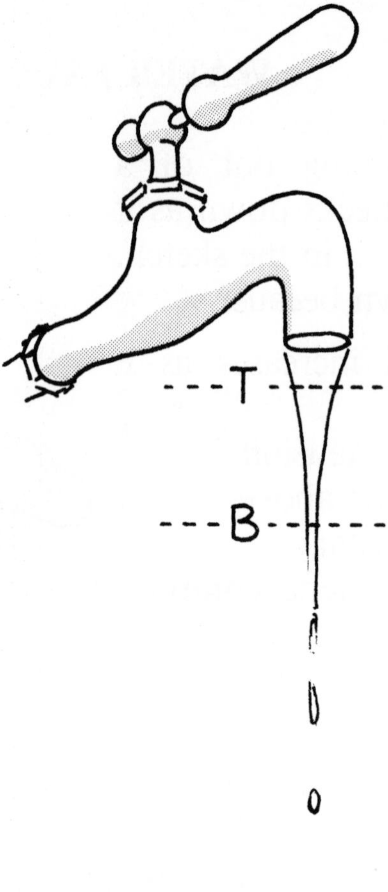

Farther down, the thin stream may break up into little drops. This beading does not have to do with the motion of water and is entirely a surface-tension effect—the total surface area is reduced. The same thing happens when you try to lay a long thin line of glue on a flat surface—and the line of glue is certainly not in motion.

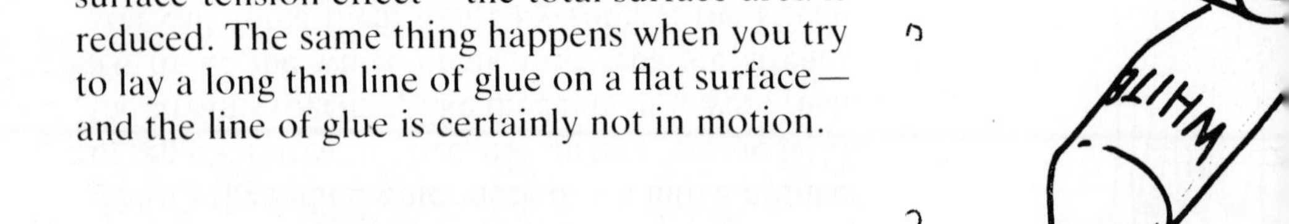

BERNOULLI SUB

A toy submarine drifts with water flowing in a pipe of varying widths. Changes in speed cause the shifting of a heavy mass suspended by springs inside the sub as shown. As the sub drifts from regions A to B, and then to C, the mass shifts

a) backward in going from A to B, then forward from B to C
b) forward in going from A to B, then backward from B to C
c) not at all

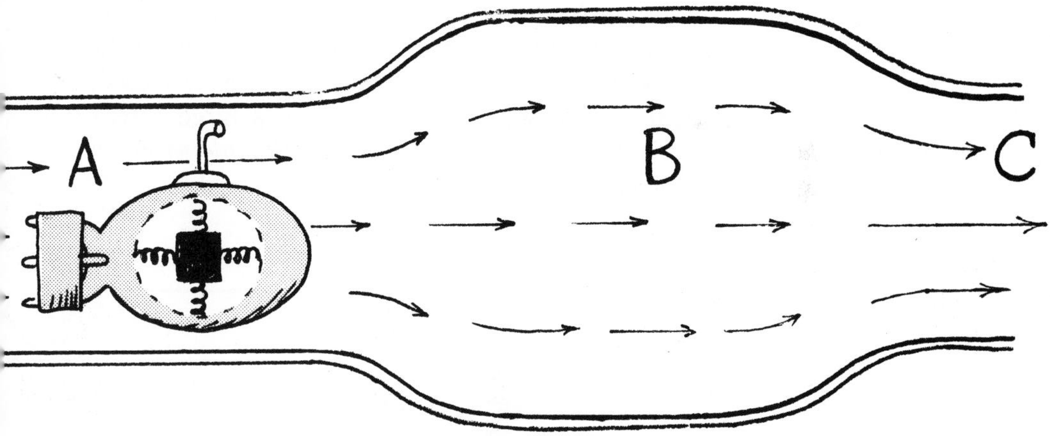

ANSWER: BERNOULLI SUB

The answer is: b. Picture yourself in the submarine in place of the suspended mass. When you move from the narrow region A to the wide region B the water slows down and you pitch forward. You press on whatever is in front of you, in this case the spring, which you compress. Similarly, pressure is applied to the water in region B that is in front of you.

When you leave the wide region B and accelerate into the narrow region C you will pitch backward. You will press on whatever is behind you and you compress the rear spring. Similarly, the surrounding water presses on the water behind and puts pressure on the water in region B. So the slower-moving water in region B has pressure put on it from both ends.

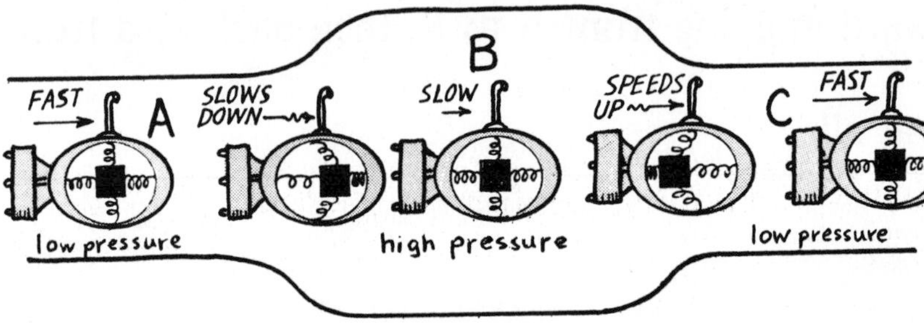

This story is exactly reversed for a narrow section of pipe. The second sketch illustrates why the water pressure is reduced in the narrow part of the pipe. So we see that water pressure is increased when its speed is decreased, and reduced when its speed is increased.* This is true to a large extent for gases as well as liquids and is called *Bernoulli's Principle* after its discoverer more than 250 years ago.

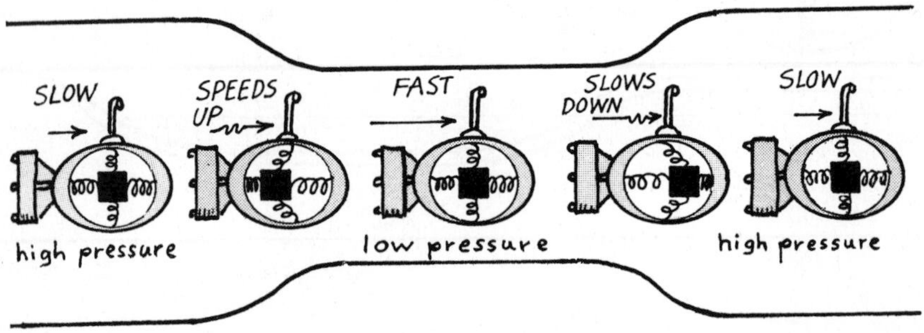

*The friction effects with water are practically nil. If we were forcing honey through the pipe, we would have a different story. The honey pressure would be high when forced into the narrow section and low when it comes out. Friction robs the honey of its pressure.

In the third sketch we replace our sub with a similarly-constructed Bernoulli Blimp that flows with the air over and under the airfoil of an airplane wing. Can you see by the displacement of the suspended mass that the pressure is increased against the bottom of the wing section and decreased along the top?

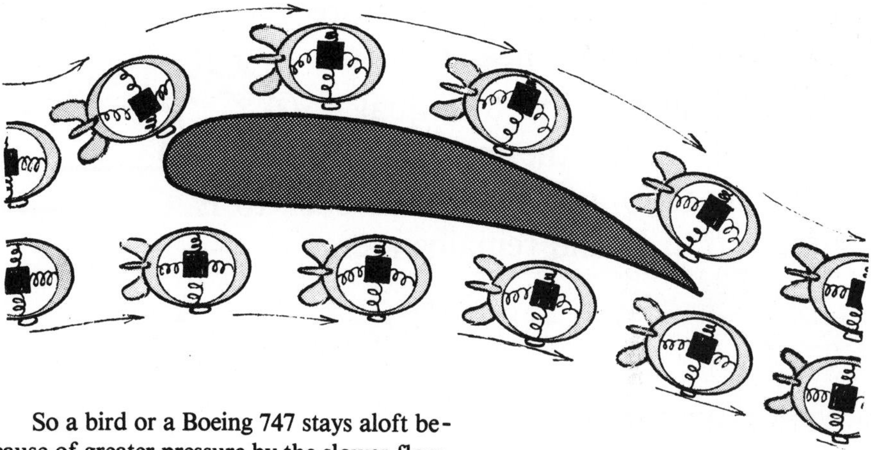

So a bird or a Boeing 747 stays aloft because of greater pressure by the slower flowing air beneath the wing than by the faster flowing air above the wing. Daniel Bernoulli lived in the 18th Century and, of course, preceded the advent of the Boeing 747. But what did birds do before the time of Bernoulli?

COFFEE RUSH

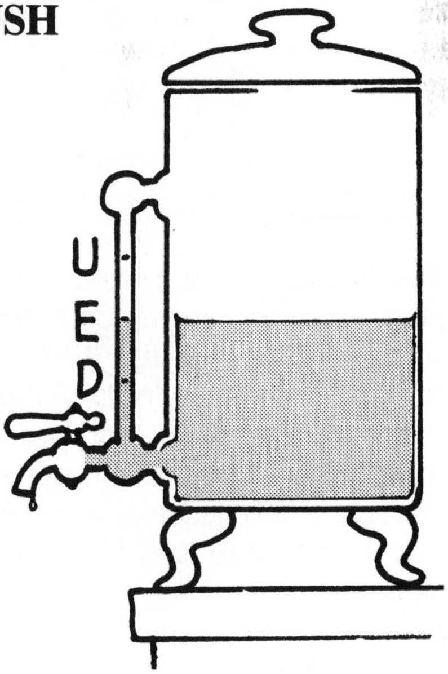

In front of a restaurant coffee maker is a glass tube called a sight pipe. The level E of the coffee in the sight pipe is equal to the level of the coffee inside the machine. But when the faucet is opened and a current of coffee runs out of the spigot, the level of coffee in the sight pipe

a) remains essentially at E
b) surges up to U
c) suddenly drops to D

COFFEE RUSH (continued)

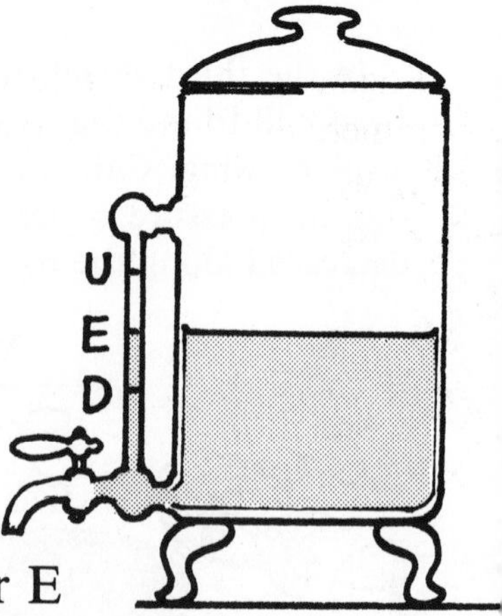

When the faucet is closed and the current of coffee is suddenly stopped, the level of the sight pipe

a) remains momentarily at D before gradually returning near E
b) returns immediately to equal the level inside the machine near E
c) surges up to U and then drops back to E
d) surges up to U and remains there

ANSWER: COFFEE RUSH

The answer to both questions is c. We expect when the coffee flows in the spigot that the level drops to D because of the Bernoulli effect. The reduced pressure in the spigot supports less coffee in the sight pipe.

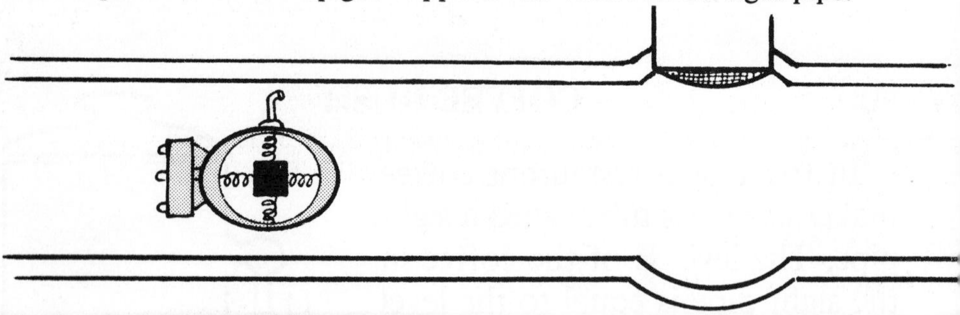

But why does it surge to U when the flow suddenly stops? Consider our little submarine again. When it is brought to a sudden stop the suspended mass slams forward. The mass of the coffee does the same. The pressure is momentarily increased — thus the surge to level U in the sight pipe.

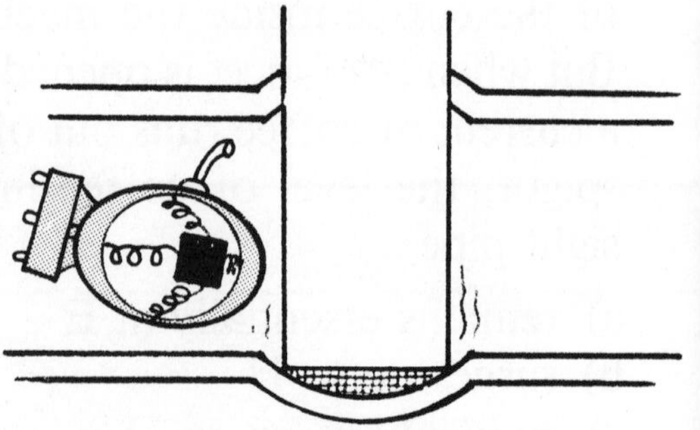

Engineers call this effect a "water hammer," and you sometimes hear it bang when you close a faucet suddenly. Needless to say, it is hard on plumbing. That's why many plumbing codes call for the installation of a short dead-ended piece of vertical pipe behind each faucet. If the faucet is suddenly closed the water surges into the stand pipe and the air at the top of the pipe cushions the water hammer.

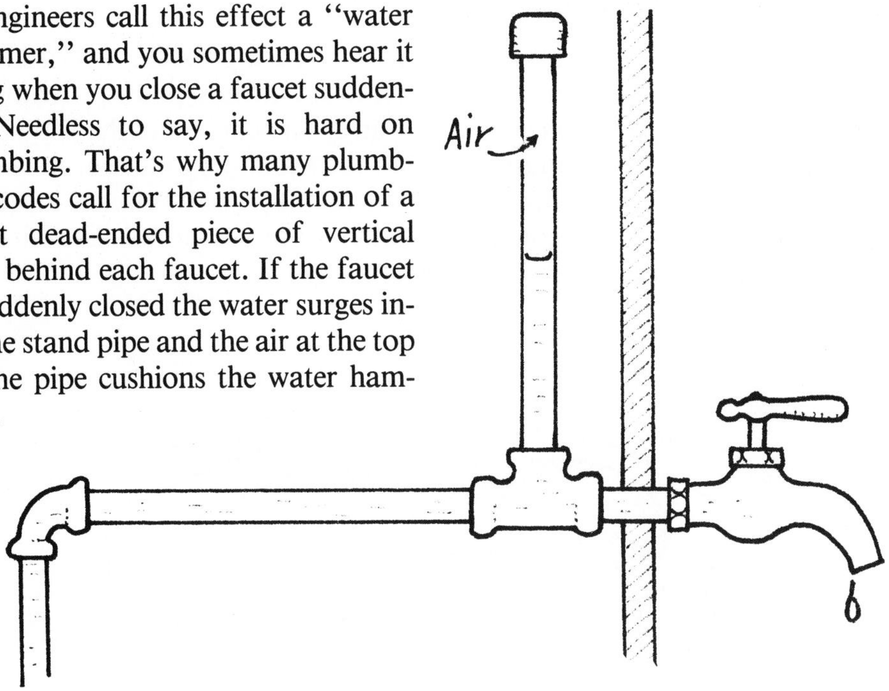

SECONDARY CIRCULATION

A large current of fluid is moving to the right in the big pipe. The current in the small pipe is moving

a) to the right
b) to the left
c) neither way

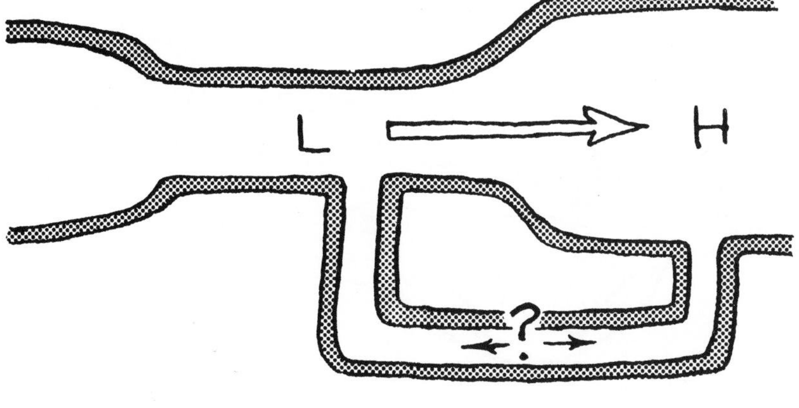

ANSWER: SECONDARY CIRCULATION

The answer is: b. As previously explained, the pressure at **L** is lower than the pressure at **H**, and the current in the little pipe flows from the region of high to the region of low pressure. The current in the little pipe is called a secondary circulation. Often the secondary current does not even need a little pipe to flow in. It can creep up the boundary layer on an airplane wing. Then you have really serious trouble! The airflow becomes turbulent and the wing stops lifting.

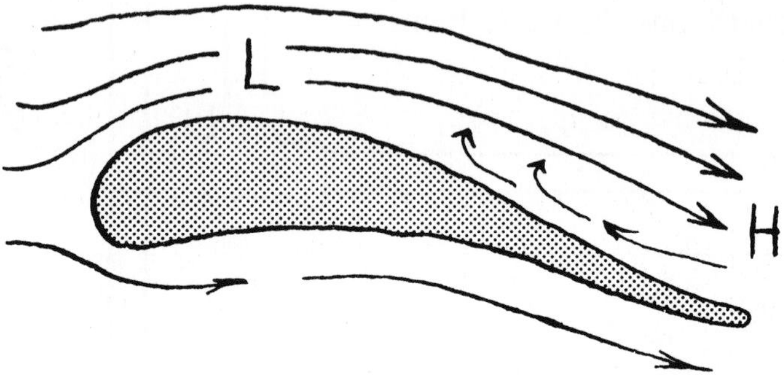

BACKWATER

The main current in a creek is flowing to the right. The small current behind the rock is flowing to the

a) right
b) left
c) neither way

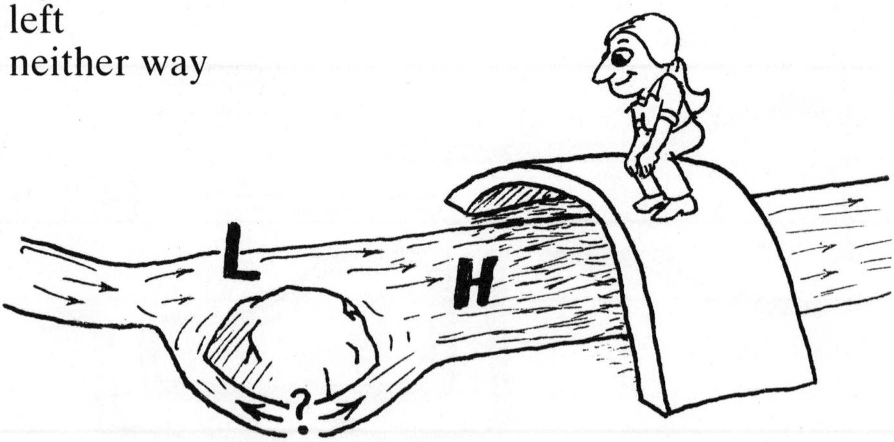

ANSWER: BACKWATER

The answer is: b. The creek is flowing rapidly where it is narrow at **L** and slower where it is wider at **H**. But how does the water get rid of its speed? Water in a pipe gets rid of its speed by flowing from low to high pressure but water in a creek can hardly change pressure. The water in the creek gets rid of its speed by flowing uphill! The water at **H** is a little higher than the water at **L**. Put your eye down close to a creek and check it out for yourself. The water flows back around behind the rock from the high water at **H** to the low water at **L**. Water flows downhill, unless it has a lot of speed, and the water behind the rock doesn't have much speed.

Often the backwater does not even need a little passage to flow in. It can creep along the boundary layer of a rock. Such backwaters start the turbulent eddy wake that forms behind boulders in a fast river. When people float through rapids (in lifejackets) they are occasionally pinned behind boulders by a pair of such backwaters.

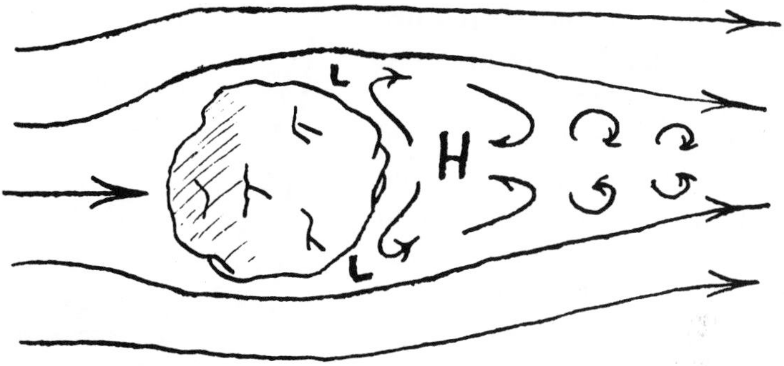

UNDERTOW

If a fast-running river makes a particularly sharp turn you will most likely find an undertow in the water on the

a) inside of the turn at **I**
b) outside of the turn at **O**

ANSWER: UNDERTOW

The answer is: b. To see why, get a glass of water and make the water spin (with a spoon). The circling water is a "model" of the river turn. The water is thrown to the outside of the turn because of centrifugal force. Were it not for this force, the water level at **O** could not remain above the water level at **I**. And were the water not circling there would be no centrifugal force. While the water on the top of the river or top of the glass of water is moving fast, the water on the bottom of the river or glass is dragging on the bottom and so hardly moving. (Remember. near the earth's surface there may be little wind while a few hundred feet up there is a strong wind.) Since the water on the bottom is hardly moving there is little or no centrifugal force on the bottom water. Now the water at **O** is deep and the water at **I** is shallow, so there is more pressure on the water under **O** than under **I**. This pressure difference forces the water on the bottom to flow from **O** to **I**. If the water on the bottom were moving, centrifugal force could counteract this pressure difference. But the bottom water is hardly moving. So as the river turns, or the glass of water spins, a secondary circulation is set up. The secondary circulation carries the water from **O** to **I** on the bottom, then up from the bottom to top at **I**, then from **I** to **O** on the surface, and back from the top down to the bottom under **O**. That down leg at **O** is the undertow. In a river you can sometimes see little whirlpools in the water near **O**. The whirlpools form where the water is being sucked down. You might also see the water boiling up near **I**, usually a bit downstream from the whirlpools.

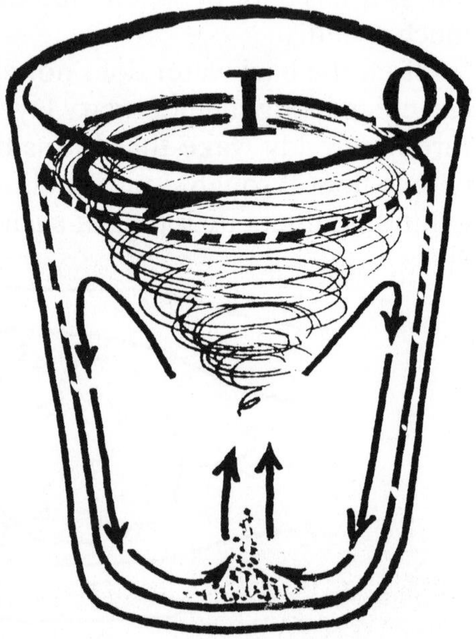

You can make this secondary circulation visible by watching the leaves in a glass of tea. (My father drank hot tea in a glass.) When the leaves get waterlogged and sink to the bottom, the secondary circulation will sweep them up and pin them under **I**. To avoid confusion and really see what is happening, use only one leaf.

MOONSHINE

Professor Mellow is syphoning a bit of his brew into a bucket. A necessary condition for the operation of a siphon is that

a) there be a difference in atmospheric pressure at each end
b) the weight of water in the discharge end exceed the weight of water in the intake end
c) the discharge end be lower than the intake end
d) all of the above

ANSWER: MOONSHINE

The answer is: c. Some people think the syphon works because of the difference in air pressure between the intake and discharge end of the syphon tube. Not so. If the difference in atmospheric pressure between the ends of the tube was responsible for the operation of the syphon it would run in the opposite direction — atmospheric pressure is slightly greater at the lower end!

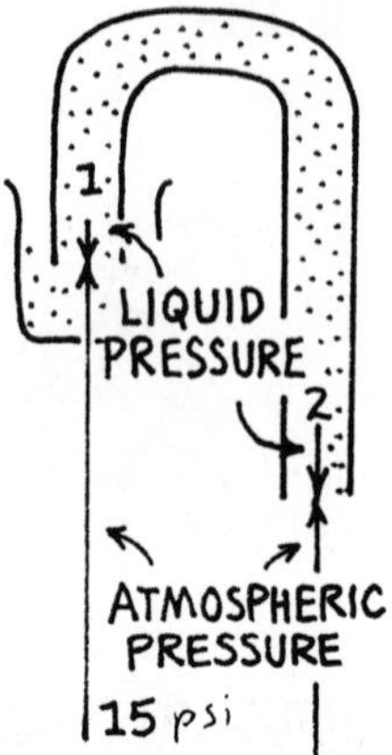

The syphon runs because of a pressure difference, but not because of any differences in atmospheric pressure. At each end of the tube are two pressures to consider: the downward pressure due to the enclosed liquid and the upward pressure due to the atmosphere. We consider in our sketch the case where the discharge end is twice as long as the intake end for a tube initially filled with brew. We have drawn the upward atmospheric pressures equal (although strictly speaking the lower end is slightly greater) and we call these 15 pounds per square inch (psi). Since liquid pressure is proportional to depth of liquid (rather than weight of liquid) twice as much liquid pressure occurs at the bottom of the discharge end compared to the intake end — so in our sketch we make the downward arrow at the discharge end twice as long as at the intake end. We see that the downward liquid pressure is less than the upward atmospheric pressure. As an illustration, let's suppose the liquid pressure at the intake is 1 psi and at the twice-as-long discharge end is 2 psi. The net result is 14 psi pressure at the intake end and only 13 psi at the discharge end with the fluid flowing from the region of higher pressure to the region of lower pressure. Note that the net pressure at each end of the tube is in an upward direction, and that this moves the fluid clockwise just as the similarly-pushed see-saw shown in the sketch rotates clockwise.

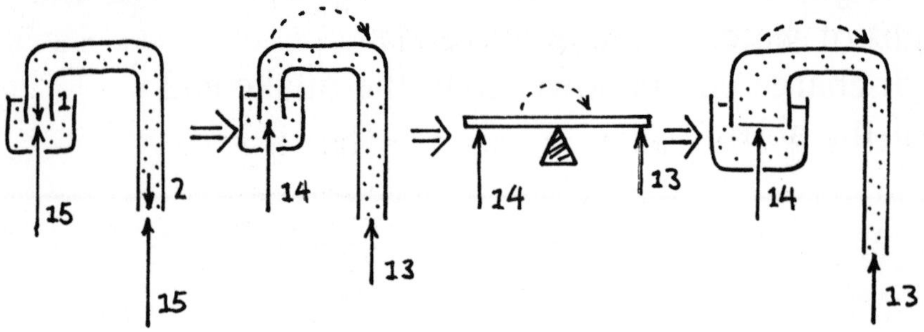

Note that the operation of the syphon does not require the tube to be of uniform thickness. If the intake side is much wider so that the weight of liquid in it is greater than the weight of liquid in the discharge side, the flow is nonetheless in a direction from higher pressure (not weight!) to lower pressure. And for each end of the syphon tube the net pressure will be upward and will be greater on the shorter or higher end.

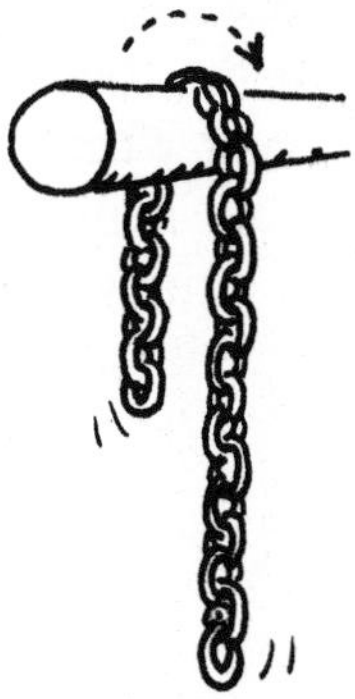

The running of a syphon is much like the sliding of a chain draped over a smooth rod. If the chain is draped so that its midpoint is over the rod and both ends hang at equal lengths, it doesn't slide. But if one end hangs lower than the other end, the longer end falls and pulls the shorter end upward. The rate of motion depends on the relative lengths of the hanging sides. So too with the syphon. But liquid is not a chain. So why does the liquid not just split apart at the top of the tube and run out of each arm? Because if it did split it would leave a vacuum in the empty space so formed, and the outside air pressure would immediately push more liquid up the tube to fill the vacuum. Atmospheric pressure, however, can only push liquid so far. It will push mercury up 30 inches, and water 34 feet. If the liquid is water and the syphon is taller than 34 feet it *will* split on top. To build a taller syphon you would have to increase the outside air pressure. The role of atmospheric pressure on syphon behavior is simply to keep the tube filled with liquid.

FLUSH

The flushing of a toilet depends on the principle of

a) the aspirator
b) buoyancy
c) the syphon
d) centripetal force
e) the hydraulic ram

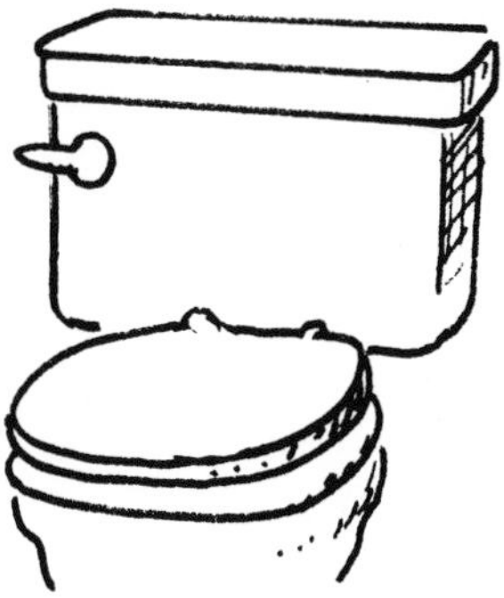

ANSWER: FLUSH

The answer is: c. A toilet, like a sink, or bathtub, or washtub has a gooseneck in its sewer pipe. The purpose of the gooseneck is to prevent sewer gas from coming up the sewer pipe into your house. It also catches dropped diamonds. When the water is released into the toilet it fills the gooseneck and the gooseneck becomes the syphon. In this way the contents of the toilet bowl are flushed and when the level of the bowl lowers to a point where air again fills the gooseneck, syphon behavior ceases.

Incidentally, many people do not know that you can make a toilet flush by simply pouring a bucket of water into it! The function of the reservoir tank is simply to supply water.

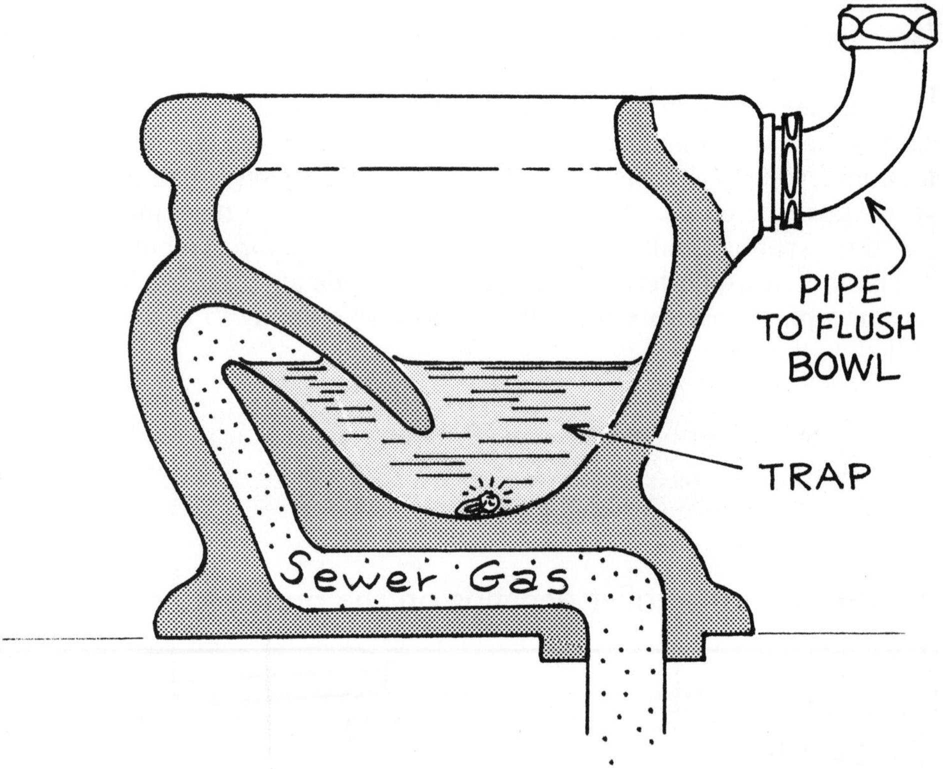

THE PISSING BUCKET

This was a favorite question of the noted hydrologist, George J. Pissing, and is still often asked of graduate students during oral exams. Consider a bucket of water with two holes through which water is discharged. Water can be discharged from a hole "B" at the bottom of a bucket which is some distance "d" below the water surface, or it can be discharged from a downspout which starts at the top "T" and has its opening at the same distance "d" below the water surface. If we neglect any friction effects the water coming out of hole "B" has

a) more speed than the water coming out of the downspout
b) less speed than the water coming out of the downspout
c) the same speed as the water coming out of the downspout

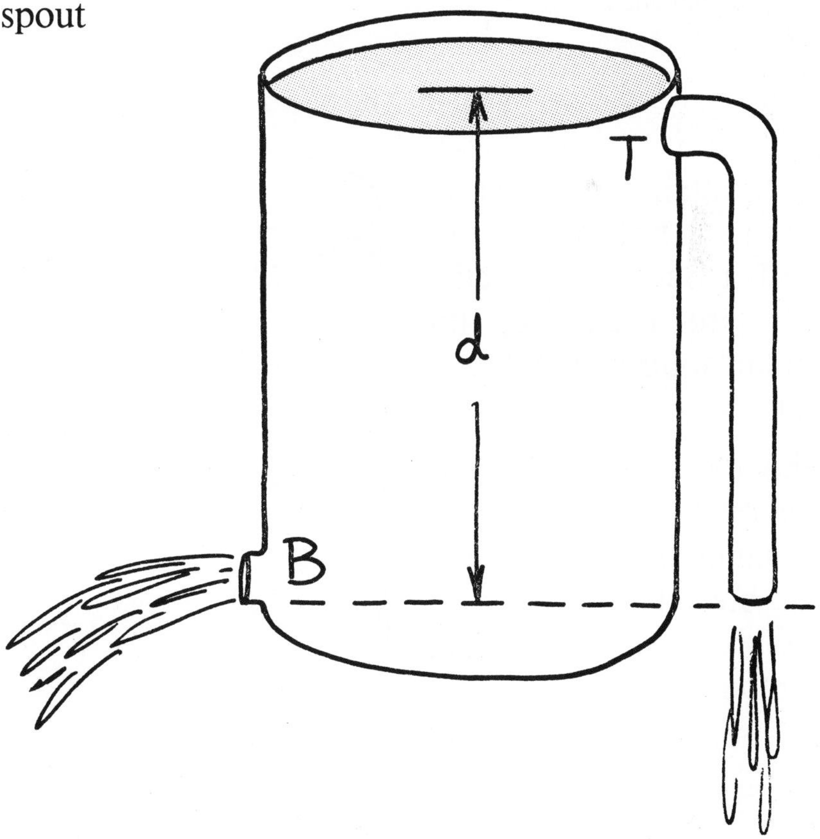

ANSWER:
THE PISSING BUCKET

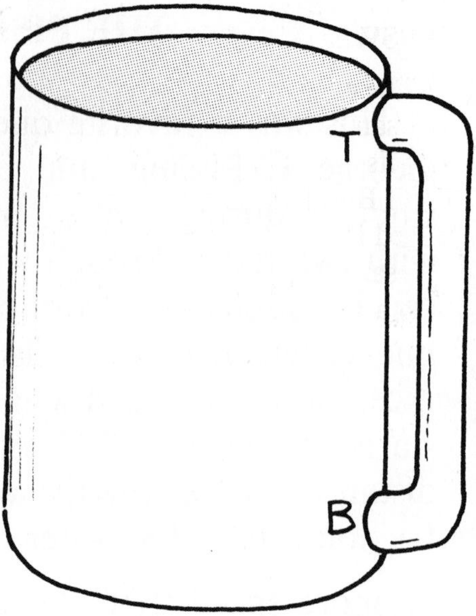

The answer is: c. The speed of the water depends on the pressure head, or depth below the free surface. The pressure heads for both outlets are the same, so water speeds are the same. Or we can look at the problem this way: Move the spout around and attach it to the hole as sketched. If the water comes out of the downspout fastest it would overwhelm the current from the hole and force its way back into the bucket at B, so you would have a perpetual motion machine with water running from T to B.

If the water comes out of the bottom hole fastest you also have perpetual motion with water flowing the other way. To avoid perpetual motion (and conserve energy), the water flowing out of both the hole and the downspout must have equal speeds.

FOUNTAIN

A fountain spout is attached to a hole at the bottom of a bucket. If friction effects are negligible, this fountain squirts water to a height

a) above the water level of the bucket
b) equaling the water level of the bucket
c) less than the water level in the bucket

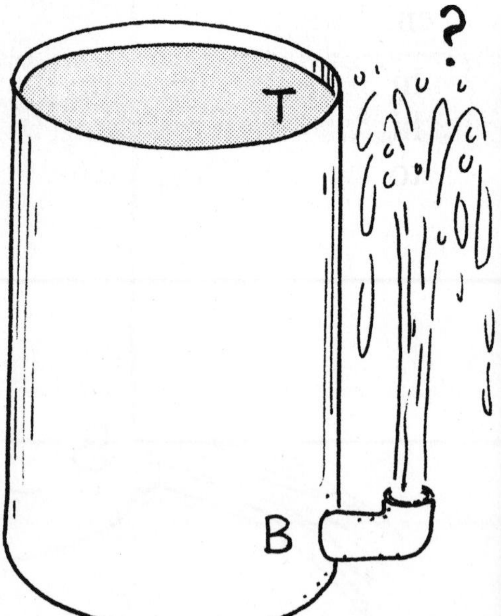

ANSWER: FOUNTAIN

The answer is: b. The speed of the water coming out of the hole at "B" is the same as the speed of the water coming out of the downspout — which has its speed by virtue of falling from "T" to "B". (Remember THE PISSING BUCKET?) So the water shoots up from the fountain with the same speed it would have if it were to fall from the top. If it starts upward with the same speed it would have gained by falling, it goes back to the top — like a perfectly bouncing ball.

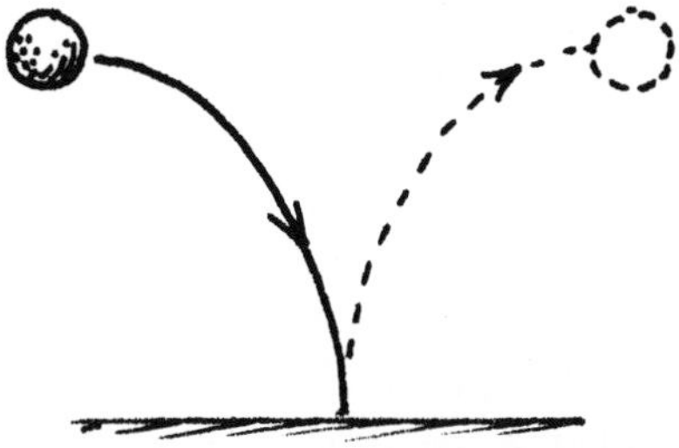

BIG SQUIRT LITTLE SQUIRT

Two fat pipes are attached directly to the bottom of a water tank. Both are bent up to make fountains, but one is pinched off to make a nozzle, while the other is left wide open. Water squirts

a) highest from the wide open pipe
b) highest from the pinched off pipe
c) to the same height for each pipe

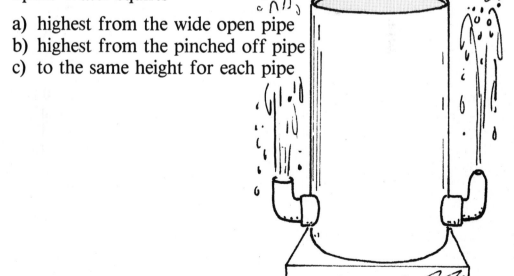

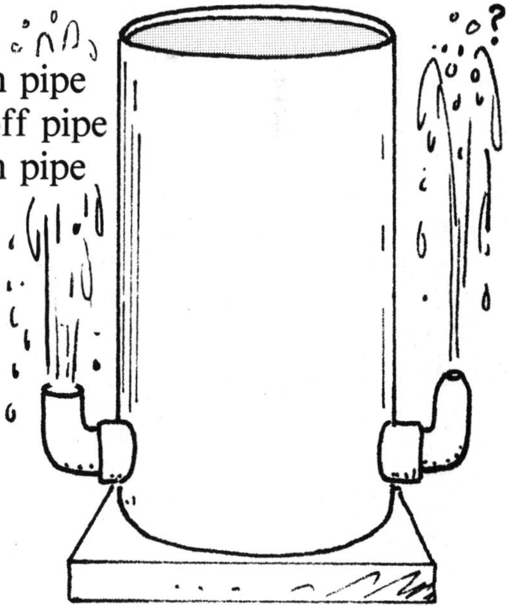

ANSWER: BIG SQUIRT LITTLE SQUIRT

The answer is: c. Remember the FOUNTAIN: the water squirts back up to the water level in the bucket and the size of the squirt hole does not even come into the story. Everyone knows, however, that putting a finger over the end of a garden hose makes it squirt much further. After all, why would anyone buy a nozzle if it did not help the water squirt further? But have you ever attached a hose DIRECTLY to a water tank? You will be in for a little surprise if you do. The water squirts just as far without a nozzle as it does with one.

Why does a nozzle make water squirt further when you put it on your hose at home? It is because the pressure at the end of your home hose has to do with more than the depth of the water tank to which it is attached. It also depends very much on the speed of the water which flows through miles of pipe. The faster the water moves through the pipe, the greater is the friction that cuts down on the pressure at the output end. (The friction in rusty pipes is even greater.) If the speed of the water is lessened, the friction also lessens and the pressure at the output is increased. Of course, if the water is shut off and does not flow there is no friction, so you get full pressure — but no water output. When the water is flowing, putting your finger over the end of the hose reduces the number of gallons per minute that flow through the pipe so the water in the pipe goes slower. This results in less friction in the pipe and consequently more pressure left in the water when it arrives at the end of the hose — so it will squirt your kid sister better. How about that!

Water friction is not like dry friction. Water friction depends very much on speed. Slide your hand back and forth on the table and the friction does not change much with speed. Next, push your hand back and forth in a bathtub full of water. The water friction is almost zero if you move slowly, but gets so strong as you move faster that it limits the speed of your hand. This whole business of water friction is almost exactly replayed in the story of a dead electric battery. When a battery dies its internal resistance goes way up. Like a rusty water pipe, the electrodes or plates become corroded — which impedes the flow of electric current. If very little or no current is drawn from a dead battery it will show full voltage (pressure) because the resistance only cuts the voltage when the current flows. (Many people are amazed to see a dead battery read full voltage.) But if you try to draw a large current from the dead battery, which is called putting a load on the battery, the voltage (pressure) drops because all of it is being used just to force the electric current through its own guts which are full of resistance. We'll learn more about electricity later.

MORE QUESTIONS (WITHOUT EXPLANATIONS)

You're on your own with these questions which parallel those on the preceding pages. Think Physics!

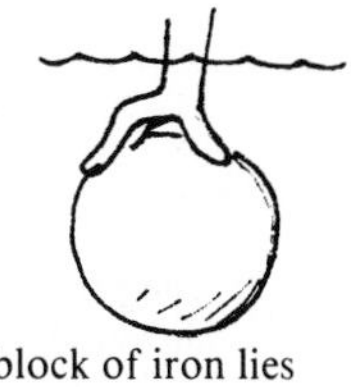

1. About how much force would you have to exert to hold a one-cubic foot beach ball under the water?
   a) 32 lb   b) 64 lb   c) 128 lb

2. A cubic-foot block of solid balsa wood floats on water while a solid cubic-foot block of iron lies submerged in water. The buoyant force is greatest on the
   a) wood   b) iron   c) . . . is the same on each

3. A boat loaded with scrap iron floats in a swimming pool. If the iron is thrown overboard into the pool, the water level at the side of the pool will
   a) rise   b) fall   c) remain unchanged

4. If a ship in a closed canal lock springs a leak and sinks, the water level at the side of the canal lock will
   a) rise   b) fall   c) remain unchanged

5. An empty container is pushed open-side downwards into water. As it is pushed deeper, the force required to hold it beneath the water
   a) increases   b) decreases   c) remains unchanged

6. A high-altitude helium-filled weather balloon rises in the atmosphere until
   a) atmospheric pressure up against its bottom equals atmospheric pressure down against its top
   b) helium pressure inside equals atmospheric pressure outside
   c) the balloon can no longer expand
   d) all of these   e) none of these

7. A pair of identical containers are filled to the brim with water. One has a piece of wood floating in it, so its total weight is
   a) greater   b) less   c) the same as the other container

8. Suppose at the end of a party your dork host serves you a "one-for-the-road" mixed drink and you notice that the ice cubes are submerged at the bottom. This indicates that the mixture
   a) produces no buoyant force on the ice cubes
   b) is not displaced by the submerged ice
   c) is less dense than ice
   d) is upside down

9. If the top part of an iceberg that extends above the water line were somehow suddenly removed, a resulting consequence would be
   a) a decrease in the density of the iceberg
   b) a decrease in the buoyant force acting on the iceberg
   c) an increase in pressure at the bottom of the iceberg which would force it up to a new equilibrium position
   d) all of these   e) none of these

10. Small amounts of mercury and water are splashed on a dry tabletop. Which balls up the most?
    a) Mercury   b) Water   c) Both the same

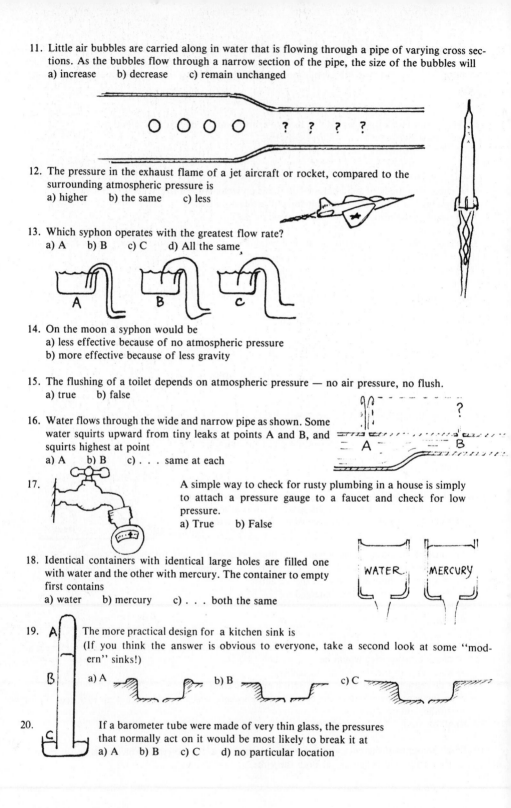

11. Little air bubbles are carried along in water that is flowing through a pipe of varying cross sections. As the bubbles flow through a narrow section of the pipe, the size of the bubbles will
a) increase b) decrease c) remain unchanged

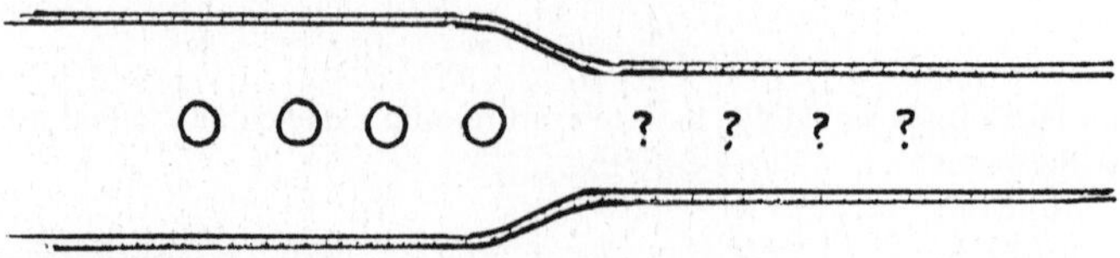

12. The pressure in the exhaust flame of a jet aircraft or rocket, compared to the surrounding atmospheric pressure is
a) higher b) the same c) less

13. Which syphon operates with the greatest flow rate?
a) A b) B c) C d) All the same

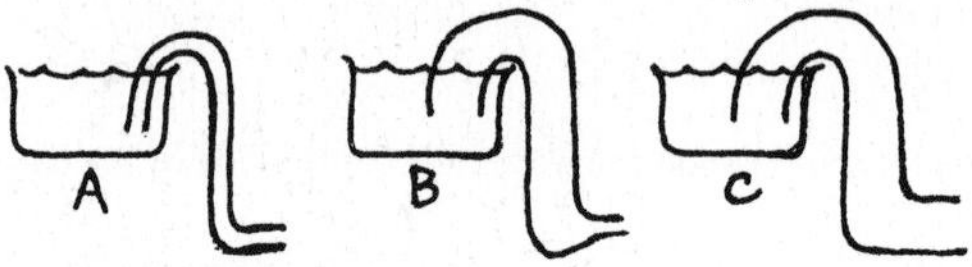

14. On the moon a syphon would be
a) less effective because of no atmospheric pressure
b) more effective because of less gravity

15. The flushing of a toilet depends on atmospheric pressure — no air pressure, no flush.
a) true b) false

16. Water flows through the wide and narrow pipe as shown. Some water squirts upward from tiny leaks at points A and B, and squirts highest at point
a) A b) B c) . . . same at each

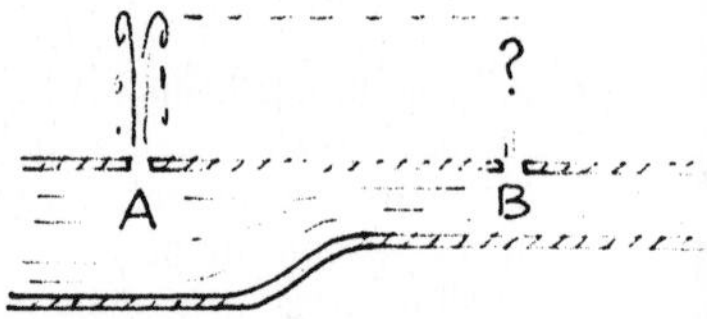

17. 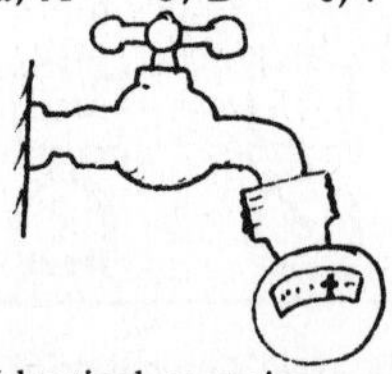

A simple way to check for rusty plumbing in a house is simply to attach a pressure gauge to a faucet and check for low pressure.
a) True b) False

18. Identical containers with identical large holes are filled one with water and the other with mercury. The container to empty first contains
a) water b) mercury c) . . . both the same

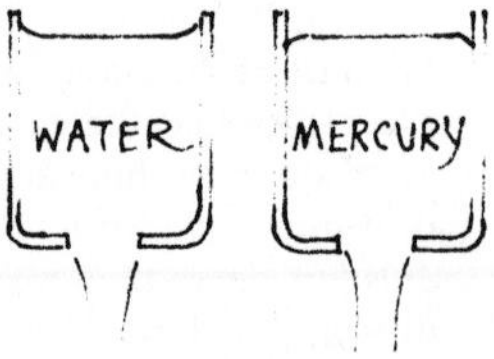

19. The more practical design for a kitchen sink is
(If you think the answer is obvious to everyone, take a second look at some "modern" sinks!)

a) A  b) B  c) C 

20. 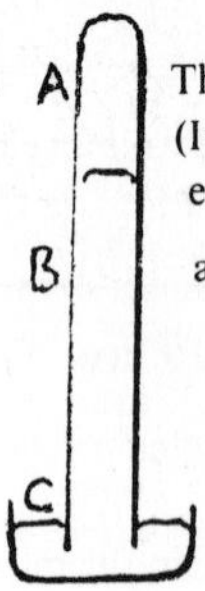

If a barometer tube were made of very thin glass, the pressures that normally act on it would be most likely to break it at
a) A b) B c) C d) no particular location

Heat

Mechanical things are visible, even fluids are visible, but heat is not. Heat is an invisible agent. Of course, invisible does not mean undetectable. Heat can be "seen" with your fingertips. Physicists have gone on to understand many things that cannot be seen, but heat was the first intangible to be treated as a "real" thing; the first invisible thing to be visualized through the mind's eye. And what was the vision? The vision was the graveyard of energy.

The problem is not that you don't know the answer.
The problem is that you don't know the question.

COMING TO A BOIL

You are bringing a big pot of cold water to a boil to cook some potatoes. To do it using the least amount of energy you should

a) turn the heat on full force
b) put the heat on very low
c) put the heat at some medium value

ANSWER: COMING TO A BOIL

The answer is: a. If you put the heat on very low you could run the stove forever without getting the water to boil. Heat is going into the pot from the stove. Some stays in the pot and some escapes as heat radiation, hot vapor, and hot air. The heat that stays in the pot eventually makes the water boil. The escaped heat is waste. The more time it takes to get the water boiling, the more time the heat has to escape. The more time heat has to escape the more energy is lost and wasted.

A similar sort of situation exists when a rocket shoots itself up. The rocket engine should fire at full force so as to get as much momentum as possible into the rocket in the least possible time. Why? Because the rocket is continously losing momentum, due to gravity, just as the pot is continously losing heat. If the rocket engine is throtled back the rocket could spend all its energy just hovering and never go up.

BOILING

The water is now boiling. To cook the potatoes using the least amount of energy you should

a) keep the heat on full force
b) turn the heat way down so the water just barely keeps boiling

ANSWER: BOILING

The answer is: b. The temperature of the water is 100°C if it is boiling vigorously or barely boiling, and all the potato cares about is the temperature of the bath. But it takes more energy to keep the heat on full force, so turn it down. Doesn't all that extra heat going into a full force boil make the water a little hotter? Not a bit! Put a thermometer into vigorously boiling and barely boiling water and see which is hotter. If you don't have a thermometer run a potato cooking race between a vigorously boiling and barely boiling pot. If the extra energy put into the vigorously boiling water does not go to cook the potato, where does it go? It goes out of the pot as steam. Now if you put a cover on the pot will you reduce the energy needed or the time needed to cook the potato? You will reduce both. You will reduce the time and energy required to boil the cold water, and save energy but not time while the water is boiling.

KEEPING COOL

The refrigerator in your house probably uses more energy than all the other electric appliances in your house combined (except electric water heaters and air conditioners). Suppose you have taken a carton of milk out of your refrigerator and used some. It is most energy efficient to

a) immediately return the milk to the refrigerator
b) leave it out as long as possible.

ANSWER: KEEPING COOL

The answer is: a. The longer the milk stands out the warmer it gets. The warmer it gets the longer the refrigerator has to run to get it cool again. The longer the refrigerator runs the more energy it eats.

TO TURN IT OFF OR NOT

On a cold day suppose you must leave your house for about one-quarter hour to go shopping. In order to save energy it would be best to

a) let the heater run so you will not have to use even more energy to reheat the house when you return
b) turn your thermostat down about 10°, but not turn it off
c) turn off the heater when you go
d) . . . it makes no difference as far as energy consumption is concerned, whether you turn your heater off or let it run

ANSWER: TO TURN IT OFF OR NOT

The answer is: c. turn off the heat. When it is cold outside, your house is always losing heat. If it were not losing heat you would only need to heat it once and it would stay hot indefinitely. The heater must replace all the heat which is lost. How much heat is lost? That depends on how well the house is insulated and how cold it is outside. The greater the difference in temperature between the inside and outside of the house, the greater is the rate of cooling (this is Newton's law of cooling: Rate of Cooling $\sim \Delta T$). Keeping the house hot while you are gone results in a greater rate of heat loss than if the house is cooler. The hotter a house is compared to the outside temperature, the faster heat is lost. Of course, if there were no temperature difference there would be no loss and no need for heating.

We can visualize this by thinking of the house as a leaky bucket, where the level of water in the bucket is like the temperature in the house. The higher the level of water in the bucket, the greater is the pressure at the holes and the faster it leaks. So more water per minute is required to maintain a high level of water than a lower level. It is easy to see, then, that we save water if we "turn the level down." Do we save more water by turning it off altogether, even for a short time? A little thought will show that less water overall will be required to refill the bucket after a complete shutoff than would be required to maintain the same level with its corresponding rate of leakage. When empty or near empty it fills quickly because inflow is greater than leakage. Filling levels off at a depth where inflow equals leakage.

So just as less water would be required to refill the leaking bucket than to maintain a constant level, less heat is required to reheat a home that has cooled than to maintain it at a higher temperature than the outside.

To save energy, turn off your lights when you don't need them, and turn off your heater when you leave your home!

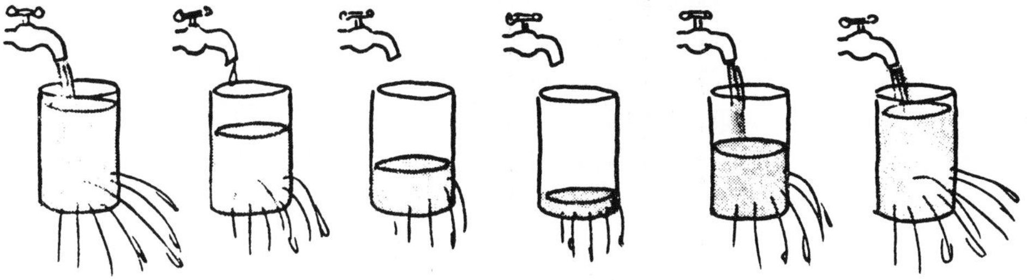

WHISTLING TEA KETTLE

One tea kettle is heated directly over a stove flame and another is set upon a heavy piece of metal which is directly over a flame. After they begin to whistle you turn off the stove.

a) The kettle heated directly over the flame continues to whistle, but the kettle resting on the metal stops promptly
b) The kettle on the metal continues to whistle for some time, but the one heated directly stops promptly
c) Both stop whistling in about the same amount of time

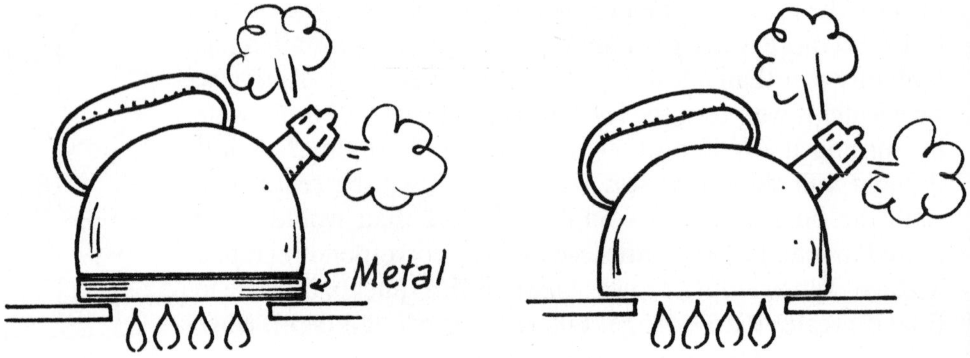

ANSWER: WHISTLING TEA KETTLE

The answer is: b. This question might trip up a "good" physics student who might reason that metal has less heat capacity than water and so will provide less energy. But the essential thing here is that the metal is hotter than the water in the kettle. It must be hotter if heat is to flow from the metal into the kettle and the metal remains hotter for a time after the stove is turned off. So, during that time heat is still forced into the kettle and the kettle continues to whistle. If there is no metal the heat supply stops when the stove goes off and so the whistle stops too.

EXPANSION OF NOTHING

A metal disc with a hole in it is heated until the iron expands one percent. The diameter of the hole will

a) increase
b) decrease
c) not change

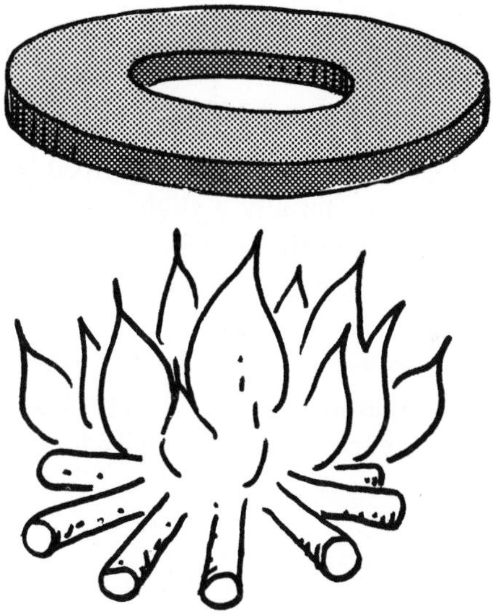

ANSWER: EXPANSION OF NOTHING

The answer is: a. The hole is nothing, but nothing expands, too. There is no way to avoid it. Every dimension of the ring expands in proportion. To visualize the expansion, suppose a photo of the ring were made and the photo was then enlarged one percent. Everything in the photo would be enlarged, even the hole.

Or look at it this way: uncurve the ring and open it up so that it forms a bar. It does become thicker when heated; but it also becomes longer. So when curved back into a ring we see that the inner circumference as well as the thickness is larger.

It's quite easy to see that the hole becomes larger upon expansion if we consider a square hole in a square piece of metal. Break it into square segments as shown, heat and expand them, then put them back together. The empty hole expands every bit as much as the solid metal.

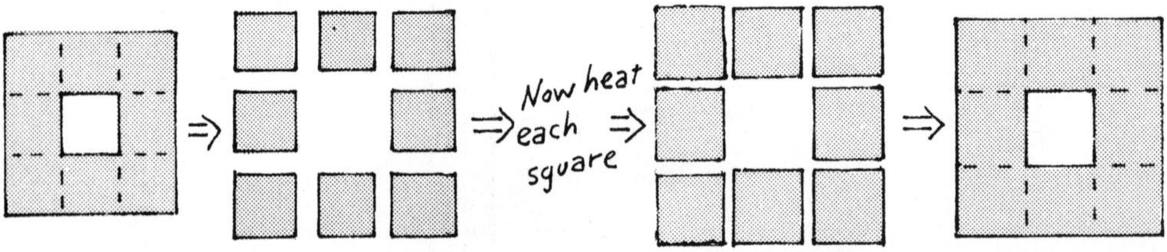

Blacksmiths used to shrink fit iron rims on wooden wagon wheels by simply heating a rim that was initially a slight bit smaller than the wheel. When heated and expanded, the rim was simply slipped onto the wooden wheel. When cooled, a snug fit was provided that required no fasteners of any kind.

The next time you can't open the metal lid on a stubborn jar, heat the lid under hot water or by momentarily placing it on a hot stove. The lid, inner circumference and all, will expand and be easily loosened.

TOUGH NUT

A nut is very tight on a screw. Which of the following is most likely to free it?

a) Cooling it
b) Heating it
c) Either
d) Neither

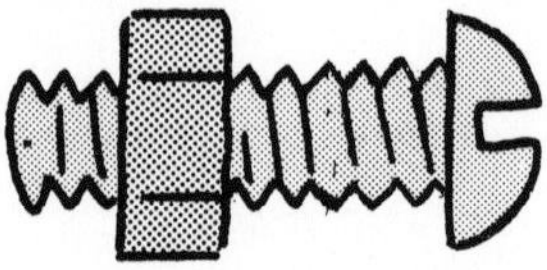

ANSWER: TOUGH NUT

The answer is: b. Think about EXPANSION OF NOTHING. The screw and the nut are not completely in contact with each other. There is a small space between the screw and the nut. For the very tight nut the problem is likely that this space is too small. How can it be increased? By heat. Heat will make everything larger. The nut will expand, the screw will expand, and most importantly, the space between them will expand. So to loosen the nut, heat it — even though the screw expands also.

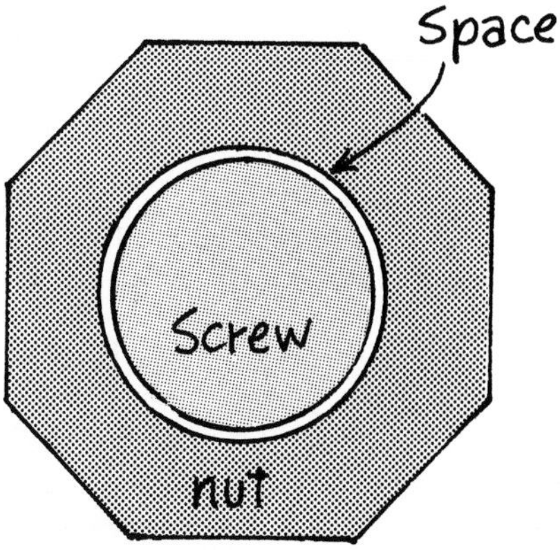

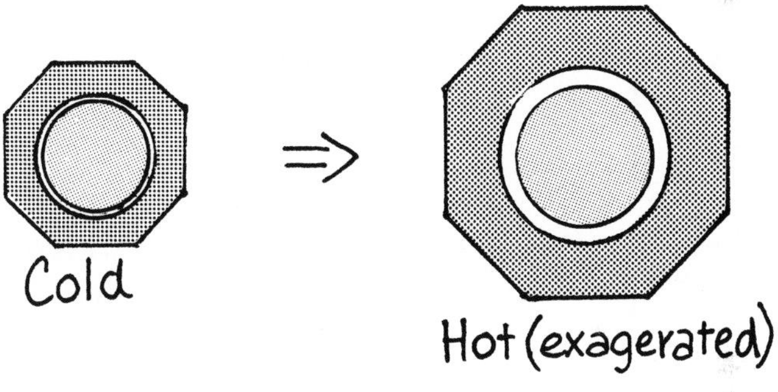

DEFLATION

If the volume occupied by some air is decreasing, then the temperature of the air must be

a) increasing
b) decreasing
c) . . . you can't tell

ANSWER: DEFLATION

The answer is: c.

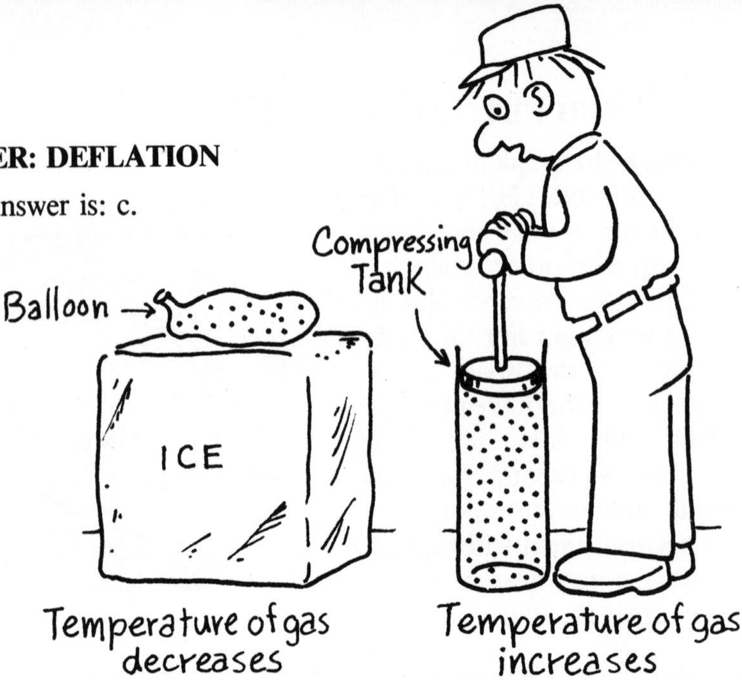

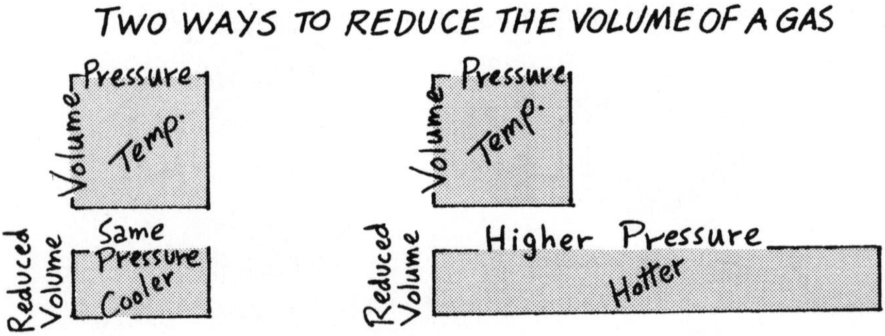

When thinking about this question the vision of a balloon put in a refrigerator might come into your mind. In this case the balloon's shrinking volume goes with decreasing temperature. But another person will flash on the air being compressed in a pump or piston engine, in which case shrinking volume means increasing temperature. The change in air temperature depends on more than just its change in volume. You must also know how its *pressure* is changing. Air temperature is a joint function of volume *and* pressure — knowing one is not enough. If volume goes down and pressure goes down then temperature goes down, or even if volume goes down and pressure does not change then temperature goes down. But if volume goes down a little and pressure goes up a lot then temperature goes up. What is a little or a lot? Well, if the volume of a gas is reduced by half and the pressure is doubled then the temperature does not change. But if the pressure increase is more than double the temperature goes up; if less than double temperature goes down. We say, temperature is proportional to pressure multiplied by volume, $T \sim PV$.

WANTON WASTE

One of the most extravagant wastes of electric energy you can see is at many supermarkets. Cold food is stocked in the following kinds of refrigeration cases. Which kind of case is most wasteful? Which is most conservative?

a) Horizontal case with sliding cover doors
b) Horizontal case without cover
c) Upright case with door
d) Upright case without door

ANSWER: WANTON WASTE

The worst case is: d. The best is: a. Cold air is denser than warm air so it falls to the floor. When you open an upright refrigerator the cold air literally falls out and new warm air comes in to takes its place. But if the upright box has no door the cold air falls out continously. Have you ever noticed how cold your feet get when you walk near an upright doorless refrigerator in a supermarket? That is wasted refrigeration energy on the floor and you are paying for it. The best refrigeration case opens only on top. That way cold air can't fall out. And it has a door so the cold air can't make contact with the warmer air outside.

Cold air is denser than warm air if they are both at the same pressure, in this case atmospheric pressure. But if the hot air is under higher pressure than the cold air it might be denser than the cold air. You must be careful about just saying or thinking things like "cold air is denser than warm air."

INVERSION

You are camped by a mountain lake. The smoke from your breakfast campfire goes up a ways and then spreads in a flat layer over the lake. After breakfast you are hiking to higher elevations. At those elevations the temperature will probably be

a) cooler
b) warmer

ANSWER: INVERSION

The answer is: b. What has caused the layer of smoke is an inversion. The air near the lake is chilled. Perhaps cold water in the lake cooled the air or perhaps the cool air just rolled down into the bottom of the valley at night. Remember, cold air is denser than warm air and sinks. Above the cold air is warmer air, and the smoke demonstrates this. The hot smoky air goes up through the cool air because hot air rises. It rises until it gets to the warm air above. If the warm air is warmer than the smoky air, the smoke has no reason to rise further. It just spreads out under the warm air. So if you hike up above the lake you will be in warm air.

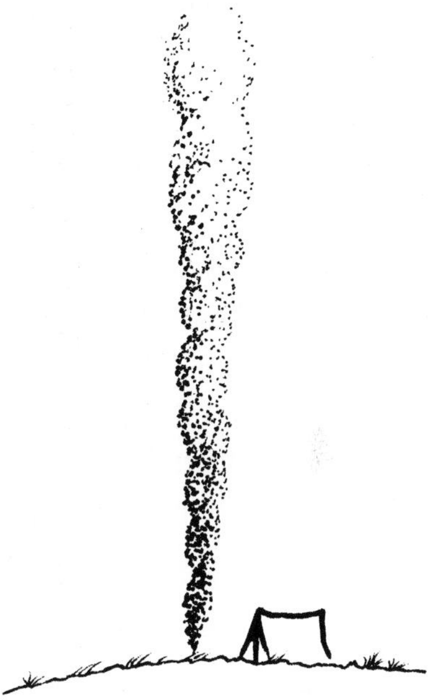

Usually if you build a fire the smoke just goes up and up and up. That means the air gets colder and colder and colder as it gets higher, and the air is always colder than the smoke. When the upper air is warmer—rather than colder, as it should be—the situation is called an inversion.

Inversions are sometimes found in coastal valleys near cold oceans. For example, at Los Angeles cool air from the cold Pacific Ocean comes in under hot air from the Mojave Desert. Smog and smoke from Los Angeles get trapped under this inversion. Often you can see a layer of yellow stuff over the city—just like the layer of smoke over the lake. For the same reason you can also see a layer of yellow stuff over the south end of San Francisco Bay.

MICROPRESSURE

Smoke is made up of numerous tiny ashes. If you could measure the air pressure in a space as small as a smoke ash you would likely find

a) it varies from place to place at any instant of time — different parts of a room have different pressures
b) it varies from instant to instant at any place in a room — the pressure fluctuates with time
c) . . . both a and b
d) except for varying weather conditions and drafts, the air pressure in a room is constant and does not fluctuate from instant to instant or from place to place — even in a very small volume of space.

ANSWER: MICROPRESSURE

The answer is: c. The air molecules are distributed at random in space so you cannot expect exactly the same number will be in every little volume of space. As the molecules rush about there will be little "gang-ups" from time to time and from place to place. In a large volume of space the effect of the little gang-ups hardly counts, but in small volumes it represents a real fluctuation in pressure — when the molecules gang-up the pressure in the little space goes up. Suppose the air pressure on the left side of a little smoke ash suddenly goes up — the little ash will get a push to the right. Later it will get pushes from other directions as the molecules bunch up at random on various sides of the ash.

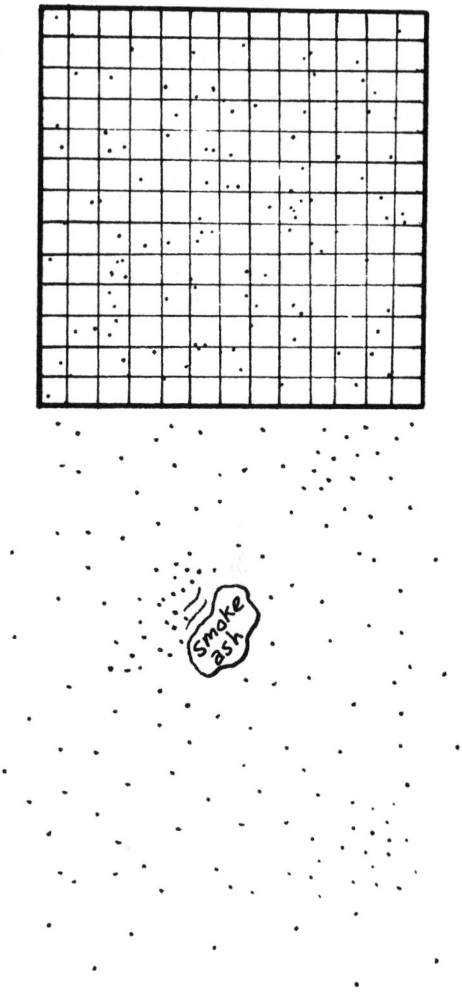

If you put cigarette smoke in a small box with a glass window and look into the box with a microscope you see the smoke particles zigzag like drunks as they float in the air. The air molecules, too small to be seen — even through the best microscope — are rushing about hitting the "big" smoke ash and so forcing it to "dance." This dance is called Brownian motion (after the first scientist to see and report it). Actually, individual "hits" do not affect the smoke particle very much, but when it gets a lot more hits on one side than on the other there is a visible effect.

OUCH

The physics teacher puts his hand in the hot steam escaping from the pressure cooker and cries ouch. But if he lifts his hand a few inches he finds the steam is cool. This is because the steam cools as it expands.

a) true
b) false

ANSWER: OUCH

The answer is: b. As a explained in DEFLATION, just because a gas expands it does not necessarily cool. But has the steam expanded in the few inches between the cooker and the teacher's hand? No. It is at atmospheric pressure as soon as it escapes from the cooker, so the steam did all the expanding it was going to do before it got out of the pot. Then why does the steam get cool a few inches above the cooker? Because it mixes with cool air. And by the way it should be called vapor, not steam—steam must be at least 100°C at atmospheric pressure.

Does that mean if you could get rid of the air, the steam would escape from the pressure cooker and not cool? Yes. If you removed all the air from a sealed room and then allowed the steam from a pressure cooker to expand into that room the steam would not cool. This is called a free expansion. If you think the steam would cool, you must have some place in mind where the steam molecules could lose some of their kinetic energy. But if the room is sealed (and insulated) no energy can escape.

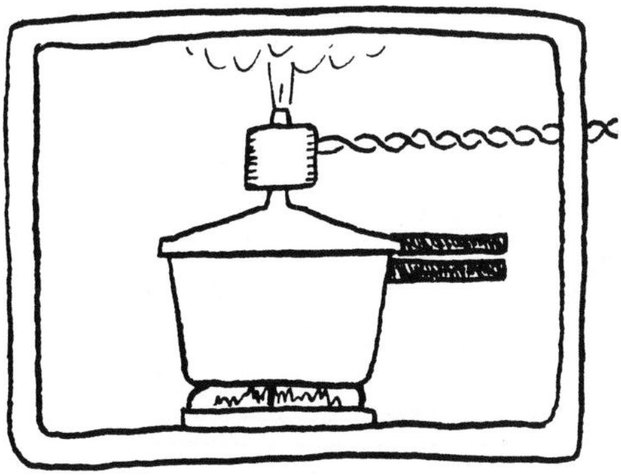

Suppose you got a toy steam turbine or steam engine and made the steam run through the turbine as it escaped from the cooker. Would that cool the steam? Yes. Yes, if the eletric energy generated by the turbine got out of the sealed room. What if the electric energy were just used to run a heater inside the sealed room? The heater would reheat the steam to its original temperature. Exactly to its original temperature? Yes, exactly.

RARE AIR

Two tanks of air are connected by a very small hole. Inside the tanks is some rare air—that is, air in which there are so few molecules that the molecules are much more likely to collide with the tank walls than to collide with each other. One tank is maintained at the temperature of crushed ice. The other tank is maintained at the temperature of steam.

a) The air pressure in the tanks must eventually equalize, regardless of the temperature difference.
b) The air pressure in the cold tank will be higher than the pressure in the hot tank.
c) The air pressure in the cold tank will be lower than the pressure in the hot tank.

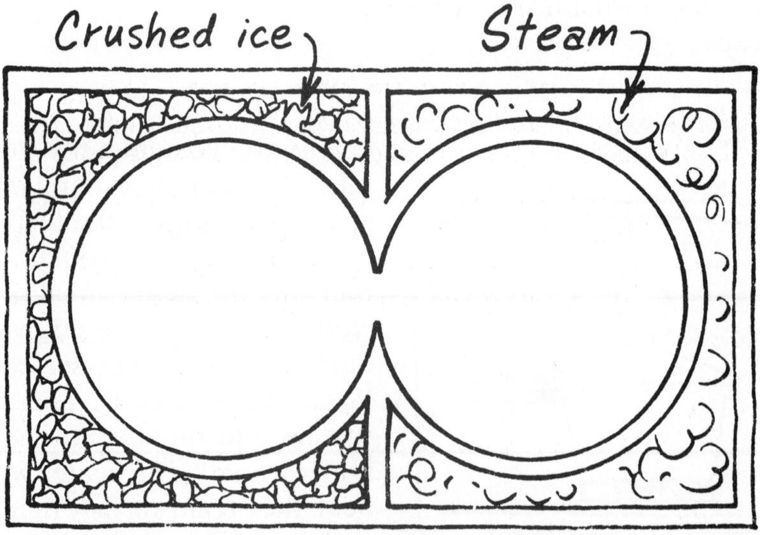

ANSWER: RARE AIR

The answer is: c. Common sense suggests that a pressure difference cannot be maintained if the tanks are interconnected. But common sense is founded upon our experience and our common experience is with dense air, that is, air in which the molecules are so packed that they hit each other much more often than the walls. Let us go down into the molecules' world.

The molecules in the hot tank will be moving faster than those in the cold tank. Some from the hot tank will go through the hole to the cold tank and some from the cold tank will go through to the hot tank. The number passing from the hot to the cold tank in a given amount of time must equal the number passing from the cold to the hot tank during the same time, or else all molecules would end up in the hot or the cold tank. So the rate at which molecules strike the walls in the tanks must be equal. However, the molecules in the hot tank are moving faster. Since air pressure depends on the rate at which molecules strike a unit area multiplied by the molecules' momentum, it must be that, regardless of the hole, the air pressure in the hot tank is higher than the pressure in the cold tank.

How then can we explain our common experience, which is that the air pressure equalizes when tanks are connected, regardless of temperature differences? The pressure equalization comes about when the air is sufficiently dense for the air molecules to affect each other a lot and we can no longer suppose the air molecules seldom hit each other.

Now, if we must include the effect of the molecules hitting each other, as well as hitting the walls, heat conduction between air molecules must be taken into account. All the molecules in the vicinity of the hole come to approximately the same temperature or speed. So the molecules near the hole in the hot tank are a bit cooler than those further in the hot tank, and the molecules near the hole in the cold tank are a bit hotter than the rest of the molecules further in the cold tank. When the cold, slow-moving molecules near the hole in the hot tank collide with the faster molecules further inside the tank, the slow ones get pushed back. Some are pushed all the way out of the hot tank and into the cold tank. That reduces the pressure in the hot tank and increases the pressure in the cold one. If the gas is dense enough this goes on until the pressures in the two tanks equalize.

HOT AIR

Warm air rises because

a) The individual hot air molecules are moving faster than the cool ones and so are able to shoot up higher.
b) The individual hot molecules find it more difficult to penetrate the dense air below than the less dense air above.
c) Individual hot molecules do not tend to rise; only large groups of hot molecules tend to rise as a group.

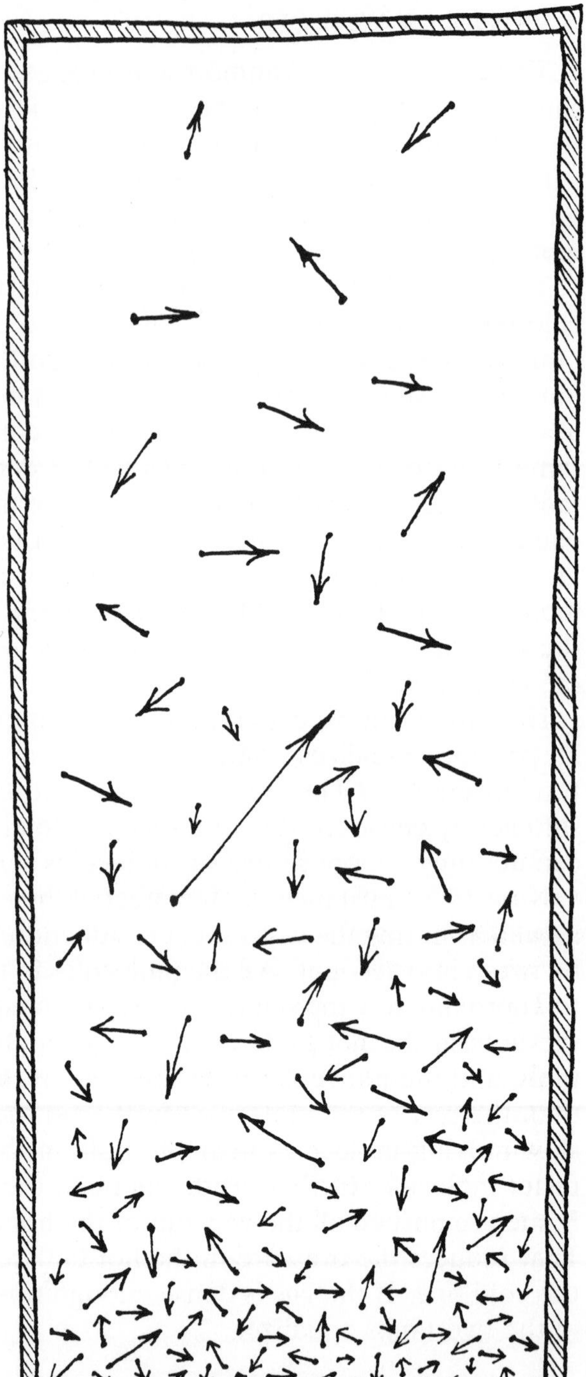

ANSWER: HOT AIR

The answer is: c. The mean free path of an individual air molecule is only about a 400,000th part of an inch before it hits another molecule—so it can't shoot far. Even if it could shoot up it would lose its speed as it went up. That is, going up would cool it.

Individual hot (fast) molecules may find it more difficult to penetrate the dense air below than the less dense air above but so also would the cool (slow) molecules find it more difficult to penetrate the dense than the less dense air.

Hot air rises because a given volume of hot air, at a given pressure, contains fewer molecules and therefore less weight than the same volume of cool air at the same pressure. That is, it is the density difference between hot and cold air that enables the hot air to float up like a bubble. It only makes sense to talk about density when you deal with large groups of molecules—not individual molecules.

Suppose you had a room in which the air temperature was the same from top to bottom. That means the average speed of the molecules at the top and bottom of the room must be the same, though some individual molecules will always be moving faster or slower. And suppose individual hot molecules did tend to rise—the hot ones of course are the fast ones. Then after a while the air on top would get hot and the air on the bottom would get cold. That is, the warm air would separate itself into hot air and cold air.

But that is contrary to real-life experience. In real life hot things cool and cold things warm. If you could even find something warm which would of its own accord separate into something hot and cold you could do miraculous things such as will be described in FREE-LOADER.

There is an important moral to all this. It is that trying to explain observable phenomena in terms of what individual molecules do is very difficult business. In the Kinetic Theory of Gas—the Molecular Theory of Gas—a little knowledge is dangerous. The molecular theory or atomic theory of matter has been around since ancient Greece and it is conceptually appealing. But it is so fraught with difficulty that it drove the man who finally established its validity, at the beginning of this century, Ludwig Boltzmann, to frustration and suicide.

DID YOU SEE THAT?

The luminous path left in the sky by a shooting star or meteor sometimes remains bright for several seconds, but the luminous streak of a lightning bolt vanishes in a fraction of a second. The reason for this has to do with the fact that

a) a meteor is more powerful than lightning
b) the meteor is hotter than the lightning
c) the lightning bolt is electric, while the meteor is not
d) the meteor is high in the atmosphere where the air pressure is low, while lightning is low in the atmosphere where air pressure is high
e) . . . the statement of fact is in error. A lightning bolt lasts several seconds, while the meteor's luminous trail disappears in a fraction of a second

ANSWER: DID YOU SEE THAT?

The answer is: d. A shooting star is usually a little cosmic grain of sand from space which runs into the earth's atmosphere. It goes through the air so fast that it knocks electrons off air atoms, which makes a plasma. Plasma is the name for gas or air atoms which have had electrons knocked off them. It is a mixture of the partially naked atoms and free electrons. The old name for plasma was ionized gas. Within a second or so the free electrons rejoin the atoms and give off the energy which was required to knock them free in the first place. The energy given off during the rejoining is the source of the light in the meteor's luminous trail.

A lightning bolt similarly makes a plasma, where electrons are knocked off the atoms by the electrons which are flowing in the electric current in the lightning.

Now the meteor is high in the atmosphere, perhaps twenty miles up. Up there the air pressure is low, which means the air atoms are far apart, so it takes a second or so for a free electron to find an atom it can recombine with and give off its energy. But the lightning bolt is low in the atmosphere, perhaps only a mile or two high. Close to the earth's surface the pressure is high which means the air atoms are close together, so it takes a free electron only a fraction of a second to locate an atom it can recombine with. So the plasma turns back to regular air in a fraction of a second. Turning air to plasma soaks up energy from the lightning. Letting the plasma turn back into air gives off energy as light, heat, and sound.

There is more energy in most lightning bolts than in most meteors and the energy in a lightning bolt is given off much more rapidly, so there is more power in the lightning discharge than in the meteor. Also, the lightning bolt has a blue tint, while the meteor trail has a yellow tint — which means the lightning plasma is hottest. The lightning plasma is created by electric energy and the meteor trail plasma is created by the kinetic energy of the meteoroid. But regardless of how the plasma is created, the amount of time it takes to turn back into regular air depends only upon how long it takes the free electrons to locate atoms to recombine with.

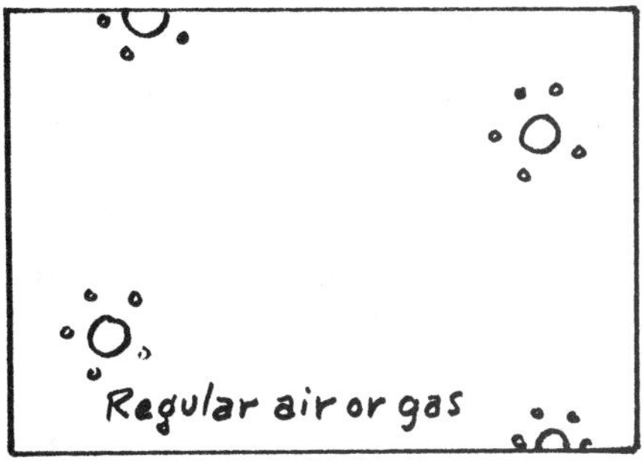

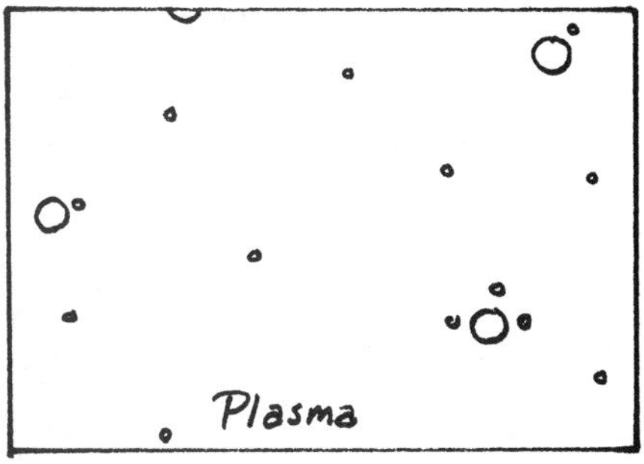

HOT AND STICKY

The weather in New Orleans and along the whole Gulf Coast is quite hot and humid in the summertime. In such climates the most comfortable time of day is

a) just after sunset when the temperature is dropping slightly
b) just after sunrise when the temperature is rising
c) at no particular time on the average

ANSWER: HOT AND STICKY

The answer is: b. The main cause of discomfort in tropical climates is humidity. When sweat evaporates from your skin it carries away heat and so cools you. But if the air is very humid it is already quite full of water vapor and is not able to take any more. So the sweat just lays on you. The number of grams of water a cubic meter of air can hold depends on its temperature. Hot air holds more water. At sundown the air cools and its ability to hold water is reduced, so this is not the time it can help soak up your body sweat. As the air cools further water vapor condenses as night dew forms over the surroundings.

At sunup the air warms and is able to hold more water again. Water evaporates back into the air. The air soaks up the morning dew and your sweat too, making you feel dry and cool — but not for long. Soon it is hot and muggy again.

Why does evaporation cool? Because the molecules in a liquid have a variety of speeds. All the molecules in 20° water, for example, are not at 20°. Some are at 30°, some at 20°, and some at 10°. Twenty is just the average. Which molecules evaporate first? The fast-moving or hot ones, like the 30° ones, which lowers the "class average." That leaves the 20° and 10° ones and so the average drops to something like 15° (depending on the relative numbers of different temperature molecules involved). So when the faster moving or hotter molecules leave, the average temperature of those left decreases.

Incidentally, the main reason for hair on your head is to allow your sweat to stick to your head. If the sweat just ran off your head it would not cool it — the sweat must evaporate from your head if cooling is to take place. In your head is your computer and you must keep the computer from overheating. Ever notice how your ability to think and concentrate is cut by a fever? Other parts of your body covered by hair are so covered because of special cooling requirements. The hair is like a wick or cooling rag. It holds the sweat in place until it evaporates.

SO WHEN ITS UNCOMFORTABLY HOT, DOES IT HELP OR HINDER TO WIPE THE SWEAT FROM YOUR BROW ?

CELSIUS*

Water boils at 100°C and freezes at 0°C at sea-level atmospheric pressure. At higher pressure, water will boil at a

a) lower temperature and ice melt at a lower temperature
b) lower temperature and ice melt at a higher temperature
c) higher temperature and ice melt at a higher temperature
d) higher temperature and ice melt at a lower temperature

*Some people love to change names — so they changed the meaningful name CENTIGRADE (a scale graded into 100 parts) to CELSIUS, in honor of the man who first devised the centigrade scale back in 1742.

ANSWER: CELSIUS

The answer is: d. Go to a high altitude and water boils at a lesser temperature (like 90°C at 10,000 feet) and all evaporates without ever reaching 100°C. That's why it's difficult to cook an egg by boiling in the mountains — the water just does not get hot enough. If the pressure gets low enough water will boil at room temperature. This is easy to demonstrate by putting a dish of water in a vacuum tank and pumping out the air. On the other hand, if the pressure is high enough water will not boil even if its temperature is way above 100°C. This is what happens in a steam boiler where the superheated water doesn't boil because of the high pressure. Also, water in a pressure cooker or water at the bottom of a geyser may be above 100°C and not be boiling.

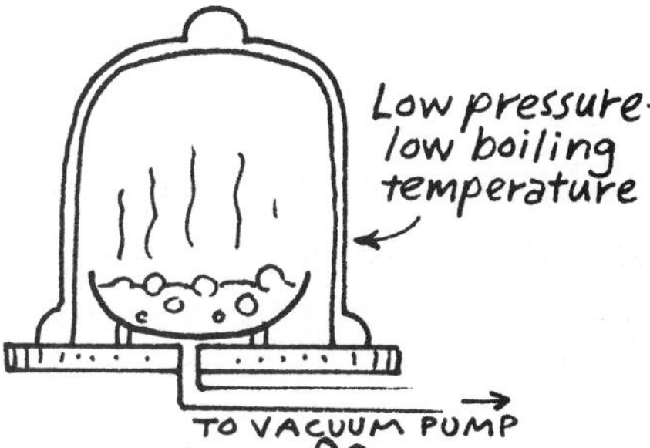

Ice can be made to melt even if its temperature is below 0°C if it is put under pressure. How? By putting a heavy object like a rock on the ice.

Why does ice melt more easily with high pressure while water boils more easily with low pressure? One explanation is simply this: the volume of ice decreases when it melts into water and pressure helps the compression, while the volume of water increases when it boils into steam and pressure hinders the expansion.

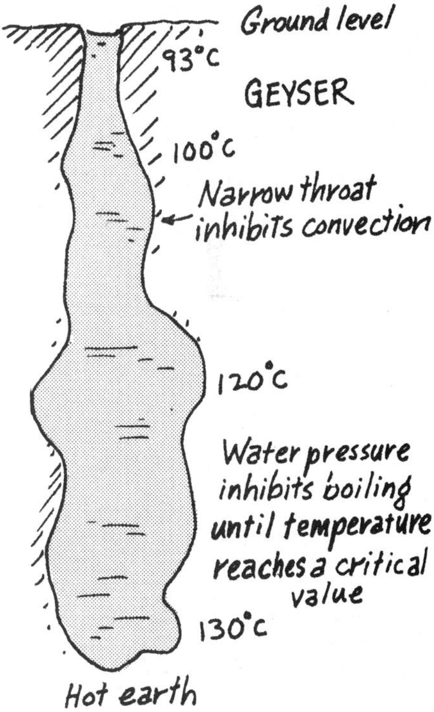

NEW WORLD, NEW ZERO

Suppose you wake up in a "new world." In the new world you make some measurements of the pressure in a tank of gas at various temperatures. A graph of your data looks like this:

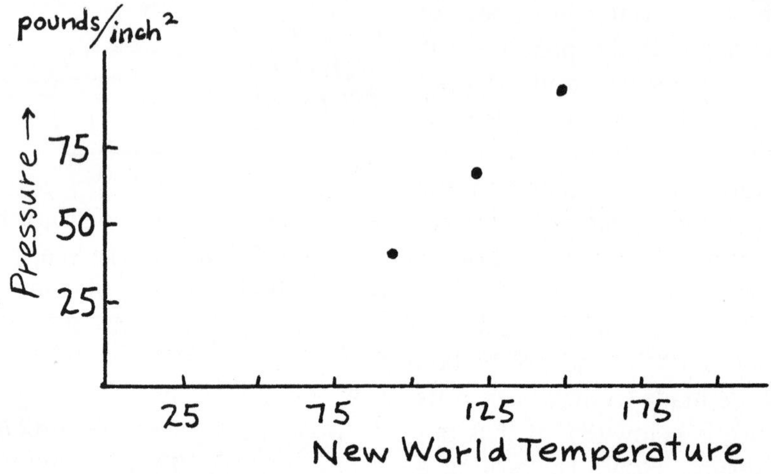

Approximately what temperature is Absolute Zero in the new world?

a) Zero New World Degrees
b) 25 NWD
c) 50 NWD
d) 75 NWD
e) 100 NWD

ANSWER: NEW WORLD, NEW ZERO

The answer is: c. If an overweight person weighs 150 lbs and loses one pound per week every week what will be the person's weight in 150 weeks? Well almost the same situation exists with the change in gas pressure when temperature is changed, and that led people to the idea of Absolute Zero. For each degree loss in temperature a gas confined in a tank loses a certain amount of pressure. If that goes on very long the gas must lose all its pressure. The temperature at which a gas loses all its pressure is called Absolute Zero. To find the zero pressure temperature just draw a line through the points and see where it hits the zero pressure line. It hits halfway between 25 and 75 New World Degrees and that makes Absolute Zero 50 NWD. In our (old) world Absolute Zero is minus 273 Celsius degrees. Now before the 150 weeks are up something is likely to happen to the person who loses a pound per week and similarly something is likely to happen to the gas before it gets to Absolute Zero. It might liquify or even freeze. But the important idea is that at room temperature all gases, oxygen, hydrogen, nitrogen, etc. behave as if their pressure would vanish at minus 273°C.

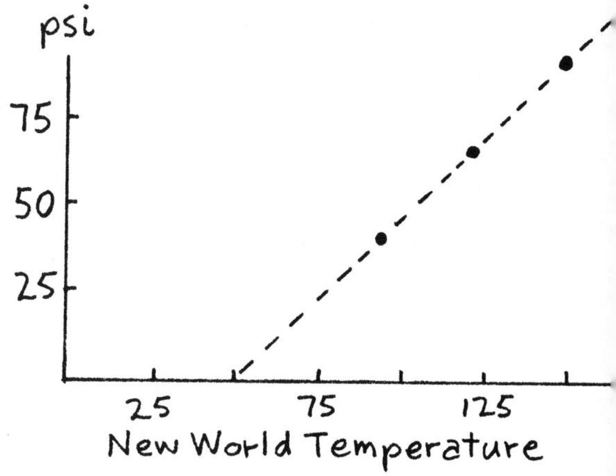

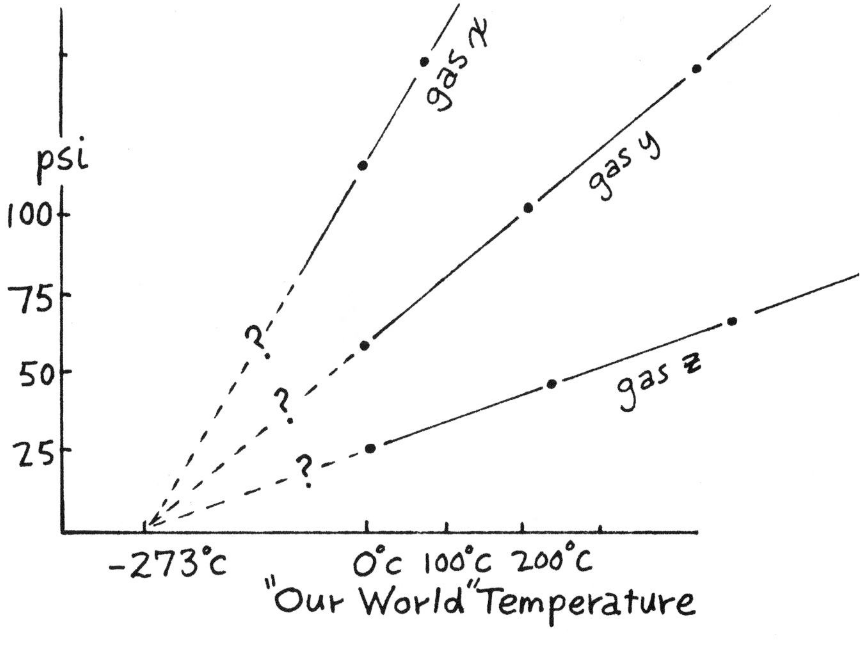

EQUAL CAVITIES

A block of metal with a white surface and a block of metal with a black surface of the same size are each heated to 500°C. Which radiates the most energy?

a) The white block
b) The black block
c) Both radiate the same

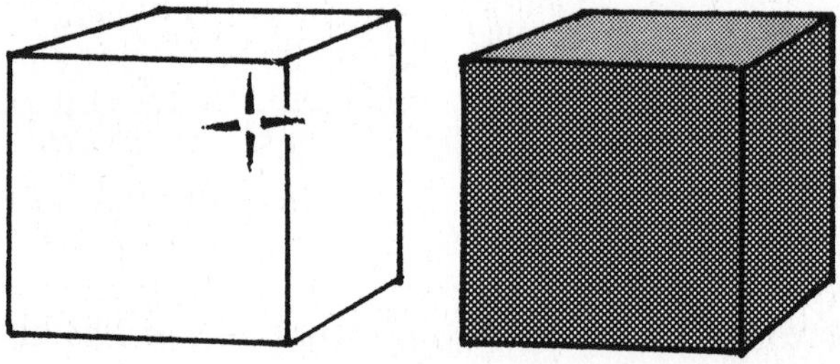

Now consider a cavity that is cut in each block of metal as shown. Again both are heated to 500°C. From which block does the most energy come *out of the cavity?* The most energy is radiated from the cavity in the block of

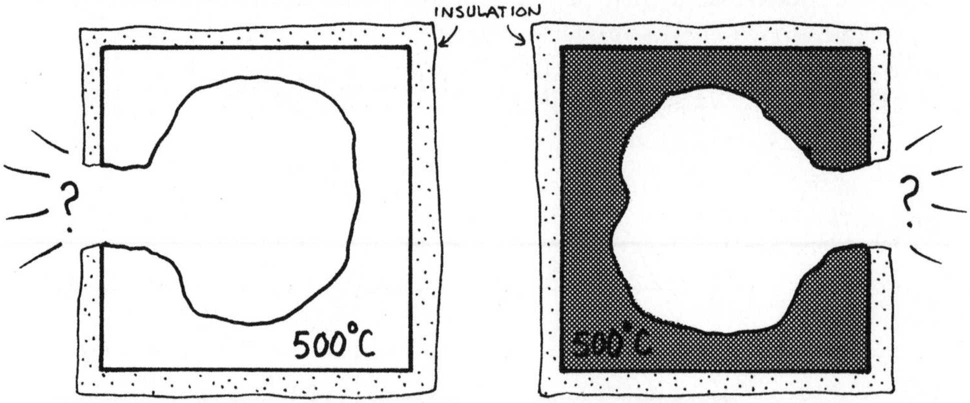

a) white metal
b) black metal
c) . . . both are the same

ANSWER: EQUAL CAVITIES

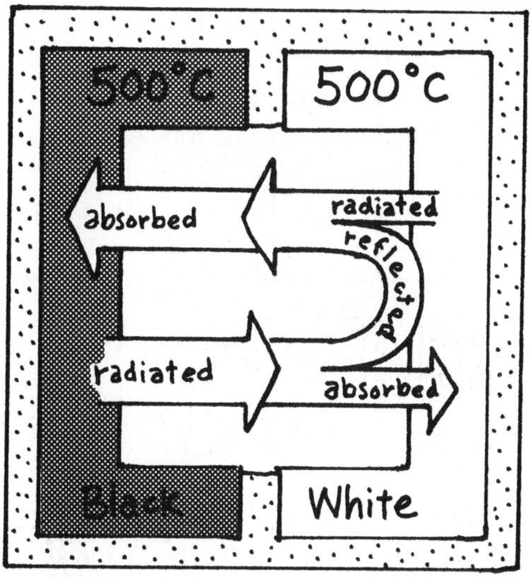

The answer to the first question is: b. Suppose you have an insulated sealed box heated to 500°C. Half the box is lined with a black-surfaced metal, the other half with white-surfaced metal. The two metals have no contact. They can exchange heat only by radiation. Some heat radiation flows from the black towards the white and some from the white towards the black. The two flows must be equal, for if they were not the side sending out more heat would soon be cooler than the other side. And a net flow of energy by its own accord would be going from a cool to a hot place — which is impossible. If the surface is perfectly black, all the heat radiation that hits it is absorbed, so if its temperature stays constant, an equal amount of heat must be radiated by it — so the surface absorbs heat at the same rate as it emits it. We see that a good absorber must also be a good radiator. Now the white surface reflects a great deal of the radiation incident upon it and absorbs very little — in turn it also radiates very little. A good reflection is a poor radiator. The flows between the white and black surfaces are equal because reflection by the white surface compensates for its lesser radiation.

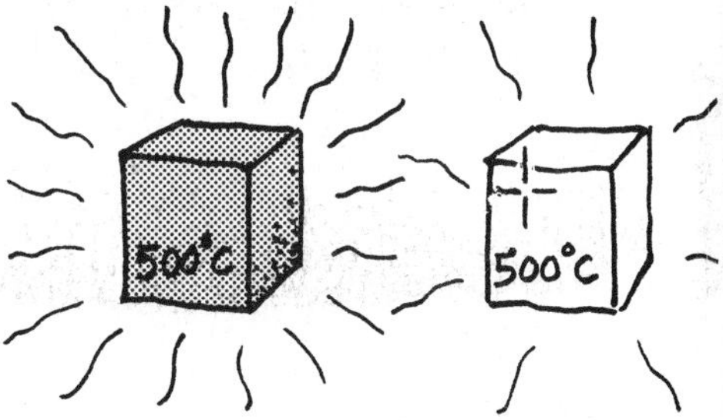

So we conclude black metal at 500°C radiates more heat than white metal at the same temperature. That is why good radiators are painted black.

Now if we bugger up the otherwise white surface, it won't be such a good reflector. It must then absorb more radiation. If we mess it up enough it will be as poor a reflector as the black surface and that means it must then absorb as much radiation as the black surface. So it acts like the black surface — which means it must radiate like the black surface. What have we done to change the white surface? We have put a lot of scratches and pits on it. As the scratches and pits get deeper they act as little cavities which trap the radiation which goes into them. Much of the radiation that goes into the cavities can't be reflected and so is absorbed. Cavities act as radiation traps. Cavities in silver, gold, copper, iron or carbon are in effect all black. Picture a house on a sunlit day with an open window. The open window is a cavity. It does not matter what color the inside room is painted (silver, gold, etc.). From the outside the room looks black.

So the answer to the second question is: c. If cavities in silver, gold, etc., are all equally good absorbers then they must be equally good radiators. If brought "hole-to-hole," two cavities at the same temperature must remain

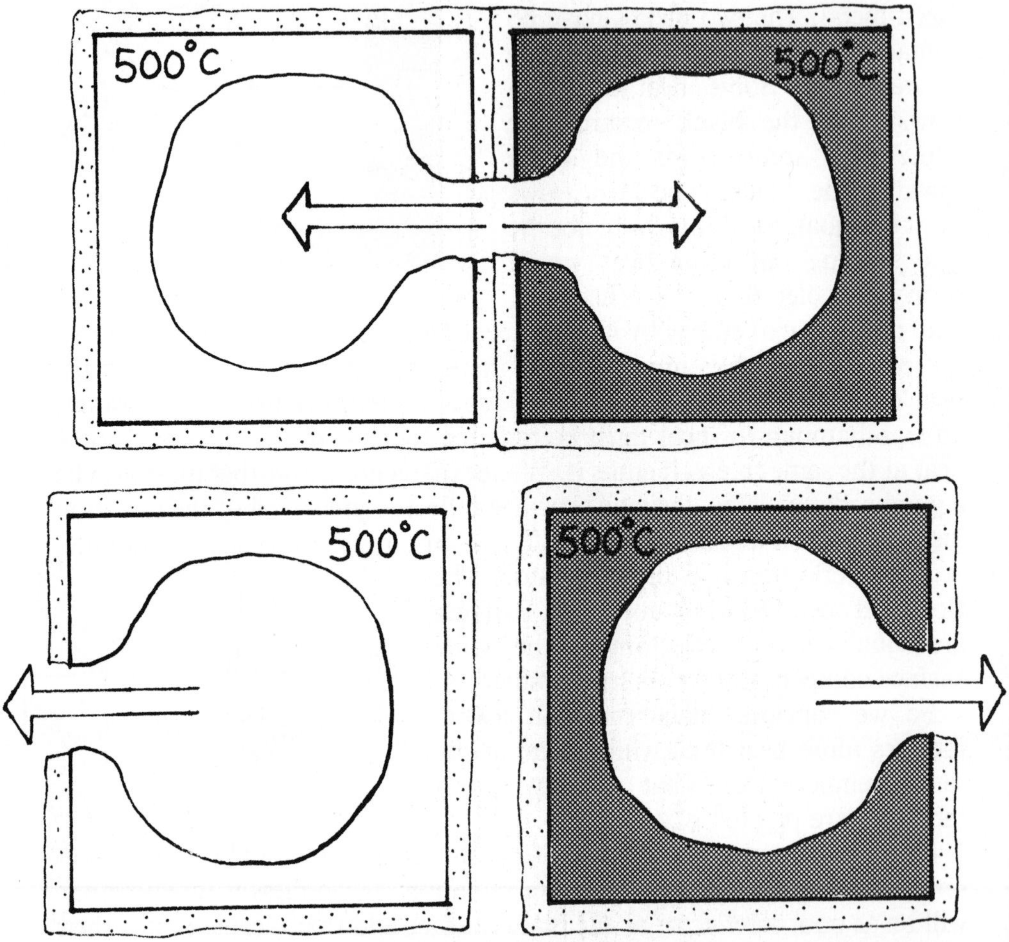

at the same temperature. This means that all the radiation energy absorbed when entering must be re-radiated. (Don't confuse radiation with reflection!) If the radiation flow from the cavity in the white metal is to equal the flow from the cavity in the black metal, then each cavity must send out the same amount of heat radiation.

HOUSE PAINT

The best color to paint your house is a

a) dark color like brown

b) light color like white.

c) The color you paint your house is purely a question of artistic taste.

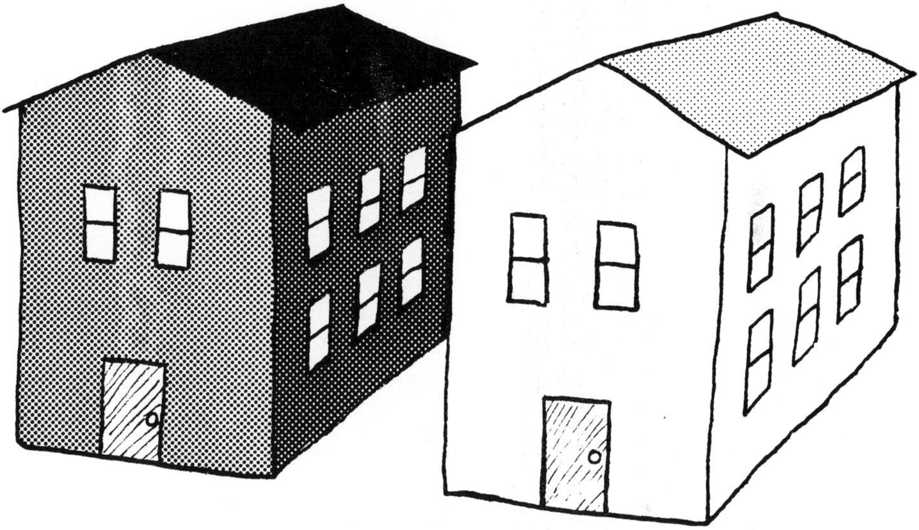

ANSWER: HOUSE PAINT

The answer is: b. There are many reasons: 1) White reflects best and keeps your house cool during the day. 2) White radiates least and so keeps your house warm at night. 3) White lasts longest because it reflects rather than absorbs light and absorbed light destroys paint—and the wood under it. Compare the paint on the north and south sides of your house if you doubt this. 4) If there is only a small space between your house and the neighbors' house so little light gets into your side windows, then painting your house and the neighbors' house white will reflect more light into the space between houses and into your windows and you can save on electric lights. Painting the inside of your house white also saves electric lighting. Don't forget to paint your roof white, or aluminum. Now you know why the astronauts wore white when they walked on the moon.

HEAT TELESCOPE

Make a heat telescope by putting the bulb of a thermometer in a paper coffee cup lined with aluminum foil. On a cool, dry, clear night point the telescope at the sky and after a few minutes read the thermometer. Then point it at the earth for a few minutes and read the thermometer again. Your results indicate

a) the sky is hotter than the earth
b) the earth is hotter than the sky
c) both the sky and the earth have the same temperature

ANSWER: HEAT TELESCOPE

The answer is: b. At night the earth radiates back to space the heat it picked up during the day (if it didn't, every day would be hotter than the day before!). When the heat telescope points down it catches some of that radiation (infrared radiation) on its way from the earth back to space. But the telescope doesn't catch this radiation when it points up.

Incidentally, the temperature of outer space has been measured to be 4 degrees above absolute zero. It takes more than a thermometer and a coffee cup to measure this low-temperature radiation, and the first experimenters to do it earned a Nobel Prize for their efforts.

SMEARED-OUT SUN

Turn on your imagination and suppose somehow the glowing disc we call the sun was smeared out into a larger and larger disc. As it gets larger suppose the intensity of each little part gets less so that the total energy we get from the whole disc stays the same as its size grows. Now suppose the disc were smeared out all over the sky so there would be no distinction between night and day. We would still receive the same total amount of energy that we received before the smear out, even though it would be received uniformly over all 24 hours at any given location rather than just during the daylight hours. If this were to happen, then the earth's average temperature would

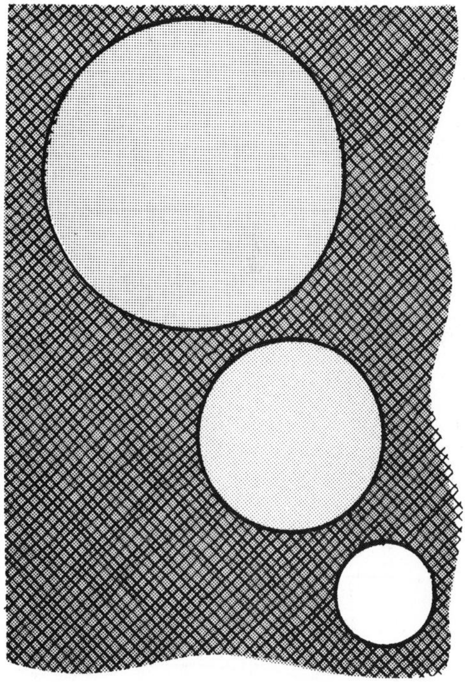

a) increase b) decrease
c) remain the same

ANSWER: SMEARED OUT SUN

The answer is: c. The earth would receive the same total amount of energy after the smear out, and the temperature of the earth must be just high enough so that as much energy is radiated back into space as is picked up from the sun. So the earth's temperature would not change even if the sun were smeared out. Incidentally, even though the earth's temperature would not change, the sun's temperature would. Spreading or diluting the same energy over a larger area results in a lower temperature, so the temperature of the disk would go down as its size goes up. As its temperature decreases the sun would turn from a yellow hot to orange hot to red hot.

IF THE SUN WERE SMEARED OUT

If the sun's disc were smeared out all over the whole sky the earth's atmosphere would

a) circulate faster than it does now so there would be more wind, rain and thunder.
b) circulate slower than now.
c) not circulate at all.

ANSWER: IF THE SUN WERE SMEARED OUT

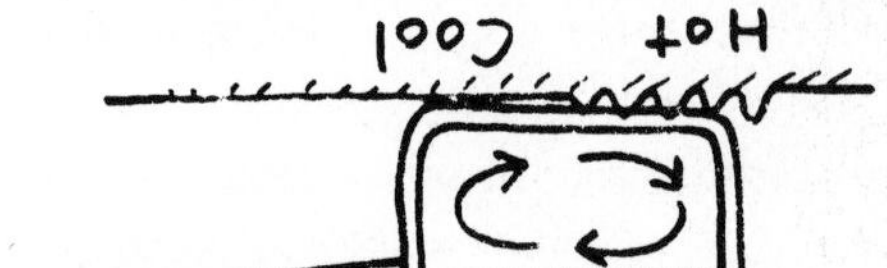

The answer is: c. The earth's atmosphere circulates because the earth is unevenly heated. For example the equator is hot and the poles cool. If one side of a cooking pot is hot and one side cool, soup in the pot will circulate for the same reason. But if the sun were smeared over the whole sky, the equator would be no hotter than the poles. If there were no temperature difference there would be no circulation, no wind, no rain, no thunder. In fact, there would be no heat-driven organized motion at all on earth. No life! No life, even though the earth would be getting as much energy as it ever did! So sunlight energy by itself is not enough — it takes more than sunlight energy to make things go on earth. The cool night sky is no less important than the sun itself in making life go. Not because it prevents the earth from overheating but because it fulfills the vital requirement of converting heat to organized motion — a *temperature difference.* There must be a temperature difference if heat energy is to do work. Heat can only make things go when it is on the way from a warm place to a cool place. Some people say energy is the ability to do work. But that is not always so. If all the world and the surrounding sky were at one temperature, no matter how high or low it was, none of the heat energy in the world, no matter how much there was, could be converted to work.

FREE LOADER

Consider the following idea: A ship heats its boilers and propels itself without the use of coal or oil in the following way. It pumps in warm sea water; extracts heat from that sea water; concentrates the extracted heat in its boilers; and discharges the cooled seawater back into the ocean. The discharged water may be ice if enough heat has been taken from it. Now ask yourself two questions. First Question: Does this idea violate conservation of energy?

a) yes b) no

Second Question: Could this idea be made to work?

a) yes b) no

ANSWER: FREE LOADER

The answer to the first question is: b. It does not violate conservation of energy because the heat in the boiler is that which is supposed to have come from the warm sea water. No energy is created — it is simply transferred from one place (the water) to another (the boiler).

The answer to the second question is also: b. If it could be done, it would be done — we simply find that this sort of thing can't occur in our world. Our collective life experience, after all, is our guide to the laws of physics. We call the nonoccurrence of this process the Second Law of Thermodynamics. Heat tends always to flow from a hot place to a cooler place. Heat by itself will not flow from the warm sea water into the much hotter boiler. That would be like a ball freely rolling uphill. Heat could be forced to go from the cooler place to the hotter place — that's what happens in a refrigerator — but it takes energy to force it from the cool to the hot place and the energy required to force it would more than equal the energy which could be obtained from the boiler.

The ship's world is the surface of the sea. If all the world is at one temperature, no matter how high, none of the heat in the world, no matter how much there is, could be converted to work (remember IF THE SUN WERE SMEARED OUT).

GULF OF MEXICO

Another FREE LOADER idea is to generate power as follows. Water on top of the Gulf of Mexico is quite warm, but deep down the water is cold. The plan is to heat some gas with warm water from the top so it will expand, and then cool the gas with water from the bottom so it will contract. The gas is alternately expanded and contracted so it drives a piston back and forth. The moving piston is attached by conventional means to an electric generator to make electricity.

a) This idea might be made to work.
b) This idea could never work.

ANSWER: GULF OF MEXICO

The answer is: a. Offhand this sounds like the FREE LOADER idea, but there is one vital difference. It does not just take heat from the warm top of the sea. It also provides a cool place to dump the heat. This idea does not force heat to go from a warm place to a hot place, rather it allows heat to go from a warm place to a cool place. It is the temperature difference between the top & bottom of the sea that makes the conversion of heat energy into work possible.

There is presently serious interest in this concept. A commercial demonstration of the Ocean Thermal Energy Conversion (OTEC) concept is scheduled for 1984. One such proposed power unit is shown to the right.

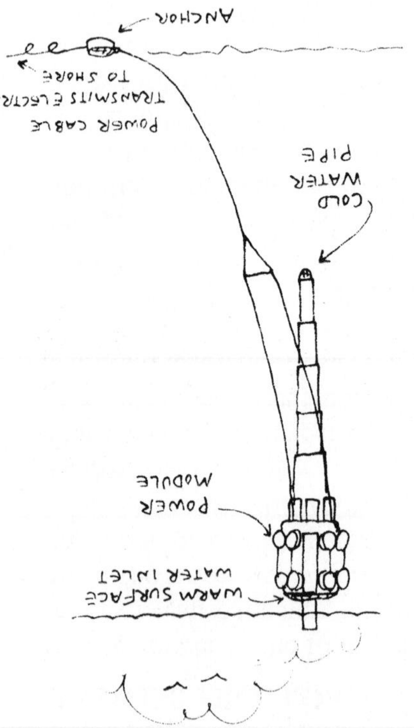

QUARTZ HEATER

Some types of electric heaters like "Quartz Heaters" are more efficient than old-fashioned electric heaters.

a) True
b) False

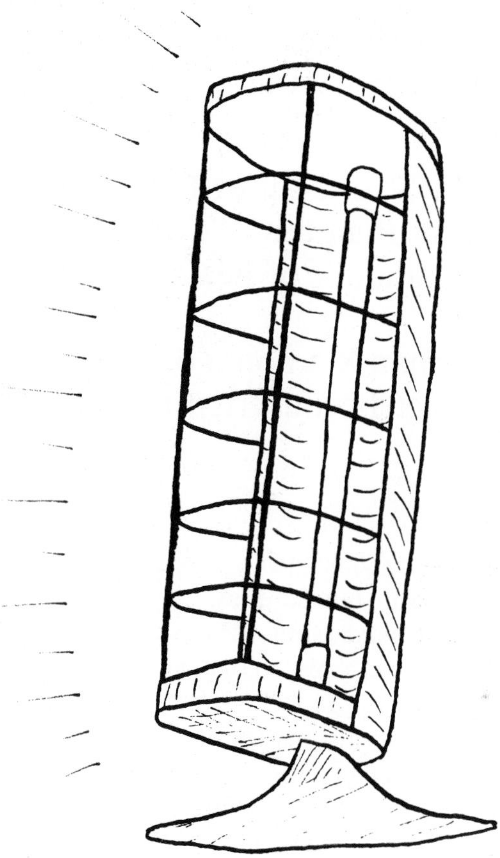

ANSWER : QUARTZ HEATER

The answer is: b. Due to advertising, a lot of people think quartz heaters are more efficient, but it is not so. If you pay the electric power company for a kilowatt-hour of energy (and your meter is working right) you get a kilowatt-hour of energy, and if you feed the energy to a heater it turns it all to heat. If some of the energy were somehow wasted, ask yourself where it could go. A motor wastes energy as heat, but a heater can't make anything but heat.

Now a fan might spread the heat over a room more evenly or a reflector might focus the heat, but they have been used on electric heaters for decades. So why are quartz heaters such a fad? The power of advertising.

ALL ELECTRIC HOME

If a certain amount of fuel (oil, gas or coal) is burned in your house stove it would produce X amount of heat. Now, if that same amount of fuel is burned in an electric generating plant, and all the electricity so generated is used to heat your house by means of an electric stove, the electric stove would

a) produce more than X amount of heat, because electricity is more efficient than gas;
b) produce exactly X amount of heat, because of conservation of energy;
c) produce much less than X amount of heat, because heat can never be completely converted to electricity.

ANSWER: ALL ELECTRIC HOME

The answer is: c. Near most electric power generating plants you see cooling towers or warm water being discharged into a river, lake or bay. The reason for this is that heat energy cannot be completely converted to electric energy. Some of the heat energy must go to waste. (At hydroelectric power generating plants, however, there is negligible waste because except for small friction losses the mechanical energy of falling water is completely converted into electric energy.) Why cannot the heat that is discharged in the towers or rivers be recycled and put back into the power plant's boiler? Because heat, by itself, will not flow from a cool place to a hot place — and the boiler is always much hotter than the waste heat. Well, why not use a heat pump to force the waste heat back into the boiler? Because the pump requires energy to operate. How much energy? At least as much energy as the power generating plant produced while making the waste heat! So, there would be absolutely no electric power left to sell.

Well, why does there have to be waste heat in the first place? Because in a steam engine or turbine the gas must expand as it pushes on the engine's piston or turbine's blades. As it expands it cools. If it could be expanded until its temperature dropped to absolute zero all the heat energy could be converted to work. But in reality, it can get no cooler than the rest of the outside world, which is about 300 degrees above absolute zero. So, you cannot get all of the energy out of the heat.

Well, how about this idea? You can expand the steam until it turns to water and then just put the hot water back into the boiler. How can there be any waste doing that? You may think there is no waste because you think you have a closed cycle, but you don't. First, some energy comes out as the steam does work pushing on the piston as it expands, but that is just what you want, so that is okay. Now comes the waste. The steam expands until its temperature drops to 100°C — then the pressure inside the engine is the same as the external atmospheric pressure. It can expand no more, but it is not water yet. It is still 100°C steam and you cannot simply put a large volume of low pressure steam back into a high pressure boiler. You must first reduce the volume of the steam by turning it into water. But to condense the 100°C steam into 100°C water you must remove its latent heat of condensation. As steam turns to water its temperature does not change, but heat — a lot of heat — must come out of it. That heat cannot be returned to the boiler because the temperature of that heat is only 100°C and the temperature of the boiler is much higher. The latent heat of condensation becomes waste heat. Too bad. Why must the temperature of the boiler be above 100°C? Because the pressure of 100°C steam does not exceed atmospheric pressure.

When you pay for all electric heating you pay both to heat your house and the rivers, sea and sky.

SOMETHING FOR NOTHING

If you put ten joules of electric energy into an electric heater, out will come ten joules of heat. Is there any way in actual practice that you can get *more* than ten joules of heat from a device if your electric input is only ten joules?

a) Yes, if you are clever enough you can get more than ten joules of heat output from only ten joules of electric energy.
b) No way! You cannot get more than ten joules of heat from ten joules of electric energy.

ANSWER: SOMETHING FOR NOTHING

Believe it or not, the answer is: a. Think about an air conditioner in a window. Outside it is hot; inside, cool. Electricity goes into the air conditioner and the air conditioner draws heat out of the house and dumps it outside. How much heat is dumped outside? If the machine draws in 9 joules of heat and takes 10 joules of electricity to work (a very poor air conditioner) it must expel 19 joules of heat outside. Come winter, the outside is cool and you want the inside to be warm. So put the air conditioner back in the window, but turn it around so the former outside part is inside. In go the 10 joules of electricity to work the machine. The machine draws in 9 joules of heat from the cool side, which is outside in winter, and it must expel 19 joules of heat on the warm side, which is now inside. A backwards air conditioner is called a heat pump.

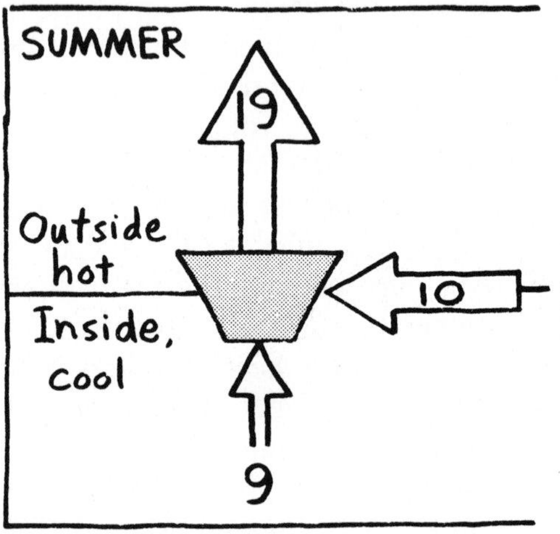

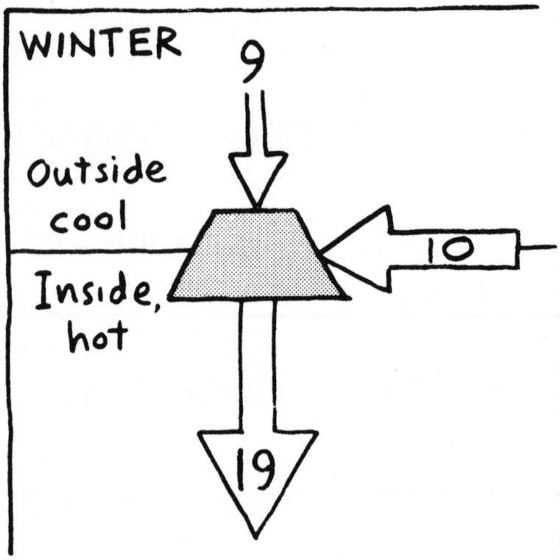

Did the heat pump really get something for nothing? In a way, yes. In a way, no. You see heat can be made, as in a toaster, or moved as in an air conditioner. The heat pump moves heat. Heat by itself moves from hot to cold places. But with a pump (and energy to run the pump), it can be moved from cold places to hot places.

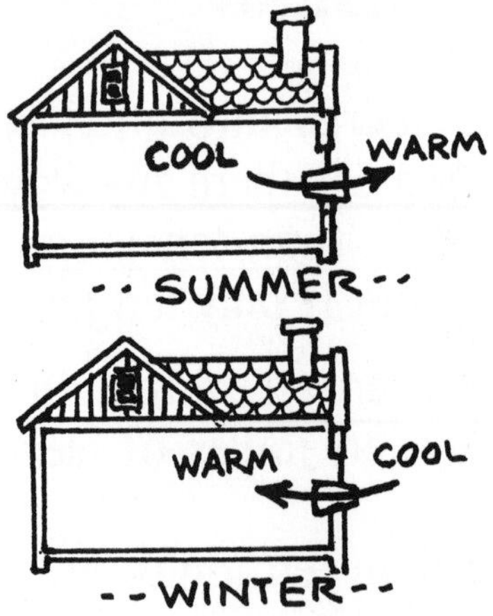

IS THERE A COOL CORNER ANYPLACE?

Some people have reason to think the temperature of the whole universe is about 4 K* due to heat released from the cosmic fireball at the creation of the universe (the "Big Bang"). If that is so, is it possible under any circumstances for a small part of the universe to be cooler than 4 K?

a) Yes, some part might be made cooler.
b) There is no way any part could be made cooler.

ANSWER: IS THERE A COOL CORNER ANYPLACE?

The answer is: a. The whole state of Louisiana might be 95°F. in July, but with a good air conditioner, you can bring a room down to about 65°F. So some laboratory can also have its temperature taken below 4 K. To do this takes energy, which must ultimately come from the stars like our sun. So ironically the reason that we can make some parts of the universe cooler than 4 K is because some parts, like the stars, are much hotter than 4 K.

HEAT DEATH

The Heat Death of the universe refers to a time in the distant future when the whole universe

a) runs out of energy.
b) overheats.
c) freezes.
d) . . . none of the above.

*K (degrees Kelvin); 4K = −269°C.

ANSWER: HEAT DEATH

The answer is: d. The *Heat Death* refers to a time when the whole universe comes to the same temperature. It might be hot or it might be cold or it might be just right. It doesn't matter what it is just as long as it is the same all over. The universe is now separated into hot and cold places. The stars are hot; the space between the stars is cold. But each day the stars get a little cooler as they radiate their energy — and each day the place the energy goes to gets a little warmer. Sooner or later the temperature difference must vanish. As far as energy is concerned the situation described in SMEARED OUT SUN will then become a reality.

Energy is usually described as "the ability to do work," but that is not a good definition. After the "Heat Death" the universe still has all its energy. But the energy, which is all at the same temperature, has no ability to do work. It has potential to do work only *if* a cooler place could be found.

Incidentally, if you were somehow able to witness the universe after the Heat Death, you would not SEE anything. There would be no contrasts between things, just as the coals and inner walls cannot be discerned inside a blast furnace of uniform temperature.

MORE QUESTIONS (WITHOUT EXPLANATIONS)

You're on your own with these questions which parallel those on the preceding pages. Think Physics!

1. Suppose you have a quart of ice-cold water and wish it to be as cold as practical 15 minutes later, and you have to mix a couple of ounces of boiling water in it either right away or in about 14 minutes. It would be best to mix it
   a) right away   b) later   c) . . . either way as it would not make a difference

2. If glass expanded more than mercury upon being heated, then the mercury in a common mercury thermometer would rise when the temperature
   a) increased   b) decreased   c) either increased or decreased

3. If the temperature of a volume of air is decreased, then its volume must be
   a) increasing   b) decreasing   c) . . . can't tell

4. When a metal plate with a hole in it is cooled, the diameter of the hole
   a) increases   b) decreases   c) remains unchanged

5. Provided the heat input was the same, water will come to a boil faster
   a) in the mountains   b) at sea level
   And cooking food in the boiling water will be fastest
   a) in the mountains   b) at sea level

6. A can of air is sealed at atmospheric pressure and room temperature, 20° C. To double the pressure in the can it must be heated to
   a) 40° C   b) 273° C   c) 313° C   d) 546° C   e) 586° C

7. If the 20° C-air in the can is to be twice as hot, its temperature would be
   a) 40° C   b) 273° C   c) 313° C   d) 546° C   e) 586° C

8. A pot of clean snow and a pot of dirty snow are placed in the sunlight. The snow to melt first will be the
   a) clean snow   b) dirty snow   c) both the same

9. A pot of clean snow and a pot of dirty snow are placed on a hot stove. The snow to melt first will be the
   a) clean snow   b) dirty snow   c) both the same

10. When heat is added to a substance its temperature
    a) will increase   b) will decrease   c) may stay the same

11. The fluid inside the cooling coils inside the freezer compartment of your home refrigerator is near its
    a) boiling point   b) freezing point

12. Inside a warm damp cave completely sealed off from the outside world
    a) some life forms could flourish indefinitely
    b) no life forms could flourish indefinitely

13. Sunlight energy can be converted at 100% efficiency to
    a) chemical energy in plants
    b) heat energy in human-made devices
    c) both of these   d) none of these

14. A certain amount of fuel burned in your house stove produces X amount of heat. If that same amount of fuel is burned in an electric generating plant and all the electricity so generated is used to heat your house by means of an electric *heat pump*, the heat pump would produce
    a) less than X amount of heat   b) X amount of heat   c) more than X amount of heat

15. In any gas there are always places where more molecules of one kind or another spontaneously and momentarily gang up, resulting in the development of hot and cold or high and low pressure spots. So the Heat Death of the universe can never be complete.
    a) True   b) False

What distinguishes man made law from nature made law? Exceptions.
— Debra Lynn Bridges, San Francisco attorney

Vibrations

A wiggle in space is a wave; a wiggle in time is a vibration. Wiggles are difficult to put your finger on, difficult to capture. Wiggles are elusive because to exist they must extend over space or over time. A wave cannot exist in one place—it must extend from place to place. And a vibration cannot exist in one instant—it needs time to move to and fro.

Besides extending through space and/or time, waves and vibrations are peculiar in yet another way. Unlike a rock, which will not share its space with any other rock, more than one wave or vibration can exist at the same time in the same space, like the voices of people singing at the same time in the same room. These vibrations and those of a whole symphony orchestra can be captured on the single wavy groove of a phonograph record, and amazingly our ears discern the component vibrations as we delight in the intricate interplay of the various sources. We enjoy the vibes.

GETTING IT UP

Meeky Mouse wants to get the ball bearing up and out of the bowl, but the ball is too heavy and the sides of the bowl too steep for Meeky Mouse to support the ball's weight. Using only its own strength without the help of levers and such, Meeky

a) can't get the ball bearing up and out
b) can get the ball bearing up and out (But how?)

ANSWER: GETTING IT UP

The answer is: b. How? By rolling the ball back and forth. Each time the ball goes back and forth, Meeky Mouse can give a little push and so add a little energy. Eventually, there will be enough energy added together to get the ball up to the rim and over the top.

The trick depends on pushing at the right times and in the right direction. The mouse must match the rhythm of its pushes to the natural rhythm with which the ball rolls back and forth. In the jargon of physics, the natural rhythm is called the resonant frequency of vibration.*

Many things besides the ball in the bowl have resonant frequencies: Swings, electric buzzers, horns and bells, even water in a bathtub or ocean swishes back and forth at some resonant frequency.

There is usually more than one way a thing can vibrate, or resonate. For example: The ball in the bowl can resonate back and forth or it can resonate around a circle. These different ways are called the "resonant modes." Sometimes the different modes have different resonant frequencies. For example: There are several ways an antenna on a car can whip back and forth. Likewise, there are many ways a string on a musical instrument can resonate, each way with its own different resonant frequency.

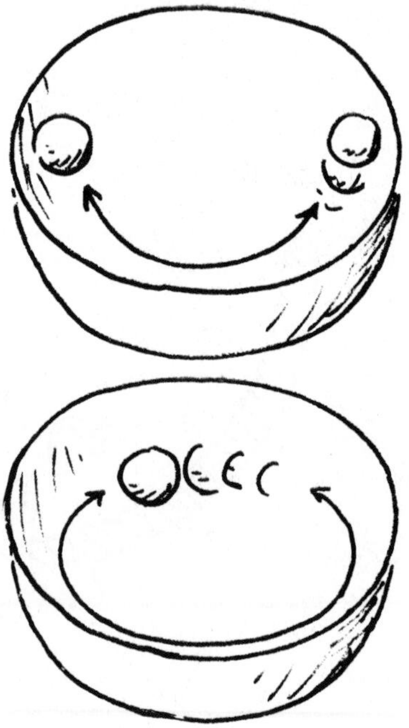

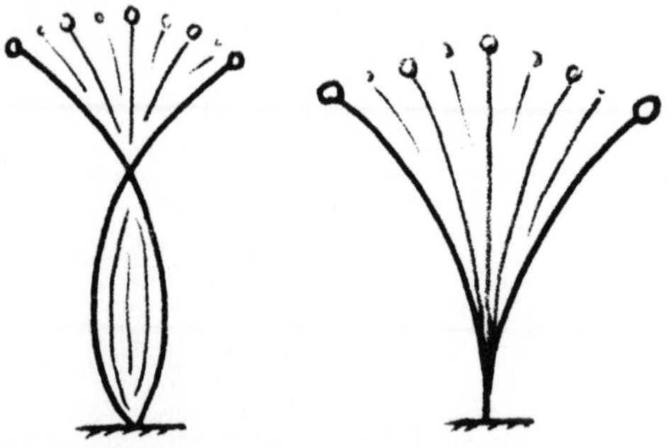

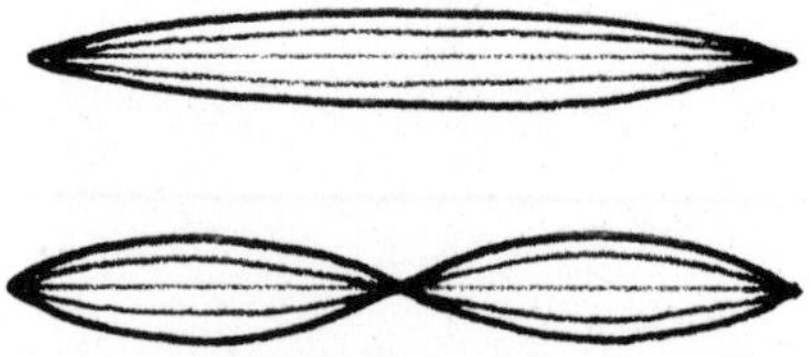

*Have you ever noticed that if you don't really understand something, but you know the "right words," people who also do not understand will often think that you do?

MIXED UP

Before the advent of the radio and telephone, long-distance messages were sent from one place to another by means of telegraph. A serious limitation of the telegraph was that only one message at a time could be successfully sent over a single telegraph wire.

a) Yes, this is correct.
b) Not so!

ANSWER: MIXED UP

The answer is: b. This is how multiple messages were sent over a century ago! An electric bell or buzzer can be tuned to ring at a high or low frequency by adjusting the spring on it. The tighter the spring, the higher the frequency. At the sending station a pair of bells A_1 and B_1 are tuned to different frequencies. At the receiving station are two more bells. One, A_2, is tuned to the same frequency as A_1 and the other, B_2, is tuned to the same frequency as B_1.

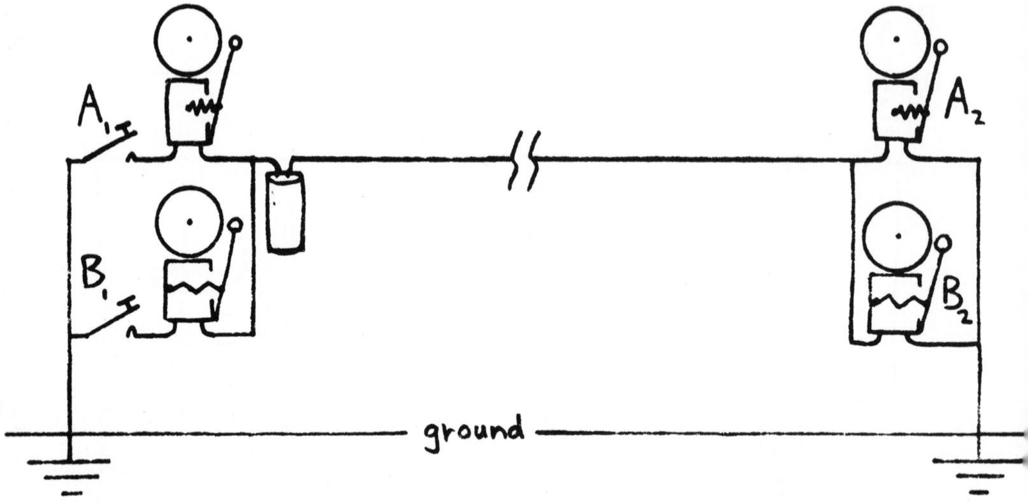

Now if you press the telegraph key at the sending station going to bell A_1 only bell A_2 at the receiving station will ring because the other bell at the receiving station, B_2, won't resonate at the frequency of the A bells.

The essence of the idea is not to simply send out a telegraph signal, say, dot dot dash (· · -), but to modulate the dots and dashes at particular frequencies. The receiver can then distinguish the signals by the modulation difference. This same means is used to tell radio stations apart. Different radio stations broadcast at specific carrier frequencies. Your radio receiver is a variable resonator, and you adjust it to resonate with the carrier frequency of your choice and thereby discriminate between the many different signals it receives.

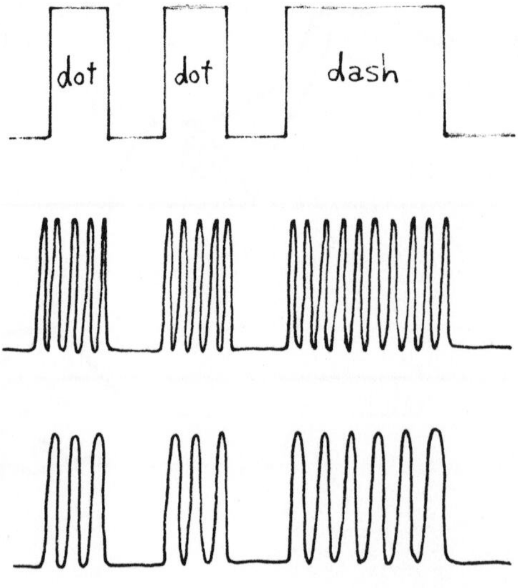

CONSTRUCTIVE & DESTRUCTIVE

Syringe A is connected to B & C by means of a glass Y and rubber tubes. When the plungers in B and C are moved, the plunger in A

a) must also move.
b) need not move.

ANSWER: CONSTRUCTIVE & DESTRUCTIVE

The answer is: b. If B and C move in or out together, then A must move. But, if B moves out while C moves in, A need not move at all. The displacement of A is the sum B + C, and the sum can be zero if B and C are opposite. Now suppose the plungers in B and C are oscillating in and out like engine pistons. If they oscillate together, the resulting oscillation of A is large. If they oscillate in opposite ways, they cancel each other and A does not oscillate.

This idea extends to waves, whether water waves or sound or light waves, all of which are produced by something that vibrates. When the effect of several vibrating things or waves comes together, you can't predict the sum effect until you first know if the things are vibrating together or vibrating in opposition.

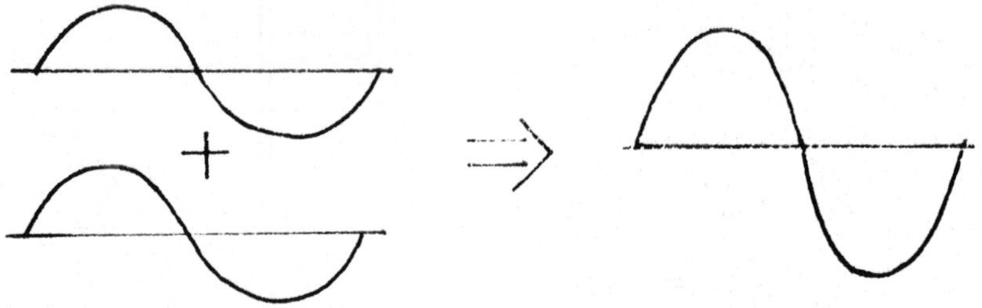

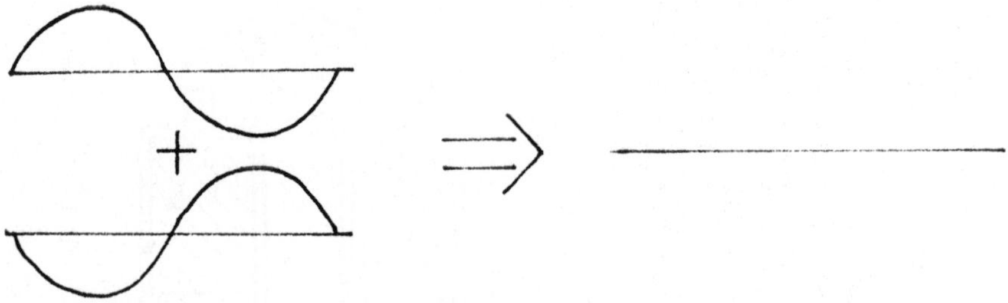

There is a scientific lingo which is used for these ideas. When things vibrate in unison, they are said to be "in phase" or "in sync." When in opposition, they are "in opposite phase" or "180° out of phase." When vibes that are out of phase superimpose they interfere destructively and cancel each other. When in phase they interfere constructively and reinforce each other. When water waves combine this way we find regions of tranquility; when sound waves combine this way we hear a throbbing, or beats; combinations of doubly-reflected light waves produce the beautiful colors we see in soap bubbles or gasoline splotches on a wet street.

GLUUG GLUUG GLUUG

You are emptying a gallon jug or gallon can. As the liquid runs out it makes a "gluug gluug gluug" sound. As the gallon becomes empty, the frequency of the sound

a) gets lower, that is: glug, gluug, gluuug
b) does not change, that is: gluug, gluug, gluug
c) gets higher, that is: gluuug, gluug, glug

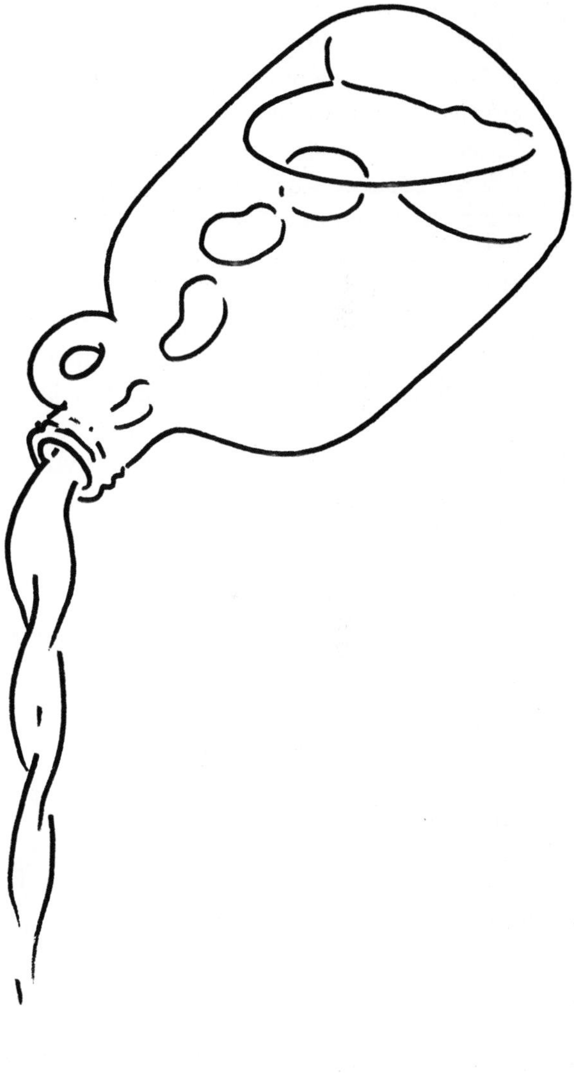

ANSWER: GLUUG GLUUG GLUUG

The answer is: a. As the liquid runs out the size of the air space in the jug grows and the large air space, or cavity, has a lower resonant frequency. Remember: big organ pipes make the low notes. The liquid flow pulses at the resonant frequency of the air cavity and that frequency decreases as the liquid spills out.

The reverse happens when you run water into a container. As the air space becomes smaller the frequency of the sound which comes out of the container increases. By listening, you can almost tell when the container is full without looking.

But why is the glugging sound frequency emitted on emptying always so very much lower than the hissing sound emitted on filling? Because when the container is emptying the water must vibrate with the air, but when the container is filling the air alone vibrates.

Imagine a little mass hanging on a spring and bouncing up and down. Then imagine a larger mass on the same spring. The larger mass bounces slower. It is harder to accelerate the larger mass. In a similar manner, when emptying the container the mass of the water slows down the vibes.

DR. DING DONG

Dr. Ding Dong rings a bell and listens to it with her new stethoscope. She moves the stethoscope all around the rim of the bell and finds

a) all parts of the rim are equally loud
b) some parts are loud and some are almost silent

ANSWER: DR. DING DONG

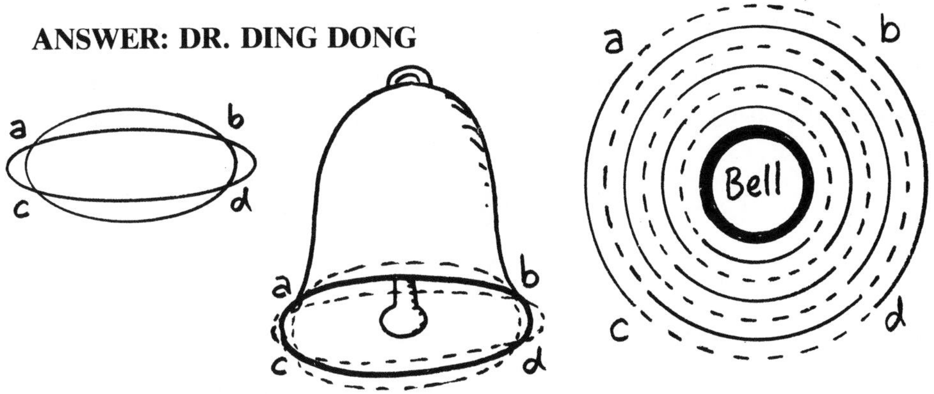

The answer is: b. When the bell is struck by the clapper, the circular rim is driven into the form of an oval, which elastically rebounds into another oval. The rim vibrates continuously from one oval form to the other as long as the bell rings. However, there are four points, *a, b, c,* and *d,* along the rim that do not vibrate. No sound comes from these points! (Can you think of a way the rim could vibrate so there would be eight silent points?) Furthermore, the silence will extend away from the bell because the sound waves that arrive in the vicinity of these points from other parts of the bell arrive out of phase and so interfere destructively. If the points of rest gradually travel around the rim, the ring of the bell is wavy.

TWANG

A guitar string is stretched from point A to point G. Equal intervals: A, B, C, D, E, F, G, are marked off. Paper "riders" are placed on the string at D, E and F. The string is pinched at C and twanged at B. What happens?

a) All the riders jump off
b) None of the riders jump off
c) The rider at E jumps off
d) The riders at D & F jump off
e) The riders at E & F jump off

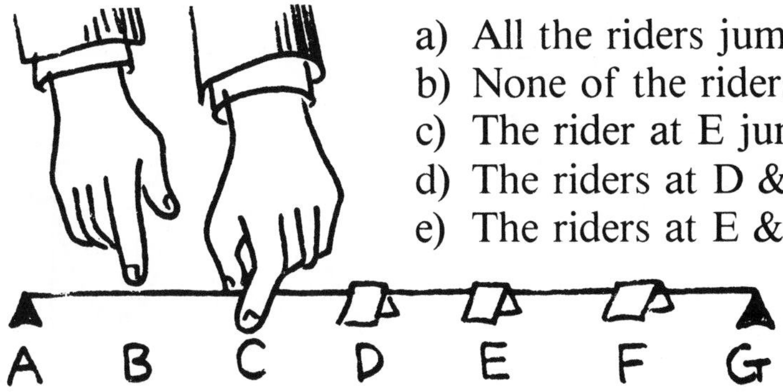

ANSWER: TWANG

The answer is: d. In this case a picture is worth 1000 words. The sketch shows how the string vibrates and which riders jump off.

CAN YOU HEAR THIS PICTURE?

Two musical notes A and B are shown, one at a time, on the oscilloscope. The note that has the highest frequency is

a) A
b) B
c) both the same

and the louder of these two notes is

a) A
b) B
c) both the same

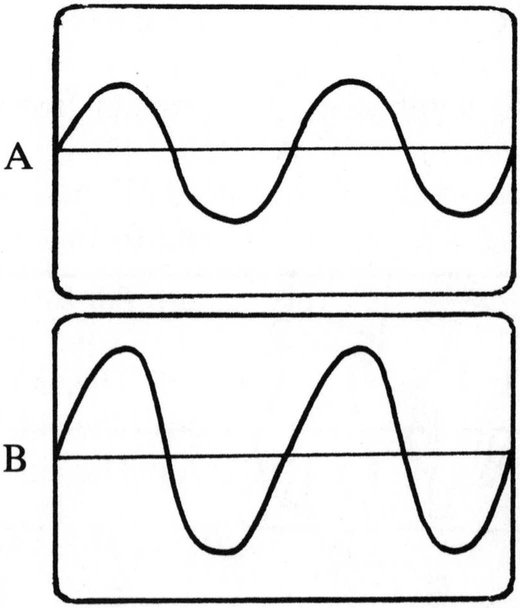

ANSWER: CAN YOU HEAR THIS PICTURE?

The answers are c and b. There are two complete wiggles (cycles) on the oscilloscope screen in each picture, so the *frequency* does not differ. But the height (amplitude) of the waves in B are greater, which indicates a greater vibratory energy. So B is the louder note.

ADDING SQUARE WAVES

These waves are called square pulse waves. When Wave I is superimposed on Wave II they add to make the wave shown in

a) a
b) b
c) c
d) d

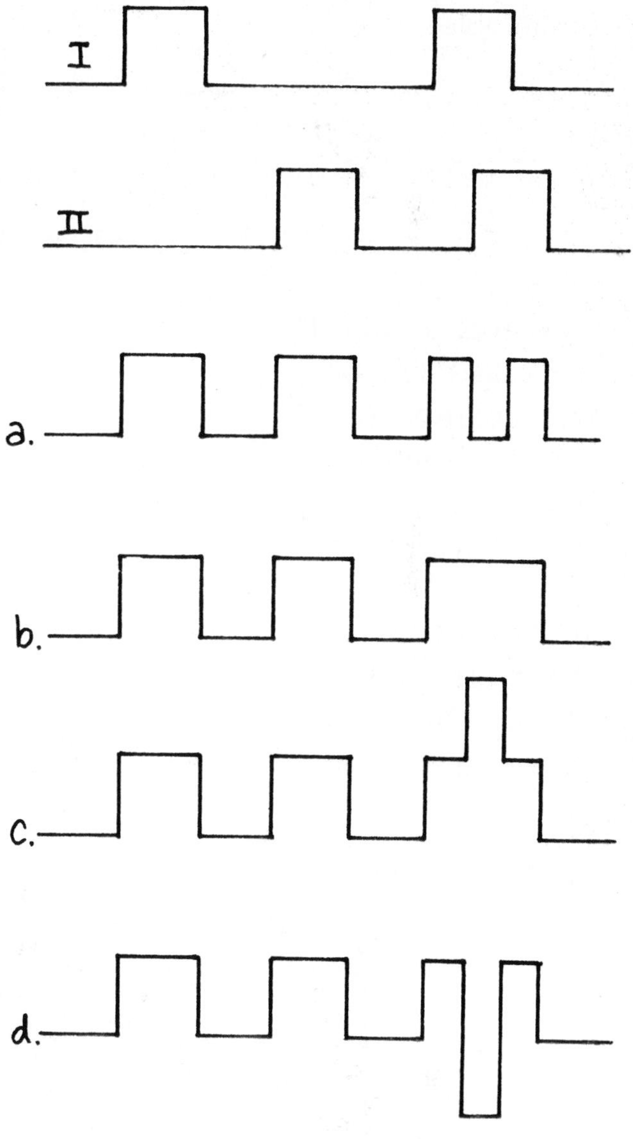

ANSWER: ADDING SQUARE WAVES

The answer is: c. The vertical displacements of waves I and II are added together to form their composite as shown in the diagram. This is a simple case because the displacements for both I and II are everywhere either zero or a single positive value.

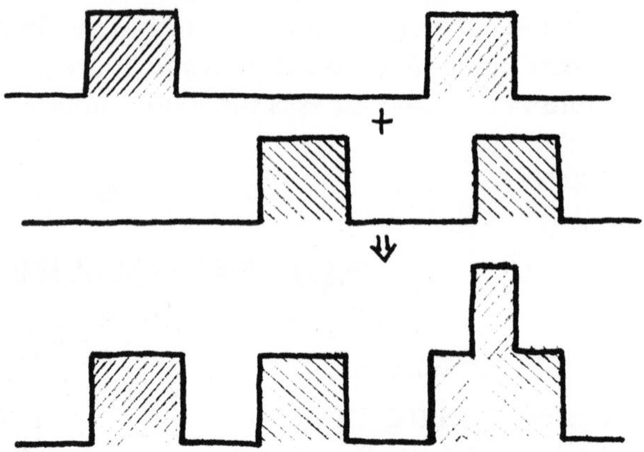

ADDING SINE WAVES

Waves I and II are called sine (sin) or cosine (cos) waves (sometimes they are called pure or harmonic waves). Wave I superimposed on Wave II adds to make the wave shown in

a) a
b) b
c) c
d) d

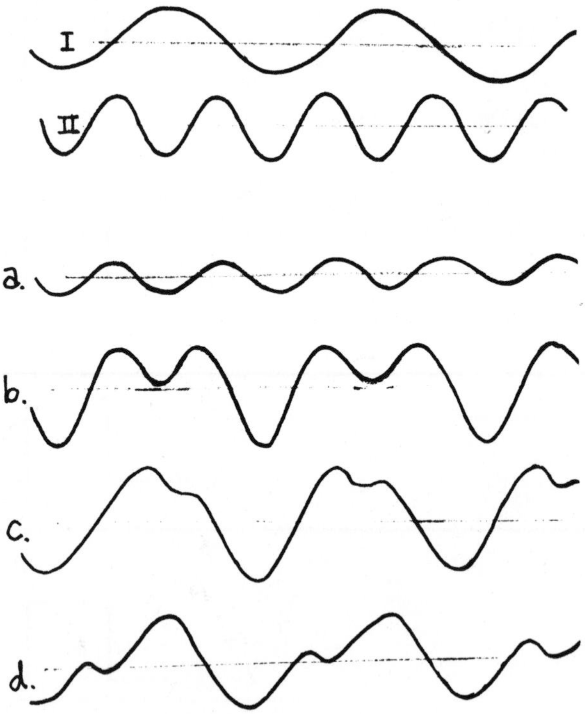

ANSWER: ADDING SINE WAVES

The answer is: b. The vertical displacement everywhere along this composite wave is simply the algebraic sum of the displacements of Waves I and II.

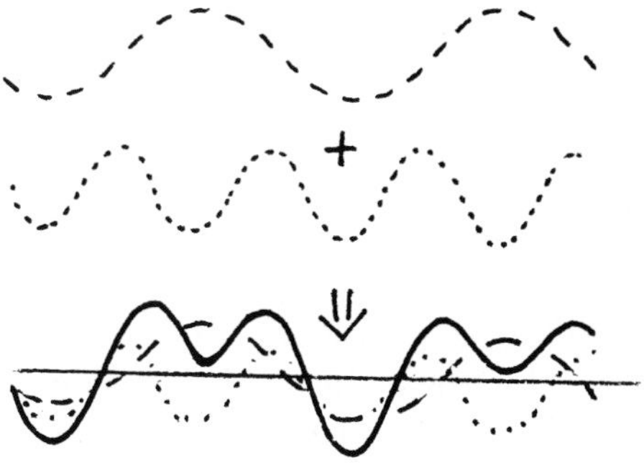

PROFILE

Wave I is a string of human face profiles. This or a wave of *any* shape can be made (built up) by adding various pure harmonic (sine) waves, such as II, III, etc.

a) Yes, this is correct
b) Not really

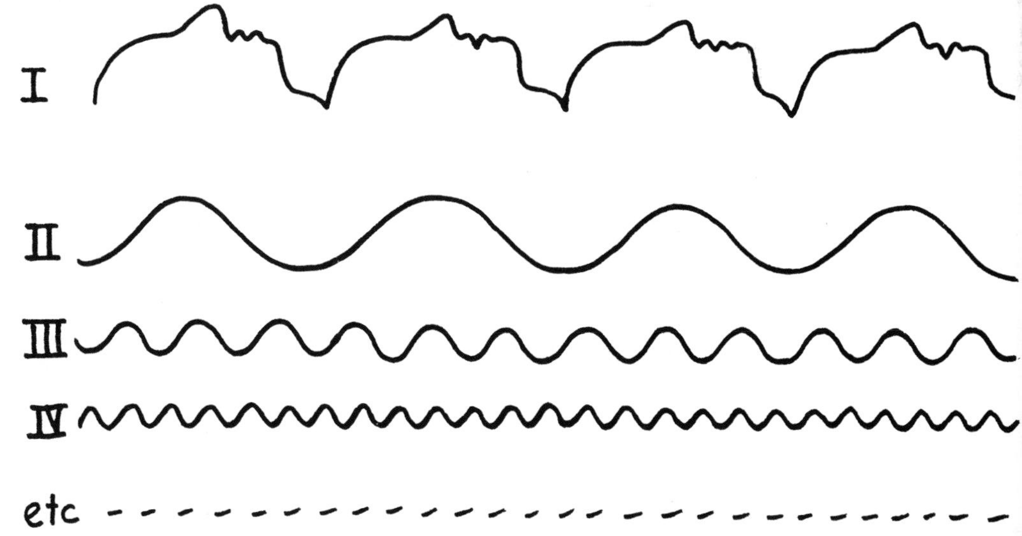

ANSWER: PROFILE

The answer is: a. Adding various pure waves will result in strange shaped waves, but Jean Fourier (who went with Napoleon to Egypt to study hieroglyphics) showed that ANY shape wave or profile can be represented as a sum of many pure waves. If the desired profile has small bumps or sharp corners, then many small (short wavelength or high frequency) waves will be required to make it. There is only one limitation on what kinds of profiles can be built from pure waves. That is that the profile must be single valued. This means any vertical line drawn through the profile must cut it at only one point, like point 1. You cannot have a profile with a hooked nose because the vertical line cuts that profile 1, 2, 3 times.

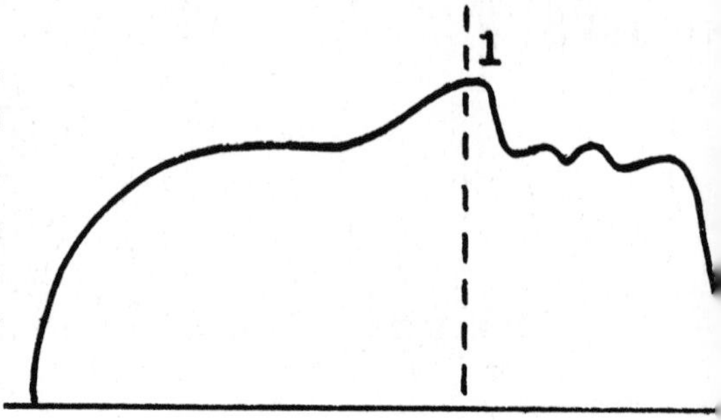

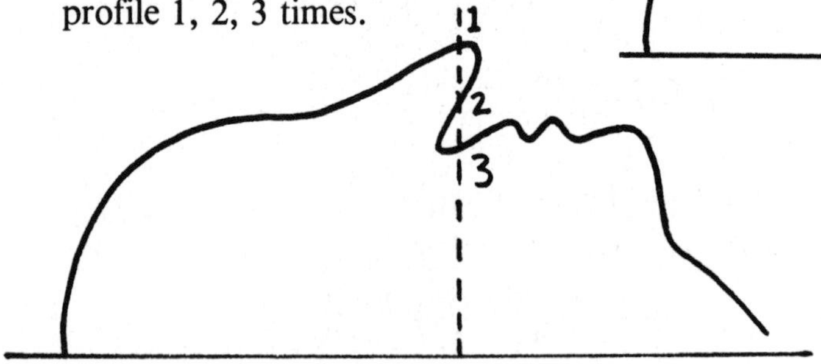

WAVES WITHIN WAVES

The square wave and the sine wave shown in the sketch both have the same frequency and wavelength. Which of these waves contains the most high frequency or short wavelength *component* waves?

a) The square wave
b) The sine wave
c) Both the same

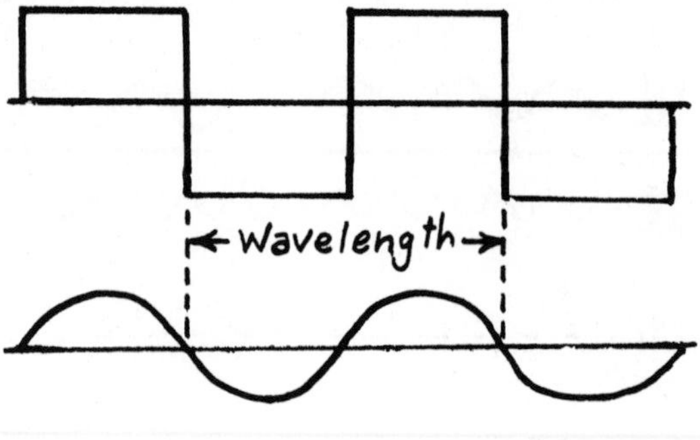

ANSWER: WAVES WITHIN WAVES

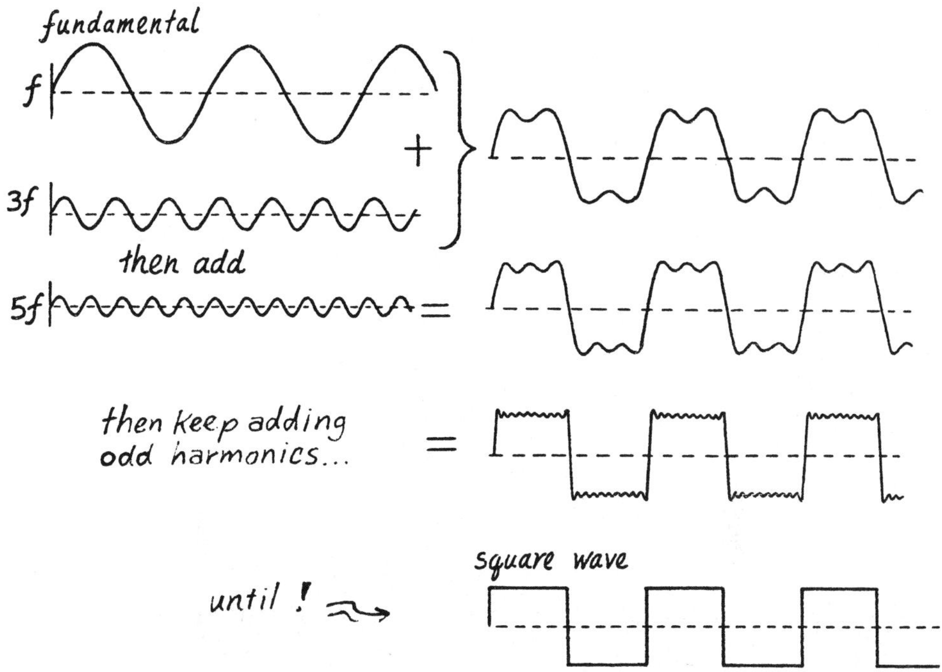

The answer is: a. The wave with the sharpest corners is composed of waves of the highest frequencies. The diagram shows how a square wave is gradually built up from many many high-frequency sine waves. If you send a square wave through an amplifier, speaker, or transmission line which is not capable of handling high frequencies, the square wave will have its high-frequency components robbed and will come out with round shoulders.

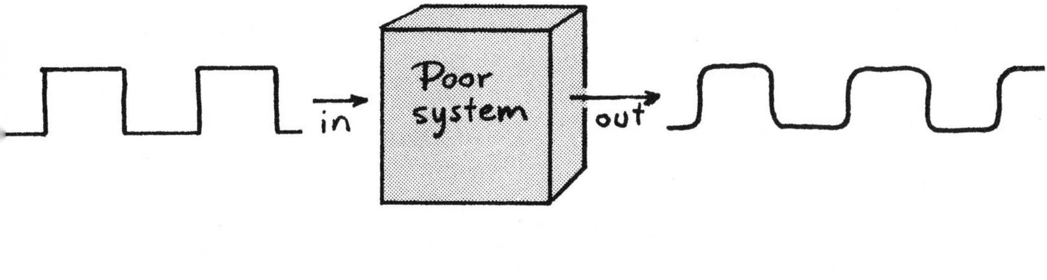

MUST A WAVE MOVE?

Must a wave move?

a) Yes b) No

Does a wave always have a wavelength?

a) Yes b) No

Does a wave always have a frequency?

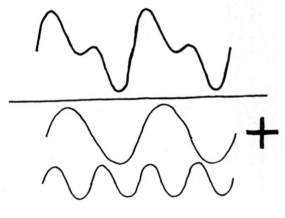

a) Yes b) No

ANSWER: MUST A WAVE MOVE?

All the answers are: b. Waves don't always move. Look at the little waves in front of a rock in a creek.

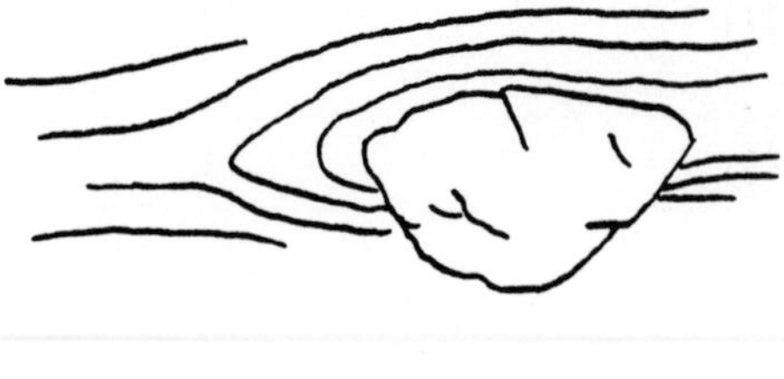

Waves don't always have a wavelength. Remember you can add two waves and get a third wave. If each of the two added waves has a different wavelength, what is the wavelength of the combined wave?

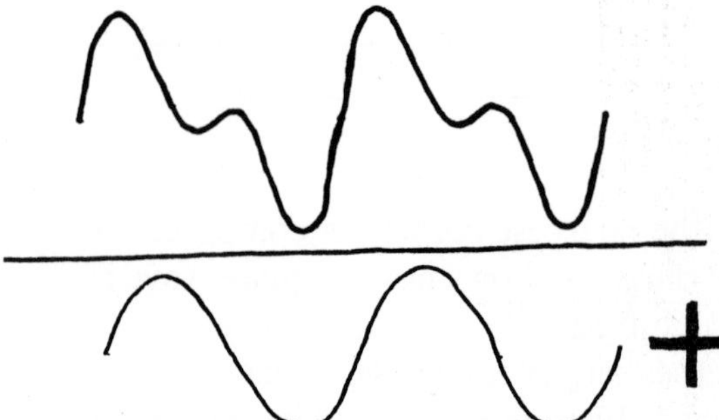

A similar argument will show that a wave does not always have a definite frequency. This uncertainty about wavelength and frequency plays a key part in quantum mechanics, where wavelength and frequency become momentum and energy—uncertain momentum and uncertain energy!

BEATS

Two different notes are sounded at the same time. Their sounds are added and the sum displayed on an oscilloscope, Display A. Then a different pair of notes is sounded and their sum displayed on the oscilloscope, Display B. We can see by the two displays that the notes shown in Display A are

A

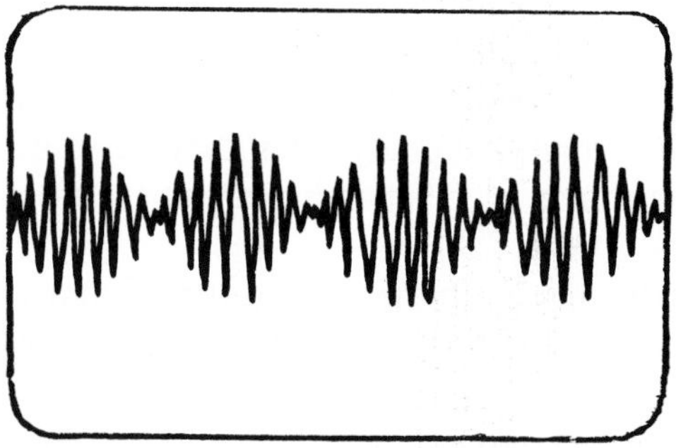

B

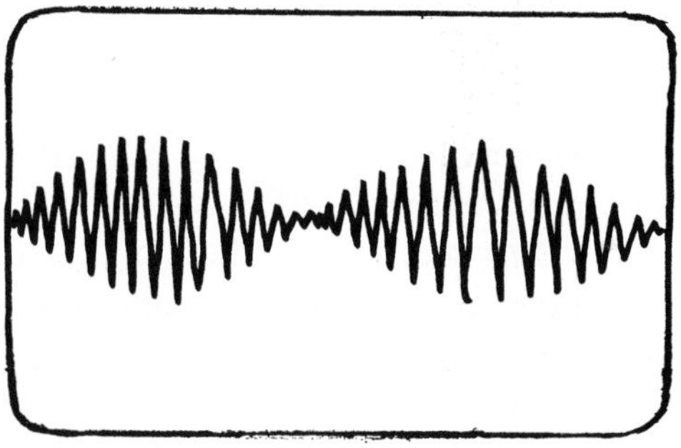

a) closer in frequency
b) farther apart in frequency
c) equally close as Display B in frequency
d) identical to the frequencies in Display B
e) . . .there is no way to tell from the displays which pair of notes has the closest frequencies

ANSWER: BEATS

The answer is: b. To understand this consider the two measuring sticks shown at the right. The marks on Stick I are spaced a bit farther apart than the marks on Stick II. Notice that the marks match at A, but farther down at B they don't match any more. yet at C they match again because II has fallen back by one whole mark. Where the marks match at A, C, and E, they are said to be in phase or synchronized (in sync). At B and D they are out of phase or out of sync.

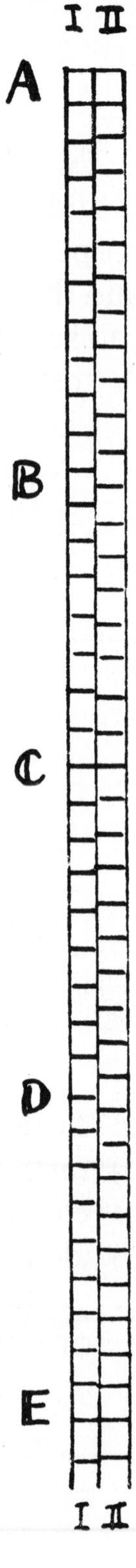

A little thought will show the greater the difference in spacings, the shorter the interval between places where matching occurs. If the spacings are only very slightly different, then matching places are fewer and farther apart. Of course if the marks on the sticks are spaced equally, then they never get out of sync — they either match or mismatch everywhere along the sticks.

Now we can take the marks on the stick to represent a sound wave. The marks themselves are the high-pressure parts of the wave (condensations) and the middle of the space between the marks are the low-pressure parts of the wave (rarefactions). We see that the frequency of Wave II is higher than the frequency of Wave I because the marks or waves on II occur more frequently than on I.

So I and II represent a pair of notes. If they were sounded together there would be *constructive interference* at A, C, and E, and *destructive interference* at B and D where high and low pressures come together and cancel. So at places like A, C, and E the sound would be loud and at B and D it would be quiet. The overall sound would pulsate or throb. The throbs are called *beats* and you can often hear them from a twin engine airplane or twin engine boat. If the engines run exactly in sync there are no beats, but if one runs a little faster than the other it has a little higher frequency than the other and beats occur. As the difference in frequency increases, the beat frequency increases. In Display A the beats occur twice as often as in Display B, so the difference in frequency of the notes fed into A are twice the difference in frequency of the pair of notes fed into B. Thus the notes that make up B are closer in frequency.

DR. DUREAU'S QUESTION*

Suppose you have just bought an excellent concert grand piano and you want it to be put in *perfect* tune. So you summon a first-rate piano tuner who tunes the piano by matching it to selected tuning forks. He sounds the piano and a fork together and listens for beats. How long will it take the piano tuner to do a perfect tuning job?

a) About an hour
b) About a day
c) About a week
d) About a month
e) About forever

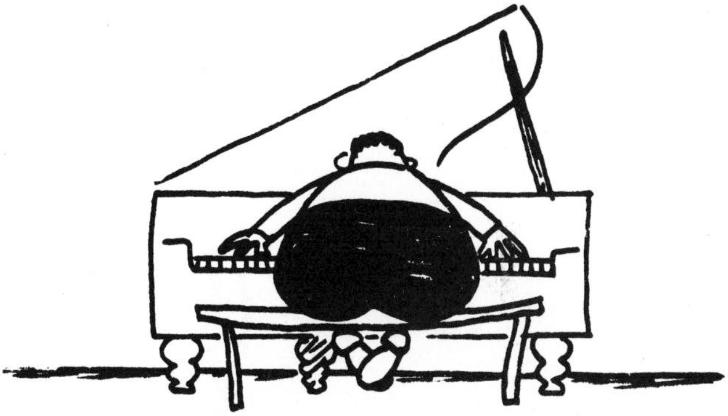

---

* Dr. Lionel Dureau was chairman of the Louisiana State University at New Orleans Physics Department, and this is one of his favorite questions. Thanks to him I broke into the teaching business, and consequently wrote this book.

ANSWER: DR. DUREAU'S QUESTION

The answer is: e. The piano tuner sounds the piano and fork together and listens for beats. As the piano string is brought closer and closer into tune with the fork, the beats get farther apart. Eventually the beats might be more than a minute apart and that is good enough. But if there are still any beats, no matter how far apart, it means the piano is not in *perfect* tune. To get it in perfect tune the piano tuner would have to have infinite patience and listen forever. But of course the sound dies out before then, so we'll have to be content with a less than perfect tuning job.

CLIP

Suppose from a complete tape of a symphony you cut out a tiny segment as shown. Can one tell from the very short clip of tape exactly what notes were being played at the moment that clip was recorded?

a) Yes, you can analyze the short clipped out piece and identify the notes exactly.
b) No, to identify the notes a long piece of tape is required. A short piece will not suffice.

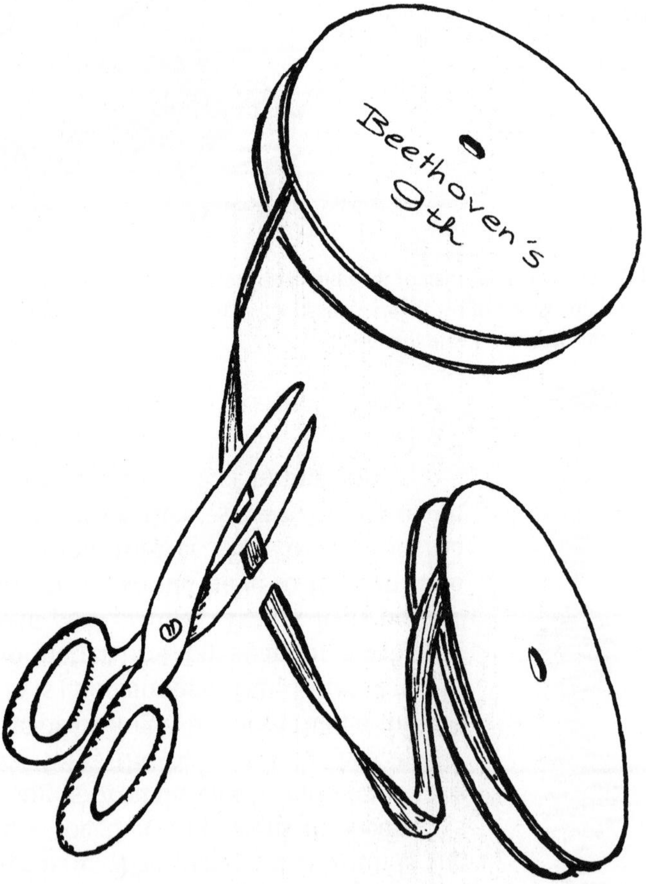

ANSWER:

CLIP

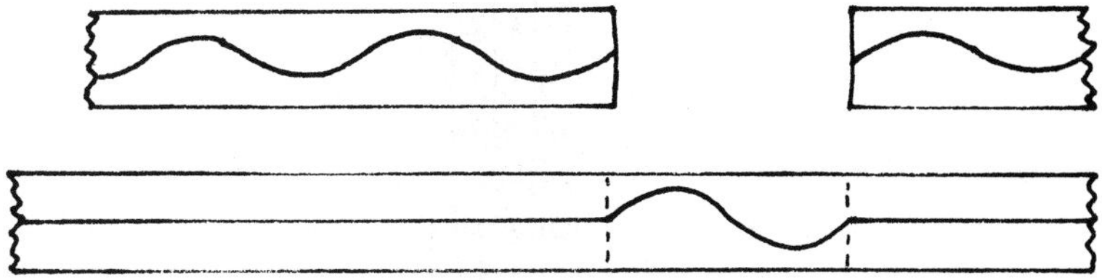

The answer is: b. If the clip were not too short you could almost tell the exact note being played, but for a very short clip you could hardly tell at all. Suppose you have a long piece of tape on which a sound is recorded as shown above, and you cut out a little segment and splice it into a long silent tape. When you play the spliced tape, you'll hear a short blip when the spliced segment goes by, but from that blip you could not possibly identify the sound. You may suppose if you could examine the blip and measure its wavelength or even a half or quarter of its wavelength you could infer the frequency of the sound.

But this is true only if the sound consists of one *pure* note. Most if not all sounds contain many notes as well as undertones and overtones of those notes. For example, in the sketch below you may look at a quarter

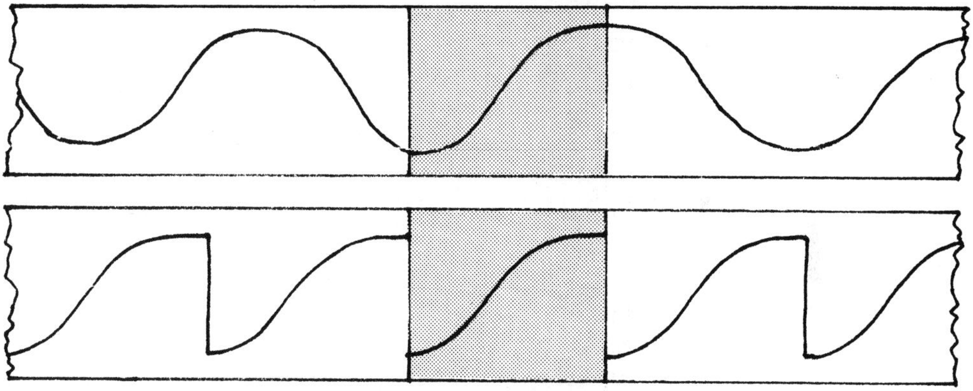

or half wavelength segment and think these two sounds were identical. But they aren't. The upper tape is of a pure note and the lower one is of a sawtooth wave that sounds quite different.

Quite apart from the theoretical problem of identifying a sound from a short segment there is a separate problem — an engineering problem. It is difficult to precisely measure a single time period or wavelength, just as it is difficult to precisely measure the thickness of a piece of paper. What to do? Stack up ten pieces, measure the total thickness and divide by ten — or stack up ten vibrations, measure the total time and divide by ten. So over a long time you can more easily measure the frequency.

This is the key to a new idea: There exist in this world various pairs of things that cannot be measured simultaneously — like the precise *frequency mix* of a sound and the precise moment of *time* during which it is sounded. This does not mean precise moments of time cannot be clipped and it does not mean precise frequency mix measurements cannot be made. It means only that you can do one or the other — but you can't do both. This new idea is called the *Uncertainty Principle* and the pairs of things are called *Conjugate Pairs*. You can measure one member of the pair as accurately as you like, but then your ability to measure the other member suffers. Or you can measure both members with moderate accuracy, but if you try to get more precise about one member of the pair you will force the measurement of the other member to become less precise.

Frequency mix and Time are a Conjugate Pair.

MODULATION

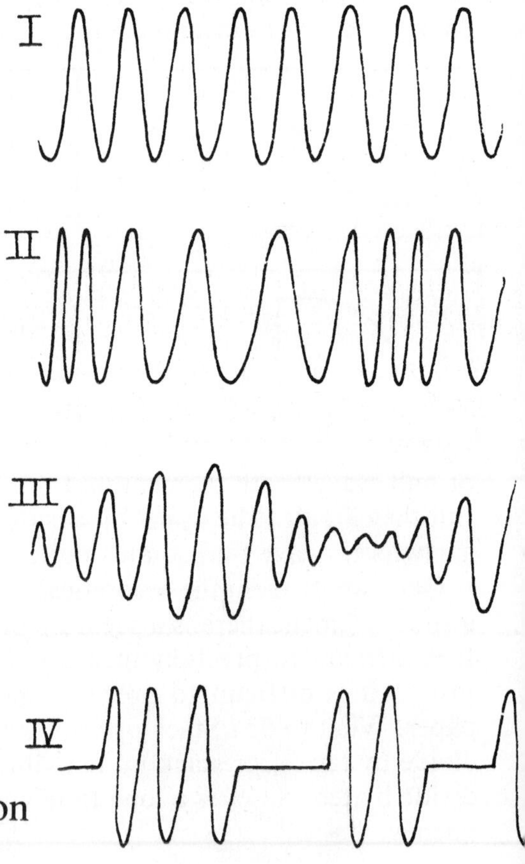

In physics the word "information" can be taken to mean to tell a story. In the sketch we see four signals, all of which represent sound or radio waves. Wave I is of one continuous frequency. Wave II has changes in its frequency. Wave III has changes in its amplitude. Wave IV pulses on and off. Which of these signals could *not* carry "information"?

a) Wave I b) Wave II
c) Wave III d) Wave IV
e) All could carry information

ANSWER: MODULATION

The answer is: a. To carry information the signal must be able to make different expressions. A wave can only make expressions by distinctively *changing* its shape — no change, no information. In physics the word "modulation" is taken to mean distinctively changing shape. There are many ways a wave can change shape. Wave II, for example, changes shape by changing its frequency or pitch. Many birds do the same. FM (frequency modulated) radio stations do it too. Another way of changing wave shape is shown in Wave III where the amplitude changes, giving variations in loudness or power — like the throbbing of two Diesel engines that drift in and out of phase. This is how AM (amplitude modulated) radio stations simulate music and voice. They vary the power of the broadcast wave. Wave IV simply turns on and off, somewhat like a barking dog. PM (pulse modulated) radio stations do the same. PM stations usually transmit code.

But Wave I contains no story. It is just a never ending hummmmmmm... The humm has no distinctive change of shape. It is a pure repetition. That means you can predict exactly what is to come. It contains no news. Now if it were to come on at a certain time or go off at a certain time that would be a different story. That would be a distinct change, or modulation. But no modulation = no story = no information.

EXACT FREQUENCY

A certain radio music station is located at 100 KHz* on your radio dial. Now suppose your radio could tune to *exactly* 100 KHz, excluding *all* other frequencies, even those very close to 100 KHz like 100.01 KHz and 99.99 KHz. On such a radio you would hear the music very clearly if it were broadcast

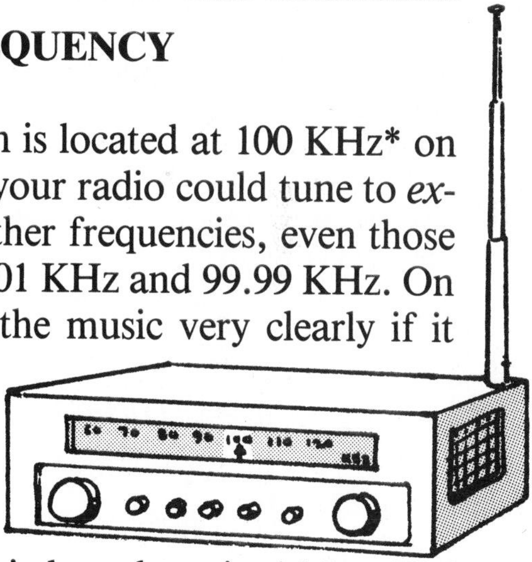

a) only in AM
b) only in FM
c) either in AM or FM
d) . . .you would *not* hear music broadcast in AM or FM

---

*Some people love to change names and units — so they changed the meaningful units CYCLES PER SECOND to HERTZ, in honor of Heinrich Hertz, who demonstrated radio waves back in 1886. So we now call the vibrational frequency of 1 cycle per second, 1 hertz (Hz). Do you think if Hertz were alive today that he would consider this an improvement?

ANSWER: EXACT FREQUENCY

The answer, perhaps to your surprise, is: d. If the radio is limited to exactly one frequency it cannot be modulated. No modulation means no information, means no music. The radio could only humm with one unchanging sound. This is true not only for frequency modulated signals, but for amplitude modulated signals as well.

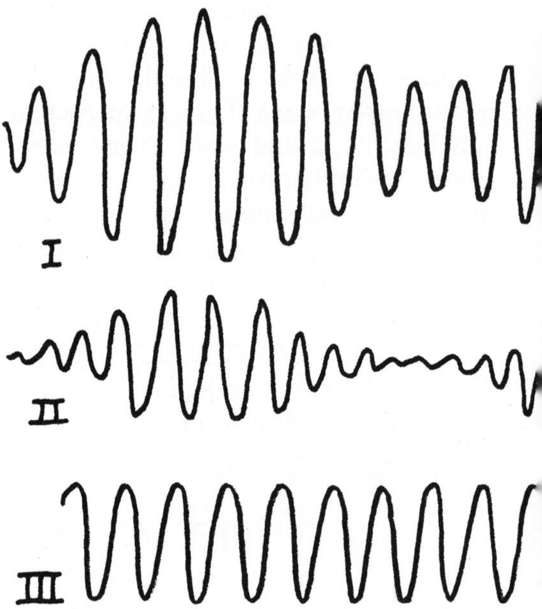

Some people might think that in amplitude modulation only a single frequency need be involved. But not so! To make an amplitude modulated wave such as I, a throbbing wave like II is added to a pure wave like III.

Now recall BEATS. The throbbing wave is made by adding two waves of slightly different frequencies. So they run in and out of sync like the sound from a pair of Diesel engines. Were it not for this slight frequency shift away from 100 KHz there would be no amplitude modulation and therefore no music.

It's interesting to note that radio engineers say *all* the information in a radio signal is carried in the sidebands. That means the frequencies just above and below100KHz. Sometimes when power must be saved, such as in transmission from distant spacecraft, the central carrier frequency at 100 KHz and one sideband is suppressed to save power. Only one side band is transmitted to earth. Back at earth the 100 KHz carrier wave and duplicate side band are added to restore the original signal. That is sort of like dehydrating food for backpacking — you can always read the water when you use it.

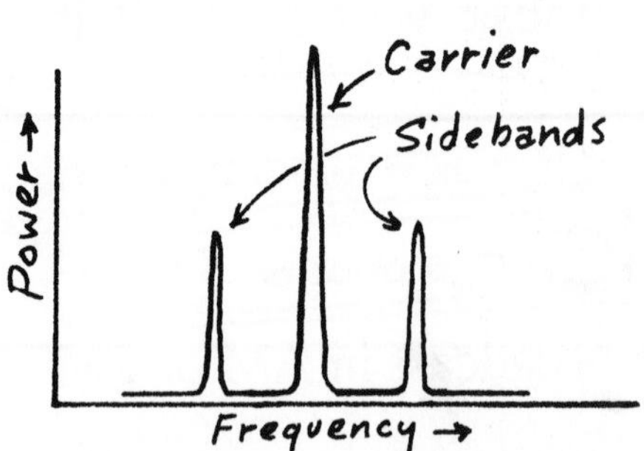

Note: While the radio must include the sideband frequencies near 100 KHz it must be carefully designed not to include too much. If the sidebands are too wide the radio will receive more than one station at the same time — a real mess.

QUICKSILVER SEA

Suppose the water in the ocean turned into quicksilver (which is about 13 times as dense as seawater). Then compared to the speed of seawater waves, the quicksilver waves would move

a) faster
b) slower
c) at the same speed

If the strength of the earth's gravity increased, ocean waves would

a) move faster
b) move slower
c) move neither faster nor slower

ANSWER: QUICKSILVER SEA

The answer to the first question is: c. Concentrate your attention on one cubic inch of water in the ocean. The forces on that cubic inch are responsi-

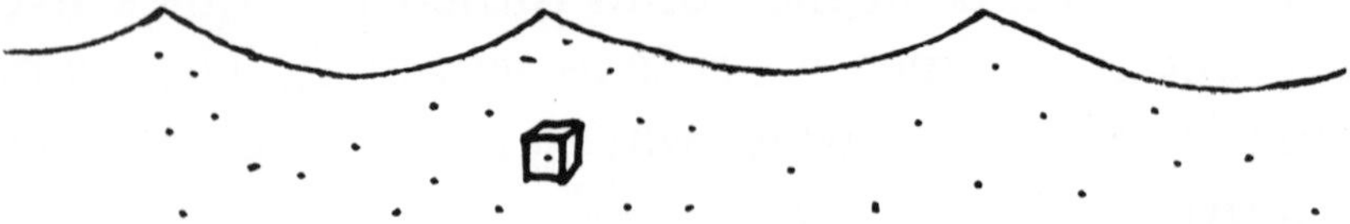

ble for its undulating motion. What are the forces on that cubic inch? The force of water pressure due to the surrounding water and the force of gravity. Now suppose the mass of each cubic inch of sea was increased 13 times — which is what would happen if the sea turned to quicksilver. The pressure forces would be 13 times larger and the gravity force (or weight) would also be 13 times larger. ALL forces on the cubic inch of water-turned-quicksilver would be 13 times larger. Does that mean the cubic inch would undulate or accelerate 13 times faster? No. Why? Because the cubic inch itself is now 13 times more massive, that is, 13 times harder to accelerate. So even though the sea turns to quicksilver the acceleration or undulation of each part of it stays exactly the same. If the motion of each part stays the same, then the total motion of the whole thing, the wave, stays the same.

The answer to the second question is: a. If gravity increases, then the weight of each cubic inch increases and the pressure force also increases. ALL the forces increase. But does the mass of the cubic inch increase? No. Increasing gravity does not increase mass. There is just more force on the same old mass and that means the cubic inch accelerates faster. If the motion of each part is faster then the total motion of the whole thing, the wave, is faster.

WATER BUG

These are the ripple waves made by a water bug on the surface of water. From the wave pattern we can see that the bug has been moving

a) continuously to the left
b) continuously to the right
c) back and forth
d) in a circle

ANSWER: WATER BUG

The answer is: c. An animal walking on snow leaves its footprints and marks its path. But on water the footprints turn into little circular ripples which don't stay put. They expand. Only one thing about the ripples do stay put — their centers. The centers of the ripples mark the position of their origin. If the water bug stays in one place all the ripples it generates have a common center. If it moves to the right the centers of the circles move to the right. This produces a crowding of ripples on the right side and a spreading on the left. You can always tell in which direction the bug has been moving by seeing in what direction the ripples are crowding together. So we see the ripples are sometimes crowded to the left and sometimes to the right, which means the bug was sometimes moving left and sometimes right.

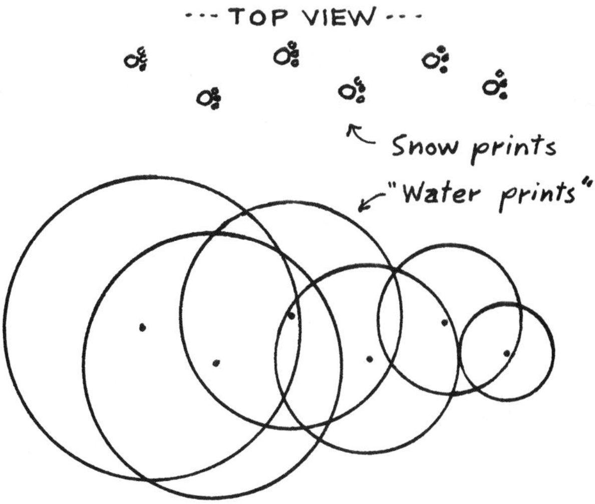

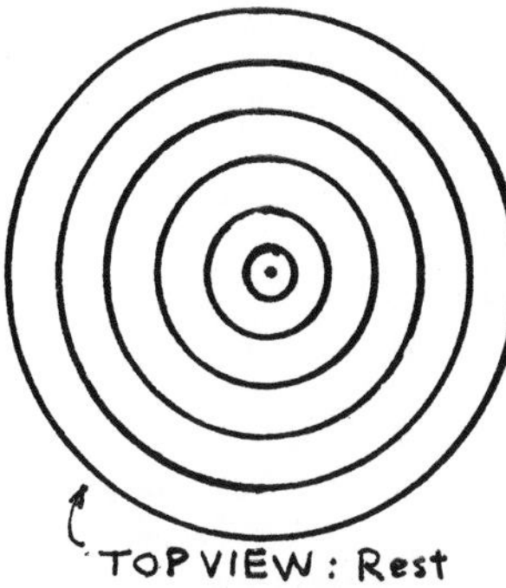

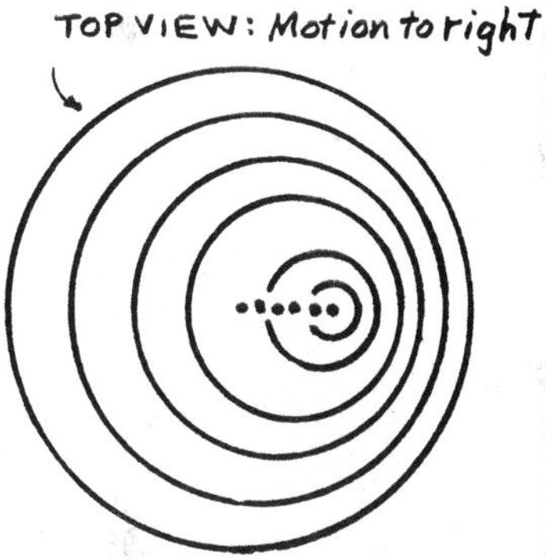

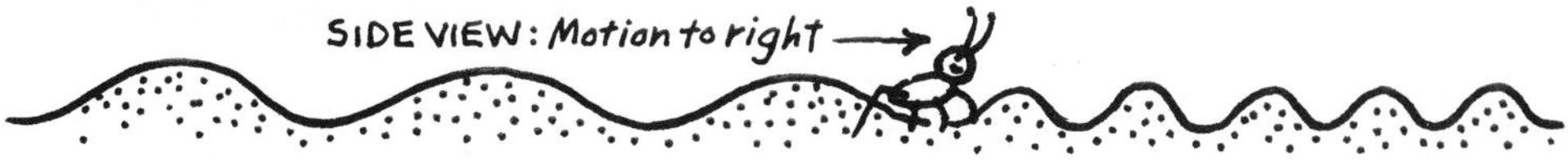

Ripples are little waves on water. Sound and light are also waves. The sound or light waves coming from a thing in motion would also be crowded together in the direction toward which the thing is moving. The spacing of sound waves determines the sound's frequency; high crowding produces high frequency. Eyes sense the frequency of light as color; high frequency is blue and low frequency is red. So, when a light is moving away from you it looks redder than it otherwise appears. We talk about the red shift in the light from stars and galaxies that are moving away from our solar system.

SOUND BARRIER

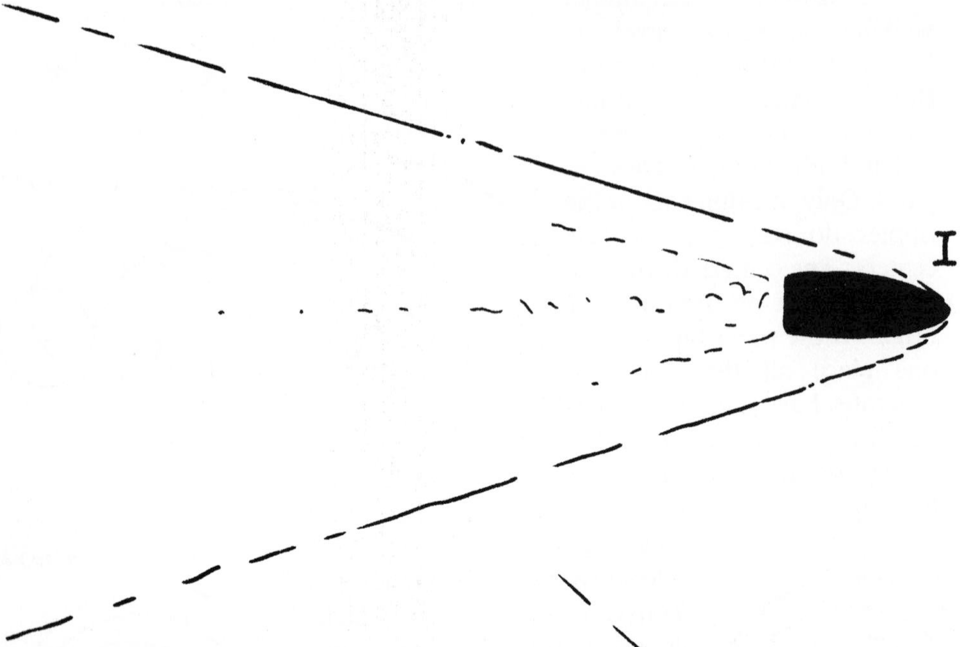

Two bullets are moving through air. From the bow waves produced by the bullets we can say for sure that

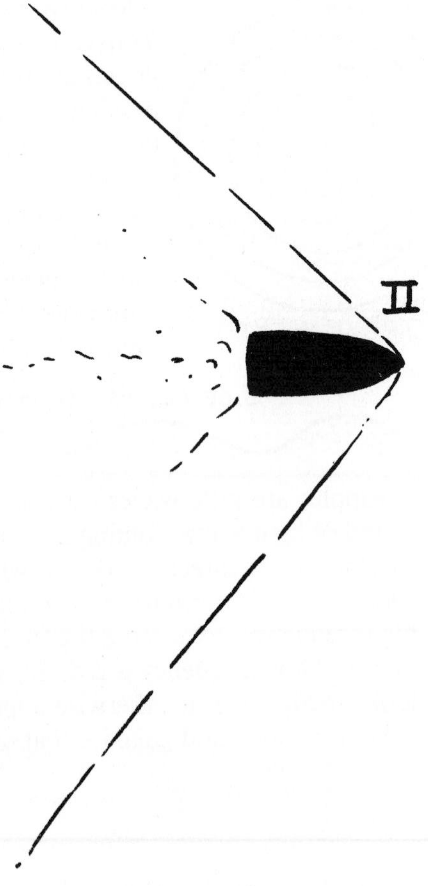

a) both bullets must be traveling faster than the speed of sound waves, and Bullet I is moving faster than Bullet II
b) Bullet I is moving faster than Bullet II, but not necessarily faster than the speed of sound
c) . . . none of the above

ANSWER: SOUND BARRIER

The answer is: a. If the water bug in our last question had moved faster than the waves it produced, the resulting pattern would not be a crowding of waves, but an overlapping of waves to form a V-shaped bow wave. The same is true for a bullet. If a bullet moves at a speed slower than the waves it produces, no overlapping occurs. Sketch *a.* If it travels just as fast as the waves, overlapping occurs only at the front of the bullet. Sketch *b.* Only when the bullet travels faster than the sound waves will overlapping occur to produce the familiar V-shaped bow wave. Sketches *c* and *d.* We see that the regions of overlapping are where the X's are. Only three pairs of X's are shown in the sketch while in practice many more occur. So the wave generated by a bullet is actually the superposition of many circular waves. Note that the faster the bullet, the narrower the V.

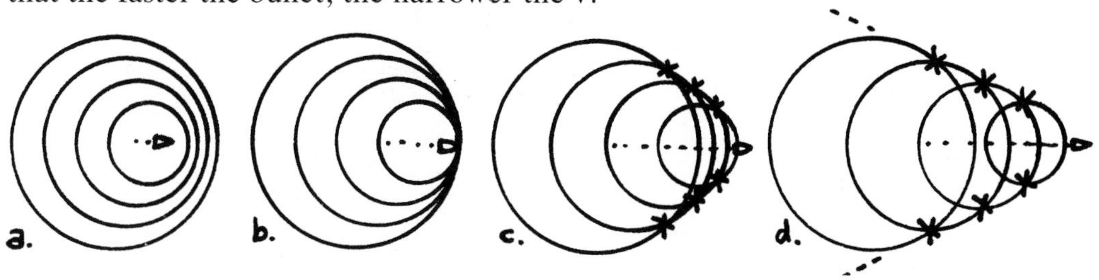

By the end of the Civil War bullets were flying faster than sound and just after the end of World War II aircraft were also flying faster than sound. Pilots experienced a great problem as they tried to fly at the speed of sound, sketch *b*, because they were flying right in their own noise. The noise built up into a wall of compressed air, a shock wave, which buffeted the aircraft and made it quite difficult and sometimes impossible to control the craft. Thus developed stories about a "Sound Barrier." When the aircraft flies faster than sound it escapes from its noise and the related difficulties. The shock wave trails behind. After a supersonic aircraft passes overhead the trailing shock wave reaches your ears as a sharp crack—the sonic boom.

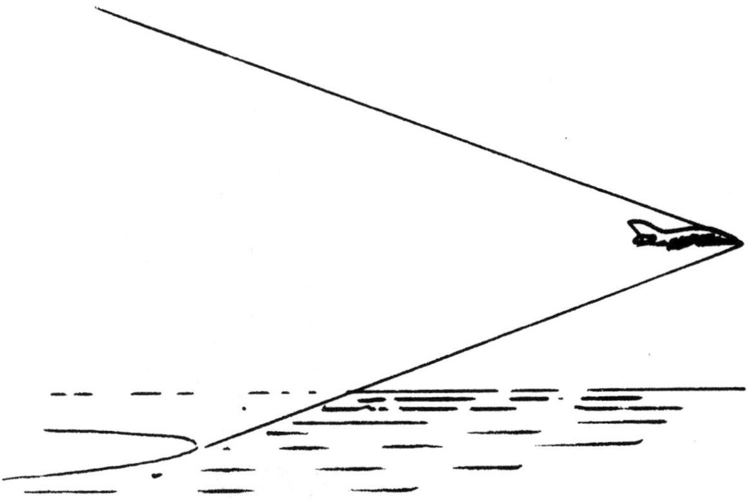

EARTHQUAKE

Many houses consist of a wooden frame structure that simply rests upon a concrete foundation or slab. In the event of a moderately severe earthquake the most frequent damage that occurs to these houses is

a) the concrete foundation shatters, but the wood frame stays intact
b) the foundation stays intact, but the wood frame collapses
c) the foundation and the wood frame both stay intact, but the frame slides off the foundation

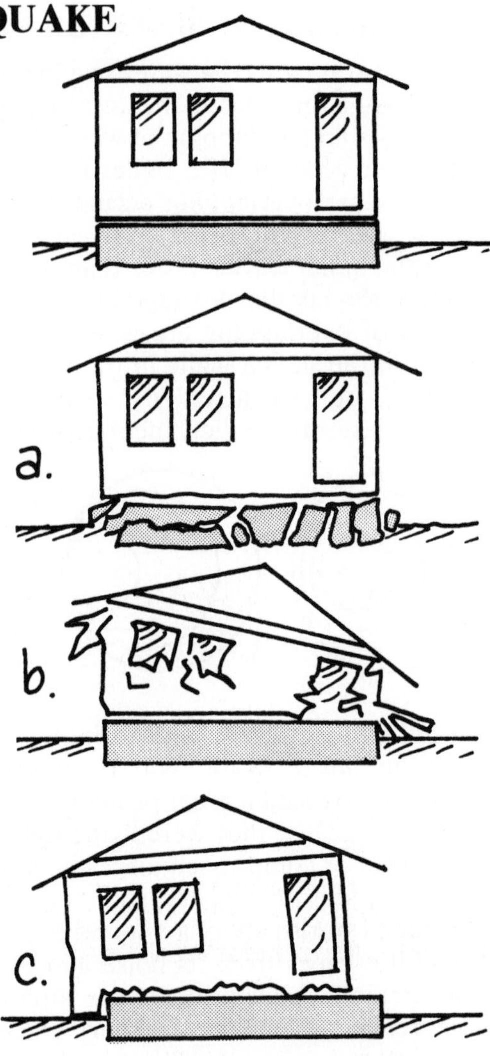

ANSWER: EARTHQUAKE

The answer is: c. It is inertia pure and simple. The concrete foundation is attached to the earth, so it shakes back and forth (and sometimes up and down) with the earth. But the house is not firmly attached to the foundation, so the foundation moves with the earth, but the house stands still! Soon the house finds nothing under it. CRASH! A $75,000 crash — and it could have been prevented by $75 worth of hold-down bolts joining the house to the concrete foundation. When you buy a house check for the hold-down bolts. Putting such bolts in all old California houses would save more lives and property than rebuilding all California schools — and at a fraction of the cost.

MORE EARTHQUAKE

The bottom of most valleys is filled with soft earth (called alluvium). In the event of a moderately severe earthquake, which house is likely to suffer the most damage?

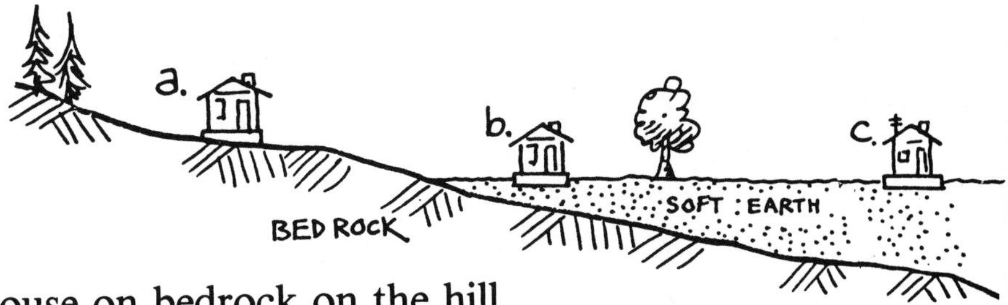

a) The house on bedrock on the hill
b) The house on soft earth near the bedrock hill
c) The house on soft earth far from the bedrock

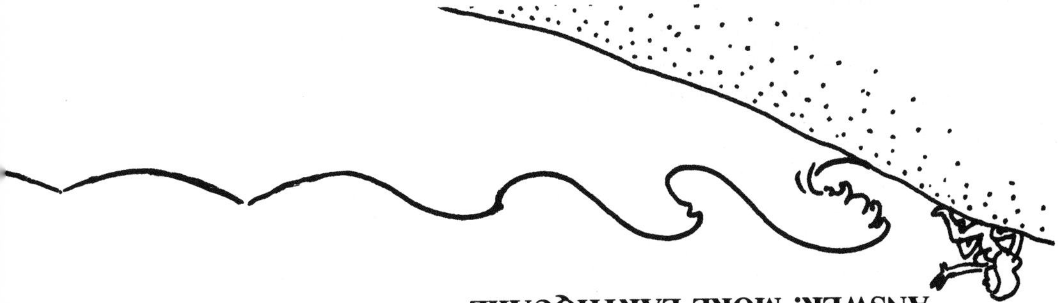

ANSWER: MORE EARTHQUAKE

The answer is: b. The earthquake wave runs through the soft earth very much like a water wave. The wave energy gets concentrated in the narrow wedge of soft earth which joins the bedrock. You can see the same thing happen if you watch a water wave come across a lake or an ocean wave hit the beach. The wave energy is concentrated in the shallow water so the wave gets high just as it hits the sand. The energy enables the wave to "run up" the beach and make wet shoes. In the case of earthquake waves there is sometimes enough energy concentrated at the edge of a valley to rupture or crack open the soft earth. After it is over it looks as though the soft earth had been plowed up.

NEAR THE FAULT

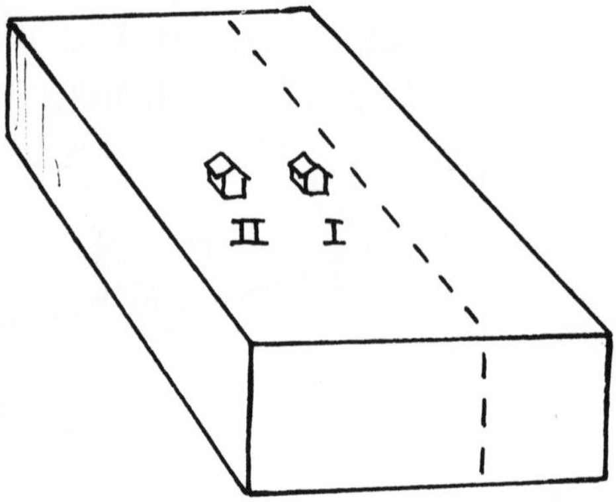

The energy in a big earthquake comes from sudden motion along a fault which is a crack in the earth's crust, sometimes over a hundred miles long. House I is one mile from the fault. House II is two miles from the fault. Which house is most likely to be harmed by the earthquake?

a) House I.
b) House II.
c) Both about the same.

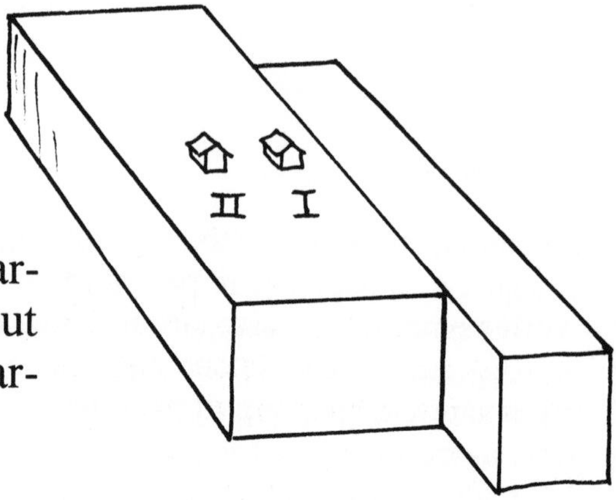

The earthquake energy arriving at House I is about how much more than that arriving at House II?

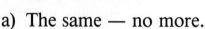

a) The same — no more.
b) Twice as much.
c) Three times as much.
d) Four times as much.
e) More than four times as much.

ANSWER: NEAR THE FAULT

The answer to the first question is c because the answer to the second question is a. The earthquake energy at both locations is essentially the same! It is not easy to see earthquake waves (though in very violent earthquakes, they might be seen), so let us visualize water waves. If you throw a stone in a lake, the waves and wave energy spread out from the stone in circles of ever increasing circumference. The energy carried by the wave is spread all along the circumference; so as the circumference grows, the energy must be more and more spread out and less concentrated, less intense. A cork floating on the water at I will bob much more than a cork at II.

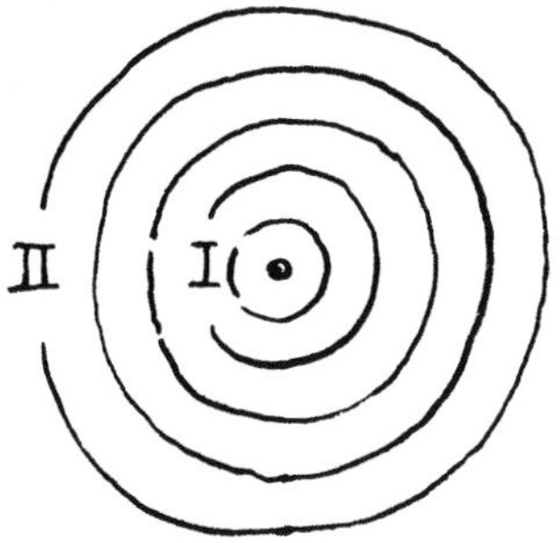

But an earthquake wave does not come from one spot like the stone. It comes from a long crack or fault. To make a wave like an earthquake wave, you must throw something long, like a log, into the lake. At the log's ends, the wave is curved and spreads out. But along the sides of the log, the wave is almost straight, and stays straight for some distance from the log. A straight wave cannot spread, and so the wave energy cannot become less concentrated or less intense as the wave moves away from the log. A cork on the water at I will bob almost as much as a cork at II. (The same sort of reasoning applies to light. Can you see why a room is more evenly illuminated by a long fluorescent tube than by a single bright bulb?)

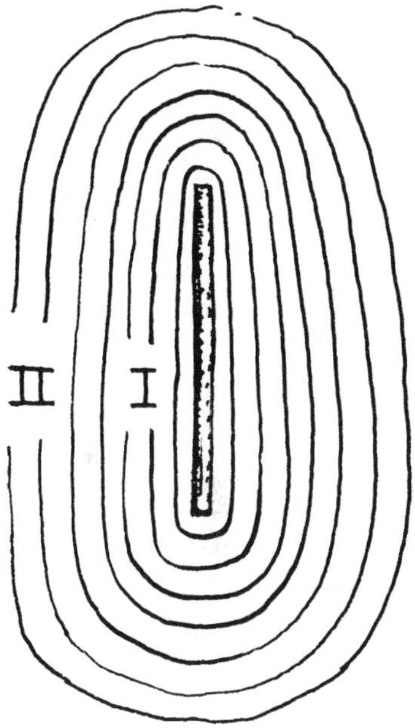

So the house near the fault is not in much more danger than the house farther away. Of course near the fault does not mean on the fault. On the fault, a house could be sheared in half. Very far away from the fault is best of all; because very far away, the waves begin to get curved and start to spread into circular waves. But near the fault, damage seems to depend more on how a house is built and upon what kind of earth it is built, than how far it is from the fault.

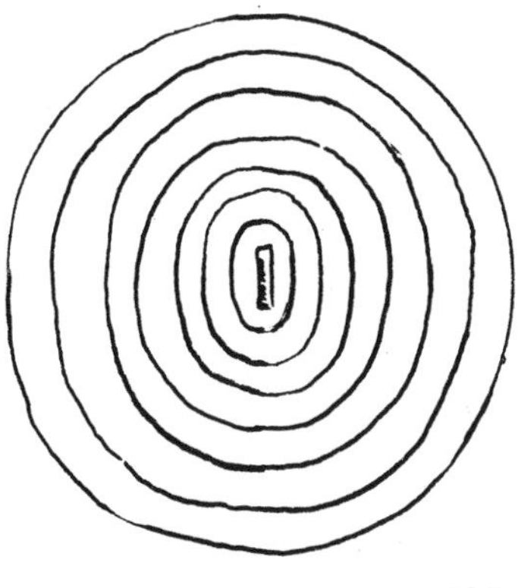

SAN ANDREAS

Being near or far from a fault is one thing, but now consider distances *along* the fault, close to and far from the epicenter. An earthquake takes the form of a rip in the earth (a shear rip — not a pull-apart rip), which begins at the epicenter and runs several miles (several hundred miles for a big one) along the fault. Cities I and II are equally distance from the fault. The earthquake will be

a) most intense at City I
b) most intense at City II
c) equally intense at each city

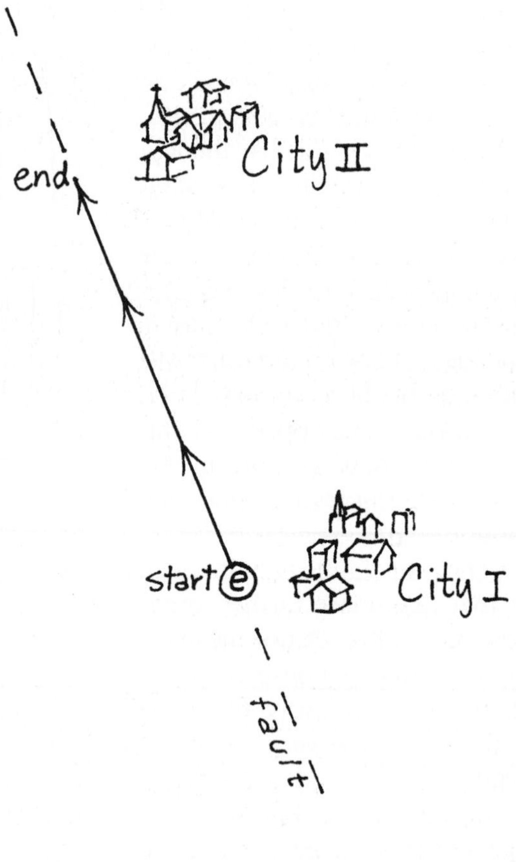

ANSWER:
SAN ANDREAS

The answer is: b. The instantaneous point of rip is the source of the earthquake wave and runs along the fault — the location of the wave source moves like a zipper. A moving source leads to a Doppler effect.

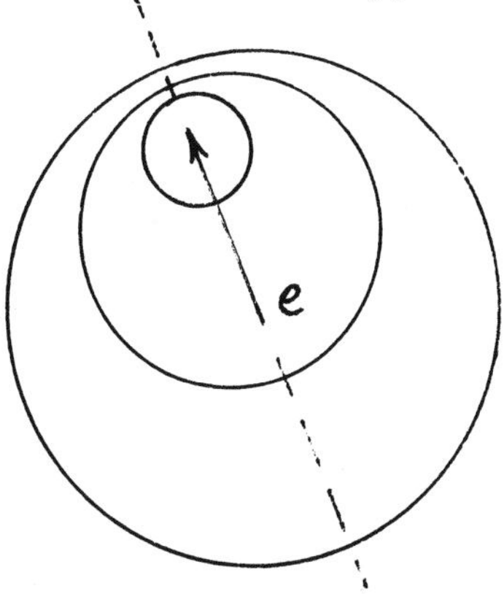

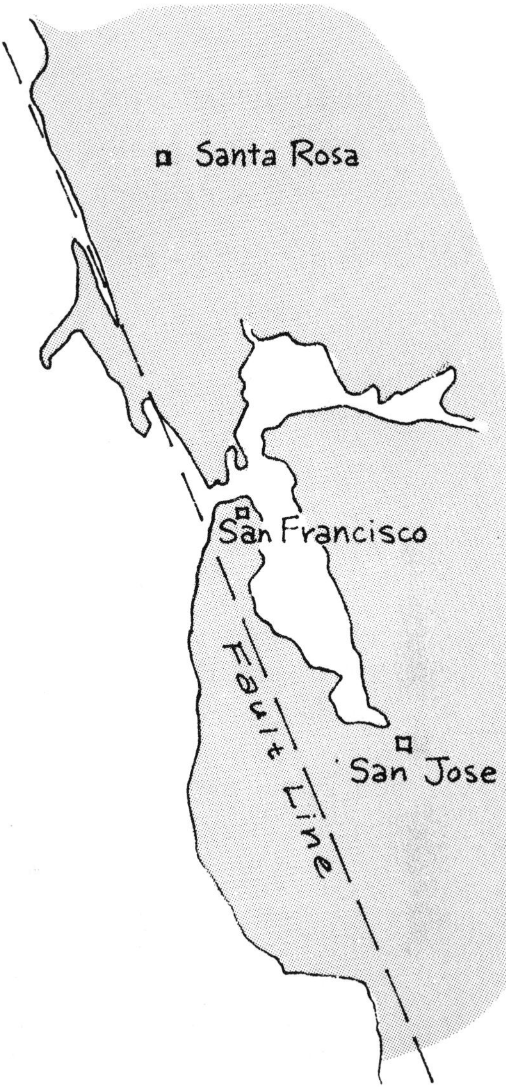

As the source starts at the epicenter and zips along the fault, it compresses the earthquake wave energy thereby increasing the earthquake's power in the direction the source is moving. In the great 1906 San Francisco earthquake, the city of Santa Rosa was damaged much more than the city of San Jose, even though San Jose was closer to the epicenter, because the rip ran along the fault from San Jose toward Santa Rosa.

How fortunate that the speed of the rip is less than the speed of the slowest earthquake wave (rip speed $\sim$ 2/3 wave speed), for were it otherwise, a shock wave could develop, which would vastly multiply the destructive power of earthquakes.

Incidentally, flip this page back and return to the answer to NEAR THE FAULT. Notice in the middle sketch that the wave pattern about the log thrown into the water is slightly narrower at the top end. It should be even more that way, for if the contact of the log with the water is analogous to the earthquake rip, then the log should drop into the water at an angle so the lower end of the log hits the water first. Then the log in effect "rips" along the water surface and generates surface waves more crowded toward the upper end.

UNDERGROUND TESTS

One of the difficulties with verification of the Strategic Arms Limitation Treaty, SALT, is that there is no easy way to distinguish between underground atomic bomb tests and natural earthquakes.

a) true
b) false

ANSWER: UNDERGROUND TESTS

The answer is: b. There are in this world two kinds of waves. There are compression (or longitudinal) waves like the wave that goes shuddering through a string of railroad cars when struck by a moving car, and there are transverse waves, like the waves in a flag. In transverse

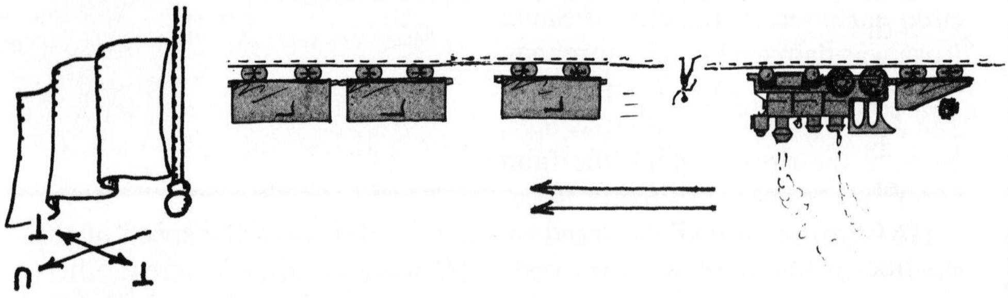

waves the displacement of the flag is in direction T which is transverse to the direction in which the wave is moving, direction U. In compression waves the displacement and the wave both move along the same line.

Earthquakes occur when a shear strain that has built up in a rock mass suddenly snaps. This snap relieves the strain and the rock mass momentarily quivers and in quivering sends out waves. Initially it sends out transverse waves which we see in the figure travel from faces **A** and **D**, compression waves from faces **B** and **E**, and rarefaction waves from faces **F** and **C**. So an earthquake sends out all types of waves.

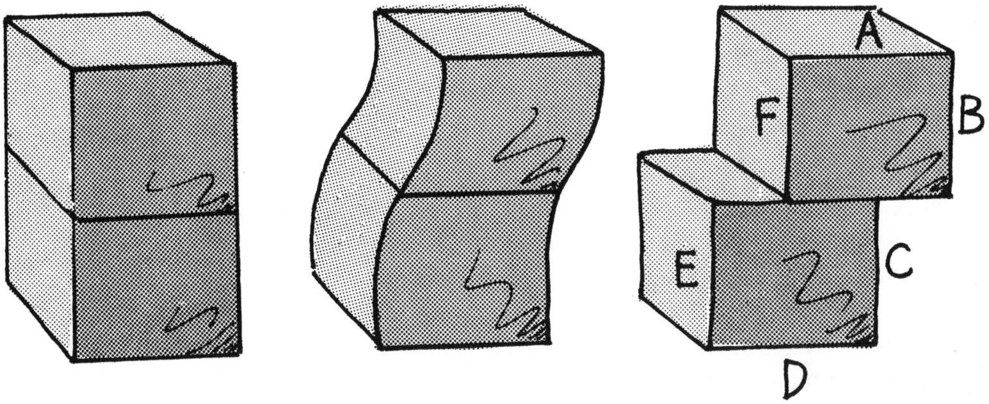

On the other hand an explosion (in air or underground) sends out only one kind of wave, compression waves. An "earthquake" with only compression waves is always a human-made earthquake. It is a dead give-away!

There are other distinctions that can be drawn between waves. For example flag waves and water waves are surface waves. Sound waves and earthquake waves are body waves because they run through three-dimensional volumes.

Sound waves are purely compression waves, and earthquakes are both compression and transverse waves. Are there any purely transverse body waves in nature? Yes. Light waves (and all electromagnetic radiation) are purely transverse. Because earthquake waves have the characteristics of both sound and light waves, they are the most complex body waves in nature.

BIO-RHYTHM

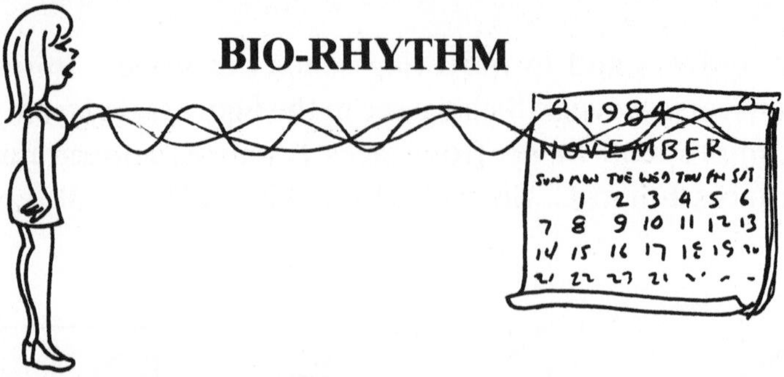

There is a popular idea that a person has certain "natural" rhythms which start at birth and run on through life, causing "up" and "down" days. You might then think that if you knew a person's date of birth and the period (or frequency) of these natural rhythms (or waves), you could figure when a person's up and down days would occur. In fact, some Hollywood types pay hundreds of dollars to have these calculations made. However, no clock runs perfectly.

How much error is in a human bio-rhythm clock? Well, let us take a well known bio-rhythm, the menstrual cycle which is about one month. About, because under normal circumstances a woman can predict the date of the onset of her next period when it is yet one month away with an error, let us say, of plus or minus one day ($\pm$ 1). Now if a woman tries to predict the onset of her period two months away, it will be more difficult. The prediction might be two days in error. But on the other hand, the errors might cancel each other, if one period comes a day early, and the next a day late.

Now suppose she tries to predict the onset of her period very far in the future, say 16 months hence. It might be hoped that there would be as many late as early periods, so the error would cancel out, and some of it usually does, but not all of it.

Little by little, the error accumulates. Suppose you throw 16 pennies. Heads represents an early period, tails a late period. You would not expect *exactly* the same number of heads and tails. On the average, there would be four more of one side than the other. If you throw 100 pennies, they seldom split 50/50 — there are usually more of one side than the other — the average is ten more of one side than the other. This number more is the error, and the average error statistically is the square root of the number of pennies thrown: $\sqrt{16} = 4$ and $\sqrt{100} = 10$. So the woman will have a 4-day average error in predicting the day of her period 16 months into the future.

Now the question: If a typical bio-rhythm is one month, with an error of plus or minus one day, and the rhythm has been running for 20 years (for a 20-year-old person), what will be the average accumulated error in predicting the date of an "up" or "down" day?

a) about zero days b) about 3 days c) about 1 week
d) about 2 weeks e) more than a month

ANSWER: BIO-RHYTHM

The answer is: d. Twenty years of twelve months is a total of 240 months. The square root of 240 is roughly 15 ($\sqrt{240} = 15.49133\ldots$). The average difference between the total number of early and late months will be 15. So at one day off per month, the cumulative error will be 15 days. Should long-range bio-rhythm forecasts be taken seriously?

MORE QUESTIONS (WITHOUT EXPLANATIONS)

You're on your own with these questions which parallel those on the preceding pages. Think Physics!

1. A note is sounded when you tap a drinking glass partially filled with water. If the glass is filled with more water, the pitch of the sound produced by tapping is
   a) higher b) lower

2. Resonance occurs when an object is forced to vibrate by external vibrations that are
   a) at a higher frequency b) at a lower frequency c) of large amplitude
   d) matched to the natural frequency of the object

3. The ringing of a bell involves a range of frequencies rather than one exact frequency. The frequencies in the range are all quite close, and eventually work their way out of phase and interfere destructively with each other as evidenced by the bell gradually falling silent. So a bell with the purest tone will ring for a
   a) long time b) short time

4. The greater the amplitude of a wave, the greater its
   a) frequency b) wavelength c) loudness d) all of these e) none of these

5. The pair of waves shown are generated at the ends of a long rope and travel toward each other. Will there be any instant at which the amplitude of the rope is everywhere zero?

   a) Yes b) No

6. These waves, I and II, are added together to produce
   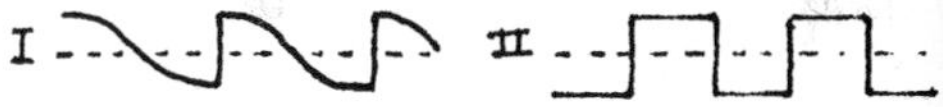
   a)  b) 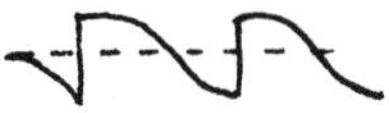 c) 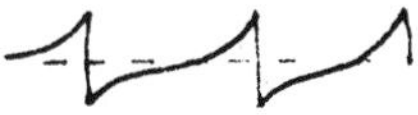 d) none of these

7. Which of these waves cannot be produced by the superposition of numerous sine waves?
   a) 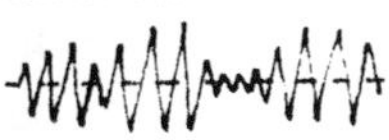 b) 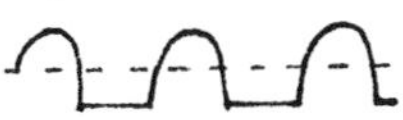 c)  d) But all can.

8. Two tuning forks are sounded, one of 254 Hz and the other 256 Hz. The resulting beat frequency will be
   a) 2 Hz b) 4 Hz c) 255 Hz

9. The waves generated by the waterbug show it to be moving
   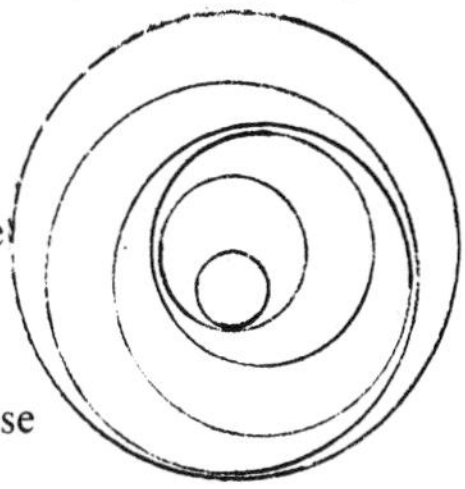
   a) back and forth b) up and down c) in circles d) none of these

10. When a sound source approaches you, there is an increase in the sound's
    a) speed b) frequency c) wavelength d) all of these e) none of these

One of the secrets of doing physics is to
keep in mind what you don't know.
The trick is to get from what you know,
to what you want to know,
without going through what you don't know.

Light

What is more familiar than light? What is less familiar than light? What is light? What is it made of? Does it have weight? Do you realize that you can't even see light? You can see the sun, you can see a bird, you can see your environment—but that is not seeing light itself! Light is some kind of "stuff" that must move to exist. If ever it should rest, even for a moment, it would cease to exist. It is really amazing that so much has been learned about such phantom-like stuff.

You can't get smart
until you realize how dumb you are.
And you can't realize how dumb you are,
until you learn how to talk to yourself.

PERSPECTIVE

A cloud casts a shadow on the ground as illustrated below. If you actually measured the size of the cloud and the size of the shadow you would find that the cloud is

a) substantially larger than its shadow
b) substantially smaller than its shadow
c) about the same size as its shadow

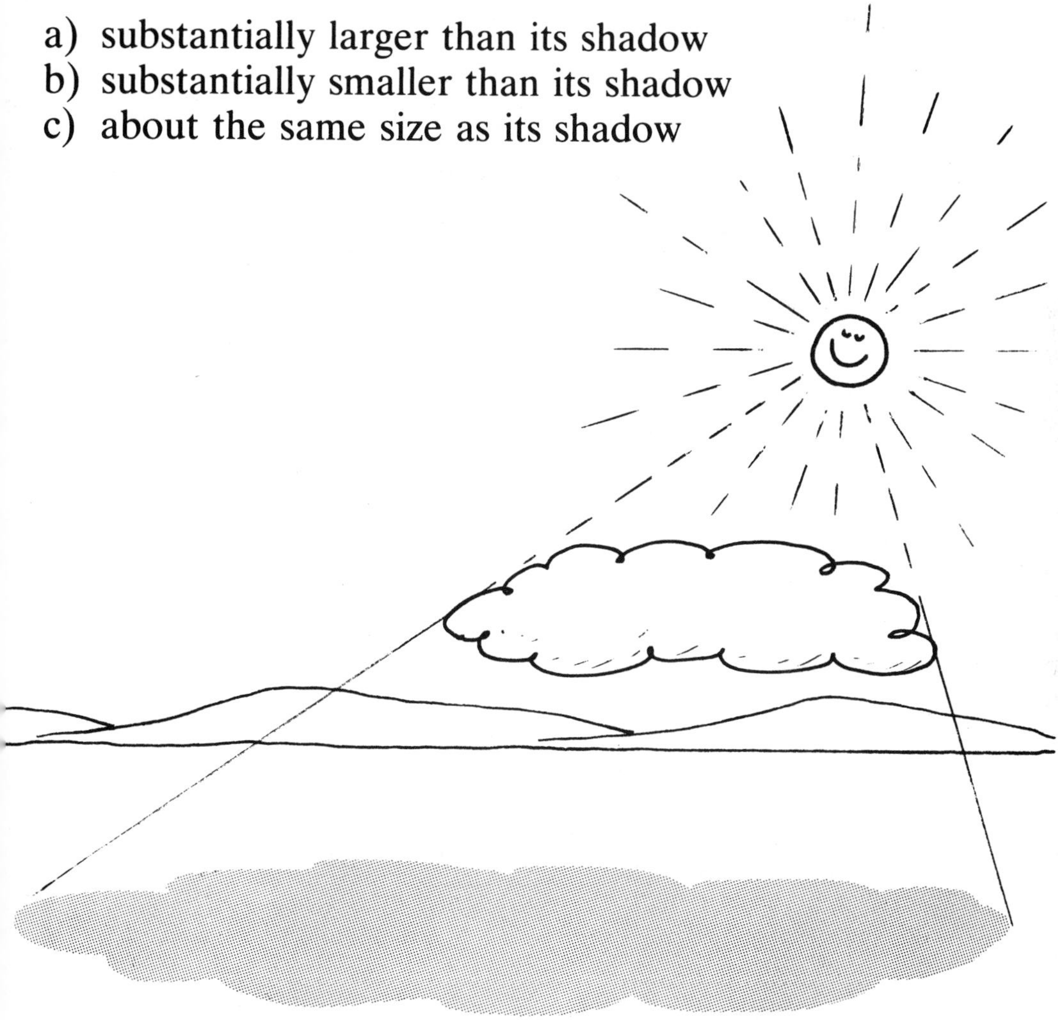

ANSWER: PERSPECTIVE

The answer is: c. The sun is so far away that the rays of light from the sun are practically parallel to each other by the time they reach the earth. Why then does it look like they are spreading out as a shaft of sunlight comes through the clouds? For the same reason distant railroad tracks appear to spread apart as they are nearer to

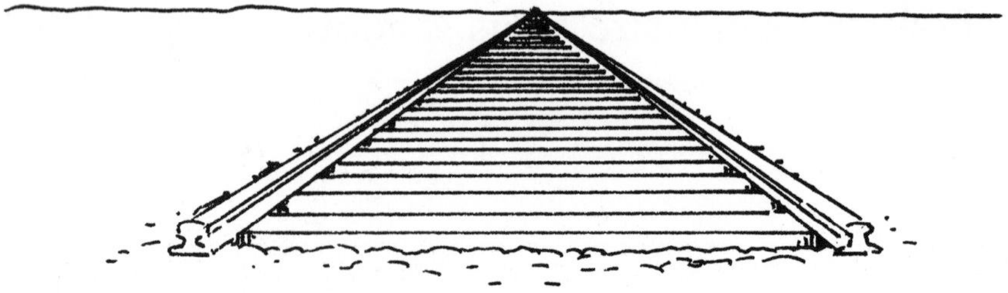

you even though they are, in fact, perfectly parallel. The drawing in question shows the cloud and its shadow between us and the sun. If the sun were in back of us and the cloud in front of us, then the shadow of the cloud would be further away and appear smaller than the cloud.

WHAT COLOR IS YOUR SHADOW?

On a clear and sunny day you are on snow and look at your shadow. You see that it is tinted

a) red
b) yellow
c) green
d) blue
e) not at all

ANSWER: WHAT COLOR IS YOUR SHADOW?

The answer is: d. The part of the snow in direct sun shows the color of the sun: yellow white. The snow in your shadow gets no direct sunlight, but is illuminated by light from the blue sky. Perhaps it is the blue color of shadows that makes people associate blue with cold.

LANDSCAPE

You are looking at two dark hills, one more distant than the other. The hill that appears a bit darker is the

a) near hill

b) distant hill

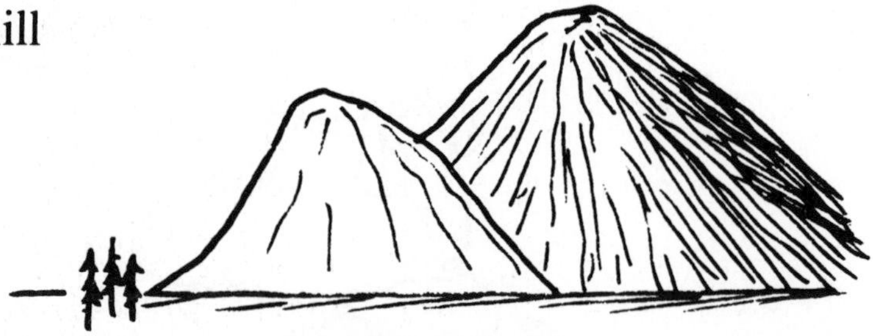

c) . . . both appear equally dark

ANSWER: LANDSCAPE

The answer is: a. The closer hill is darker. When you look at the hills, most of the light you see comes from the air between you and the hills. The air scatters light from the sky above and scatters some of it into your eyes. There is more air between you and the distant hill than between you and the near hill, and that means more air to scatter light towards you. So distant mountains appear bluish because the atmosphere between you and the mountains scatters blue light. Similarly, the sky is brighter when you look towards the horizon and darker when you look straight up (unless the sun is straight up).

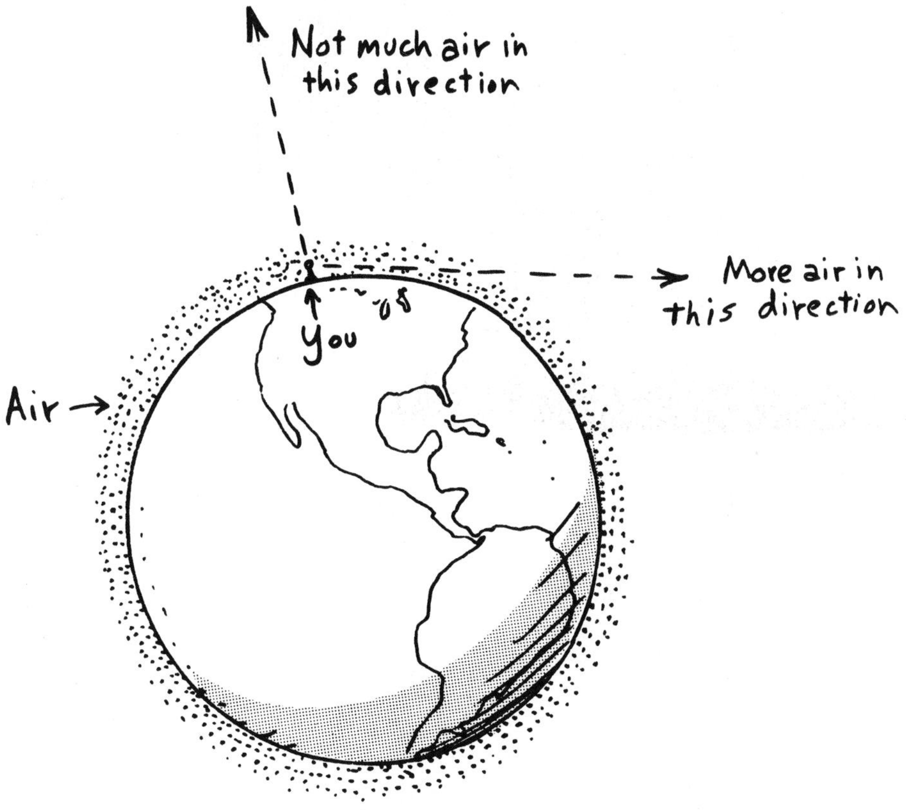

ARTIST

The sun has just gone down behind the hill. Smoke from a campfire is going up to the sky. The part of the smoke in front of the hill (part I) is visible because it is slightly brighter than the hill. The part of the smoke above the hill (part II) is visible because it is slightly darker than the sky. What are the proper color tints to apply to part I and part II?

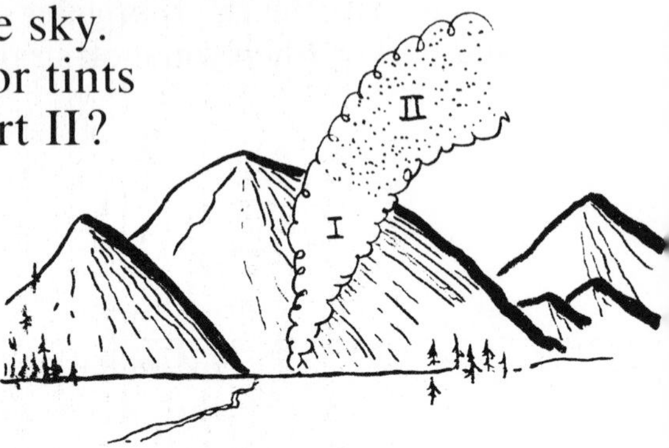

a) Tint part II blue and part I red.
b) Tint both parts red.
c) Tint both parts blue.
d) Tint part II red and part I blue.
e) There is no reason why any part need have any tint— it is up to the artist.

ANSWER: ARTIST

The answer is: d. When light hits smoke some of it goes through the smoke and some is scattered. That which is scattered is likely to be bluish and that which penetrates is likely to be reddish (that's why the sky is blue and sunsets are red). In part II of the smoke we see the fraction of the light that could penetrate through the smoke from the bright sky behind, so it is reddish. In part I no light comes from behind the smoke since a dark hill is behind. The light coming from part I is due to the smoke scattering (and reflecting) light from the sky above and in front of it. So part I is bluish.

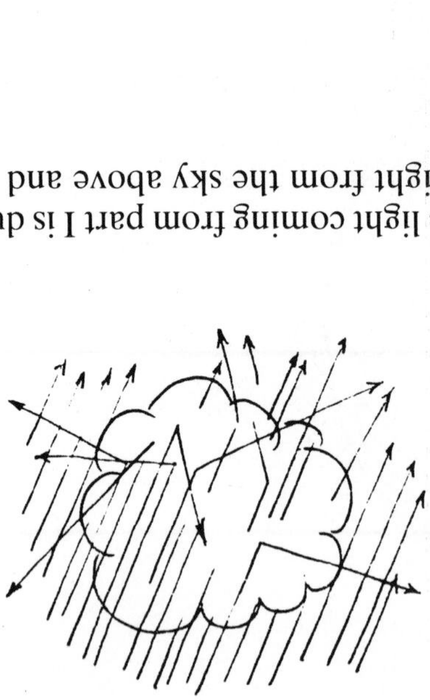

RED CLOUDS

Sometimes the clouds at sunset are very red, and sometimes they are only a little red. The very red clouds are usually the

a) low clouds
b) high clouds

ANSWER:
RED CLOUDS

The answer is: b. By looking at the sketch you can see that the sunlight scattered from a high cloud can travel through three times as much air as can light scattered from a low cloud. The longer the path through the air, the redder the light.

By the way, sometimes there are clouds at various heights. How can you tell which are high and which are low? The low ones get dark first. Can you see why? The last clouds to go dark will be the highest and therefore reddest. Check it out for yourself.

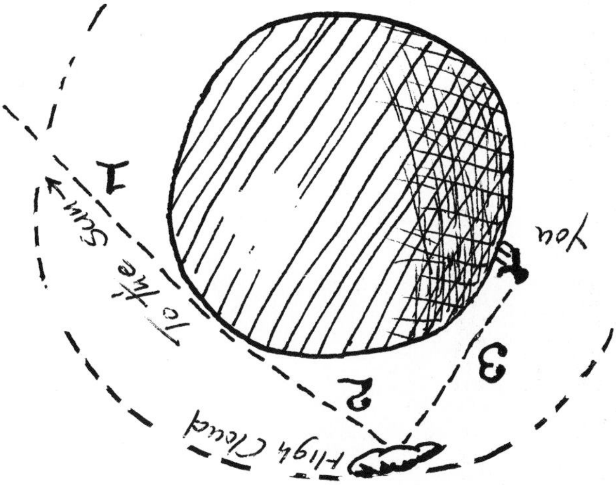

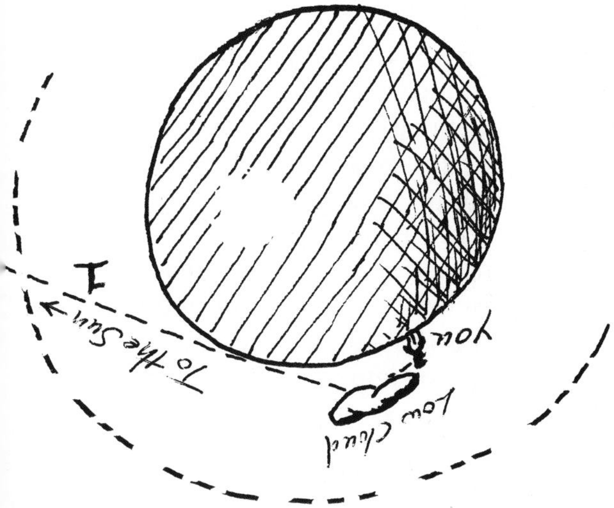

EDGE OF NIGHT

Twilight is the time between when the sun sets and when the sky gets dark. Twilight lasts longest at

a) New Orleans, Louisiana
b) London, England
c) neither; it lasts the same time at both locations.

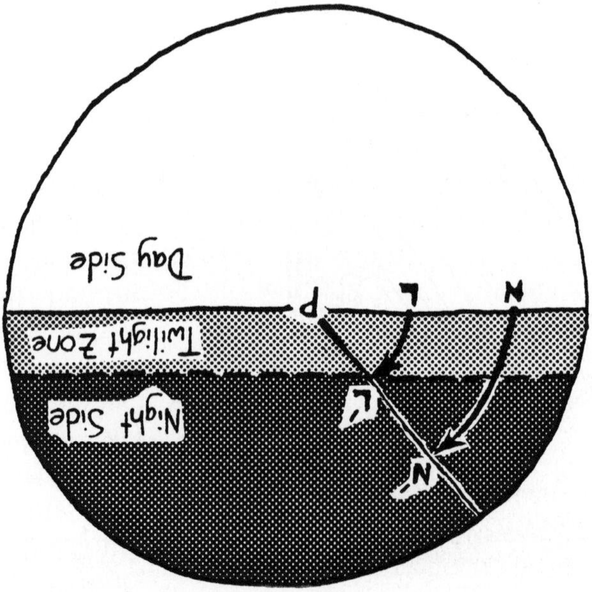

ANSWER: EDGE OF NIGHT

The answer is: b. The sketch is a view of the earth, looking straight down at the north pole, marked **P**. Half the earth is in day, half in night. The twilight zone extends around the earth on the edge of night.

Now London is closer to the north pole than New Orleans. In three hours the earth spins London from **L** to **L′**, and New Orleans from **N** to **N′** (in three hours the earth turns 45°). The sun sets at **L** and twilight ends at **L′**, so London spent all three hours in the twilight zone. But for New Orleans three hours was enough to carry it from sunset through the twlight zone and well into the night. One summer evening I took a walk in New Orleans. The next evening (after an airplane ride) I took a walk in London. The change in the duration of twilight was most amazing.

ANSWER: TWILIGHT

The answer is: c. The sketch shows three views of the earth, as it would be seen by someone far above the north pole, marked **P**. The first view shows the winter situation. The sun is below the equator so most of the northern hemisphere, including the north pole itself, is in darkness. The last view shows the situation reversed in summer. The

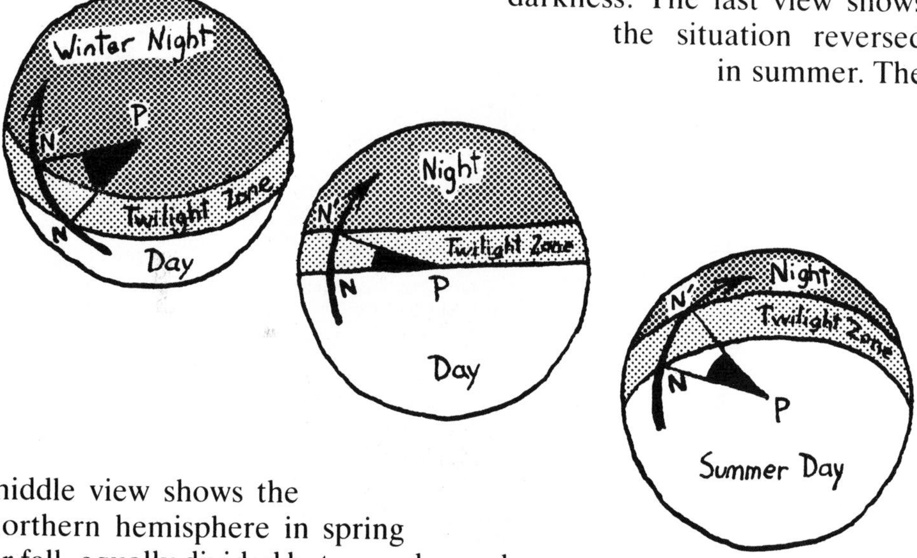

middle view shows the northern hemisphere in spring or fall, equally divided between day and night—that's why it is called the equinox.

Because the earth spins around the pole, a person standing on the planet would be carried in an arc, like that from **N** to **N′**, around the pole. As seen from the sketches, the length of arc

TWILIGHT

At any location on earth, twilight is shortest

a) in winter
b) in summer
c) between winter and summer—on the equinox

and time required to pass through the twilight zone are shortest on the equinox.

Now let me bring up something else. Does twilight last longest in the mountains or near sea level? It lasts longest at sea level. The reason there is such a thing as twilight is because if you are at **Y**, in the earth's shadow, on the dark side of the earth, the air high above you at **A** is not in the shadow. The air at **A** scatters some sunlight down to you at **Y**. When you are high up in the mountains there is less air above you to do the scattering, so there is less twilight. In Ecuador in the Andes, on the equinox the day ends "like someone switched off the light." On the moon where there is zero air, there is zero twilight.

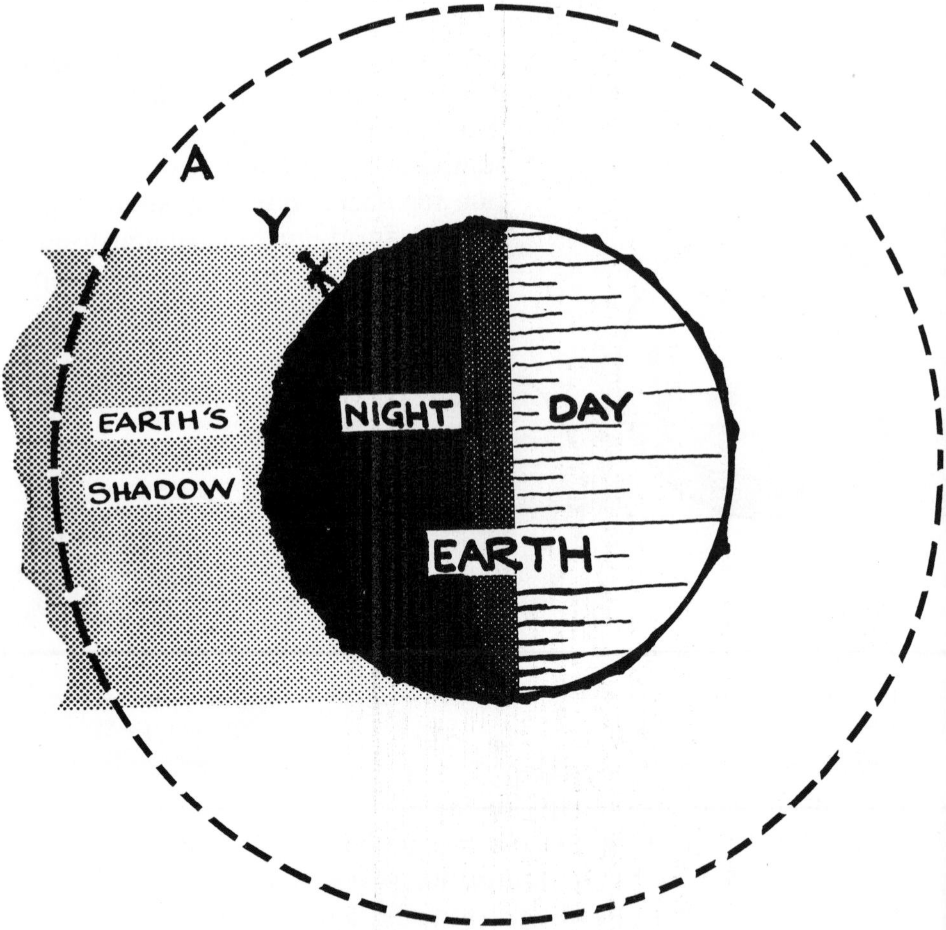

DOPPLER SHIFT

Jupiter takes about 12 years to go once around the sun, so compared to the earth it is practically motionless. Jupiter has a moon, Io, which takes about 1¾ days to go once around Jupiter. During the 6 months the earth takes to move away from Jupiter, from A to B, the moon Io as perceived from the earth orbits about Jupiter

a) more frequently than
b) less frequently than
c) at the same frequency as

it will appear during the following 6 months when the earth moves from B to A.

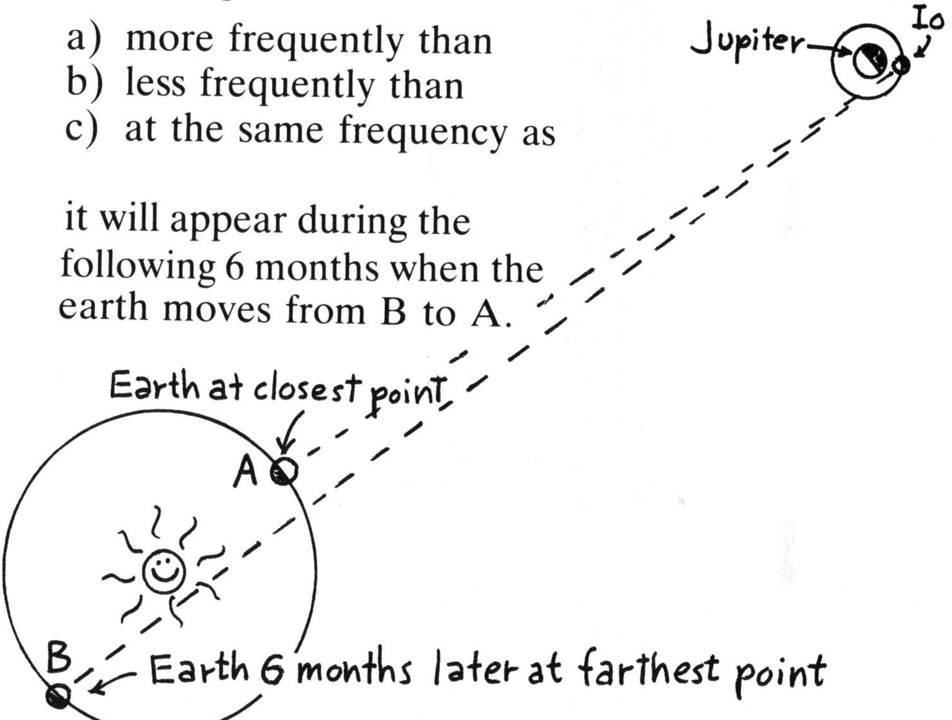

ANSWER: DOPPLER SHIFT

The answer is: b. As you move away from any periodically recurring signal its frequency appears to be lower than when you move toward it. So as the earth moves away from Jupiter each revolution of Io appears to take longer than it should. It turns out that when the earth gets to B this cumulative discrepancy adds up to 1000 seconds, a delay that has some historical significance, as we shall see in the next question.

TAURUS

High in the winter night sky resides Taurus the Bull. The horn of Taurus is a little cluster of stars, the Hyades. Over many decades the stars in the Hyades cluster are observed to be slowly converging towards a common point in the sky! Will these stars all collide?

a) Yes b) No

Does the light from these stars show

a) a blue Doppler shift?
b) no Doppler shift?
c) a red Doppler shift?

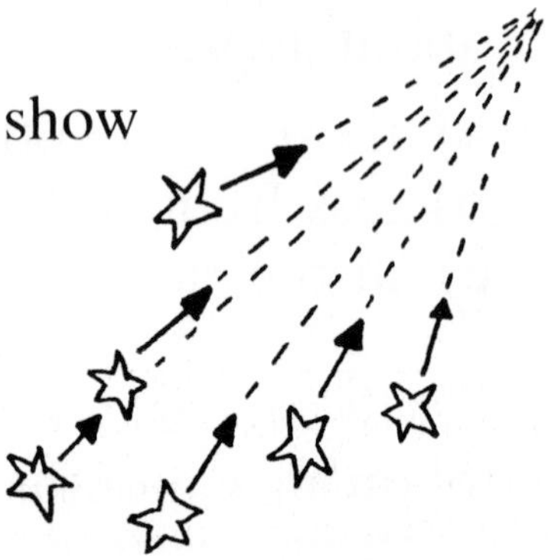

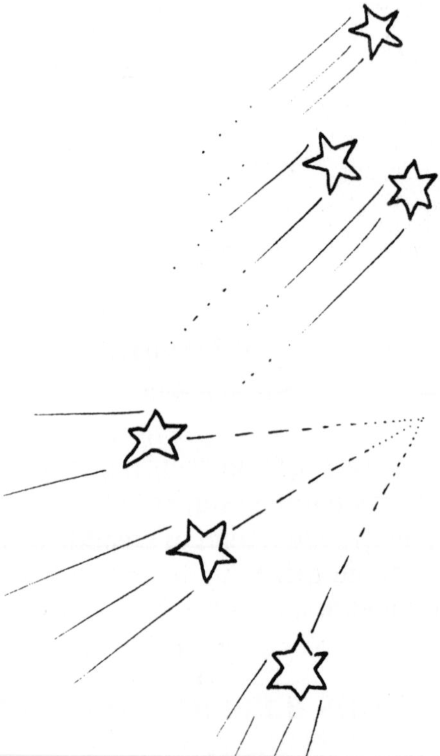

ANSWER: TAURUS

The answer to the first question is: b. It would be quite a coincidence if all the stars had agreed on a celestial suicide pact.

The answer to the second question is: c. The stars in the cluster are, in fact, moving through space parallel to each other, like a fleet of ships. But the fleet is moving away from the earth, so the parallel paths of the stars seem to converge, like the railroad tracks. Of course, we get a red Doppler shift from stars moving away from us.

SPEED OF WHAT?

Astronomers for years had noted the great regularity of the eclipse of Jupiter's major moon Io. In about 1675 the Danish astronomer Roemer noted a 1000 second delay in the eclipse of Io when viewed from the earth at its farthest orbital position from Jupiter as compared to its closest position six months earlier (see sketch). Using this information Roemer was able to calculate the speed of

a) light
b) the orbital motion of the earth
c) the orbital motion of Io about Jupiter

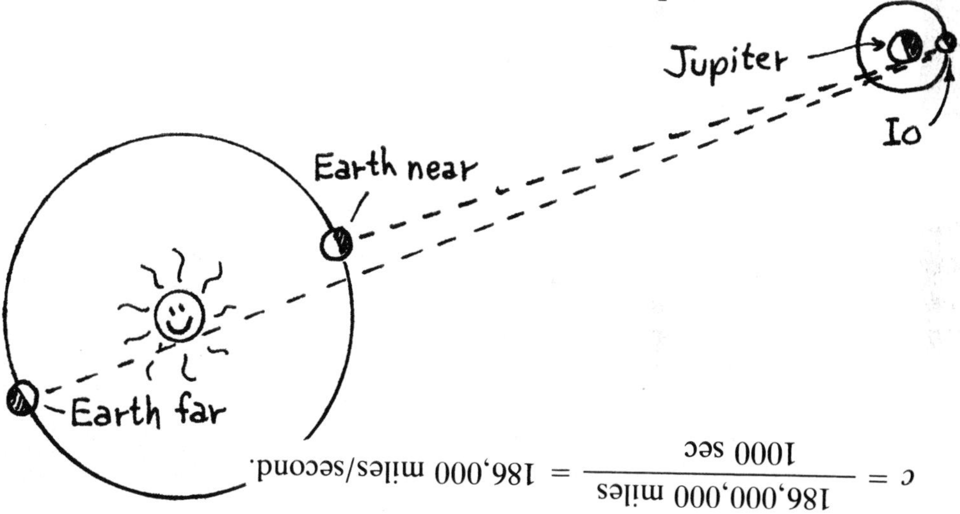

ANSWER: SPEED OF WHAT?

The answer is: a, the speed of light. The speed of anything can be calculated by simply dividing the distance traveled by the time of travel. The distance that light travels in 1000 seconds is the diameter of the earth's orbit about the sun. This is 300,000,000 kilometers (or 186,000,000 miles). So the speed of light is simply

$$c = \frac{300{,}000{,}000 \text{ km}}{1000 \text{ sec}} = 300{,}000 \text{ km/second, or}$$

$$c = \frac{186{,}000{,}000 \text{ miles}}{1000 \text{ sec}} = 186{,}000 \text{ miles/second.}$$

RADAR ASTRONOMY

Objects are detected by radar, that is, by sending out radio signals and then receiving the radio waves reflected back from the objects. In the sketch we see that a radar signal is sent from earth to a planet. The signal from which part of the planet returns to the earth first?

a) A b) B c) C

d) no marked part

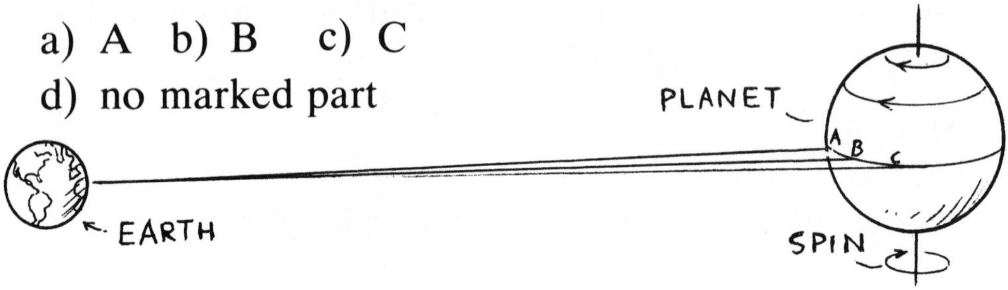

The signal from which part returns with the highest frequency?

a) A
b) B
c) C
d) no marked part

ANSWER: RADAR ASTRONOMY

The answer to the first question is: a. Point A is the point on the planet nearest to the earth, so the signal reflected from A returns first.

The answer to the second question is: c. The planet is spinning and point C is moving towards the earth faster than any other part of the planet. So the reflection from C has the largest Doppler shift.

By using the delay time and the Doppler shift jointly, radar astronomers can tell from what point on a planet the radar signal is being reflected. From this they can make "radar photographs" of the planet's surface.

TWINKLE TWINKLE

When the stars twinkle in the nighttime sky, do planets twinkle also?

a) Yes, both twinkle
b) No, only the stars twinkle

From the point of view of an astronaut orbiting the earth

a) only the stars would twinkle
b) only the planets would twinkle
c) both stars and planets would twinkle
d) neither stars nor planets would twinkle

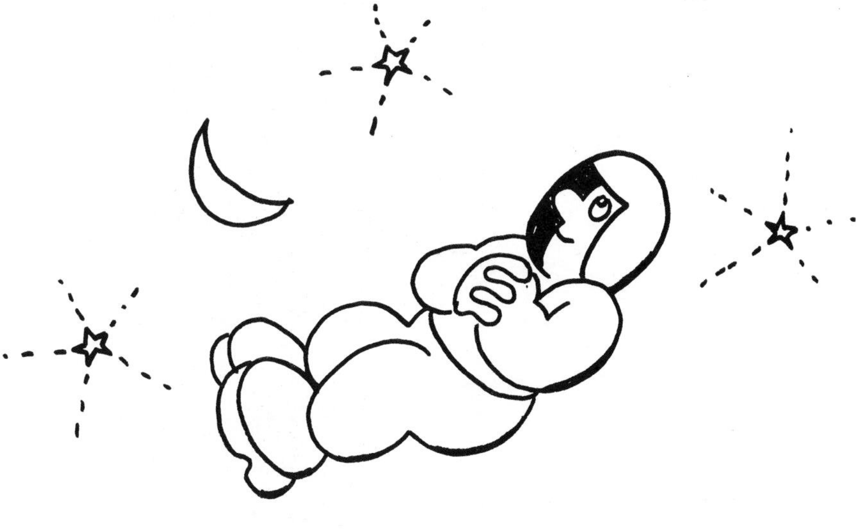

ANSWER: TWINKLE TWINKLE

The answer to the first question is: b. Twinkling is the result of variations of air densities in the earth's atmosphere. These variations are evident in pronounced form in the daytime when we see the shimmering of objects through the heated layers of air above a hot surface. Any distant light source will shimmer if enough air with its incessantly swirling currents separates it from the observer. This shimmering in the turbulent layers of the atmosphere deviates your line of sight. In the drawing we see that if your line of sight is deviated to A or B it will miss the star. But this deviation will not miss the planet because the planet is a larger target—that is, it subtends a larger angular diameter in the sky. You *can* see the disk of Jupiter with a six-power telescope. You *can't* see the disk of a star even with a six-hundred-power telescope.

As seen from and through the vacuum of space, no twinkling occurs, so the answer to the second part is: d.

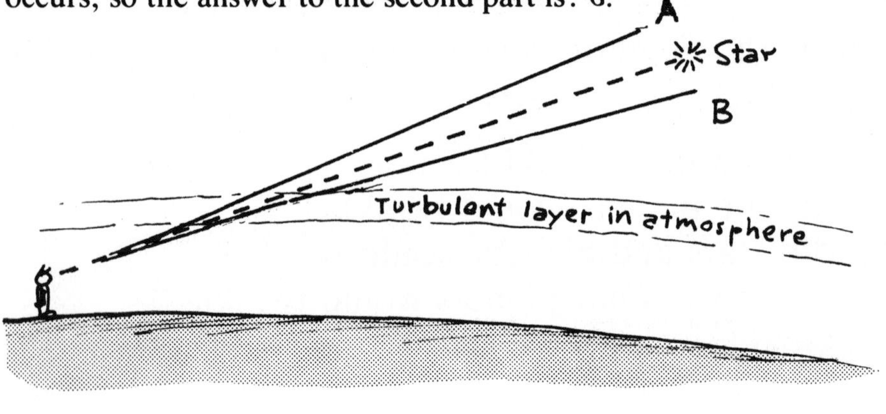

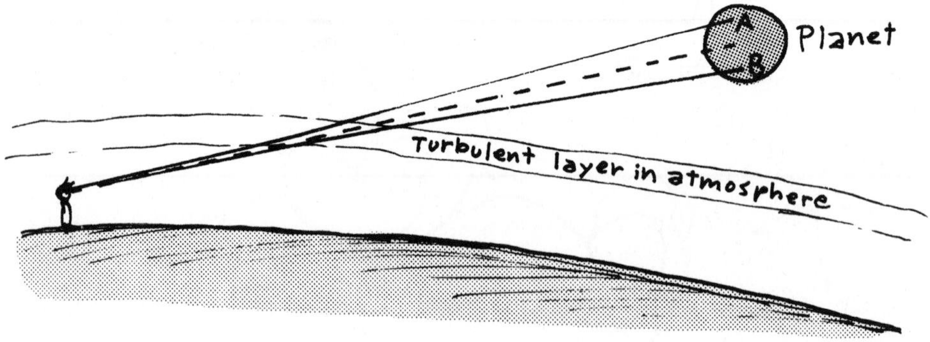

HOT STAR

Can you tell, just by looking, which are the hottest stars in the sky?

a) Yes, you can tell
b) No, you cannot tell

The hottest stars in the sky are the brightest stars in the sky.

a) True
b) False

ANSWER: HOT STAR

The answer to the first question is: a. The hot stars are blue, the cooler stars are red. Remember: as a piece of iron heats up it first glows red, then to orange, yellow, and white. If you could heat it further it would eventually glow blue.

The answer to the second question is: b. The apparent brightness of a star depends on three things: (1) how hot it is, (2) how far away it is and (3) how big it is. Thus, a big cool star that is near could outshine a distant or small hot star.

SQUEEZED LIGHT

During the last century some physicists thought of light as a gas. They thought this because they understood the properties of a gas and tried to see if that understanding would help them understand light — and it did! In particular, it allowed them to think of the "temperature" of light.

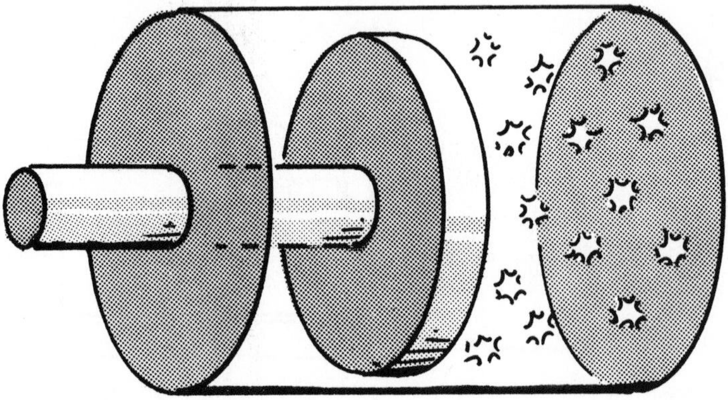

Just as gas in a cave comes to the temperature of the cave, so also light inside a glowing cave comes to the temperature of the cave. The old-timers also imagined that the light could, like a gas, be put inside a cylinder and compressed by a piston. The imagined cylinder and piston had to be completely reflecting on the inside. Just as compressing a gas increases its density and temperature, so also would compressing light increase its density and temperature. Now as the temperature of the light increases, its color

a) shifts from blue to red
b) shifts from red to blue

ANSWER: SQUEEZED LIGHT

The answer is: b. As the light was compressed, thereby increasing its temperature and density, its color would shift from red to blue. There are two ways to see why. First, by the Doppler effect: if a source that emits light approaches you the light waves are compressed and the light has a blue shift. (The shift is red if the light source runs away from you.) Now if light bounces off a mirror that is approaching you, that also produces a blue Doppler shift. The light in the cylinder bounces off the advancing piston as it is compressed and so gets a blue shift. Second, the light waves in the cylinder are quite literally compressed by the piston. Thereby red, long wave light is made into blue, short wave light.

Of course, no one compresses light in a cylinder — but interestingly enough we see that this gas model for light provides a means of explaining why blue light is hotter (comes from a hotter place) than red light. For example, blue stars are hotter than red stars.

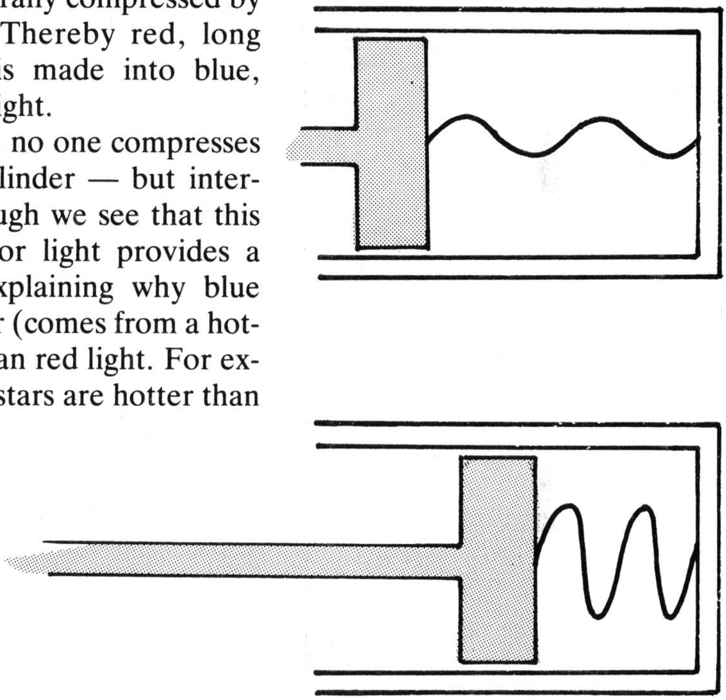

The gas model for light works here because light is composed of particles called photons just as gas is composed of particles called molecules, and the particles, regardless of name, bounce off the piston in such a way as to conserve both momentum and energy. Surprisingly, the gas model for light was developed before the photon theory. The main reason light does not act even more like a gas is because gas molecules interact strongly with each other while light photons do not.

1+1=0?

Can two light beams ever cancel each other out and result in darkness?

a) No, light can never make darkness — this would violate the Law of Conservation of Energy
b) Yes, there are various ways to add light to more light and come out with no light

ANSWER: 1+1=0?

The answer is: b. Light is a wave and waves can combine so as to reinforce each other or cancel each other. The same can be done with water waves where the crest of one wave overlaps the trough of the other wave (or sound waves or any kind of waves). This is called destructive interference. OK, but this sounds a bit like a violation of conservation of energy. If light cancels light where does the energy go? It turns out that every time you make a set-up where light cancels light at one location there is another location — usually very nearby — where light reinforces light and all the energy that is missing from the canceled location shows up at the reinforced location. This is true for sound, water and any other kind of wave.

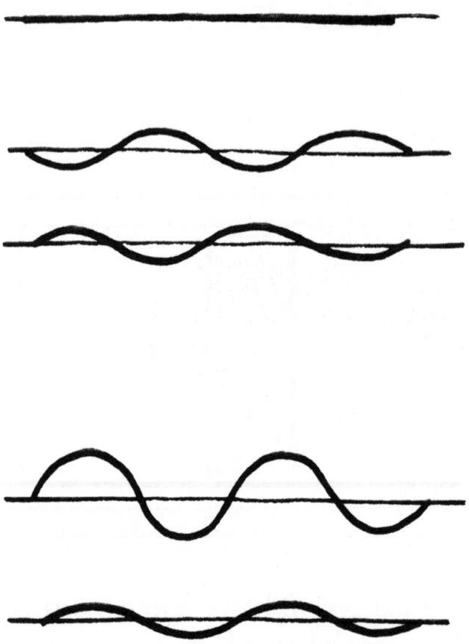

LEAST TIME

A lifeguard at L on the beach must rescue a drowning person in the water at P. Time is of the essence! Which path from L to P will take the least time? (Hint: consider the relative speeds of the lifeguard on land and in the water.)

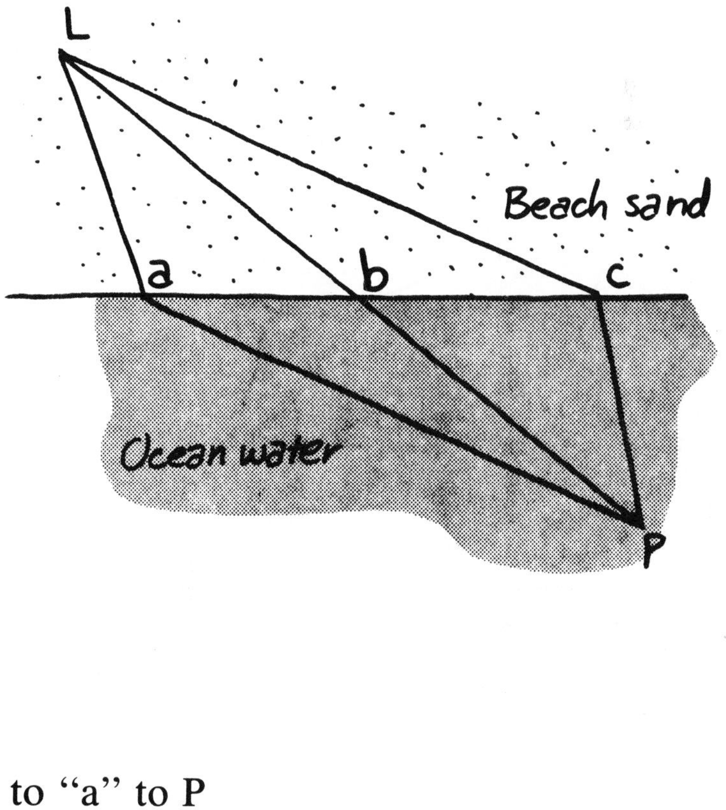

a) L to "a" to P
b) L to "b" to P
c) L to "c" to P
d) All these paths take the same time

ANSWER: LEAST TIME

The answer is: c. Why does the lifeguard not run directly from L to "b" to P in one straight line? Because the guard can go faster running over the beach than swimming through the water. The guard's speed on the L to "c" run and the reduced distance of the slow "c" to P swim more than compensate for the increased distance traveled. By running to "c" and then swimming to P the distance the guard must travel is not minimized but the *time* required for the rescue is minimized.

Suppose a dolphin was employed as a lifeguard. Which path from L to P would take the least time for the dolphin? The dolphin makes good time in the water, but can hardly move on the beach, so the dolphin goes from L to "a" to P.

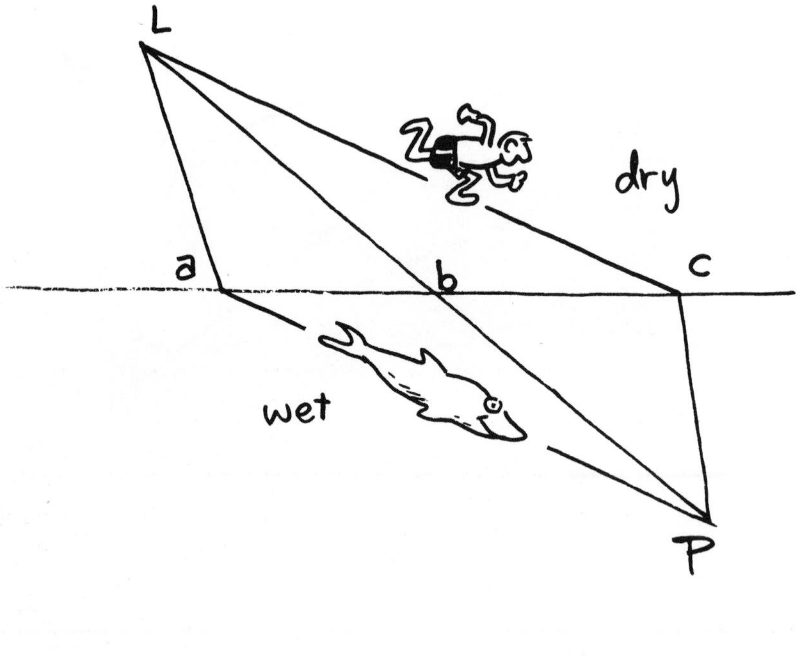

This Principle of Least Time is important in optics. It so happens that whenever light travels from one point to another it will always do so along the path requiring the least time.

SPEED IN WATER

If light takes the path of least time between two places, then the way it bends or refracts when it goes from air (or vacuum) into a transparent material suggests its speed in the material is

a) faster than its speed in air
b) the same as its speed in air
c) slower than its speed in air

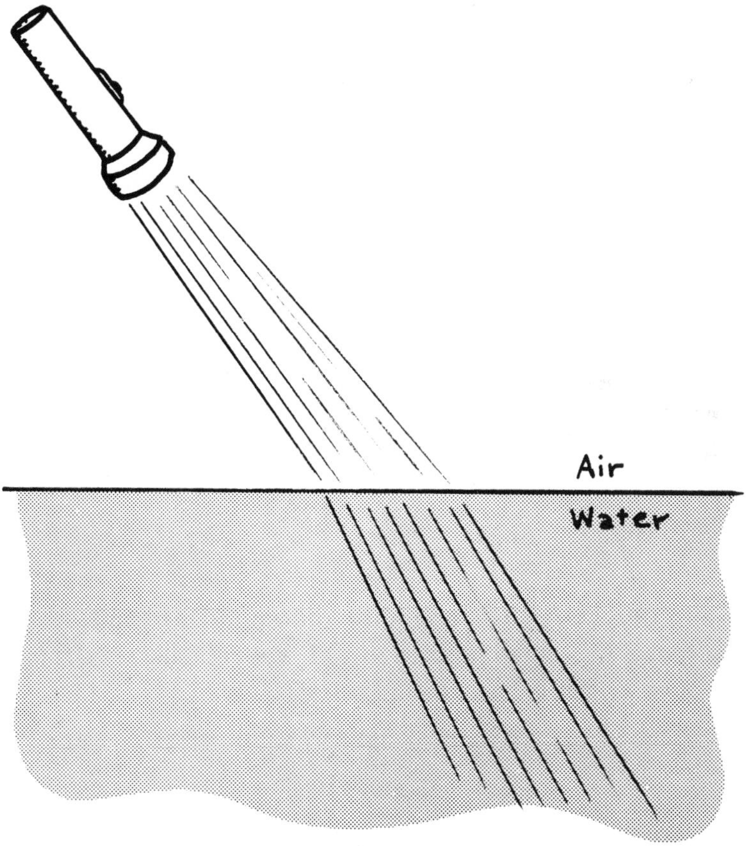

ANSWER: SPEED IN WATER

The answer is: c. Think about the human and dolphin lifeguards. The path of a light beam going into glass is like the path of the human lifeguard and not like the path of the dolphin. So if light, like the lifeguard, takes the path of least time we see that the speed of light is faster in the air and slower in the water. This is true for any transparent material.

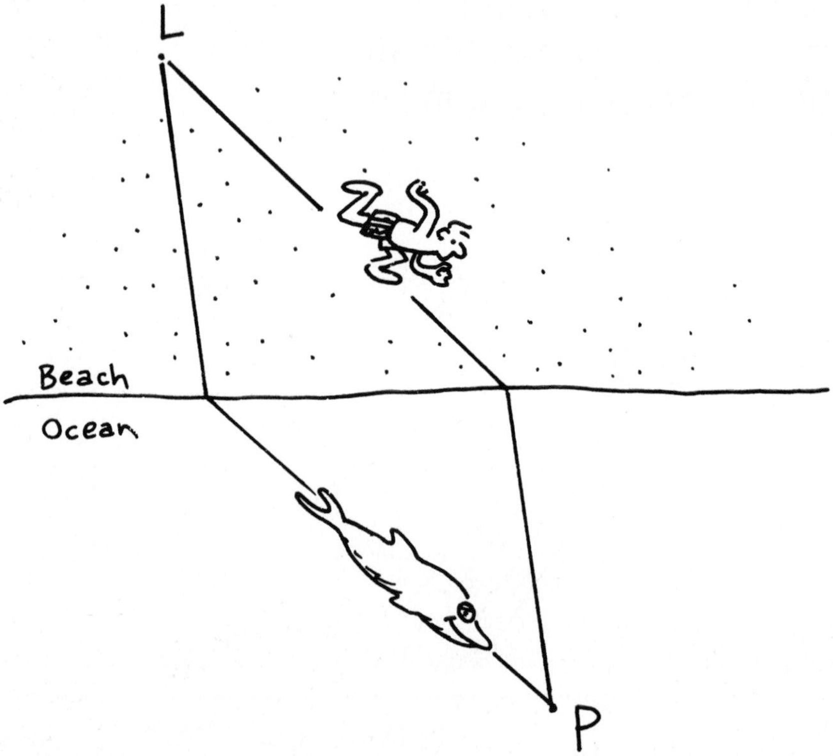

Incidentally, when we compare the speed of light in a vacuum with its speed in a particular material, we call the resulting ratio, *n*, the *index of refraction* of the material:

$$n = \frac{\text{speed of light in vacuum}}{\text{speed of light in material}} = \frac{c}{v}$$

For example, the speed of light in water is $\left(\frac{1}{1.33}\right) c$; so $n = 1.33$. In common glass, $n = 1.5$. For a vacuum $n = 1$.

REFRACTION

Suppose a pair of toy cart wheels connected by an axle is rolled along a smooth sidewalk and onto a grass lawn. Due to the interaction of the wheels with the grass, they roll slower there than on the smooth sidewalk. If the wheels are rolled at an angle onto the lawn they will bend from their straight-line path. Which of the sketches below shows the path they will take?

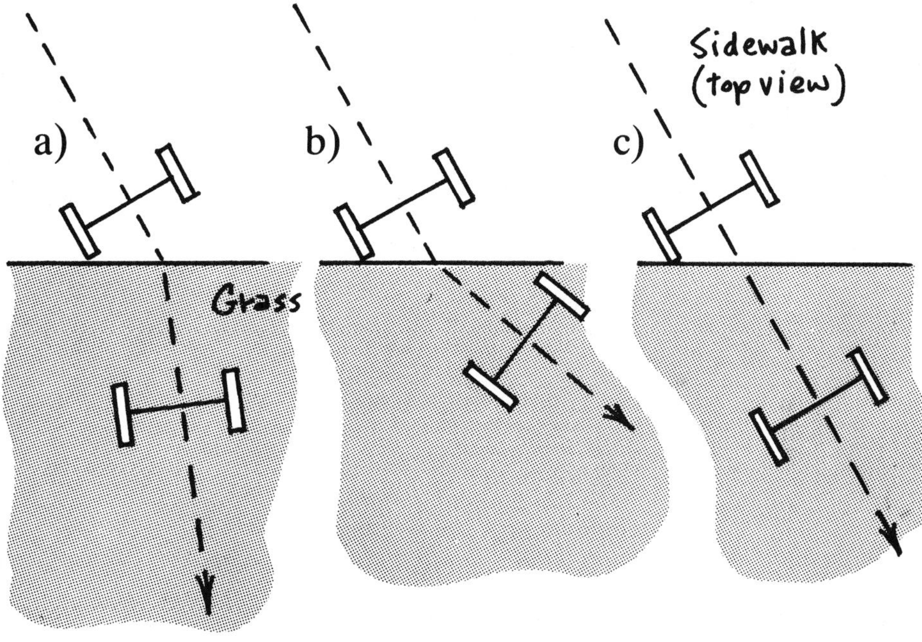

ANSWER: REFRACTION

The answer is: a. That's because the left wheel is the first to encounter the grass, where the greater interaction results in slowing. This acts as sort of a pivot, for the right wheel remains at its greater speed on the sidewalk until it meets the grass. When both wheels are in the grass they again move in a straight line. Likewise with wave-fronts of light incident at an angle upon a transparent medium.

LIGHT BEAM RACE

Three light rays start out from the candle flame together (at the same time). Ray A goes through the edge of the lens. Ray B goes through the center and Ray C through an intermediate part of the lens. Which ray arrives at the image on the screen first?

a) A
b) B
c) C
d) All arrive at the same time

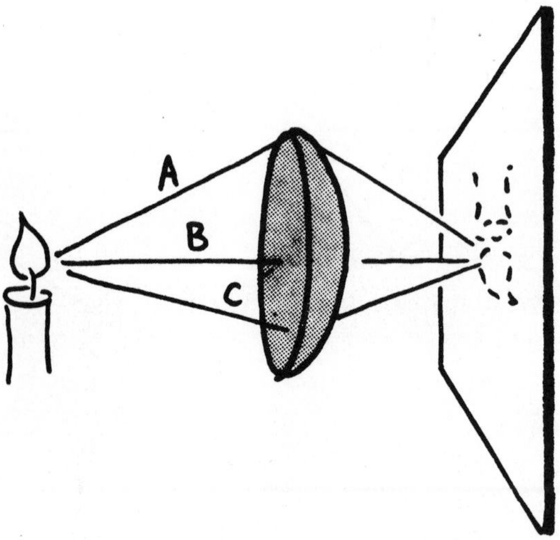

ANSWER: LIGHT BEAM RACE

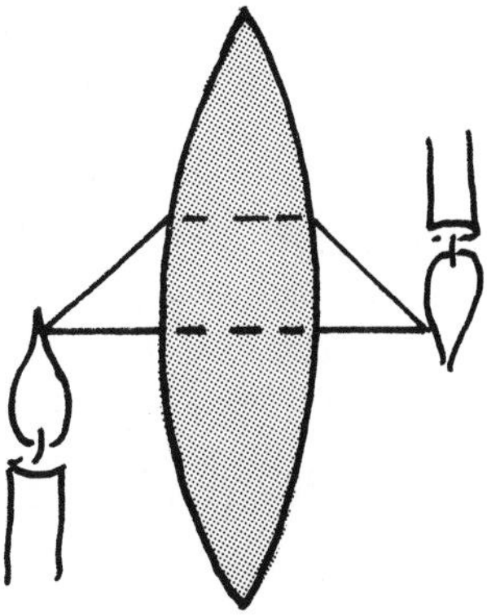

The answer is: d. The race is a dead heat. There are a bunch of reasons why this is so. We'll consider two. Reason I: There is the basic principle that a light ray going between any two points will go by the path that takes the least TIME. In our case the rays go between the candle and its image. If one path took less time than the others, all the rays would go on it. But we know all rays arrive by different paths, so no path must take less time than the others. But some paths are longer, like the ones through the edge of the lens, and some are shorter, like those through the center — so how can they all take equal time? The answer is that the longest overall paths have the shortest passages in the glass, and the shortest overall paths, like the one through the center, have the longest passages in the glass. Although the paths vary in their overall distances, they do not vary in the overall time required to follow these paths.

Reason II: Visualize the light as waves, not rays, and run an animation in your head. It should look about like this: All the waves that start together at the candle must reconverge together at the image. If a wave which starts at the candle at one particular time did not all reconverge at the image at the same time, then the wave could cancel itself out. If one path to the image took a little less time than another, then the part of the wave which took the faster path would always arrive ahead. When the waves recombine at the image they would partially or completely cancel each other. To get an image, we need reinforcement, not destructive interference, and to get that, all paths must take equal travel time.

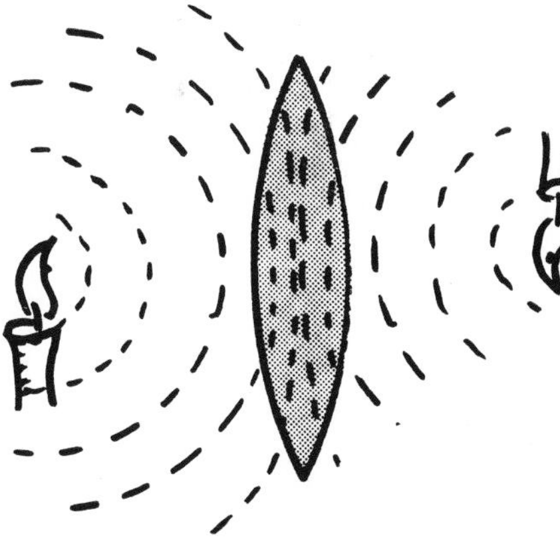

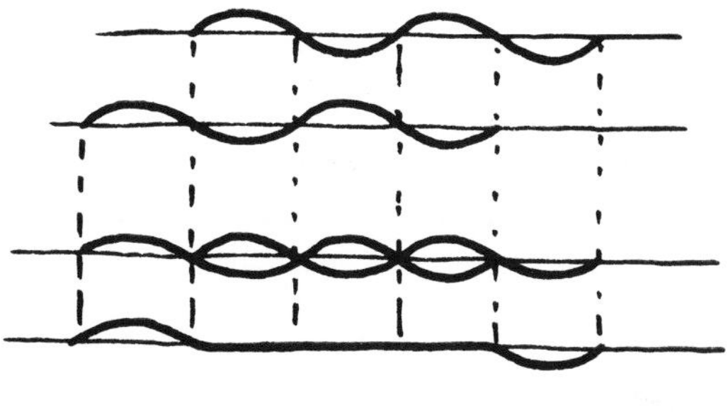

I COULD ALMOST TOUCH IT

A coin is under water. It appears to be

a) nearer the surface than it really is
b) farther from the surface than it really is
c) as deep as it really is

ANSWER: I COULD ALMOST TOUCH IT

The answer is: a. You estimate the distance of an object by using both eyes. The brain senses how much your eyes have to be crossed to converge on the object. The closer the object, the more the eyes have to be crossed. Now when water gets into the act, it bends the light rays as sketched so your eyes have to be crossed as if the coin were at position II even though it is really at position I. So the water makes it appear closer!

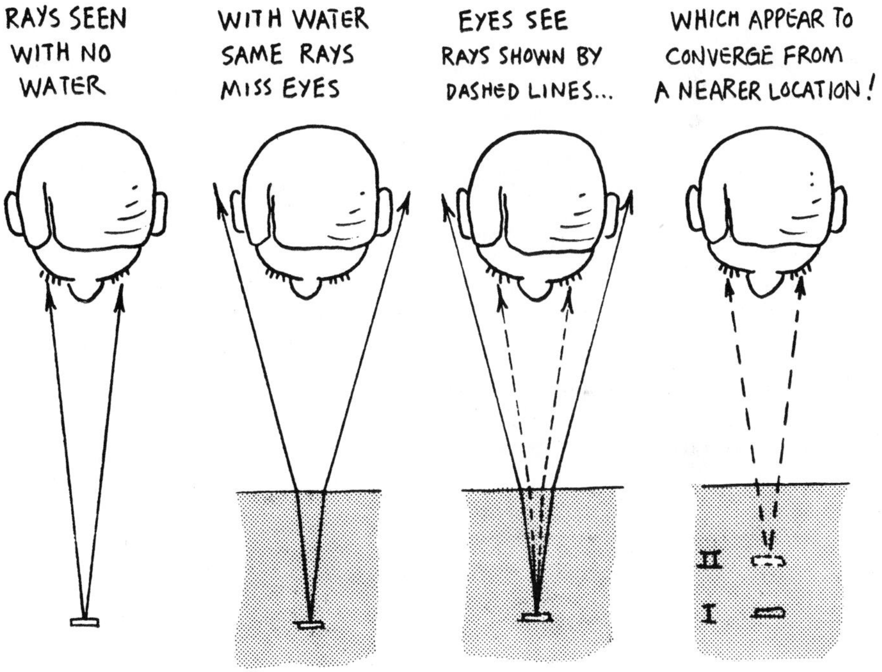
RAYS SEEN WITH NO WATER
WITH WATER SAME RAYS MISS EYES
EYES SEE RAYS SHOWN BY DASHED LINES...
WHICH APPEAR TO CONVERGE FROM A NEARER LOCATION!
II
I

HOW BIG?

When you look down into a fishbowl, the fish you see appears

a) deceptively large
b) deceptively small
c) the size it would appear if there were no water

ANSWER: HOW BIG?

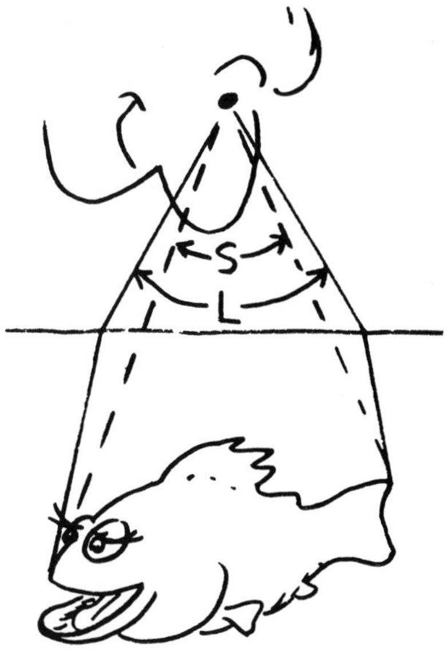

The answer is: a. You estimate the size of objects by their angular size. Without water the fish would occupy the small angle *S* but with water the light is bent (refracted) and the fish seems to occupy the large angle *L*.

Some reflecting camera lenses take advantage of this apparent change in the angular size of things by filling the space between the mirror and the film with a solid piece of glass. That reduces the size of the image on the film. Reducing the image size makes the image more intense and so reduces exposure time. It also increases the camera's field of view (we can see this by noting in the sketch that the reduced image of the candle leaves additional space for imaging rays that pass above and below the candle).

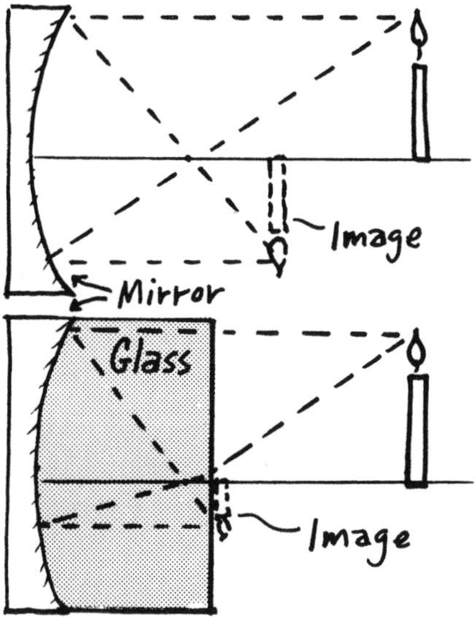

Some people think that space itself might be curved so light coming from a very distant galaxy would be bent as it passed through space (see sketch). If that is so, the distant galaxies may appear deceptively large just as the fish seemed deceptively large.

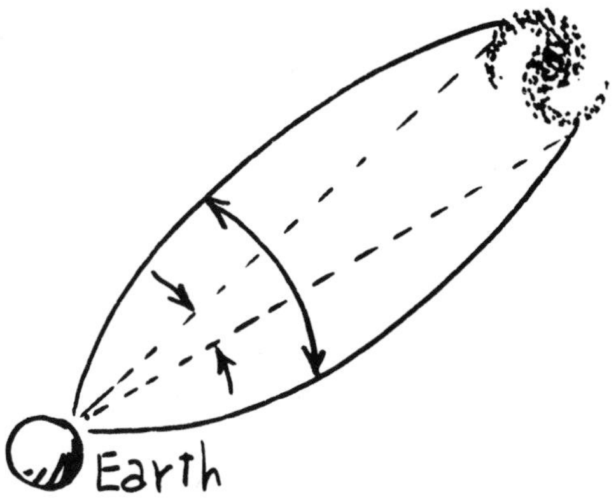

A MAGNIFYING GLASS IN THE SINK

If a magnifying glass is held under water its magnifying power is

a) increased
b) the same as it was out of water
c) decreased

ANSWER: A MAGNIFYING GLASS IN THE SINK

The answer is: c. You might find the answer to this one by actually holding a magnifying glass under water and seeing what changes occur. Try it! We know that a magnifying glass bends rays so as to magnify. It bends rays because of the curvature of the lens and because the speed of light in glass is less than the speed of light in air. It is the speed change that produces the bend. However, in water the light is already slowed down; when it goes into glass it goes still slower, but the speed change is not as much. So, under water the bend is less and so the effect of the lens is less. If the speed of light in water was as slow as the speed of light in glass then the lens would not bend rays at all, they would go straight through, just like through a window—flat glass will not focus light so windows have zero magnifying power.

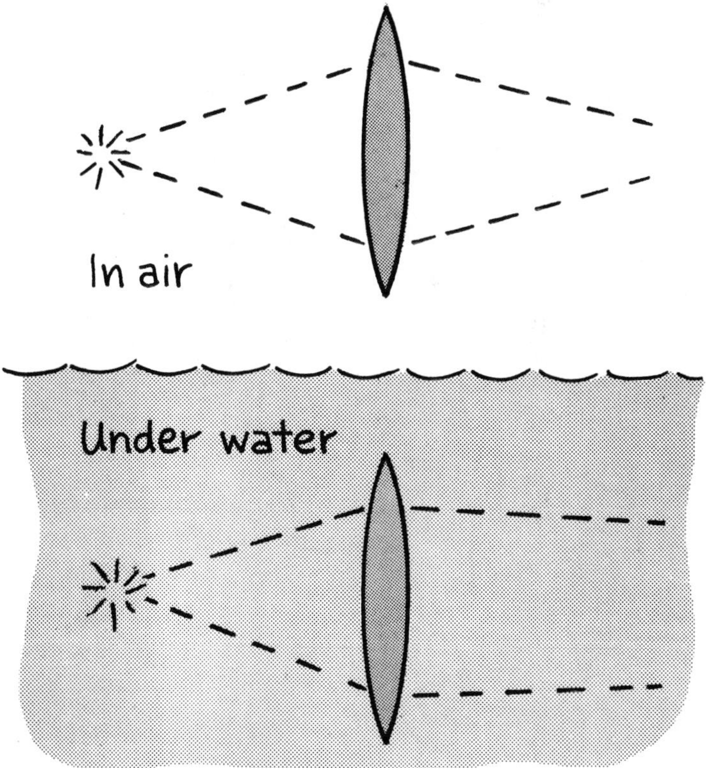

TWO POSITIVE LENSES

A positive lens is sometimes called a convex or converging lens because it is convex and causes parallel rays of light to converge to one point, called the focus.

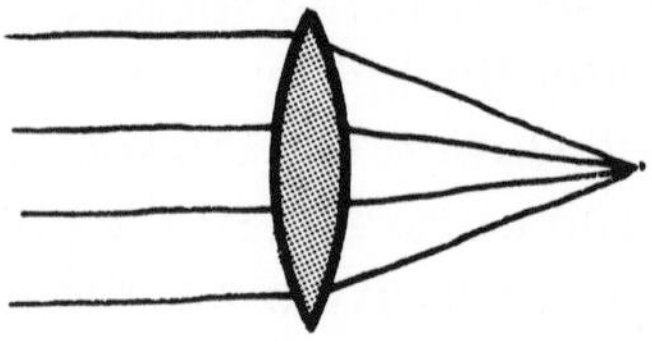

If two positive lenses are put next to each other the rays will converge:

a) more than with one lens;
b) less than with one lens;
c) the same way as with one lens.

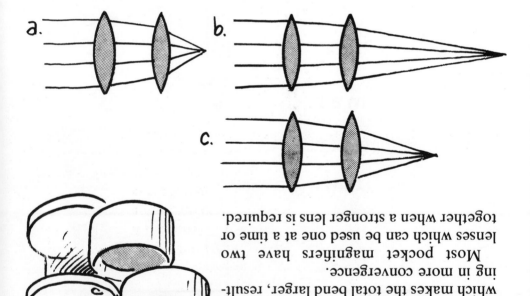

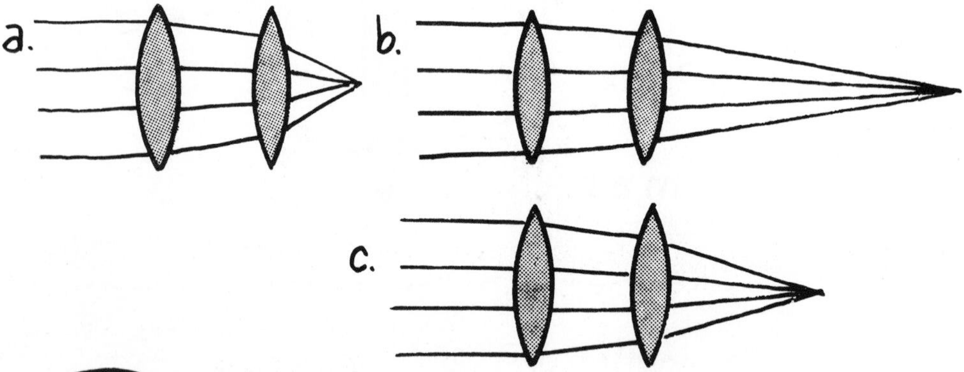

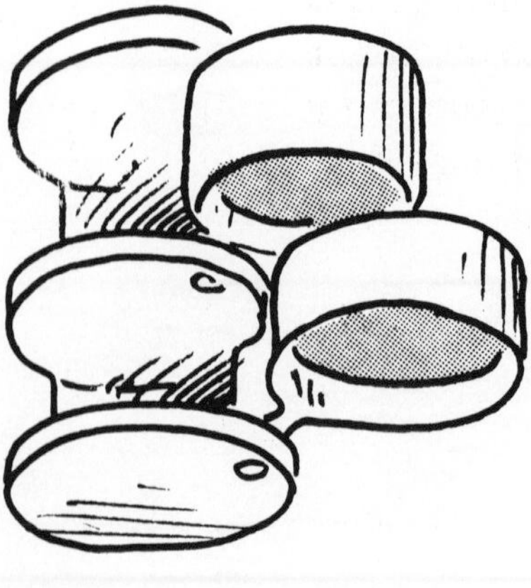

ANSWER: TWO POSITIVE LENSES

The answer is: a. The rays are bent or refracted by the lens to converge. The second lens just bends the light more which makes the total bend larger, resulting in more convergence.

Most pocket magnifiers have two lenses which can be used one at a time or together when a stronger lens is required.

FASTEST LENS

Sunlight is concentrated on a piece of paper. Which of the lenses illustrated below will most effectively set the paper on fire?

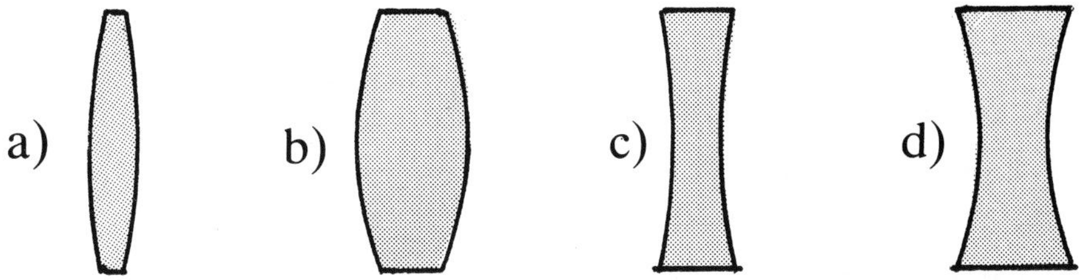

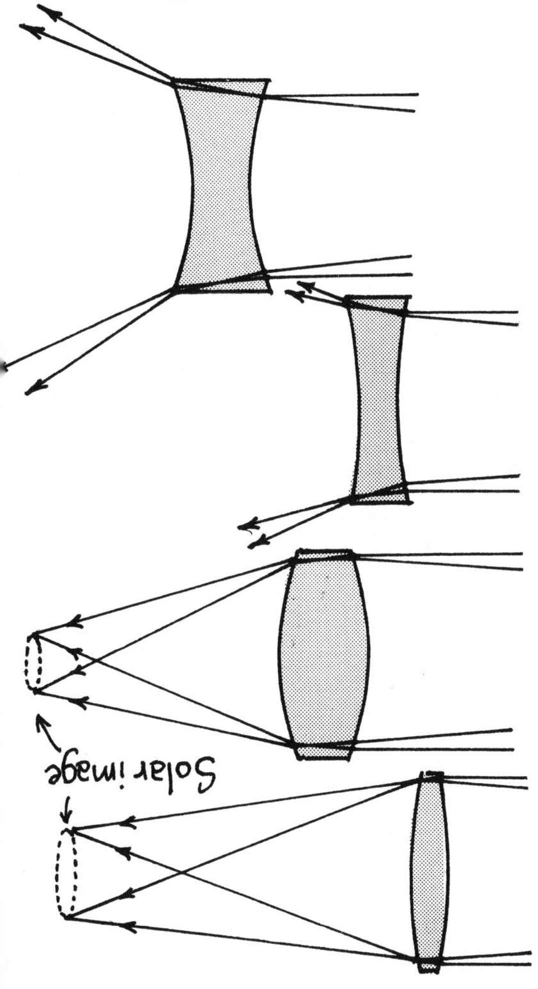

ANSWER: FASTEST LENS

The answer is: b. The negative lenses (C and D) will not cause the sunlight to converge at all, so they cannot start a fire. Lenses A and B will make the light converge, but B will bend the light the most, since it is thicker and behaves like two thin lenses stuck together, and so convergence is greater and a smaller image of the sun is produced by B than A. (Remember: a short pinhole camera makes a smaller image than a long pinhole camera.)

Now the energy in the smaller image is more crowded than in the larger image, so it will be hotter and burn the paper faster. It will also expose film faster, which is why short focal lenses are called fast lenses.

THICK LENS

Two thin converging lenses can be "melted" together to make one thicker lens which will have the effect the two thin lenses had. Thus the thicker the lens the more converging power it has. Suppose two diverging lenses are "melted" together. The resulting lens will have

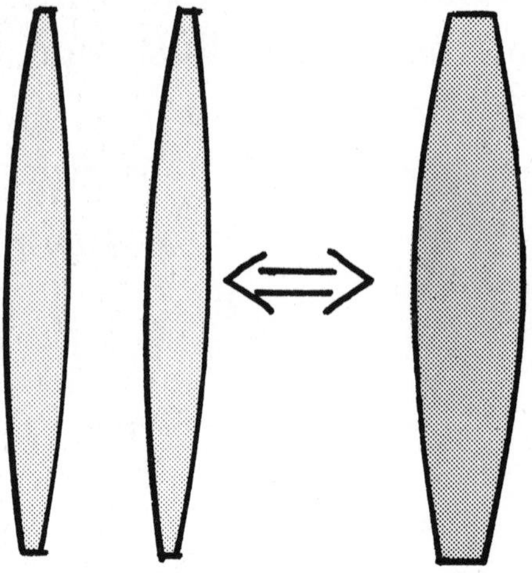

a) more diverging power
b) less diverging power
c) the same diverging power

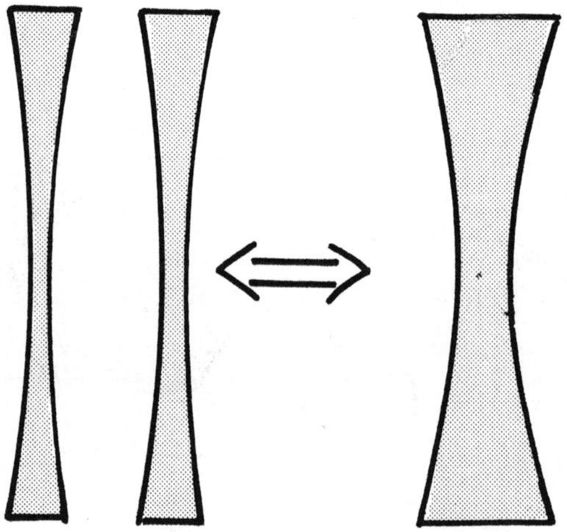

ANSWER: THICK LENS

The answer is: a. Lenses converge if they're thicker in the middle, and diverge if they're thinner in the middle. If the thickness in the middle were the same as on the edges you'd have a flat window instead of a lens. So melting two lenses together makes the difference in thickness between the center and the edge twice as great. And that makes it twice as effective in converging if you fuse converging lenses, and twice as effective in diverging if you fuse diverging lenses. Question: What happens when you fuse a converging and a diverging lens?

BUBBLE LENS

Underwater is a bubble. A light beam shines through it. After passing through the bubble the light beam

a) converges

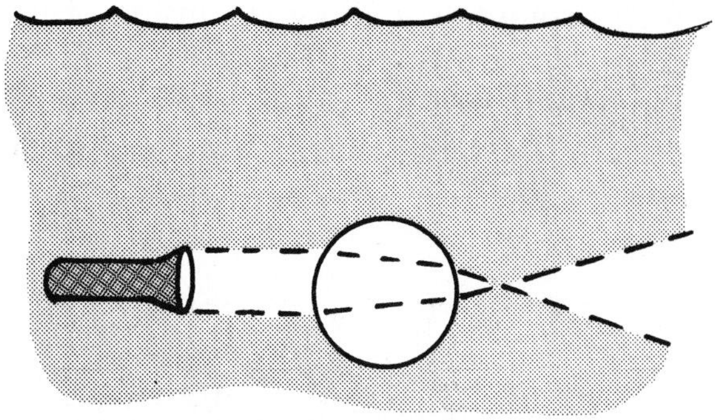

b) diverges

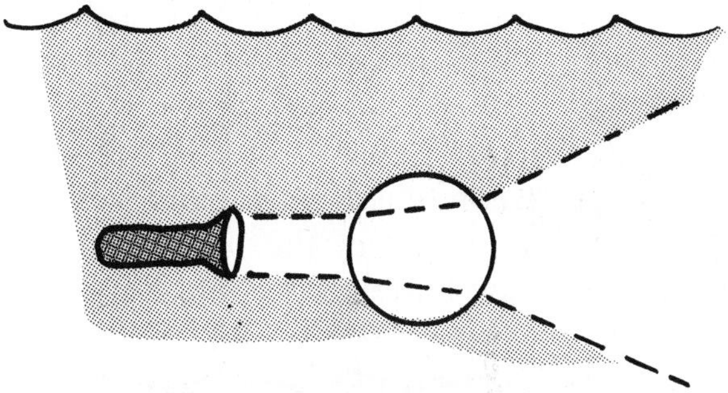

c) is unaffected

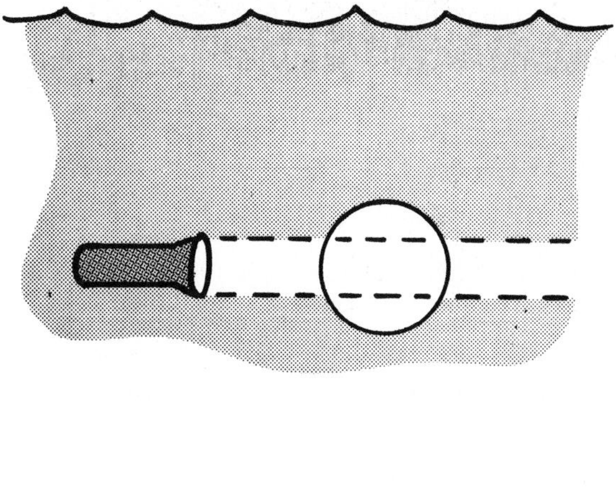

ANSWER: BUBBLE LENS

The answer is: b. There are a lot of ways to explain this, but there is also a general way to think about this type of problem. The general way goes like this. If there were a ball of water by itself in space, the ball, acting like a positive lens, would make the beam converge. Now let's think of light traveling through plain water with no bubble at all — and also let's think of no bubble as a bubble with a ball of water completely filling the inside of it. The light beam would not converge or diverge. It would go straight. So the combined effect of the bubble and the ball is a straight beam, which we might call NO effect. The ball by itself, however, would make the beam converge.

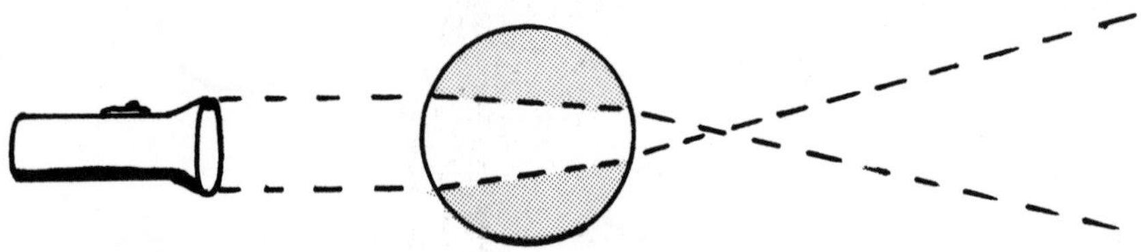

What effect combined with convergence makes no effect? Divergence. So the effect of the bubble must be to make the beam diverge. And the bubble does make the beam diverge.

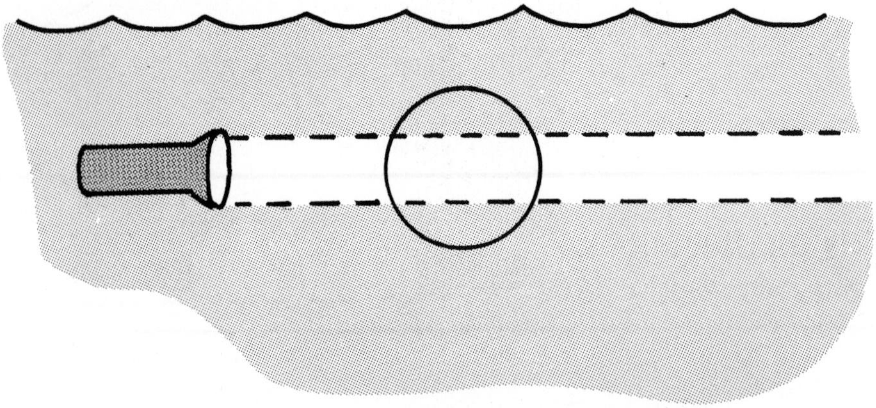

BURNING GLASS

A magnifying glass used to focus sunlight in an intense hot spot becomes a burning glass or solar collector. If the lens is made larger and/or if the focus is made more concentrated, the hot spot gets hotter. Could a lens be made bigger and bigger or could its focus be made more and more concentrated so that its hot spot would be hotter than the sun itself?

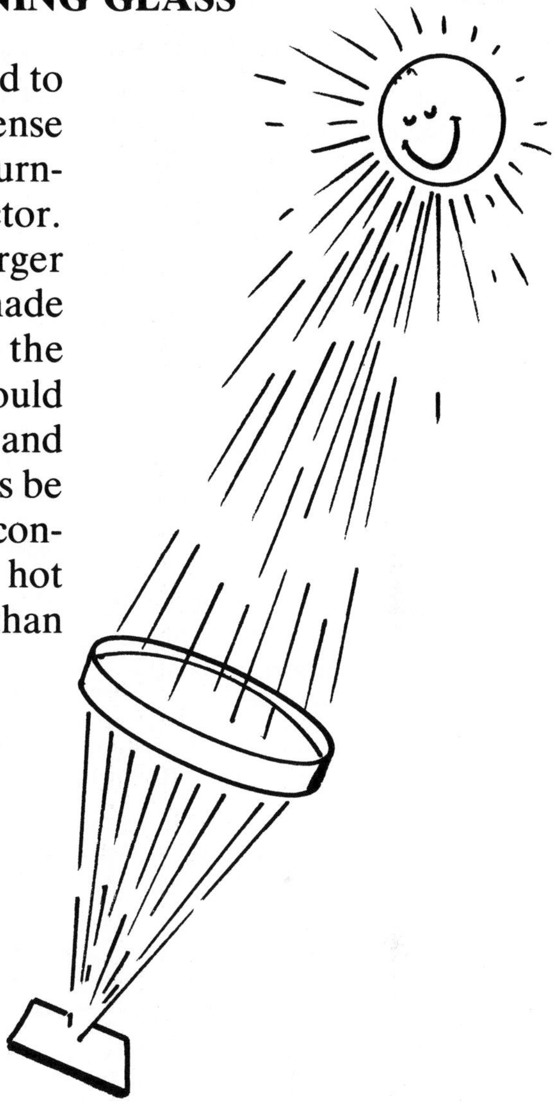

a) There is no limit on how hot the spot could get
b) The spot could not be made hotter than the sun's surface
c) The spot could never be made anywhere near the temperature of the sun's surface
d) If several lenses were used, a spot could be made hotter than the sun's surface

ANSWER: BURNING GLASS

The answer is: b. There are two ways to see the answer. First, imagine your eye at the exact focus of the lens. In every direction you look into the lens your line of sight is directed back to the sun's surface. Now suppose the lens was so large (or perhaps several lenses were used), so that no matter in which direction you looked you would see the sun's surface. Then you would seem to be completely encased in solar surface. So your temperature would be the same as the temperature of the solar surface.

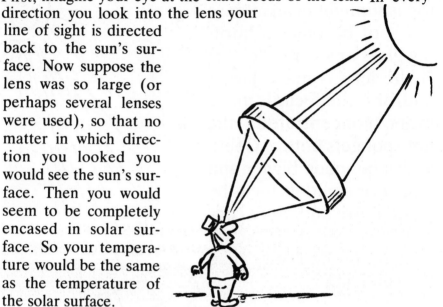

Second, imagine the spot could be made hotter than the solar surface. That means heat flowed from the solar surface, through the lens and to a place — the spot — which was hotter than the place it came from. But heat energy will not do that. By itself, heat energy always goes from a hot place to a cool place and never from a hot place to a hotter place.

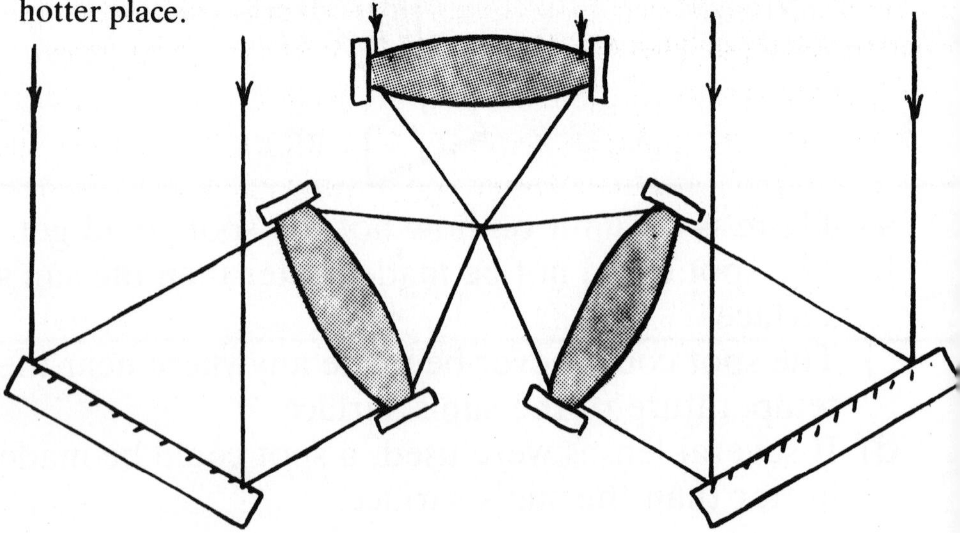

CONDENSING LENS

1. If someone said they could devise an optical system which would focus ALL the light from the large lamp filament into an image of the filament which was smaller than the filament, would the claim be fundamentally at odds with any basic law of physics?

a) Yes, the claim is at odds with a basic law of physics
b) No, the claim is not at odds with any basic law of physics

2. If someone said they could devise an optical system which would focus SOME of the light from the large lamp filament into an image of the filament which was smaller than the filament, would the claim be fundamentally at odds with any basic law of physics?

a) Yes, fundamentally impossible
b) No, fundamentally possible

3. If someone said they could devise an optical system which would focus ALL of the light from a SMALL lamp filament into an image of the filament which was LARGER than the filament, would the claim be fundamentally at odds with any basic law of physics?

a) Yes, fundamentally impossible
b) No, fundamentally possible

ANSWER: CONDENSING LENS

The answer to the first question is: a. Almost everyone who works with lights and lenses sooner or later tries to increase the intensity of a light source by condensing all the light from a lamp filament into an image smaller than the filament. But it cannot be done. Why?

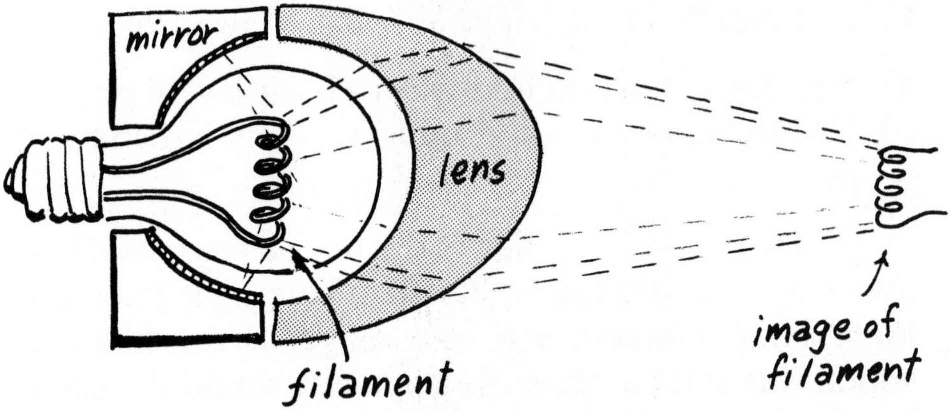

For one reason, the image would be hotter than the filament itself and according to the Second Law of Thermodynamics you cannot make heat flow from a hot place to a hotter place without using a heat pump, which requires energy to operate. A lens does not use energy so it is not a heat pump. The image of a light source can never be more intense than the light source itself.

The answer to both the second and third question is: b. The Second Law of Thermodynamics would not be violated, for in each of these cases the image is not required to be more intense than the light source.

CLOSE-UP

The first sketch shows a camera properly focused on the DISTANT MOUNTAINS. When the camera is re-focused on a very NEAR subject it will be set as shown in

a) the second sketch
b) the third sketch
c) the fourth sketch

a) 

b) 

c) 

ANSWER: CLOSE-UP

The answer is: c. As the subject moves towards the lens, the lens must move away from the film, which is at the rear of the camera. To see why, focus your attention on one part of the lens, say the upper tip. It is in effect a small prism. The prism can bend light through a certain angle, θ. So if light is headed from **A** to **B**, it is turned and ends up going to **C**.

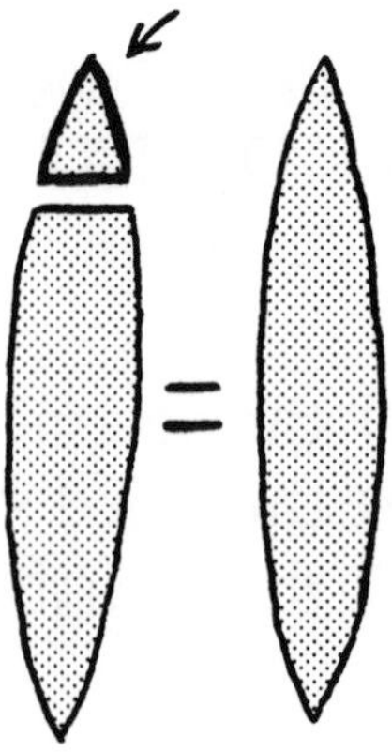

Next look at the sketch showing the subject or light source at **A**, the lens, and the film at **C**. If **A** moves closer to the lens, and the bend angle θ can't change and the light must remain focused at **C**, then the only way this can happen is for the lens to be moved away from the film, as shown in the last sketch. If **A** moved towards the lens and the lens were NOT moved away from **C**, then the bend angle θ would have to get larger, which means you would have to use a stronger lens.

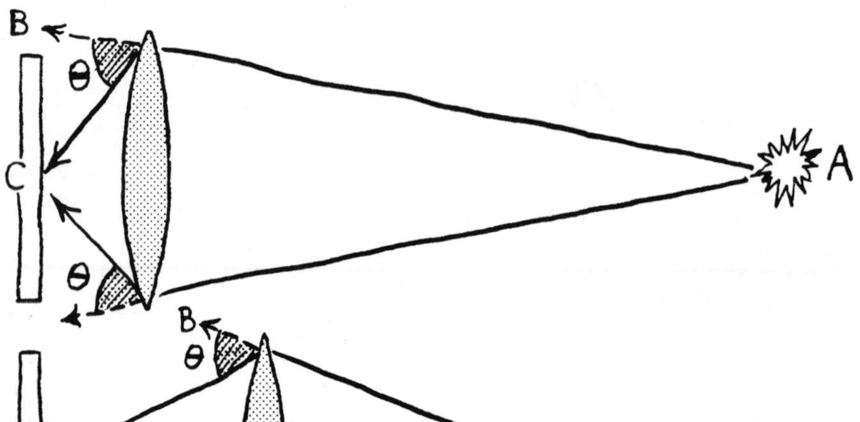

NEARSIGHTED AND FARSIGHTED

People who hold reading material close to their eyes when reading likely have elongated eyeballs where the retina is farther back than normal from the lens of the eye. This is the nearsighted eye. A farsighted eye is just the opposite — the retina is too close to the lens. If the book is held too close, the image cannot be formed in the available distance between the lens and the retina.

Now if the nearsighted person's book moves away into the distance, the image of the book moves toward the eye lens and so is not focused on the retina.

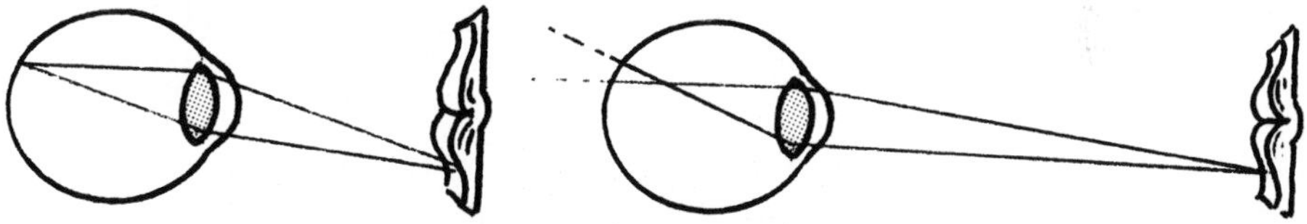

The Nearsighted Eye

To correct for these defects, the eyeglasses for a nearsighted person consist of

a) positive lenses
b) negative lenses

The Farsighted Eye

And the eyeglasses for a farsighted person consist of

a) positive lenses
b) negative lenses

ANSWER: NEARSIGHTED AND FARSIGHTED

The answer to the first question is: b. The lens in the eye is a positive (converging) lens and in a nearsighted eye brings light too abruptly to focus. If a negative (diverging) lens is added it diminishes the focusing effect of the positive lens. The rays converge more slowly. So a negative lens put in front of the nearsighted eye will cause the eye's focus to move back to the retina.

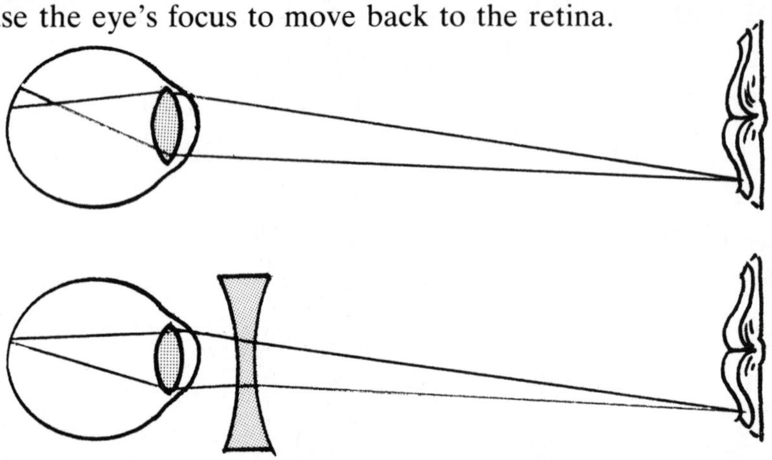

The answer to the second question is: a. In the farsighted eye the light converges too slowly and comes to focus beyond the retina. So a positive or converging lens placed in front of a farsighted eye will make rays converge faster, and converge in the small distance between the lens and retina. So the eyeglasses of a farsighted person are positive.

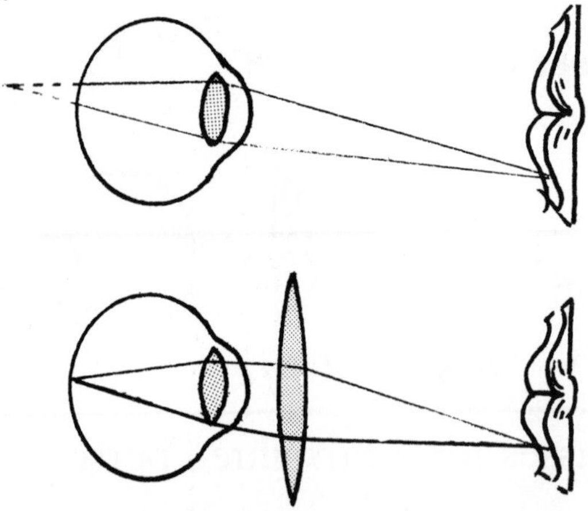

The eyeglasses of nearsighted people tend to make their eyes look smaller, whereas the eyeglasses of farsighted people make their eyes seem larger.

BIG CAMERA

Both cameras shown are the same in every way except for the diameter of the lenses. If they are used to photograph a distant object, which camera produces the biggest image of the object?

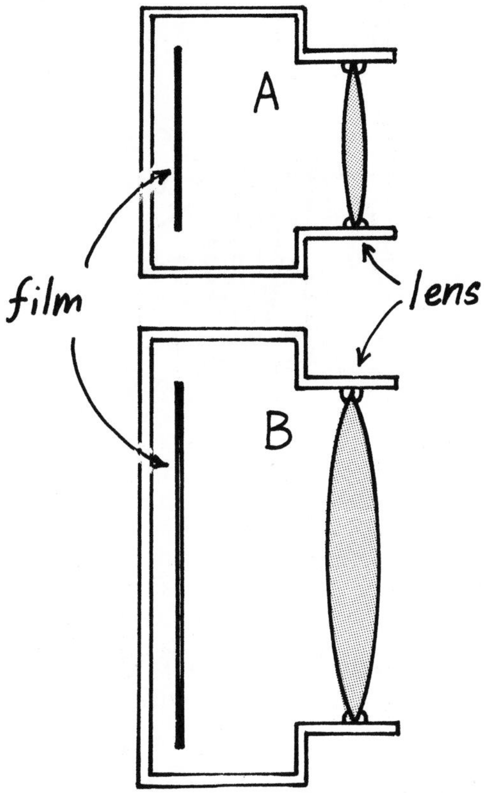

a) camera A
b) camera B
c) both the same

ANSWER: BIG CAMERA

The answer is: c. The image size depends on the distance between the lens and the film. The diameter of the camera lens has nothing to do with the size of the image. The larger diameter lens catches more light, but that goes to make a brighter image, not a larger image. You can see that the diameter of the camera lens will not affect image size because if you close down the iris diaphragm of a camera (going from *f*2 to *f*8) it effectively reduces the diameter of the camera lens, but it does not reduce the size of the image on the film.

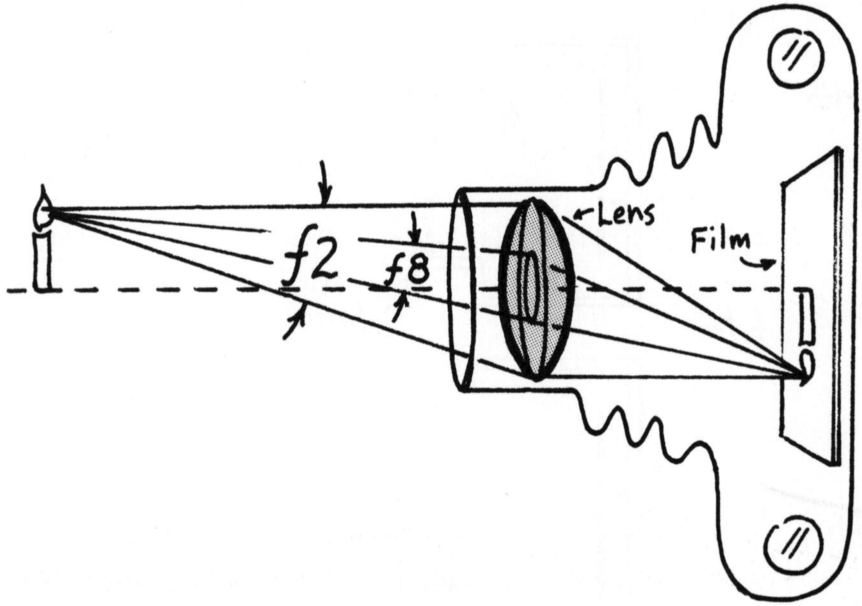

What's more, you don't even need a camera to see that this is true. Your eyes change their effective diameter without altering image size.

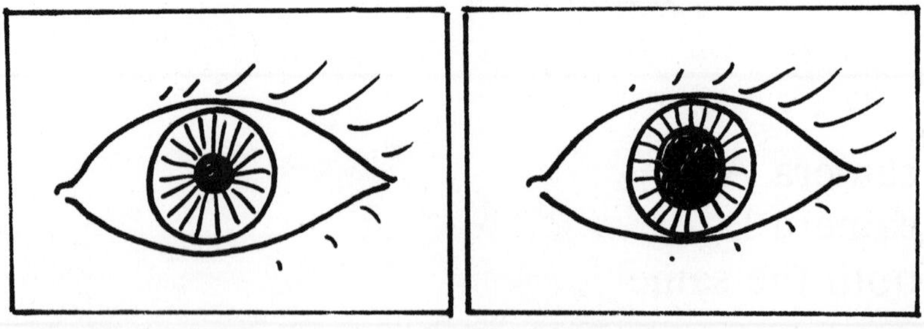

BIG EYE

In 1609 Galileo was the first to survey the heavens with telescopes of his own construction. Since then popularizers of science depict the telescope as being essentially a "big eye." Telescopes are sometimes called "big eyes" in newspaper stories. The fact of the matter is that

a) this oversimplified depiction greatly misrepresents the essence of the telescope
b) the popularizes and newspaper writers have been quite correct about this

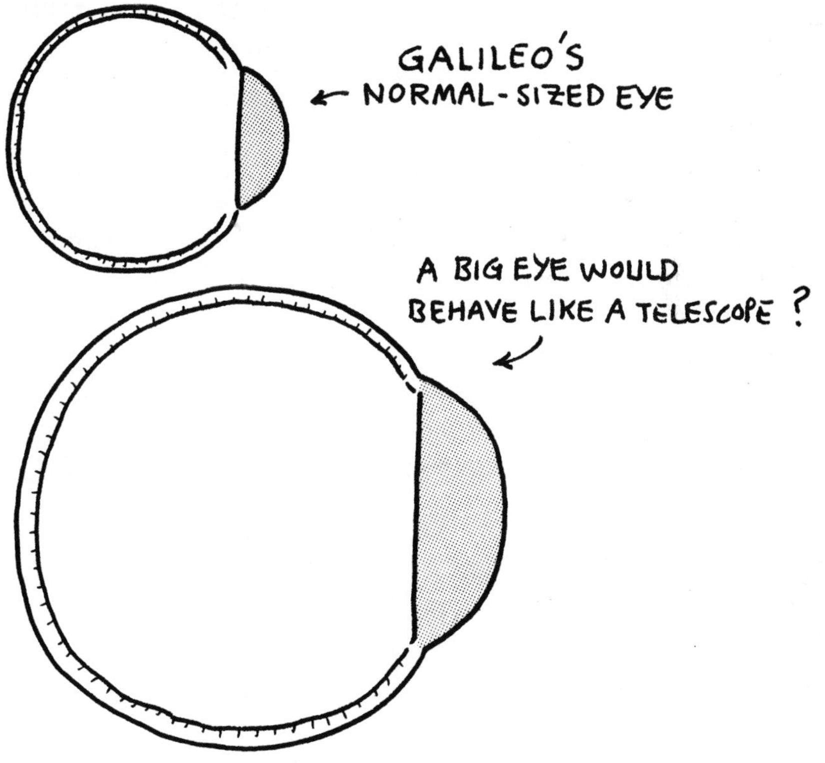
GALILEO'S
NORMAL-SIZED EYE
A BIG EYE WOULD
BEHAVE LIKE A TELESCOPE ?

ANSWER: BIG EYE

The answer is: b. It is interesting to note that Galileo knew nothing of the lens and magnification equations so commonly found in the optics sections of today's physics books. Nor did anyone else at that time. He just wanted to make a "bigger eye," one that would scale up everything he looked at.

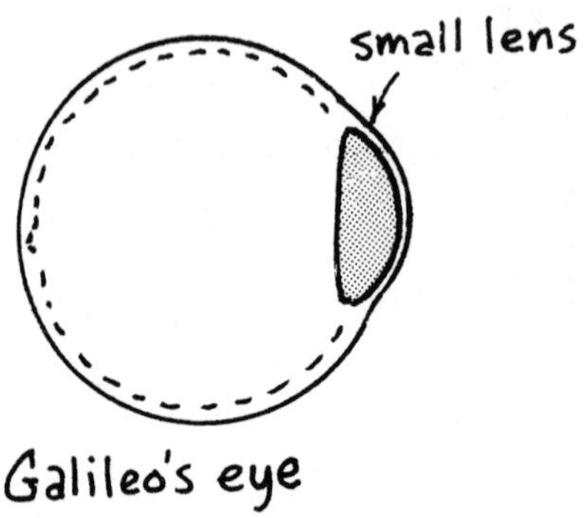

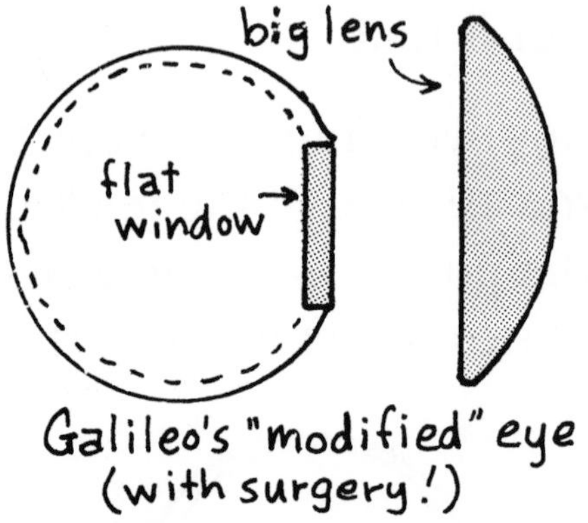

One way would be to surgically replace the little lens in his own eye with a larger one. But a larger one wouldn't fit. So he would replace the lens in his eye with a flat window, and place a larger one in front of the eye. Now, how to do this without resort to surgery? To effectively replace the little lens in his own eye with a flat window he neutralized it with a little glass lens of the opposite kind. This is the eyepiece. Then a larger lens with an outer curvature like that of the eye lens was placed in front. The resulting telescope was indeed equivalent to a bigger eye!

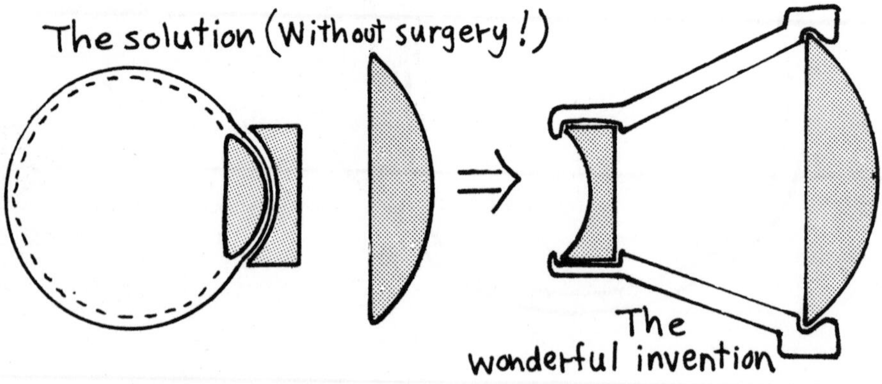

GALILEO'S TELESCOPE

Galileo's idea for a telescope is shown in sketch I and Kepler's idea for a telescope is shown in sketch II. Both will work but, except for opera glasses, Galileo's version of the telescope is seldom used. This is because

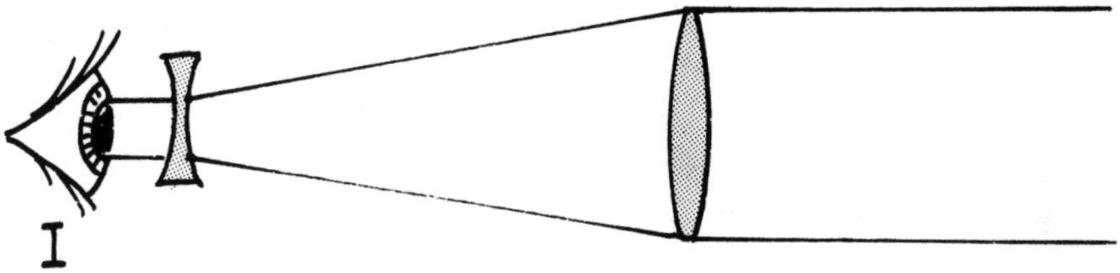

I

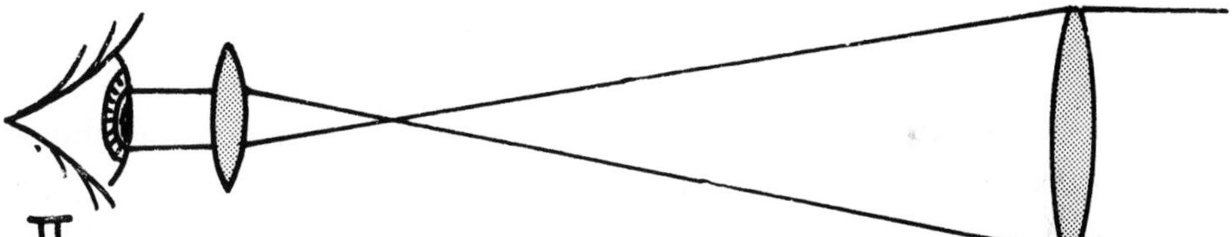

II

a) of Galileo's arrest record
b) in Galileo's version the tube is longer than in Kepler's version
c) Galileo's version spills light outside the eye's pupil
d) in Galileo's version the image is upside down

ANSWER: GALILEO'S TELESCOPE

The answer is: c. Galileo's telescope has many advantages over Kepler's. In Kepler's the image is upside down, but not in Galileo's. Also, Kepler's tube must be longer (some historians suspect Kepler never even looked through a telescope!). But Galileo's telescope has one devastating problem. Light coming from different parts of the field of view cannot all enter the eye's pupil in any one position. The telescope spills light. You must move your eye to different positions to see different parts of the field of view. Kepler's version brings the light from all over the field to one position where the eye's pupil can be parked. It has a built-in light funnel. Telescope people call the place where the eye goes the "exit pupil."

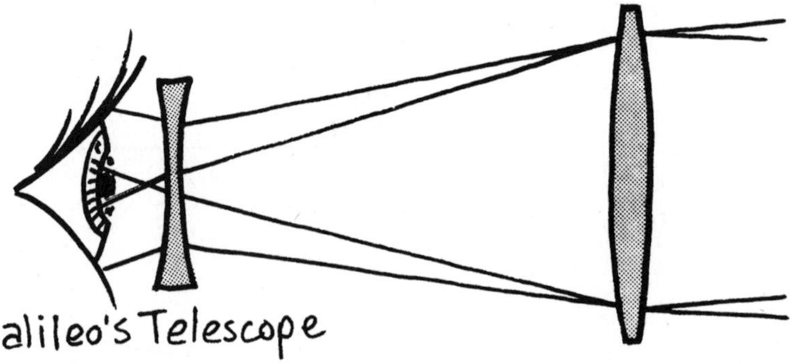

Galileo's Telescope

What is the moral of this story? The moral is that when you think about optical systems: cameras, microscopes, projectors, etc., you must not limit your thinking to the light that comes straight into the system. You must also think of the light that comes in at an angle.

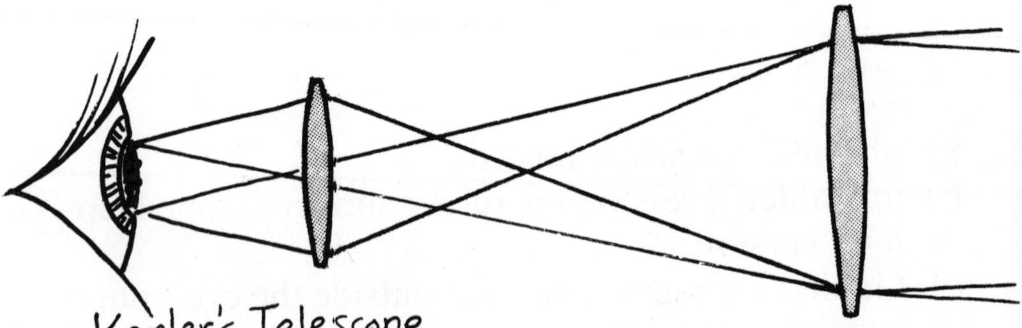

Kepler's Telescope

DROP OUT

If a beam of red light falls on a round drop of water

a) equal amounts of red light will come out of the drop in all directions
b) all the red light will come out in one direction
c) some red light will come out in all directions, but more will come out in some directions than in others
d) most of the light will go *straight through* the drop and will *not* be bent at all

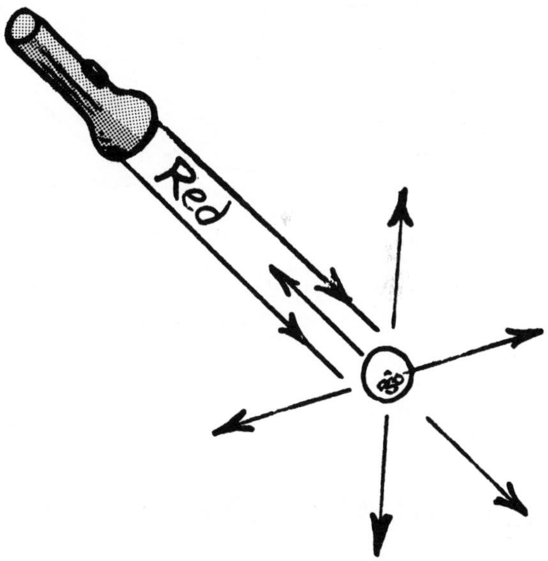

ANSWER: DROP OUT

The answer is: c. The sketch shows the beam of red light coming from the direction of ray 1 and entering the drop. The rays are refracted when they enter. Part of ray 1 is reflected from the back of the drop and part of ray 1 escapes through the back of the drop as ray 2. The reflected part emerges through the front of the drop as ray 3. Looking at the sketch we can see some of the red light emerging in almost every direction, but more comes out in the general direction of ray 3 than in other directions. So what? So the rainbow! It is this concentration of emerging light in a particular direction that makes rainbows possible. Incidentally, how much light goes straight through the drop without being bent at all? Very little. Only one ray, ray zero.

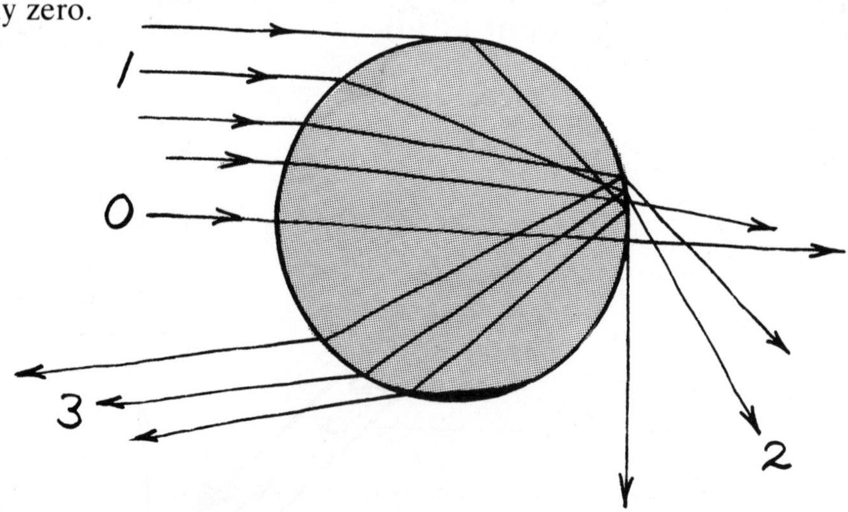

BLACK AND WHITE RAINBOW

Where you see a rainbow, your totally color-blind friend sees nothing in particular.

a) True
b) False

ANSWER: BLACK AND WHITE RAINBOW

The answer is: b. Like so many cases, a picture is worth thousands of words. You can photograph a rainbow with black and white film. The colors aren't there, but the bow certainly is! In explaining the rainbow it is common to concentrate on the colors. But the colors are secondary to the main problem: Why is there a bright arch in the sky?

Rays of sunlight encountering a raindrop emerge in many directions. But due to focusing effects caused by reflection and refraction, slightly more rays come out at particular angular regions than in others.

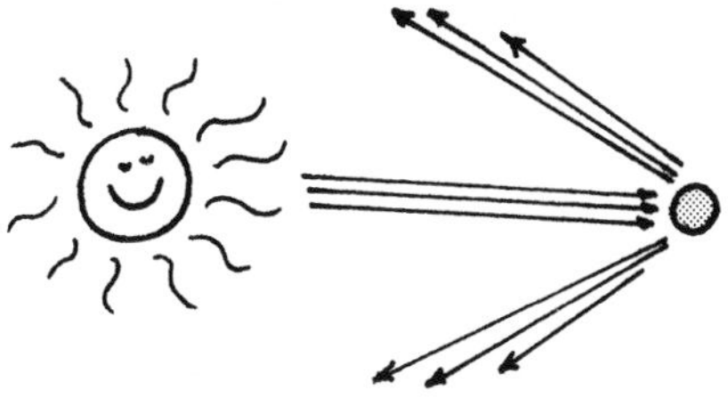

This slight preference is the key to the rainbow. Note from the sketch that virtually all the sunlight worked over by the drop is cast back into a cone. The center of this cone is exactly opposite the sun. The cone forms a bright disk of light exactly opposite the sun. As seen from ground level, only part of this disk is seen; from a high-flying plane the full disk can sometimes be seen. It is the bright edge of this disk that forms the rainbow.

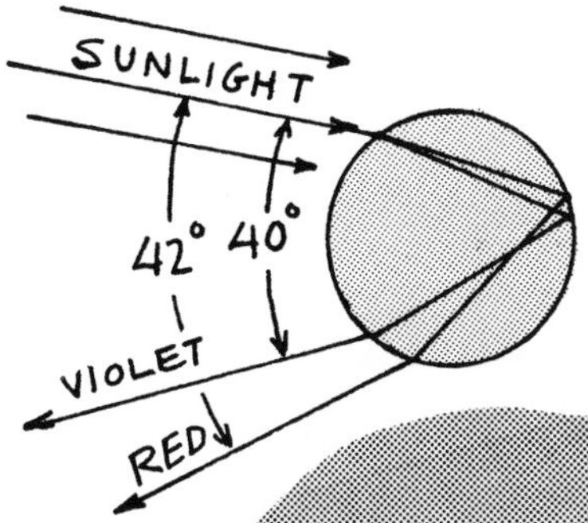

The colors of the rainbow are separated by refraction. The details of refraction depend slightly on color; different colors of light travel at different speeds in the drops and refract differently. This produces a color separation, as seen by the ray diagram.

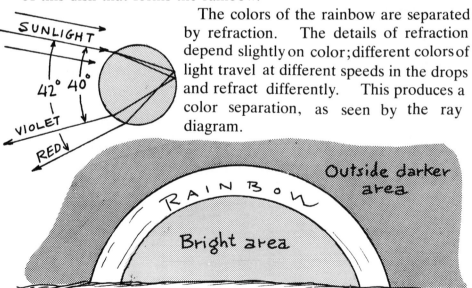

The attention of many people viewing a rainbow is so captured by the colors that they fail to notice the bright disk segment, the colored edge of which is the bow.

MIRAGE

A common mirage on a hot day is an *apparent* pool of water on a hot highway which in fact doesn't exist when approached. Such a mirage can be distinguished from a pool of water with polaroid glasses because

a) reflection from water is polarized while the mirage is not
b) the mirage is polarized while water reflection is not
c) both are polarized, but along different axes.
d) neither is polarized; they cannot be distinguished.

ANSWER:
MIRAGE

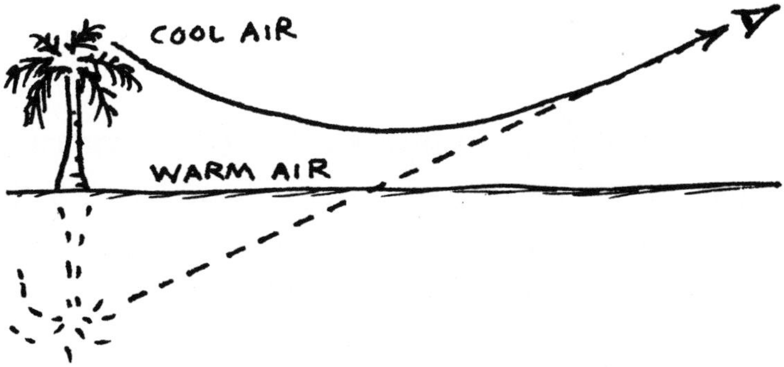

The answer is: a. Light from the mirage is bent by refraction, not by reflection. A layer of very hot air about a foot or so deep just above the highway is of reduced density which results in a slightly greater speed for light passing through it, compared to the cooler denser air farther above the road. The speed "gradient" causes the light waves to bend.

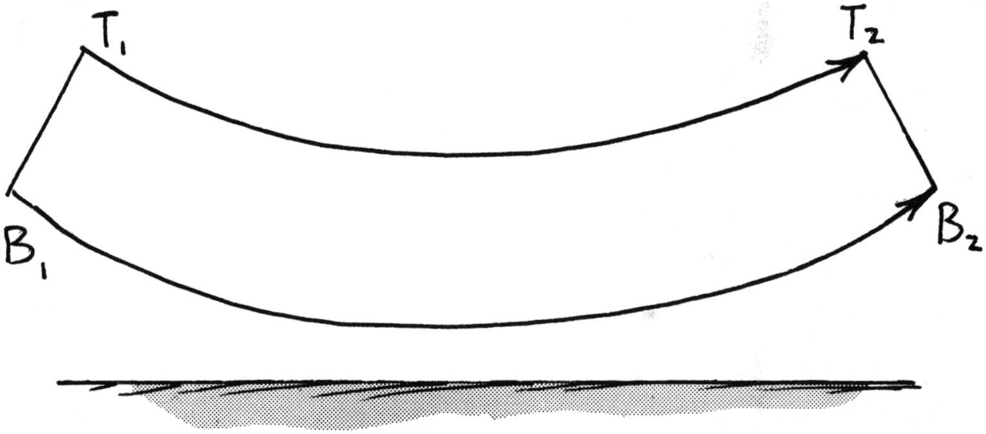

We can see how this bending takes place in Figure 2. Consider light wave T_1B_1 approaching the road from the left. The bottom of the wave, B_1, moves a farther distance to B_2 in the same time that the top of the wave goes from T_1 to T_2. Can you see how the beam bends?

Light speed depends only on the air density and not at all on the polarization of the light waves. So a mirage is no more polarized than the light that goes into making it. A mirage will look the same when viewed through a polarization filter no matter how the filter is aligned. But not so with light reflected from water. A mirage is caused not by the reflection of light, but by the refraction of light.

MIRROR IMAGE

A lovely lady holds a hand mirror one foot behind her head while standing 4 feet from a dresser mirror as shown. How far in back of the dresser mirror is the image of the flower in her hair?

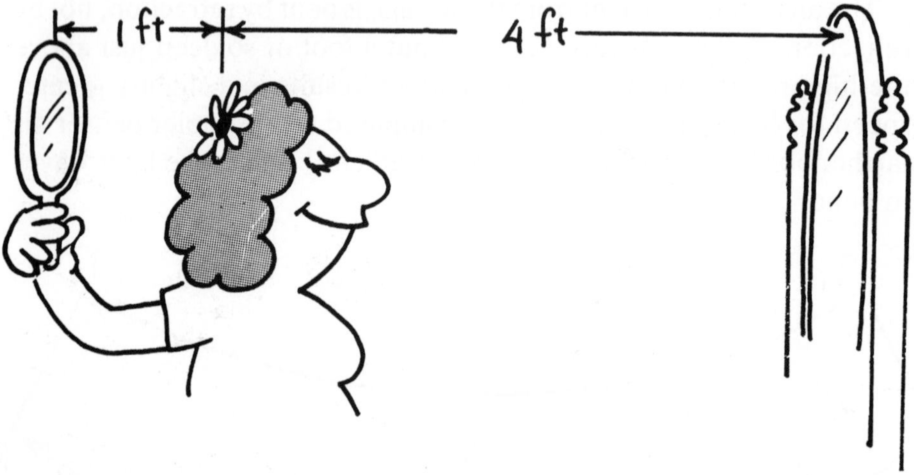

a) 4 ft b) 5 ft c) 6 ft d) 7 ft e) 8 ft

ANSWER: MIRROR IMAGE

The answer is: c, 6 feet. Why? Because the image of the flower in the hand mirror is as far behind the mirror as the flower is in front—1 foot. This puts the first image 6 feet (4+1+1) in front of the large mirror. The image is as far behind the large mirror—6 feet.

PLANE MIRROR

What must be the minimum length of a plane mirror in order for you to see a full view of yourself?

a) One-quarter your height
b) One-half your height
c) Three-quarters your height
d) Your full height
e) The answer depends on your distance.

ANSWER: PLANE MIRROR

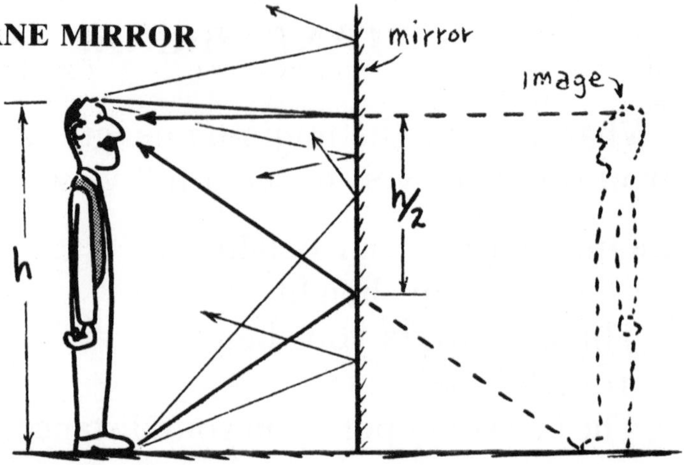

The answer is: b. Exactly one-half your height. Why? Because for reflection, the angle of incidence is equal to the angle of reflection. Consider a man standing in front of a very large mirror as shown in the sketch. The only rays of light from his shoes to encounter his eyes are those that are incident upon the mirror at a level halfway up from ground level to eye level. Rays of "shoe light" that are incident higher reflect above his eyes and rays incident lower reflect below his eyes. So the part of the mirror below the halfway level is not needed — it shows only a reflection of the floor in front of his feet. Similarly for the top part of the mirror. The only rays to reach his eyes from the top of his head are those that are incident upon the mirror halfway between the top of his head and his eye level. The part of the mirror above this is not needed. So the portion of the mirror useful for seeing his image lies halfway above his eye level to the top of his head, to halfway below his eye level to his toes — that's half his height.

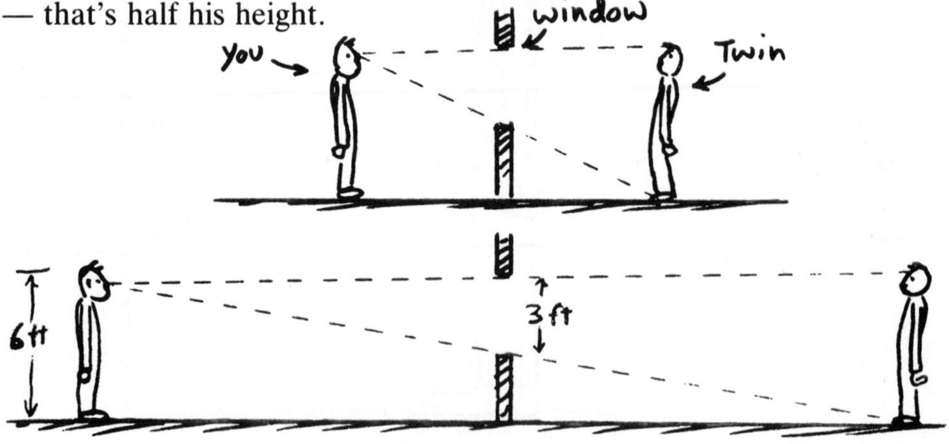

Mirrors act like windows to a world behind them. Everything in mirrorland is the mirror image of this land. The sketch shows that to see your image in mirrorland the window need be only one half your height—regardless of how near or how far you stand from the window.

The next time you view yourself in a mirror, mark it where you see the top of your head and where you see the bottom of your chin. Note that the distance between the pair of marks is half your face size and that as you move closer to or farther from the mirror the image of your face still fills the space between the marks.

FOCUSING MIRROR

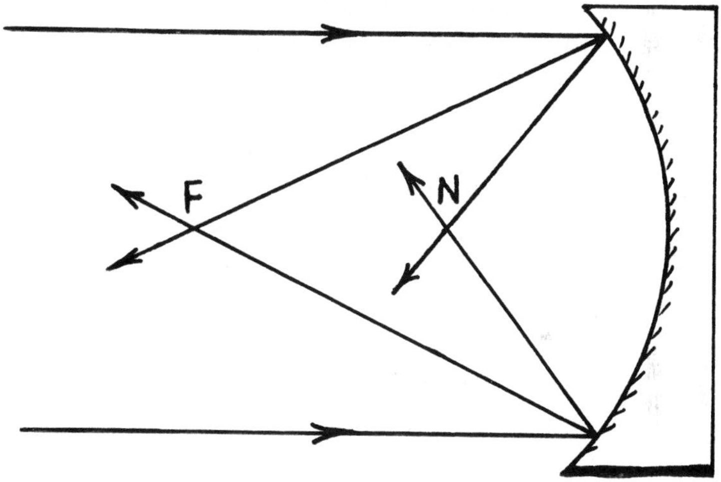

If a beam of white light falls on a curved mirror and is focused as indicated in the drawing above, then the red component of the light will be focused at

a) the near point (N) and the blue light at the far point (F)
b) F and the blue at N
c) . . . the drawing is wrong — all colors are focused at the same point

ANSWER: FOCUSING MIRROR

The answer is: c.

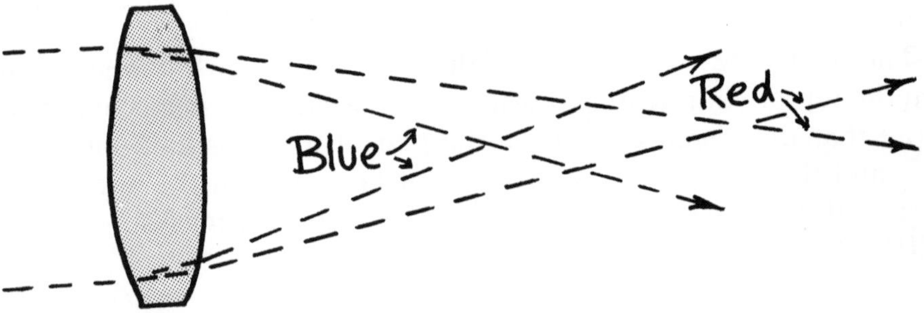

The law of reflection states that the angle of incident light upon a reflecting surface is equal to its angle of reflection — regardless of its frequency. So when a mirror focuses light it treats all the colors the same and brings them to the same focus. This is not true for a simple lens. A simple lens will bend blue light more than red and so bring each color to focus at a different place. It was to avoid this undesirable color separation (called chromatic aberration) that Sir Isaac Newton invented the reflecting telescope, which uses a mirror rather than a lens to focus light. Galileo's first telescope (a refracting telescope) used a lens but now most big telescopes use mirrors.

In 1733 an English lawyer and amateur scientist, Chester Mohr Hall, surprised the experts and devised a lens that focuses most colors in one place—it was a compound lens made of different kinds of glass, called an *achromat,* and is in popular use today.

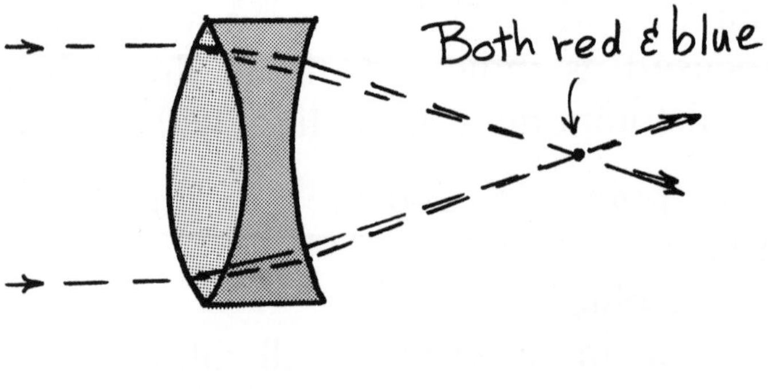

POLAROIDS

Light will pass through a pair of polaroid sheets when their polarization axes are aligned, but when their axes are at right angles to one another light will not get through. Light will not be transmitted through the crossed polaroids. Now if a third sheet of polaroid is inserted as shown between the crossed polaroids, light will

a) get through
b) not get through

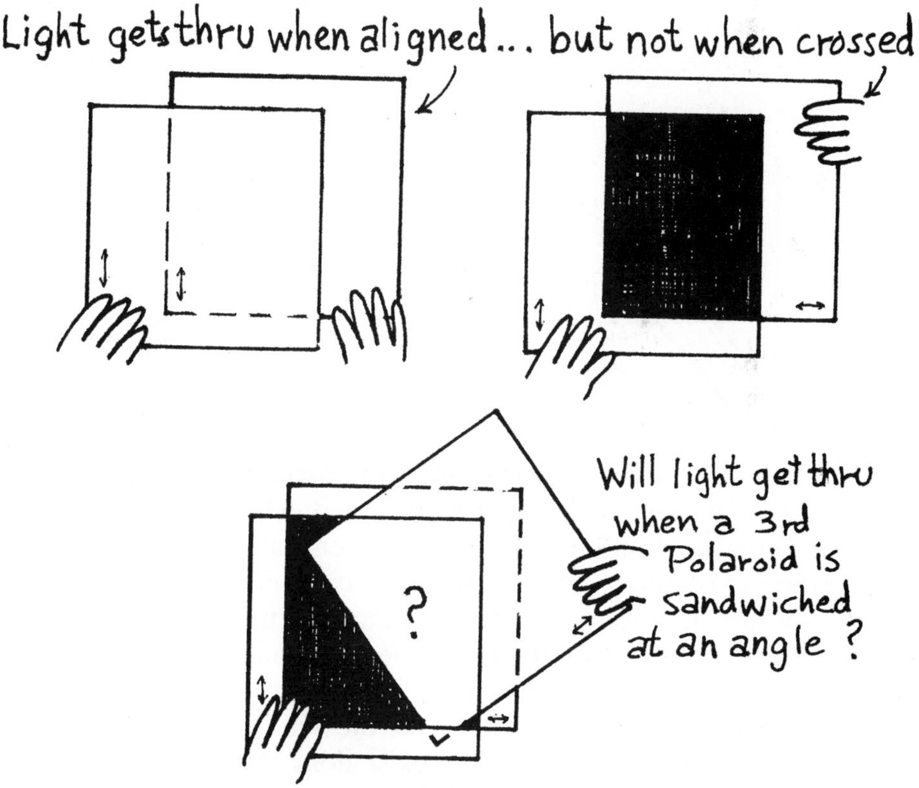

Light gets thru when aligned... but not when crossed
?
Will light get thru when a 3rd Polaroid is sandwiched at an angle?

ANSWER: POLAROIDS

The answer is: a. For a simple pair of crossed polaroids, no light passes because the axis of the second polaroid is completely perpendicular to the components that pass through the first. But when the third polaroid is sandwiched between at a non-perpendicular or non-parallel orientation, we find that light, although diminished in intensity, is transmitted by the polaroids. This is best understood by considering the vector nature of light. Light is a wave that oscillates

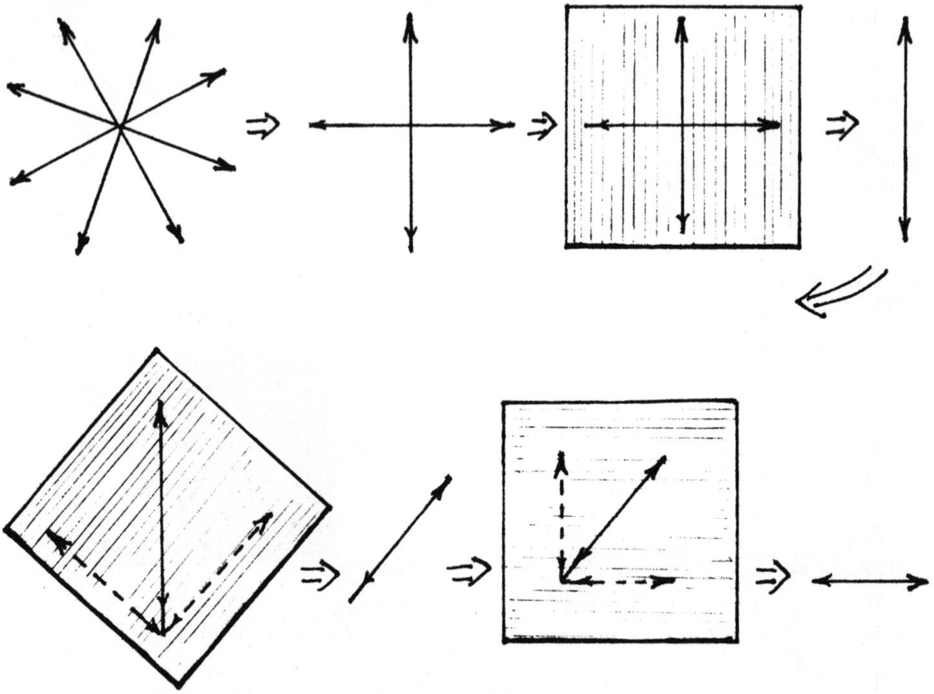

through the space in which it travels. If the oscillations are in a preferred direction, we say the light is polarized. The light that passes through a polaroid is polarized, for components at right angles to the polarization axis are absorbed rather than transmitted. The diagram shows how initially nonpolarized light incident upon the first polaroid is filtered so that only the components parallel to the polarization axis are passed. If these fell on a polaroid with its axis at 90 degrees, none would get through. But the sandwiched polaroid is NOT at 90 degrees. Components do exist along the axis of this polaroid, and these pass through and fall on the third polaroid through which components parallel to its axis are transmitted.

MORE QUESTIONS (WITHOUT EXPLANATIONS)

You're on your own with these questions that parallel those on the preceding pages. Think Physics!

1. If the atmosphere of a planet predominantly scattered low frequencies of light and more readily transmitted higher frequencies, then sunsets on that planet would appear
a) red b) white c) blue

2. As viewed on the planet just described, distant mountains in mid-day would appear tinted
a) red b) white c) blue

3. Here on earth, red clouds at sunset are illuminated by the sun from the
a) top b) bottom

4. At the same time that astronaut observers on the moon see a solar eclipse, the earth experiences a
a) solar eclipse also b) lunar eclipse c) both d) neither

5. Suppose the axis of a spinning star is parallel to the earth's axis, so that one surface of the star is spinning toward us and the opposite side is spinning away from us. The light that reaches us from the side spinning toward us is measured to have a greater
a) speed b) frequency c) both d) neither

6. The color of a star indicates its temperature. The hotter star will appear
a) red b) white c) blue

7. The famous Willow Pattern is a dark blue design on a white china plate. If the ceramic plate is heated until it glows, the pattern in the glowing plate will
a) remain a dark design on light china
b) reverse, becoming a light design on dark china

8. The stars that twinkle most in the nighttime sky are those that lie
a) nearest the horizon b) directly overhead c) both about the same

9. The average speed of light is least in
a) air b) water c) the same in each

10. The bending of light in passing from one medium to another (refraction) is primarily caused by differences in the light's
a) speed b) frequency c) both d) neither

11. The most important difference between the focusing ability of one lens and another is its
a) thickness b) curvature c) diameter

12. Different colors of light correspond to different
a) intensities b) frequencies c) speeds

13. The person who sees most clearly under water without the use of goggles or a face mask is one who is normally
a) nearsighted b) farsighted c) neither nearsighted nor farsighted

14. The size of the image formed by a positive lens depends on the
a) diameter of the lens b) distance between the lens and the image
c) both of these d) none of these

15. For the production of a rainbow, light is
a) refracted b) reflected c) both d) neither

16. A lens can be used to take light emanating from one place and bring it back to another place—its image. For a simple lens, which color is brought to focus nearest to the lens?
a) red b) blue c) the same

17. Interference is a phenomenon which can be evidenced with
a) light waves b) sound waves c) water waves d) all of these e) none

18.
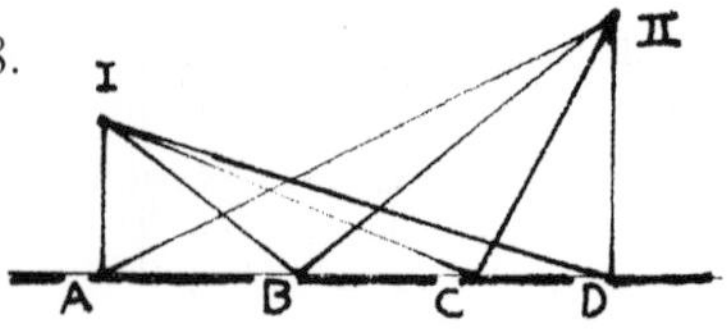

The shortest path from point I to point II is the one that meets point
a) A b) B c) C d) D
e) all are equally long

19. She walks toward the large plane mirror at a speed of 1 mile per hour. Her image approaches her at a speed of
a) ½ mile per hour b) 1 mile per hour c) 1½ miles per hour
d) 2 miles per hour e) none of these

20. She's 6 feet tall. The minimum length mirror required for her to see her full-length image is
a) 6 feet b) 4 feet c) 3 feet d) depends on her distance from the mirror

21. As the sun rises in hilly places like San Francisco, the light from the east reflects from the windows of houses on hills to the west. As the sun rises, the area of sparkling windows seems to move to houses

a) farther up the hill
b) down the hill
c) neither—the sparkling area essentially stays put

Electricity & Magnetism

Fluids, mechanics, heat, vibrations and light were all known and familiar to the ancient engineers who built the pyramids. But now we shall treat something "new"—electricity and magnetism. At the time of the American Revolution electricity and magnetism was still exotic, hard to make and of little practical use. Then the secrets of electricity and magnetism were found... and thereafter the world changed.

Spinning turbines generated electricity to take the place of whale oil and gas for lighting cities, and to run motors to ease the efforts of human and animal labor; heat came out of walls by means of wires; electric vibrations carried messages across the country and then to the moon and, at last, physicists understood the nature of light—the radiation of electric and magnetic fields.

How can a magnet attract iron without touching it?

STROKING

If something gets a positive electric charge, then it follows that something else

a) becomes equally positively charged
b) becomes equally negatively charged
c) becomes negatively charged, but not necessarily *equally* negatively charged
d) becomes magnetized

ANSWER: STROKING

The answer is: b. When you stroke a cat, the cat becomes positively charged and the brush becomes negatively charged. In so doing you do not create electricity. The electricity was already there. The cat's fur, before stroking, contained equal amounts of positive and negative electricity in *each* of its atoms, the positive in the nucleus and the negative in the surrounding electrons. The stroking only separated the negative from the positive. This is because the bristles in the brush have a greater affinity for electrons than does the cat's fur. Negatively charged electrons are transferred by friction from the cat's fur to the brush which leaves an imbalance of electric charge on both the fur and the brush. The fur is deficient in negative charge, and we say the fur is

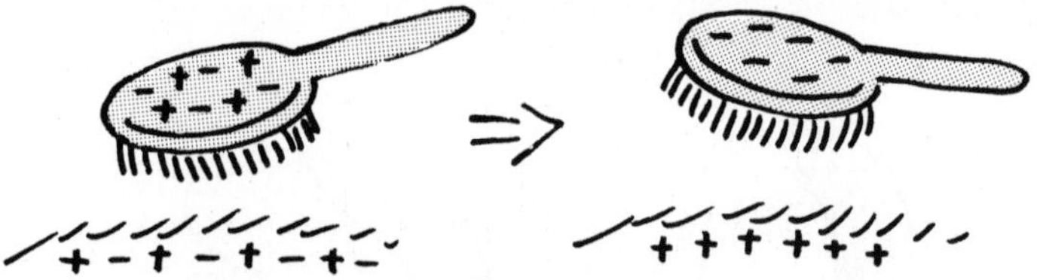

therefore positive. The excess negative charge on the brush makes it negative. So the fur and the brush are equally but oppositely charged. The energy that is expended in the brushing action is stored in the separated charges, which is evident if the brush is brought near the fur and a spark is produced.

DIVIDING THE VOID

Charging is a process of charge separation that required work. But given enough energy, can electric charge be created from nothing, say in the vacuum of empty space?

a) Yes, this is not unusual
b) No, such an occurrence would violate presently-understood laws

ANSWER: DIVIDING THE VOID

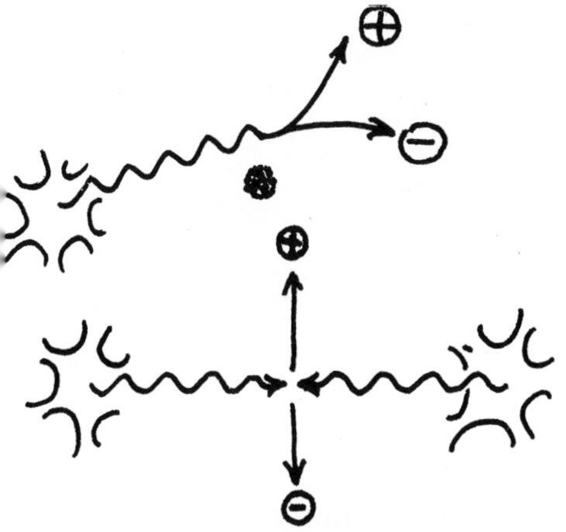

The answer is: a. If a sufficiently energetic x-ray passes *near* a piece of matter or if a pair of x-rays collide, the energy of the x-rays creates a positive and a negative electron *right in empty space.* The process has been photographed, and is a routine occurrence when high energies are involved. The positive electron is known as a *positron* or anti-electron.

How does this come about? Granted you can see where the energy comes from—the x-rays. But where do the charges come from? Now a *net* charge can never be created. That is, if there was zero charge to start with there must always be zero charge. Charge can be created only in the limited sense that equal amounts of plus + and minus − are created together, so the total effect is zero creation since the + exactly cancels the −. More accurately, we might state that charge is separated out of the void. We can picture it like this:

Suppose the vacuum of empty space is a gray void. Out of one area called **a** in the gray void we remove a bit of grayness, leaving a white area **b** and compressing the removed gray part into another gray area **c**. In this way we make **c** even grayer, or perhaps black. So in effect we have not created black or white, but have only separated black and white out of gray. So it is that + and − electric charges can be separated from the vacuum of space—in essence, dividing the void. Of course, it takes energy to do the division.

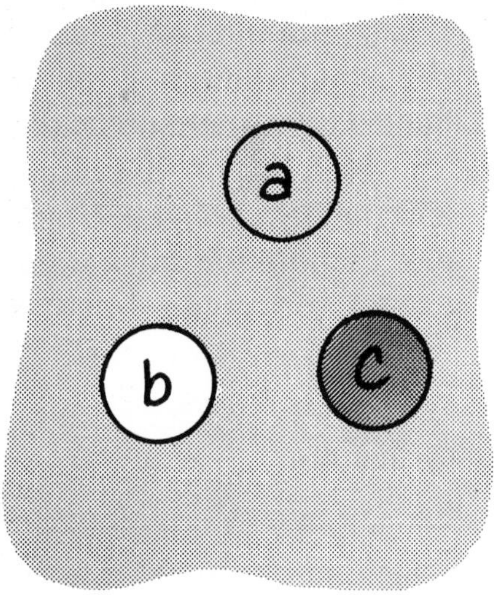

And another thing: If we take a piece of empty gray space and force the grayness to a different location leaving white behind, we produce a *strain* or *displacement* that is left in the space between the black and white areas. That strain is the *electric field* which is sometimes called a *displacement field.* If the plus and minus are not held apart, the strain will cause them to snap back together.

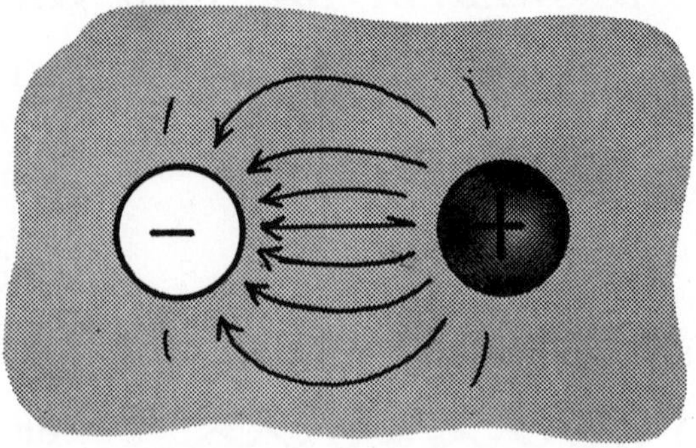

If the plus and minus are allowed to snap together you might expect the strain would be relieved and everything would return to the prior state—a plain gray void. But when the strain in the earth is relieved by the snap of an earthquake fault, does everything go back quietly to the way it was? Everything goes back, but not quietly. The strain energy escapes, thereby making the surrounding earth shiver. So also does the sudden relief of the strain between the plus and minus which makes the surrounding gray shiver. The shiver is the pulse of radiation always emitted by electron-positron annihilation.

Incidentally, this notion of creating things (two opposite things) out of nothing is not only found in physics. In the world of business a new corporation is created selling stock (or bonds). On the one hand the corporation then has money to operate with, but on the other hand it has its debt obligations to the people who supplied the money. The stock and bonds represent the debt obligation of the corporation and the debt obligation exactly balances the money gathered by the new corporation when it starts up. The gathered money is called the capital, so: capital + debt = 0.

ELBOW ROOM

Molecules in a gas resist crowding and get as far apart as possible. Free electrons also resist crowding and get as far apart as possible. When a tank is filled with gas the molecules distribute themselves more or less uniformly throughout the tank's volume, thereby giving each molecule the maximum possible distance from its nearest neighbors. When a copper ball is charged with electricity, the free electrons will distribute themselves more or less uniformly throughout the ball's volume for much the same reason.

a) True

b) False

ANSWER: ELBOW ROOM

The answer is: b. Offhand you might expect that the electrons, like the gas molecules, would spread themselves throughout the volume of the copper ball, thereby giving each electron as much elbow room as possible between itself and its neighbors. But that is not what happens. The electrons all crowd on or near the outer surface of the copper ball. Why the dramatic difference between the electron and the gas distribution? Because the gas molecules only interact with their nearest neighbors by physically hitting them. Molecules exert short-range forces on each other. A molecule has no interaction with a distant molecule on the other side of the gas tank. The molecules distribute themselves so as to maximize the distance between their immediate neighbors.

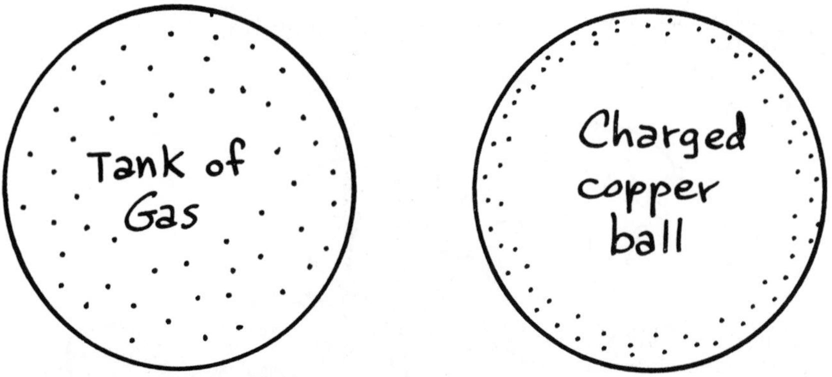

On the other hand, the electron is able to interact with distant electrons by means of its field. It can exert a force on another electron without being near to it. An electron maximizes its distance, not from its nearest neighbors, but from ALL the electrons in the copper ball. The electron ends up accepting a few close neighbors in return for getting all the other electrons as far away as possible. "As far away as possible" means on the other side of the ball. Electrons exert long-range forces on each other.

The electrons that make a metal object negatively charged are always spaced along the outer surface of the object.

MOON DUST

Before the first moon landing, several NASA scientists were concerned about the possibility that the lunar lander might be engulfed by a layer of dust hovering just above the moon's surface. Could there be some particular distance from the moon where electrically charged dust, or even electrons, could hover?

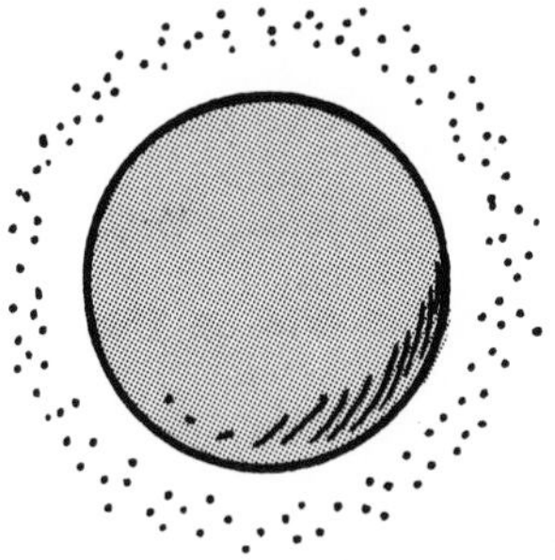

Suppose the moon had a negative charge. Then it would exert a repelling force on electrons near it. But the gravitational force of the moon exerts an attracting force on the electron. Suppose the electron is one mile above the lunar surface and the attraction exactly balances the repulsion, so the electron floats. Next, suppose the same electron was two miles above the moon. At the greater distance

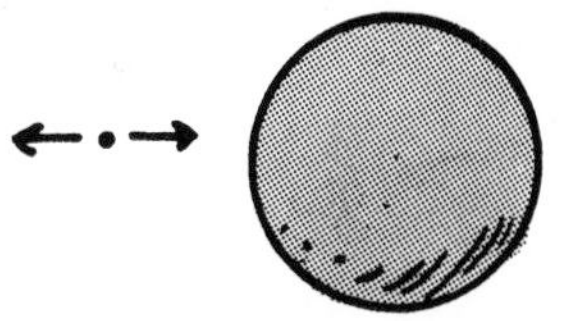

a) the gravity would be stronger than the electrostatic force, so the electron would fall
b) the gravity would be weaker than the electrostatic force, so the electron would be pushed into space
c) the gravity would still balance the electrostatic force, so the electron would float

ANSWER: MOON DUST

The answer is: c. There cannot be some particular distance from the moon where, and only where, the electric and gravitational forces balance out. Why? Because if they balance at some distance from the moon and you go to a place twice as far away, both will be reduced by the SAME factor and so they will still balance. If dust could float, due to an electrostatic charge, one inch above the lunar surface, it could float at any height and so would float right off the moon! In fact, it is impossible to suspend or levitate an object by any combination of STATIC electric, gravitational or magnetic force fields because each obeys the inverse square law.

UNDER THE INFLUENCE

Two uncharged metal balls, X and Y, stand on glass rods. A third ball, Z, carrying a positive charge, is brought near the first two. A conducting wire is then run between X and Y. The wire is then removed, and ball Z is finally removed. When this is all done it is found that

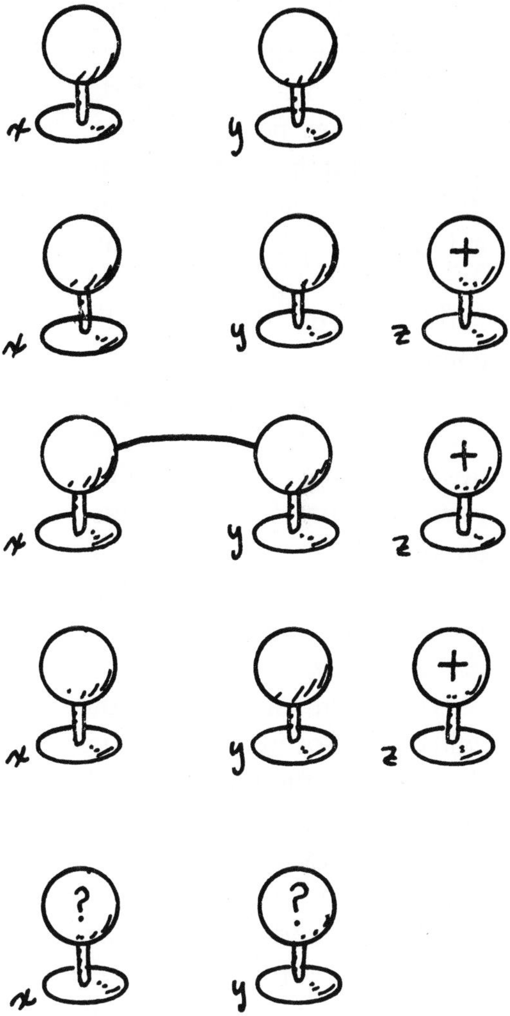

a) balls X and Y are still uncharged
b) balls X and Y are both charged positively
c) balls X and Y are both charged negatively
d) ball X is + and ball Y is −
e) ball X is − and ball Y is +

ANSWER: UNDER THE INFLUENCE

The answer is: d. The trick to this thing is to see the world in a slightly different way. True, X and Y are uncharged, but that does not mean that they do not have charge on them. They each have equal amounts of plus and minus mixed so the gross effect is zero charge. But then Z comes on the scene with its plus charge. Though Z never touches X or Y they are nevertheless under the influence of plus Z. The minus charges in X and Y are drawn towards Z. The plus charges in X and Y are repelled from Z. So one side of X becomes plus, the other side of X becomes minus. The same for Y. This split is called electrostatic polarization. When a wire is run from the minus side of X to the plus side of Y the minus charges on X can get still closer to Z, and the plus charges on Y can get farther from Z. So the minus on X moves to Y and the plus on Y moves to X. But that leaves a net plus on X and a net minus on Y. This process is called charging by electrostatic induction. It is important to note that no electric charge was created. True, ball X became plus, but ball Y became minus by exactly the same amount so the net effect is zero. What happened is that charge was separated.

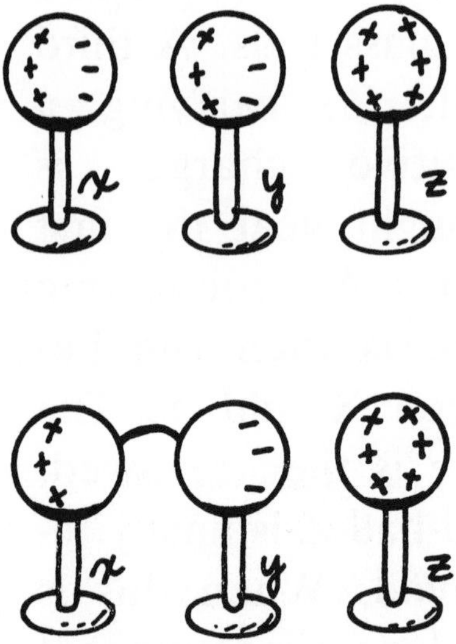

JAR OF ELECTRICITY

A Leyden Jar is an old-fashioned capacitor. Now a capacitor, or condenser as they are sometimes called, consists of metal surfaces that are separated from one another. They are storehouses of electric energy when one surface is charged + and the other surface −. Two hundred years ago capacitors were made by putting one piece of metal foil on the inside of a bottle and one piece on the outside. The bottle was called a Leyden Jar because the first were made at the University of Leyden in Holland—the Cal Tech of its day. The energy stored in a charged Leyden Jar is actually stored

a) on the metal foil inside the jar
b) on the metal foil outside the jar
c) in the glass between the inner and outer foil
d) inside the jar itself

ANSWER: JAR OF ELECTRICITY

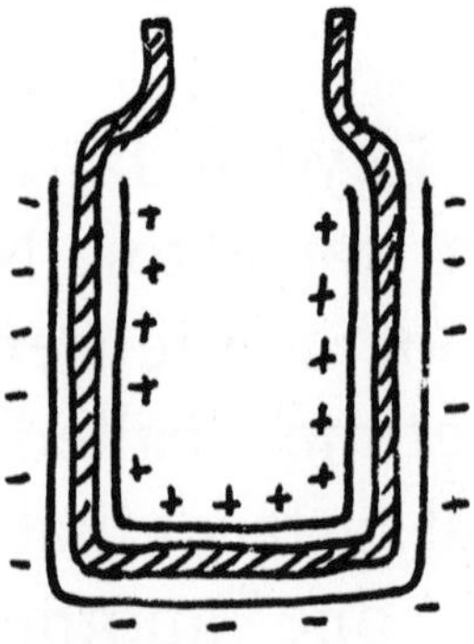

The answer is: c. A simple capacitor consists of two conducting pieces, usually metal, *close together but not in contact.* It allows the + and − electricity to get real close together, but not touching. It is the electrical equivalent of the 18th-century practice of courtship by "bundling," where opposite sexes could lie together in the same bed provided an upright board prevented any touching. In a Leyden Jar, opposite charges are prevented from touching by means of the glass bottle. Suppose the inside is charged + and the outside −. Then electric force field lines run from the + charges on the inside foil to the − charges on the outside foil. The charges mark the beginnings and ends of the force lines. So the force field is in the glass and the energy is in the force field. That means the energy is in the glass!

So a Leyden Jar is a bottle that holds electricity, the energy of which is not inside the bottle, but rather inside the glass. How do you empty the bottle? Simply connect wire leads from + and − sides of the glass.

The energy in a capacitor is always in the space between the opposite charges. From that you might suspect the amount of energy in a capacitor depends not only on how much electric charge is in the capacitor, but also on how much space is between the charges and what the space is filled with—like glass, air, or oil. We will investigate this in the next two questions.

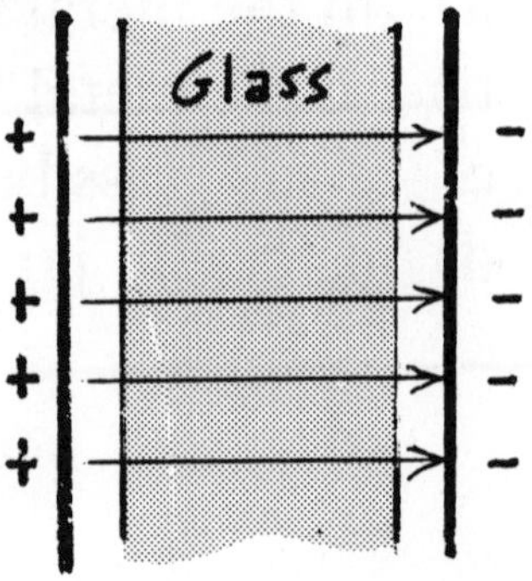

ENERGY IN A CAPACITOR

Consider a simple capacitor made of a pair of conducting plates in close proximity. Suppose the plates are appropriately charged + and − and then discharged to produce a spark. Next, the plates are charged again exactly as they previously were, only this time after being charged they are pulled farther apart. If they are then shorted out a second time, the spark produced will be

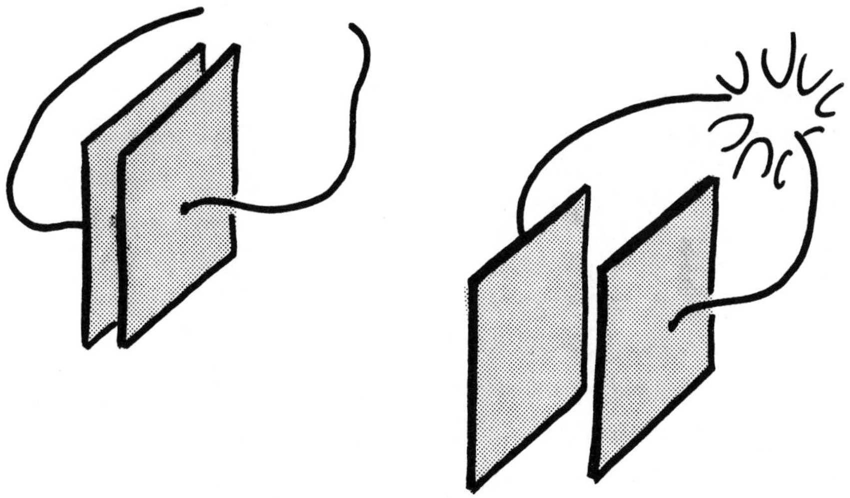

a) bigger (liberate more energy) than the first spark
b) smaller than the first spark
c) the same size as the first spark

ANSWER: ENERGY IN A CAPACITOR

The answer is: a. Where did the energy come from to make the bigger spark? Energy came from the work someone did in pulling the + plate away from the − plate. No one added any electricity to the condenser by pulling the plates apart. Instead, the work done in overcoming the mutual attraction between the oppositely charged plates in pulling them apart went into the electric field between the plates. We say the *voltage* between the plates is increased. Voltage is an electric energy potential difference, like the gravitational energy potential difference related to falling objects. In our case the electrons "fall" from the − to the + plate, so if there is more distance between the plates there is farther to fall and therefore more potential difference.

Another way to talk about this is to say the capacitance of the capacitor was decreased but the charge held constant, so the voltage increased—but that just amounts to saying what has been said in different words.

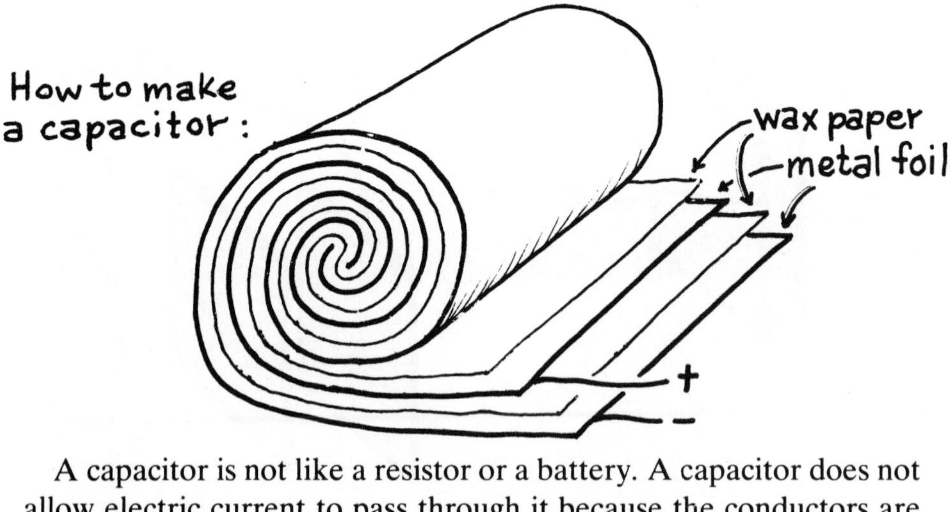

A capacitor is not like a resistor or a battery. A capacitor does not allow electric current to pass through it because the conductors are separated, so it is not like a resistor, which does allow current to pass. A capacitor does not make electric current; it must be charged, so it is not like a generator which makes current without being charged. A capacitor is not like a battery which puts out one voltage because a capacitor can be charged to have many different voltages. It is a storehouse for electrical energy.

GLASS CAPACITORS

Capacitors may have an airspace between their plates, or have glass, plastic, wax paper, or oil between the plates. In the days of Benjamin Franklin capacitors were the previously mentioned Leyden Jars—so you are now learning 200-year-old physics. If a glass capacitor is charged, but the glass between the plates is removed before it is discharged, the spark will be

a) bigger than it would have been if the glass were left in at discharge

b) smaller than it would have been if the glass were left in at discharge

c) the same as it would have been if the glass were left in at discharge

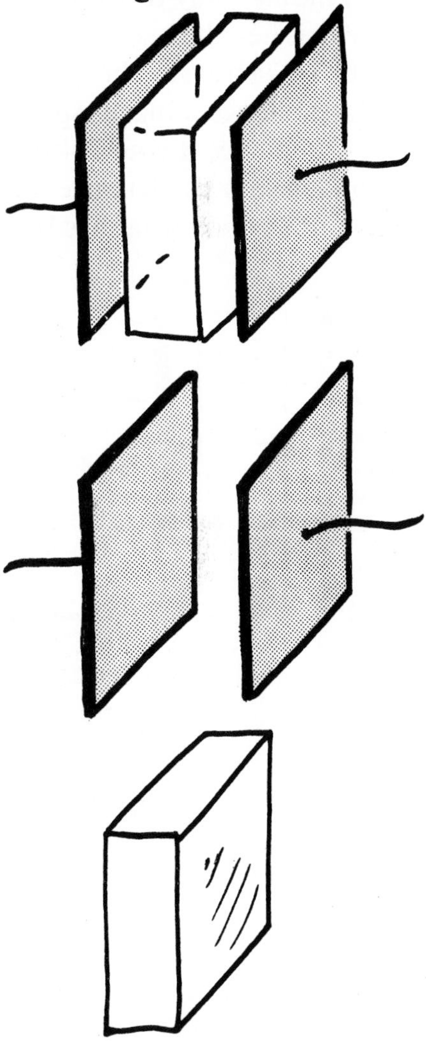

ANSWER: GLASS CAPACITORS

The answer is: a. The glass in the capacitor is polarized, so the side of the glass near the + plate becomes − and the glass near the − plate becomes +. When the glass is removed, the − charge on the glass is removed from the vicinity of the + charge on the plate and the + charge on the glass is removed from the vicinity of the − charge on the plate and doing that requires work to overcome the attraction of like charges. So work is required to remove the glass and that work shows up in the spark.

Another way to view the situation is to say the glass weakens the electric field between the plates. Removing the glass restores the field and so increases the potential difference or voltage between the plates, hence a bigger spark.

Of course, it could also be said removing the glass decreases the condenser's capacitance and therefore increases its voltage which again is saying what has been said in different words.

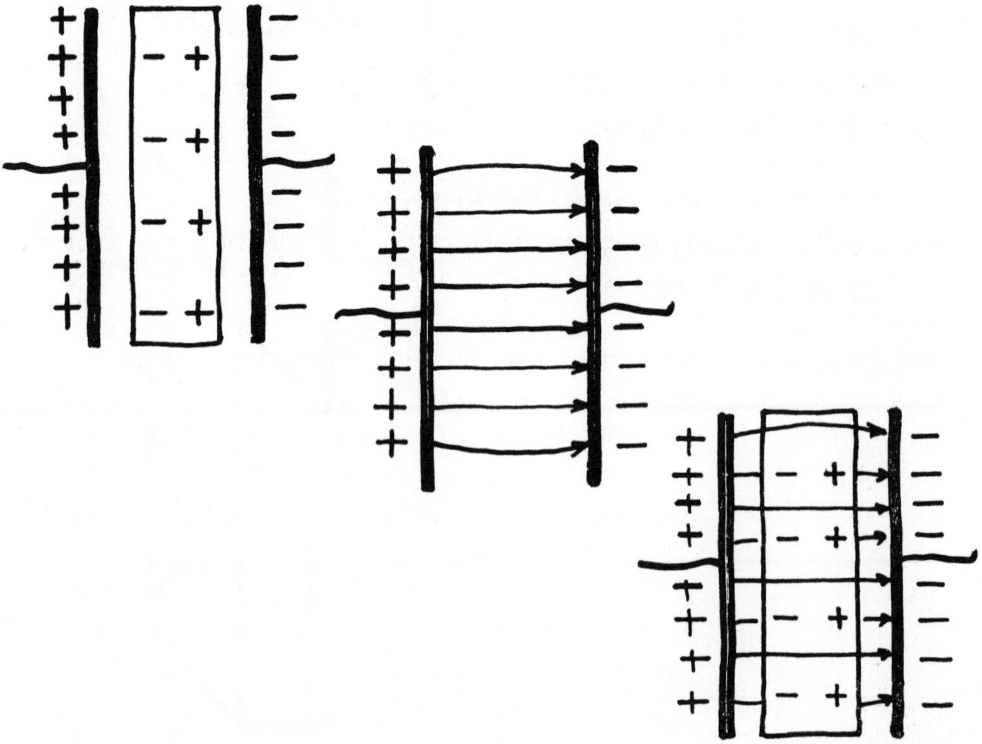

HIGH VOLTAGE

Is it likely that a situation could exist where there was a lot of voltage without also having a lot of current at the same time?

a) Yes, such situations commonly exist
b) No, such situations are not common

ANSWER: HIGH VOLTAGE

The answer is: a. Plus charge and minus charge attract each other, while plus and plus as well as negative and negative repel; like charges repel and unlike charges attract. Energy is required to separate plus and minus. This energy can be stored as long as the charges are separated, just as energy can be stored in the hammer of a pile driver as long as it is elevated. Stored energy is called *potential energy*—in the case of the pile driver we say the hammer has gravitational potential energy with respect to the ground below; in the case of the separated charge we say the charges have electric potential energy with respect to each other. When we speak of the amount of electric potential energy *per charge,* we are speaking of *voltage*. On a fully charged 12-volt automobile battery, for example, the opposite charges on the battery terminals have an energy of separation of 12 units of energy per unit of charge, or more precisely, 12 volts = 12 joules/coulomb. If the dome of a Van de Graaff generator is charged to 100,000 volts, then each coulomb of charge on the dome has a potential energy of 100,000 joules.

How about current? There is none. However, if a conducting path is provided between the locations of opposite charge, then charge will flow and we have an electric current.

HIGH AMPERAGE

The unit of electric current is the *ampere,* and electric current is often loosely called *amperage.* Can there exist any situation where there is a lot of amperage without also having a lot of voltage at the same time?

a) Yes
b) No

ANSWER: HIGH AMPERAGE

The answer is: a. The amount of current in a simple circuit depends not only on voltage, but on resistance as well. If the resistance of a conductor is very small, a tiny voltage across it can result in a large current flow. Some materials cooled to very low temperatures, called *superconductors,* have zero resistance. Tiny voltages produce enormous currents in these superconductors. In fact, the current in a superconducting circuit will continue to flow indefinitely after the voltage source is disconnected!

HIGH RESISTANCE

Most of the resistance in this circuit is in the

a) wire cord
b) light bulb

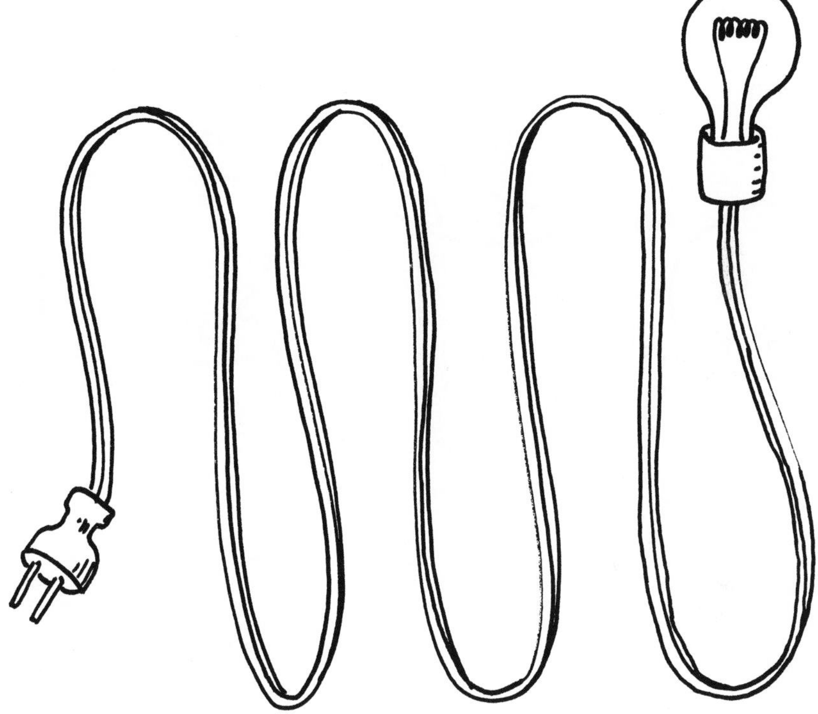

ANSWER: HIGH RESISTANCE

The answer is: b. The wire in the cord is much fatter than the wire that makes up the light bulb filament. If the cord had more resistance than the bulb, the cord would get hot and glow and the lamp would be relatively cool.

COMPLETE CIRCUITS

A simple electric circuit can be made with a dry cell, a lamp, and some wire. In which of the arrangements shown will be lamp be lit?

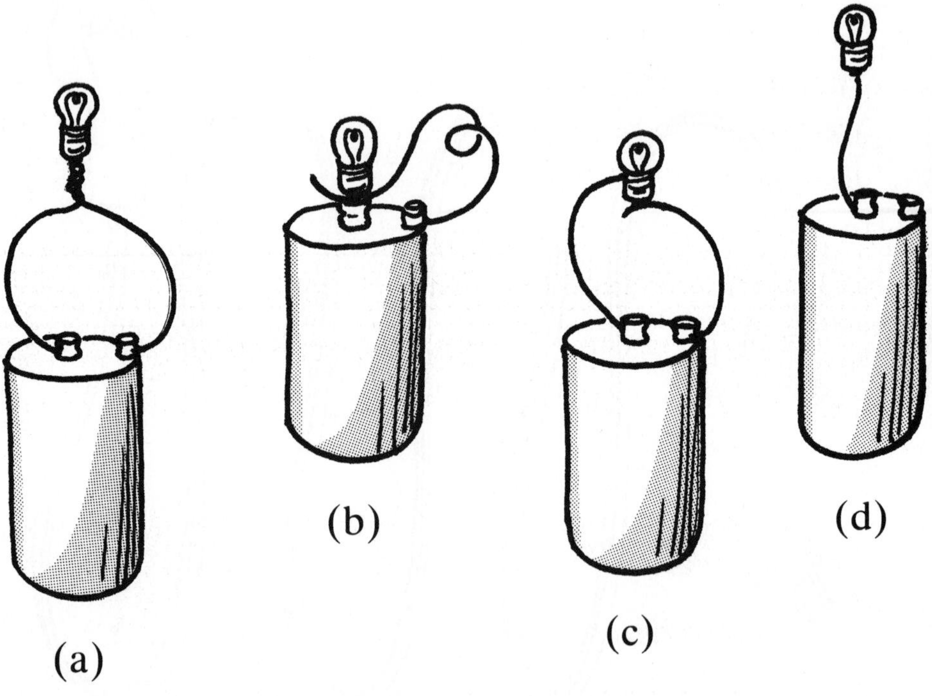

(a) (b) (c) (d)

ANSWER: COMPLETE CIRCUITS

The answer is: c. A battery is not a source of energy which fills a lamp in a sink-like fashion. There must be a "passing through" in a circuit. Current does not flow into a lamp, nor does it even flow out of a lamp. Current flows *through* a lamp, just as it does anywhere else in the circuit, including the battery itself. Investigation of the arrangements shows that only in **c** is a pathway provided for the necessary in-to and out-of connection to the lamp.

ELECTRIC PIPE

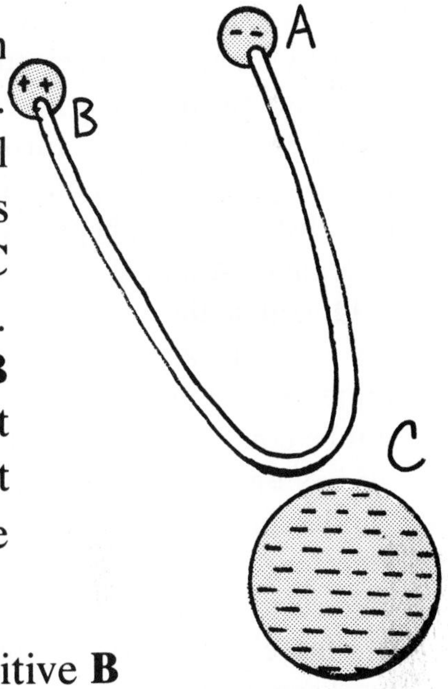

This is an important question so be careful with this one. Suppose object **A** has a small negative charge and object **B** has a small positive charge. Object **C** has a very large negative charge. Further suppose that **A** and **B** are joined by a copper wire that goes very close to but does not touch **C**. What will happen is the negative charge from **A** will

a) flow through the wire to positive **B**
b) not flow to **B** because of the repelling influence of **C**

ANSWER: ELECTRIC PIPE

The answer is: a, even though it might seem that the electrons would be pushed back by the negative charge on **C**. The problem comes from picturing the wire as an empty glass tube through which little balls called electrons flow. That picture is misleading. If it were correct the answer to this question would be b. But in fact, the current in the wire will be as if **C** were not present.

What happens is that the electrons in the wire shield the inside of the wire from the influence of **C**. It happens like this: Consider only the wire and object **C** without the presence of **A** and **B**. Positive charge is induced in the part of the wire near **C** while negative charge is induced in the part of the wire away from **C**. This induced charge distribution sets up an electric field in the wire that exactly balances the field due to **C**. This happens by a momentary electron flow in the wire until the net electric field in the wire is equal to zero. Then further flow is nil unless an additional potential difference is established; which happens with the introduction of **A** and **B**.

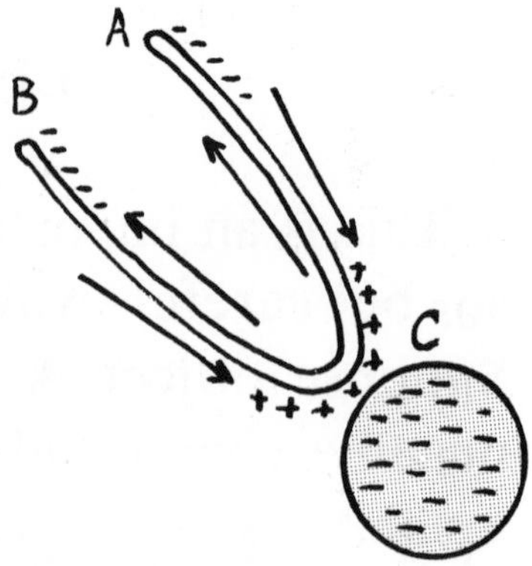

Of course there must be enough free electrons in the wire to allow for a redistribution of charge to offset the effect of **C**. But suppose there are not enough free electrons in the wire. Then the wire could not be completely shielded from the effect of **C**. In practice there are always enough free electrons in a metal wire, but there are semiconducting materials like germanium which have relatively few free electrons. This situation is taken advantage of in the making of an electric valve called a *field-effect transistor.* The device allows the flow of a few electrons to be stopped by the presence of other electrons and works as is shown in the sketch: A semiconductor bridge joins two metal wires. Normally, electrons flow from the source wire across the semiconductor bridge to the drain wire. However, if another piece of metal called the *gate,* which is very close to but not touching the bridge, is made negative it repels electrons in the bridge. That stops the current and closes the valve.

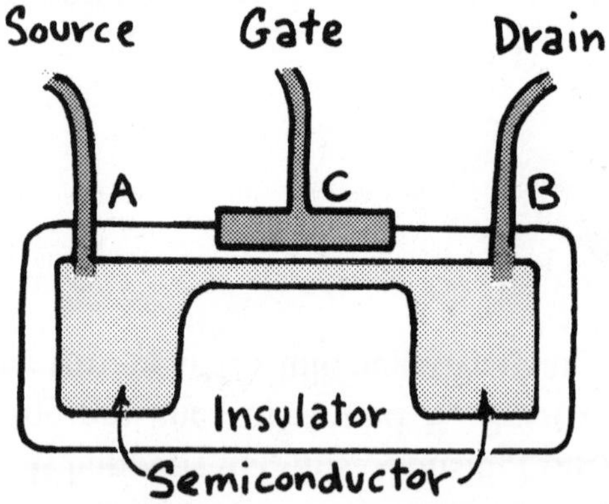

Were the bridge made of copper this electric valve wouldn't work because the top of the bridge would get positive enough to shield the material below it from the negative charge on the gate. But in a semiconductor there are not enough free electrons to make the top of the bridge sufficiently positive to shield itself.

The idea of the field-effect transistor was developed in the 1920's but wasn't put into practice until the 1960's. There is another type of transistor called a *junction* or *bipolar* transistor. That type is described in most physics books.

IN SERIES

A defective toaster may blow your house fuse if it has a short in it. But suppose you add a light bulb to the circuit, as illustrated. When it is plugged in with the bulb in the circuit it will

a) sometimes blow the fuse
b) never blow the fuse
c) always blow the fuse

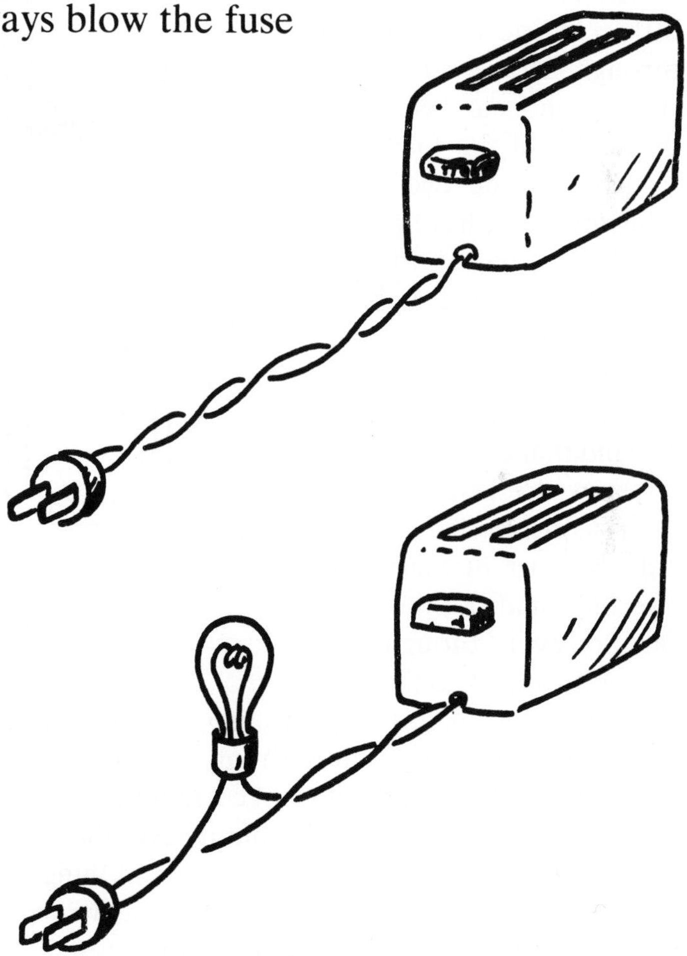

ANSWER: IN SERIES

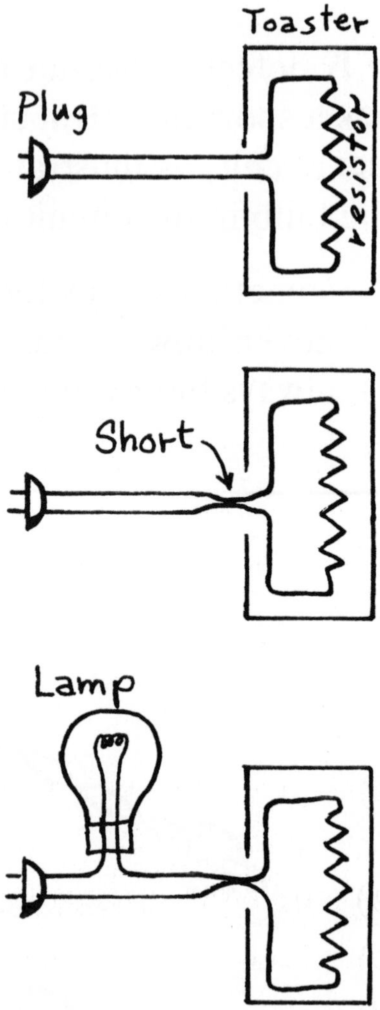

The answer is: b. The first sketch shows a good toaster. The electric current comes in one wire at the plug, goes through the heating element in the toaster and then goes out the other wire at the plug. All the current that comes in must go back out. The heating element has resistance which is like electric friction. It resists the current flow so only a little electric current can squeeze through. The resistor puts a drag on the circuit and keeps the current down, but if the toaster is damaged such that the two wires from the plug touch, we have a short circuit. This is called a "short" because it gives the current a shortcut. It does not have to go through the resistor anymore. With the resistor bypassed, the current can flow like crazy—and that blows the fuse. If it didn't blow the fuse it would probably start a fire! If a light bulb is now added, the current must flow through the bulb and the bulb has resistance. So even though the current does not have to go through the toaster's resistor it still has to go through the light bulb's resistor. And the bulb's resistance stops the current from running wild. Of course, the light bulb hinders the flow of current through the toaster and will not let it get as hot as normal, but it saves the fuse.

Some old-timers who experiment with electrical things purposely put a light bulb in the circuit so that if something goes wrong and there is a short, the current will not run wild and blow a fuse.

WATTS

The power, called wattage, delivered to an electric device (a power saw, for example) can be increased by increasing

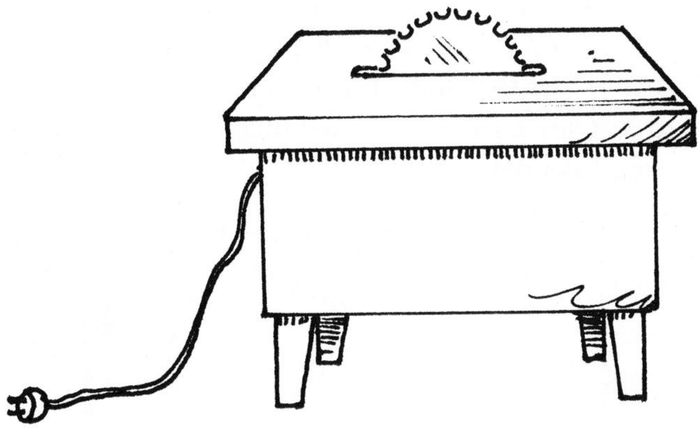

a) the current flow (amperage) it draws but not the potential difference (voltage)

b) the voltage applied to it but not the amperage

c) the amperage or the voltage

d) none of these is correct.

ANSWER: WATTS

The answer is: c. When a power saw is plugged in, the applied voltage is 110 volts. That is the most voltage available to it. How does it increase its power output when a big piece of wood is shoved into it? By increasing the amperage it draws. If you overload the saw and slow it down it will draw excess amperage which can be evidenced by the dimming of lights elsewhere in the circuit. This is like the drop in house water pressure when someone opens a big water faucet.

Can you think of a case where voltage is increased without changing amperage? Think of a battery in series with a light bulb. Then think of two batteries in series with two light bulbs. The voltage and power output are doubled but since the load resistance is also doubled, the amperage in the circuit is unchanged.

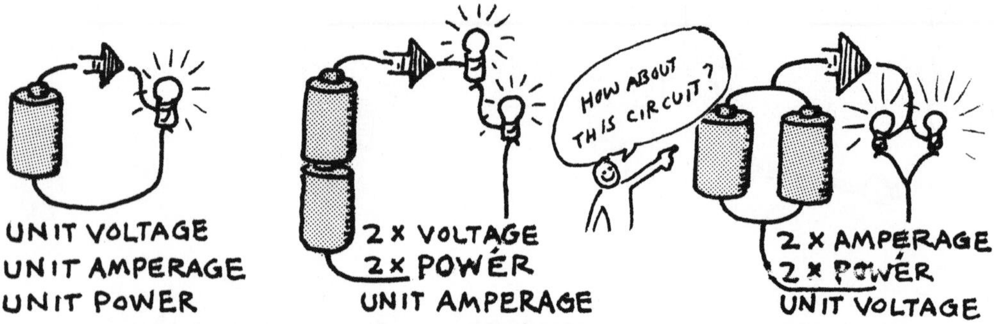

The delivered power can be doubled by doubling either the current flow or the voltage applied to the electric device. In general

$$(\text{power}) = (\text{wattage}) = (\text{voltage}) \times (\text{amperage})$$

This idea is not unique to electricity. It applies, for example, to water wheels. The power output of a water wheel depends on the product of two things. First on the diameter of the wheel, which measures the potential difference the water falls through. That is like the voltage. And second, it depends on the number of gallons per hour of water going over the wheel—the current. That is like the amperage.

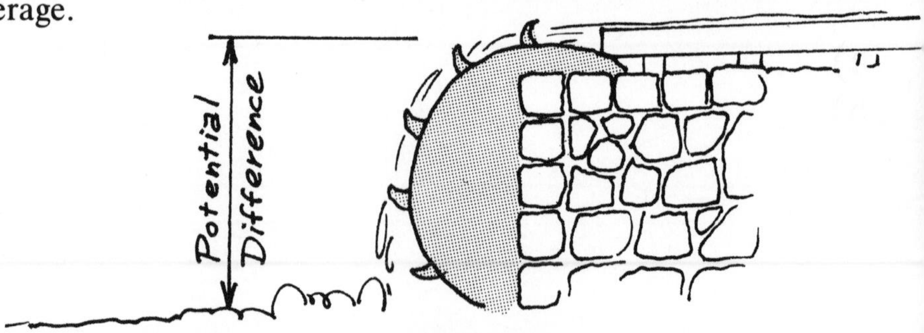

PUTTING OUT

First one light is connected to a battery. Then two are connected in series to the same battery. When two are connected, the battery puts out

a) less current
b) more current
c) less voltage
d) the same current

A

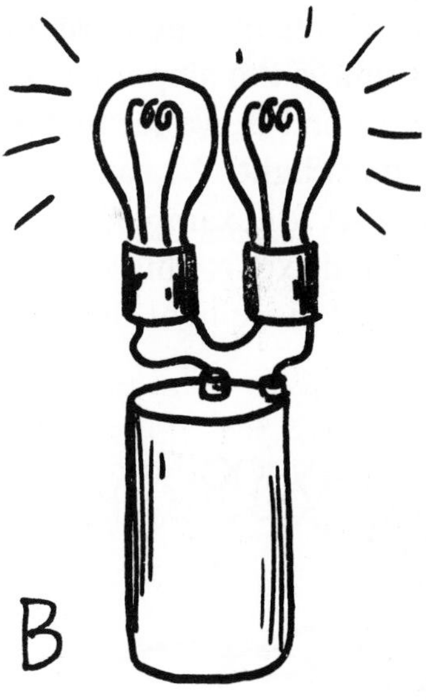

B

Which arrangement puts out the most light?

a) A
b) B
c) both the same

ANSWER: PUTTING OUT

The answer is: a. The battery provides a voltage (like six volts or twelve volts) which is like a certain pressure. The voltage forces a flow of charge (current) through the light bulbs which resist the flow. The wires also resist the flow, but the resistance in the bulbs is much, much greater. Two identical bulbs connected in series have twice as much resistance as one bulb. As the resistance doubles only half as much charge flows. The current is one-half.

The situation here is quite like that inside a human body. As blood arteries become obstructed their resistance to the blood current increases and less blood flows. If resistance is doubled only half as much blood flows. But the body can't get by with half as much blood current and demands more current. So the heart pumps with more pressure (high blood pressure) to force more blood through the resistance in the arteries. The heart is like the battery: it provides the pressure or voltage. But a battery can only put out some fixed maximum voltage pressure. A heart, unlike a battery, can put out more pressure if it is required to do so, but an overworked heart is the result.

The answer to the second part is: a. We have established the fact that **A** puts out the most amperage, and since the power the battery puts out is the amperage × voltage, and since voltage is the same in both **A** and **B**, we see that **A** puts out the most power—and therefore the most light. We can exaggerate this and see the idea more clearly: suppose, for example, that 50 bulbs were connected in series—at best we could expect only a faint red glow from the filaments. A single bulb puts out more light than multiple bulbs in series.

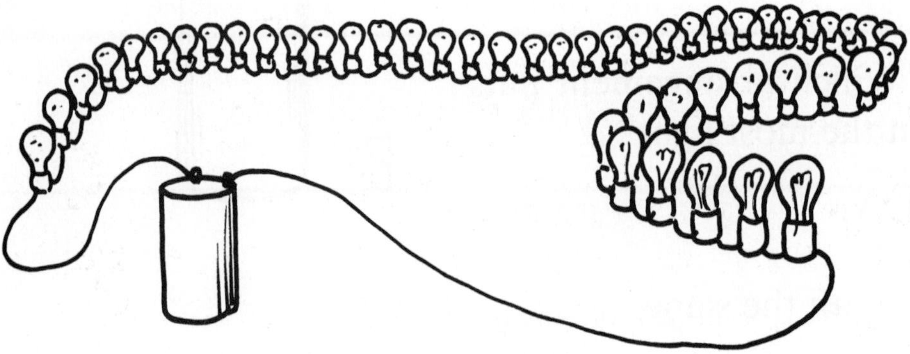

TROLLEY

A street car has one trolley.

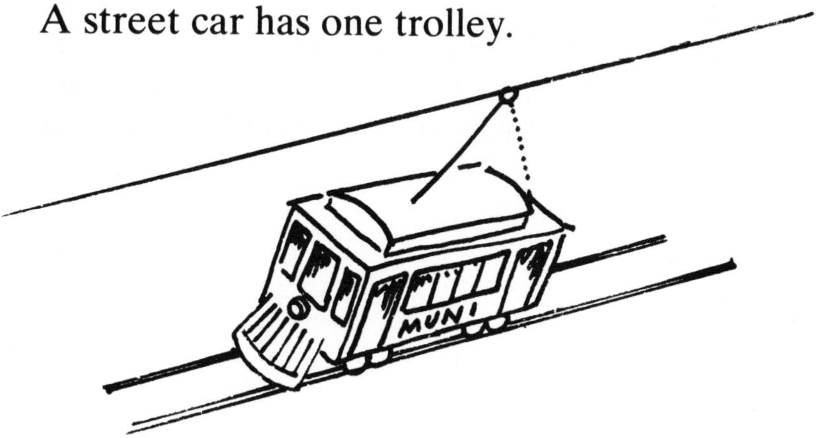

An electric bus has two trolleys.

This is because the

a) extra trolley on the bus is a backup for increased reliability.
b) bus is AC while the street car is DC
c) bus is DC while the street car is AC
d) bus draws more current than the street car
e) street car uses its wheels as one of its trolleys

ANSWER: TROLLEY

The answer is: e. The electric generator that powers the street car has one side "put to earth." Current from the generator goes out to the car by means of the one trolley wire, but comes back to the generator through the earth. A bus rides on rubber wheels, so it cannot return electricity through the earth. Therefore, two wires and two trolleys are necessary.

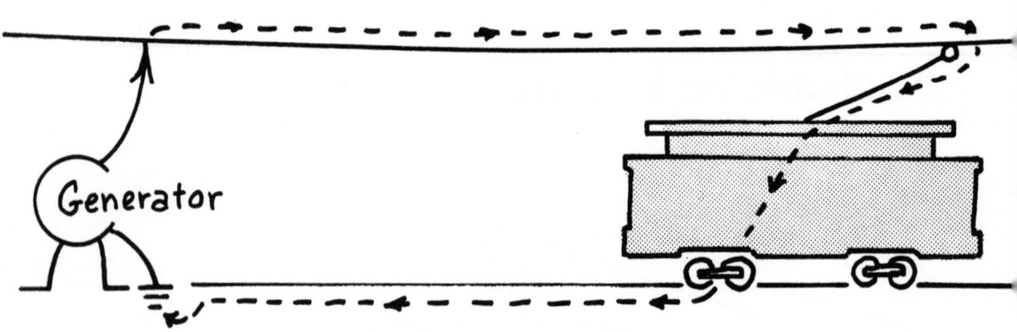

GROUNDED CIRCUIT

Will the light go on if the circuit is grounded as illustrated?

a) Yes
b) No

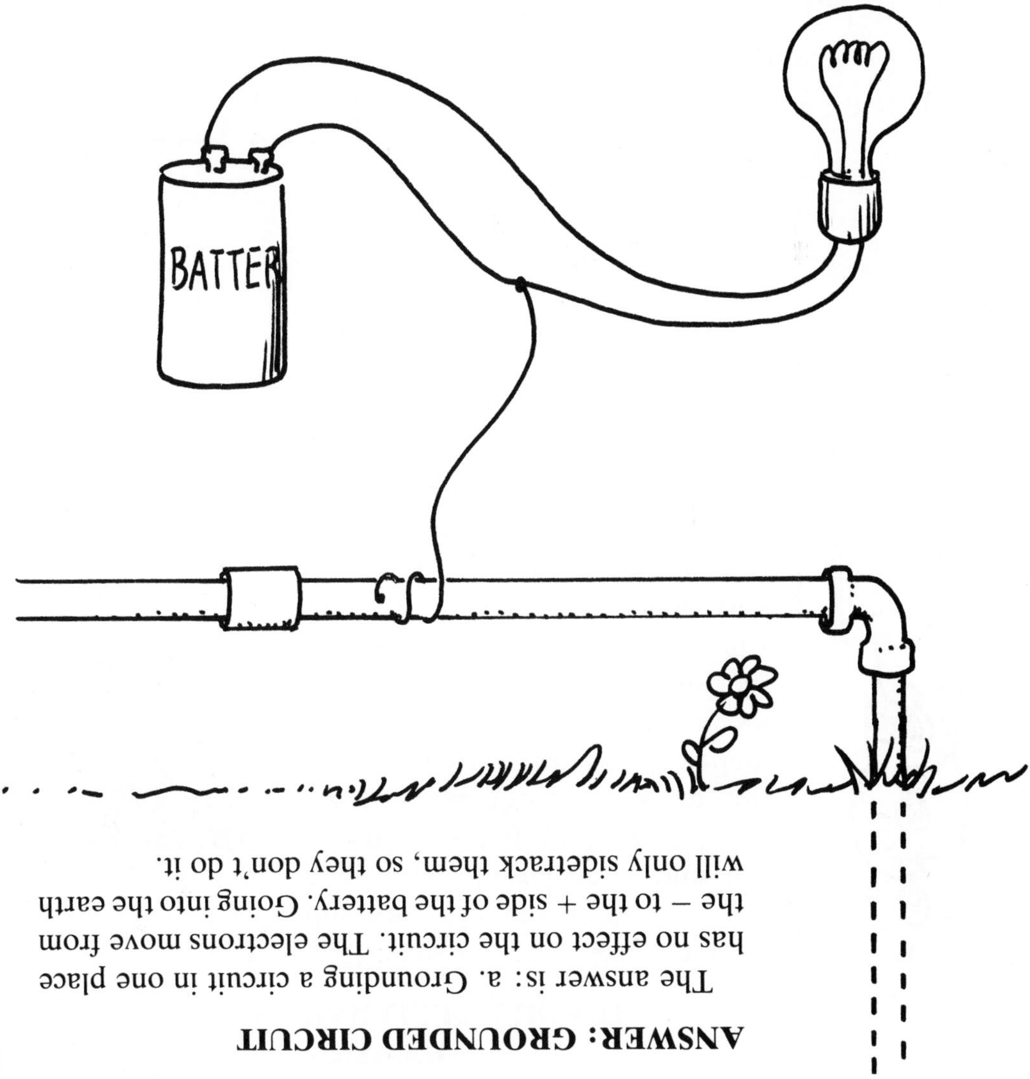

ANSWER: GROUNDED CIRCUIT

The answer is: a. Grounding a circuit in one place has no effect on the circuit. The electrons move from the − to the + side of the battery. Going into the earth will only sidetrack them, so they don't do it.

PARALLEL CIRCUIT

In the circuit shown at the right, the voltage drop across each resistor is

a) divided among the three resistors
b) dependent on the overall resistance
c) the same

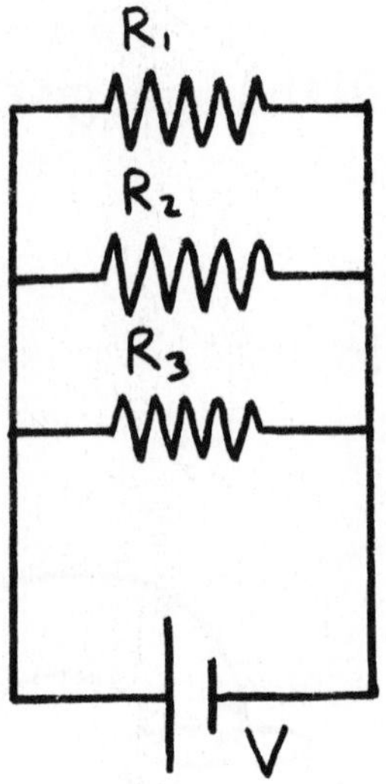

ANSWER:
PARALLEL CIRCUIT

The answer is: c. We can see it this way. The circuit shown in the question might well be a representation of the situation shown below at the left. All three bulbs are connected across the same voltage source. If the cell is 1.5 volts, we see that 1.5 volts is across each bulb. But is the situation any different if the bulbs are connected as shown at the right? No. If the resistance of the connecting wires is negligible, the two cases are the same. In a circuit, the voltage across branches connected in parallel is the same.

THIN AND FAT FILAMENTS

Light bulbs **A** and **B** are identical in all ways except that **B**'s filament is thicker than **A**'s. If screwed into 110-volt sockets,

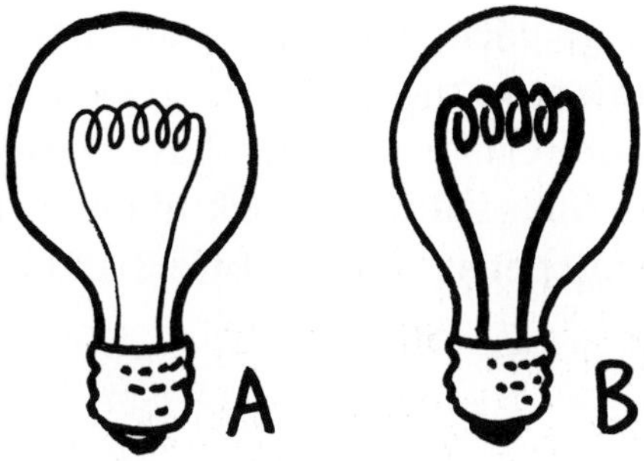

a) **A** will be the brightest because it has the most resistance
b) **B** will be the brightest because it has the most resistance
c) **A** will be the brightest because it has the least resistance
d) **B** will be the brightest because it has the least resistance
e) Both will have the same brightness

ANSWER: THIN AND FAT FILAMENTS

The answer is: d. The brightest light is the one that consumes the most energy per second. The energy consumed depends on how much charge falls through how much potential difference or voltage difference. The voltage difference across each bulb is 110 volts since each is screwed into a 110-volt socket. So the only distinction between the bulbs is how much charge goes through them per second, that is, how much current goes through each. The thick filament offers less resistance than the thin filament so more current goes through the thick one. After all, the thick one can be thought of as several thin ones side by side. Therefore, the bulb with the thick filament uses the most energy per second (energy per second = power) and is the brightest.

IN BED

You have a nice new thick blanket which is a good heat insulator, and a thin old blanket which is a poor heat insulator. It is a cold night and you need both blankets. You will be warmest if you

a) put the good blanket on top to keep the cold out of the bed, and put the poor one next to you.
b) put the good blanket next to you to keep the heat in, and put the poor one on top.
c) do it either way; it doesn't matter which blanket goes where.

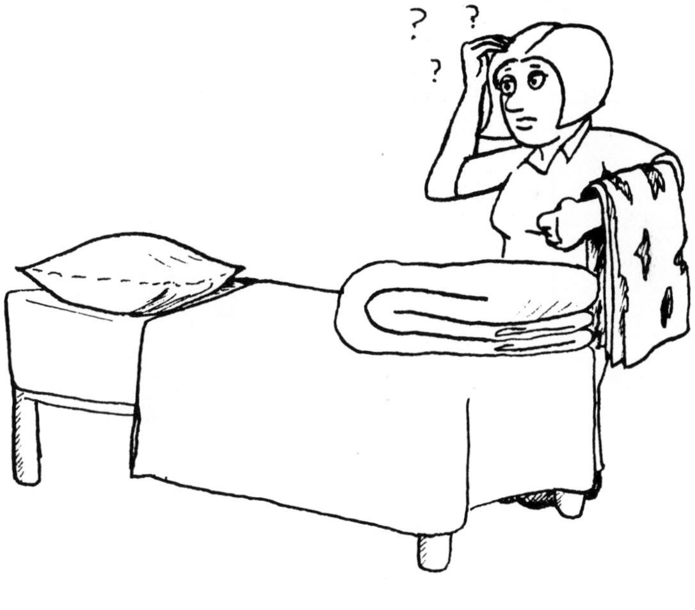

ANSWER: IN BED

The answer is: c. The blankets are in series. That is, the heat must go through both before it escapes (or the cold must come through both before it enters the bed). Heat flows from hot to cold places as electricity flows from high to low voltage. The blankets are heat insulators in series. Heat insulators in series act just like electric resistors in series. Suppose you have a big resistor and a little resistor in series (light bulbs can be used as resistors). Wil! more or less electricity flow through them if you interchange their order? Of course there is no difference. So understanding something about electricity lets you understand heat flow. Actually, heat flow was understood first and it helped people understand the flow of electricity.

HIGH VOLTAGE BIRD

Will this bird get a shock sitting on a bare high-voltage line?

a) Yes!
b) No

ANSWER: HIGH VOLTAGE BIRD

The answer is: b. You might think that a voltage high enough to compensate for the high resistance of the bird, say 20,000 volts, would produce a damaging current in the bird. But 20,000 volts refers to the voltage of the whole length of wire with respect to the ground. Although the bird on the wire would also be at 20,000 volts, there is no part of its body that isn't; there is no voltage *difference* across its body. Current is made to flow in a conducting medium when there is a voltage difference across the conductor—no voltage difference, no current. Now if the bird extended its wings so as to touch a neighboring wire at a different voltage—ZAP! Power lines are strung sufficiently far apart so as to be beyond the wingspans of birds so they aren't short-circuited.

ELECTRIC SHOCK

What causes electrical shock—current or voltage?

a) current
b) voltage
c) both
d) neither

ANSWER: ELECTRIC SHOCK

The answer is: c. Electric shock occurs when current flows in the body; if there is no current there is no shock. So it may seem that answer **a** should be correct. But what causes the current that causes the shock? A voltage is responsible for the current. So an impressed voltage and the resulting current are the causes of electrical shock. One could rightfully quibble that answer **a** is correct because shock is directly associated with current alone, irrespective of what the cause of the current is. Or one could as well argue that answer **b** is correct on the grounds that an impressed voltage is the cause of the shock whatever the role of the intermediate current. For example, we don't see warning signs: "DANGER—HIGH AMPERAGE," but rather, "DANGER—HIGH VOLTAGE." So give yourself a point whether you chose answer **a**, **b**, or **c**. But if you picked answer **d**, you struck out!

HIGH VOLTAGE BIRD AGAIN

Suppose a bird stands with its legs bridging the light bulb in the circuit shown. In this case, the bird will likely get

a) a shock if the switch is open
b) a shock if the switch is closed
c) a shock if the switch is either open or closed
d) no shock at all in any of these cases

ANSWER: HIGH VOLTAGE BIRD AGAIN

The answer is: b. When the switch is open all the wire on one side of the switch is at say 12 volts, and all the wire on the other side is at zero volts. The bird is all on one side so there is no voltage difference across the bird. Now close the switch. Current flows and is forced through the light-bulb resistance. Some of the current takes a detour and flows through the bird. The bird receives a shock.

In a circuit the voltage difference is always across the part of the circuit which presents an impasse to the current. When the switch is open the impasse is the switch. When it is closed, the remaining impasse is the resistor. The voltage difference is from one side of the resistor to the other, and the unfortunate bird has her legs spread across that voltage difference.

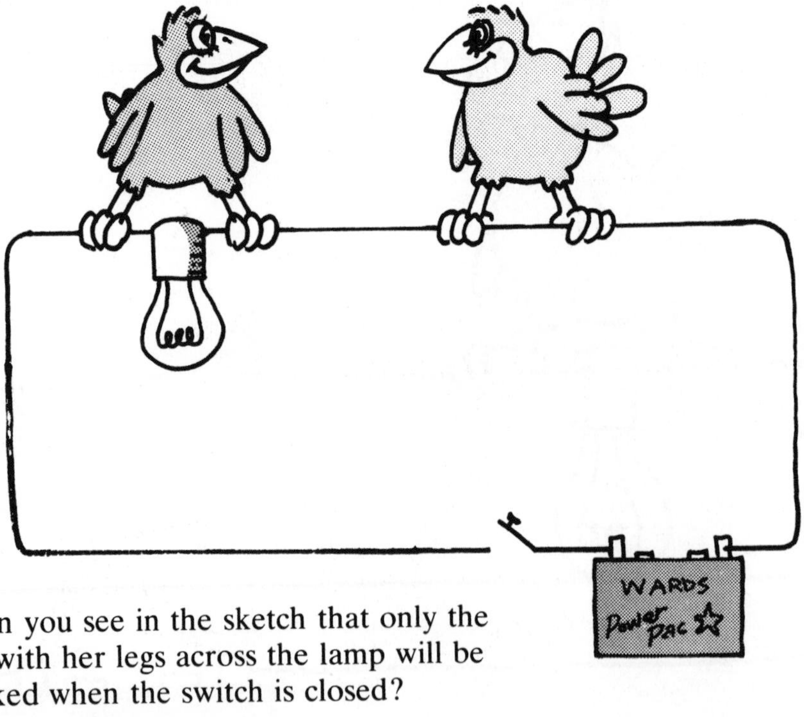

Can you see in the sketch that only the bird with her legs across the lamp will be shocked when the switch is closed?

ELECTRON SPEED

When you turn the ignition key in an automobile, you complete a circuit from the negative battery terminal through the electric starter and back to the positive battery terminal. This is a DC circuit and electrons migrate through the circuit in a direction from the negative battery terminal to the positive terminal. About how long must the key be in the ON position for electrons starting from the negative terminal to reach the positive terminal?

a) A time shorter than that of the human reflex turning a switch on or off
b) ¼ second
c) 4 seconds
d) 4 minutes
e) 4 hours

ANSWER: ELECTRON SPEED

The answer is: e. Although the electrical signal travels through the closed circuit at about the speed of light, the actual speed of electron migration (drift velocity) is much less. Although electrons in the open circuit (key in OFF position) at normal temperatures have an average velocity of some millions of kilometers per hour, they produce no current because they are moving in all possible directions. There is no net flow in any preferred direction. But when the key is turned to the ON position the circuit is completed and the electric field between the battery terminals is directed through the connecting circuit. It is this electric field that is established in the circuit at about the speed of light. The electrons all along the circuit continue their random motion, but are also accelerated by the impressed electric field in a direction toward the end of the circuit connected to the positive battery terminal. The accelerated electrons cannot gain appreciable speeds because of collisions with anchored atoms in their paths. These collisions continually interrupt the motion of the electrons so that their net average speed is extremely slow—less than a tiny fraction of a centimeter per second. So some hours are required for the electrons to migrate from one battery terminal through the circuit to the other terminal.

COULOMB EATER

When an electric motor is doing work, or when an electric toaster is toasting, more coulombs of electricity must run into it than come out.

a) True
b) False

An electric generator

a) generates coulombs of electricity
b) takes in as many coulombs of electricity as it puts out

ANSWER: COULOMB EATER

The answer to both questions is: b.

An electric motor or toaster does not consume electricity. It consumes energy. An electric generator does not generate electricity. It generates electric energy. The same number of coulombs of electricity that enter a motor or toaster must leave that motor or toaster—but they leave "pooped." What does "pooped" mean? In this case it means at a lower voltage. Think of steam going through a steam engine. All the steam comes out of the engine, but it comes out at a lower pressure than it had when it went in. Similarly, the coulombs lose voltage in the motor or toaster and gain voltage in the generator. The energy in a coulomb of electricity depends on its voltage:

$$\text{Energy} = \text{voltage} \times \text{coulombs}$$

so zero voltage means zero energy. Coulombs—any number of coulombs—with zero voltage still have zero energy.

ELECTRONS FOR SALE

Estimate the number of electrons annually that pass through to the homes and business establishments of a typical American city of 50,000 inhabitants.

a) None at all
b) About the number of electrons that exist in a pea
c) About the number of electrons that exist in the Great Lakes
d) About the number of electrons that exist in the earth
e) About the number of electrons that exist in the sun

ANSWER: ELECTRONS FOR SALE

The answer is: a. None at all. It is a common misconception that electrons flow from generating plants through power lines and into the outlets of consumers. Electric power in a typical American city is AC, which means that electrons do not migrate through power lines, but simply vibrate to and fro 60 times each second in the lines. Power lines are not conduits for electrons, but for ENERGY. When you plug an appliance into an AC outlet, energy flows from the outlet into the appliance and sloshes free electrons back and forth that are already in the conducting elements of the appliance. The power company supplies the energy if you supply the electrons.

If you receive an electric shock from an AC circuit, remember that the electrons making up the current in your body were there all the time. It is electrical energy that comes out of the wire to pass through you and into the ground. Not electrons.

ATTRACTION

A comb can sometimes be given an electric charge by running it through your hair. The charged comb will then attract small pieces of paper. Will the charged comb also be attracted to magnets?

a) Yes, it will be attracted to magnets.
b) No, it will not be attracted to magnets

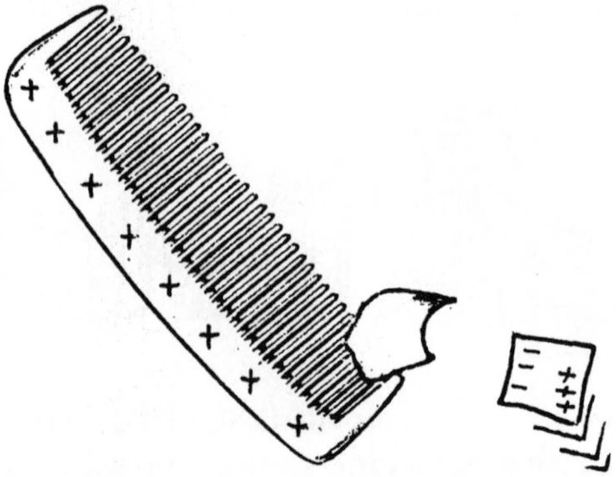

ANSWER: ATTRACTION

The answer is: b. Magnetic attraction and electric attraction are different things. There is a connection between them, but they are nevertheless different—as we shall see.

CURRENT AND COMPASS

If a current-carrying wire runs directly over a magnetic compass, the needle of the compass will

a) not be affected by the current
b) point in a direction perpendicular to the wire
c) point in a direction parallel to the wire
d) tend to point directly to the wire

ANSWER: CURRENT AND COMPASS

The answer is: b. The magnetic field lines circle the current in the wire as shown. The needle of the compass then orients itself parallel to and along the magnetic field lines. Therefore the needle is perpendicular to the current.

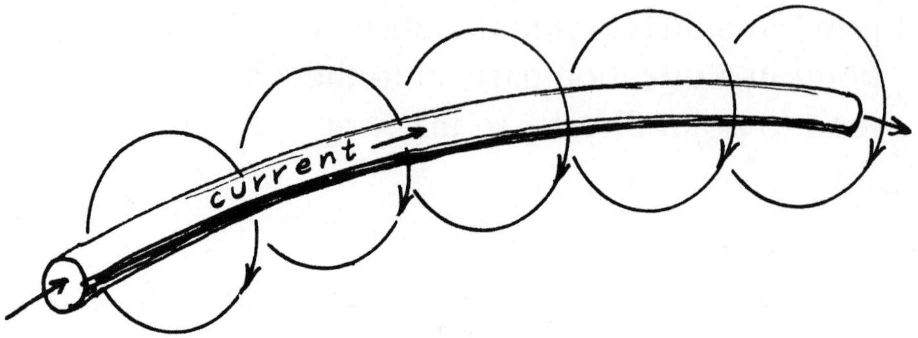

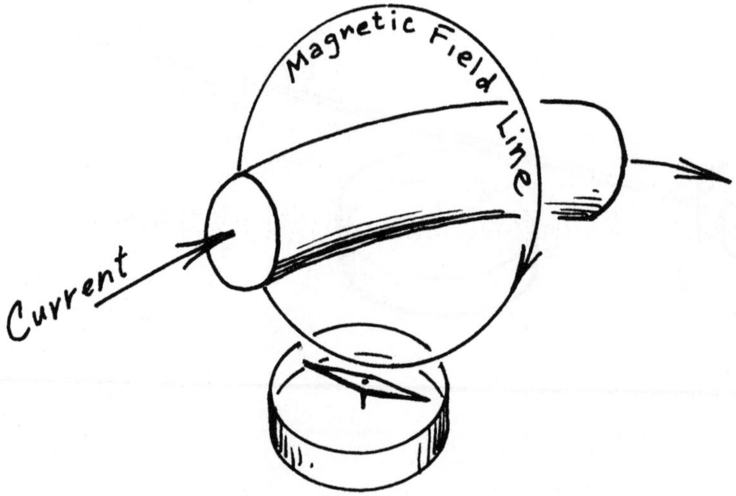

ELECTRON TRAP

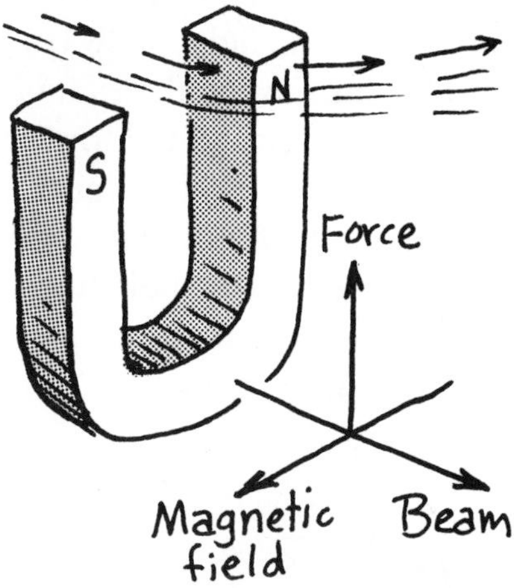

A charged particle experiences a force when it moves through a magnetic field. The force is greatest when it moves perpendicular to the magnetic field lines. At other angles the force lessens and becomes zero when the charged particle moves along the field lines. In all cases the direction of the force is always perpendicular to the magnetic field lines and the velocity of the charged particle.* In the sketch we see electrons curving as they move through the field of the small magnet. The curved part of their path is small because their time in the field is brief—they quickly pass into and out of the field. Now if their motion is all the time confined in a uniform magnetic field and they move perpendicular to the field lines, their paths will be

a) parabolas 
b) spirals 
c) complete circles 
d) straight lines 

---

* You may have noticed that we have just stepped into three-dimensional space. True, we were always in 3-D space, but as far as the physics we have discussed up to this point, space might just as well have had only two dimensions. Up to here you were never forced to visualize anything in 3-D. The game of pool, for example, illustrates all the laws of mechanics, but you don't need three-dimensional visualization to understand pool. But electromagnetism is different. Unlike pool, the ideas of electromagnetism will not fit into two-dimensional space—three dimensions of space are essential. Perhaps future findings will not even fit into 3-D space!

ANSWER: ELECTRON TRAP

The answer is: c. The force on an electron (or any charged particle) in a magnetic field is always perpendicular to the particle's direction of motion, just as the radius of a circle is always perpendicular to its circumference. So the force on the charged particle is radial and the particle is swept into a circle. Thus, the particle becomes trapped in the magnetic field. If it is moving at an angle greater or less than 90° with respect to the field lines, the "circular" path extends into a helix. That's because its component of motion along the field lines persists without interacting with the field.

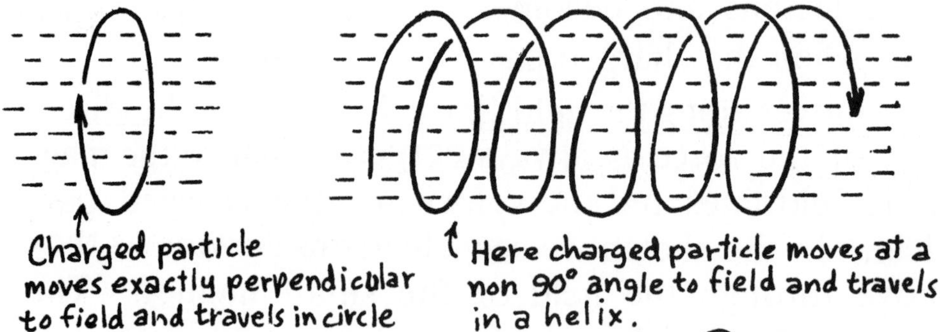
Charged particle moves exactly perpendicular to field and travels in circle
Here charged particle moves at a non 90° angle to field and travels in a helix.

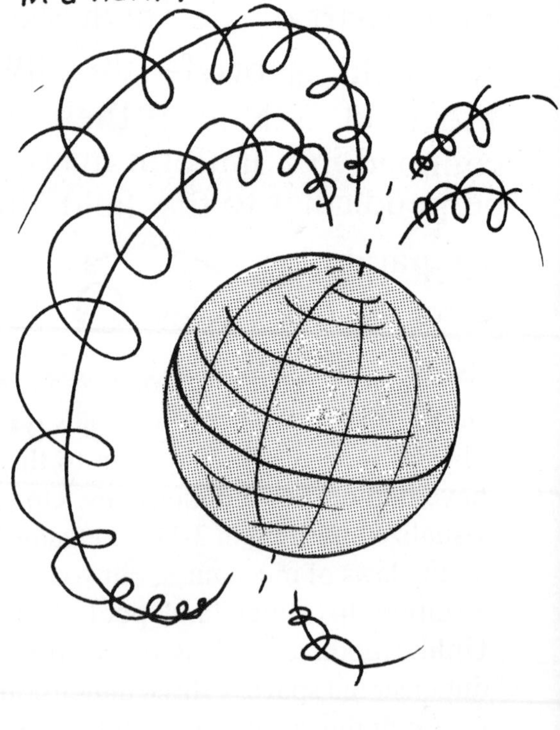

Electrons (and protons) in space become trapped in the earth's magnetic field and follow helical paths along the field lines. The cloud of magnetically trapped particles that surround the earth is what we call the Van Allen Radiation Belts. The crowded field lines at the poles act like magnetic mirrors and the charged particles bounce from pole to pole. Sometimes they dip into the atmosphere and we see the aurora borealis, or "northern lights."

ARTIFICIAL AURORA

Hydrogen bombs have been detonated above the earth's atmosphere and have induced an ARTIFICIAL aurora borealis, "Northern Lights," as the particles emitted by the explosion enter the earth's atmosphere. If such a device were exploded high above the earth's North magnetic pole, at what place or places on the earth could the artificial aurora be seen?

a) the North magnetic pole
b) the equator
c) the South magnetic pole (south of Australia)
d) at both the North and South magnetic poles
e) at both poles and also at the equator

ANSWER:
ARTIFICIAL AURORA

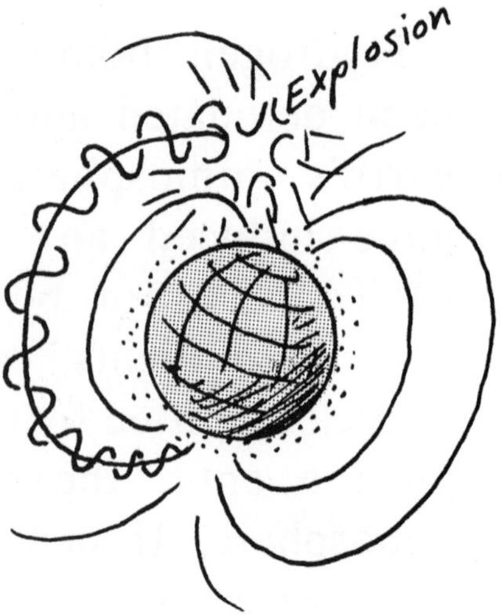

The answer is: d. The electrons and protons released by the explosion would spiral along the lines of the earth's magnetic field. Above the North magnetic pole the lines run almost straight up and down, with the down end going directly to the North magnetic pole. The up end curves up and around the planet and goes into the South magnetic pole. So some electrons will spiral into the atmosphere near the North pole and some will spiral around to the South pole.

No bomb has been detonated above the pole, but one was detonated above Johnston Island in the Pacific Ocean in the early 1960's (Project Starfish). The magnetic field lines above Johnston Island led near to Hawaii so the Hawaiians were able to witness an auroral display as the particles from the bomb re-entered the earth's atmosphere. Most Hawaiians had never seen an aurora before.

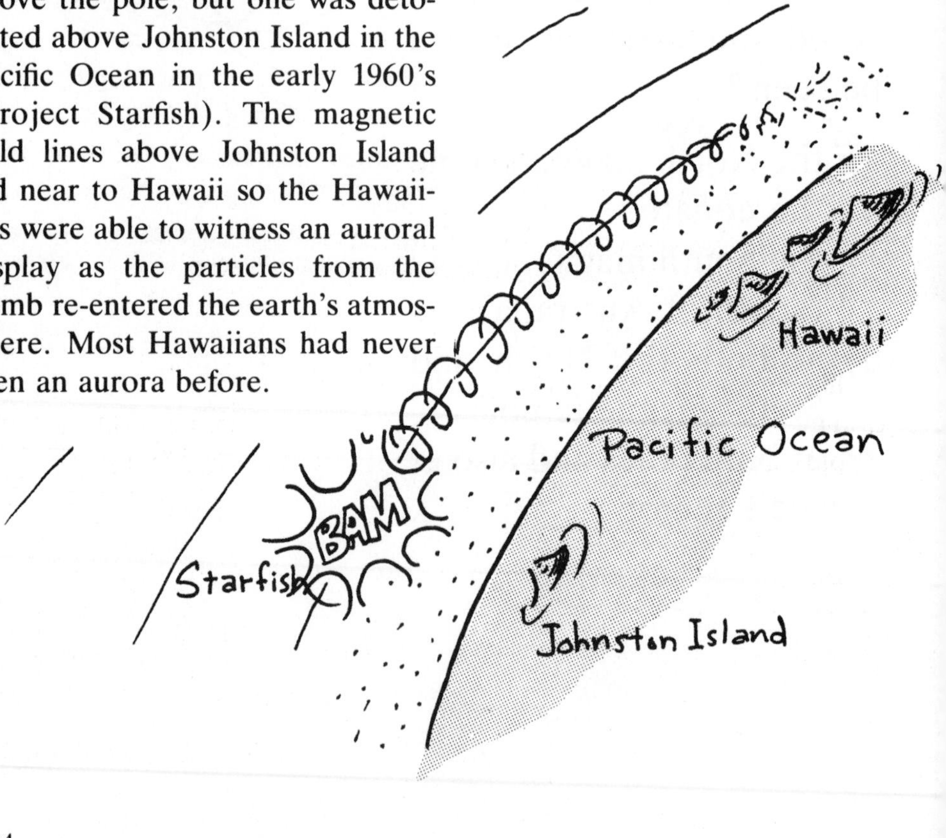

IRON FREE

Is it possible to make a magnetic field without the use of iron?

a) Yes
b) No

ANSWER: IRON FREE

The answer is: a. A magnetic field is associated with moving electric charges. A current-carrying wire is surrounded by a magnetic field. This is evident when a wire carrying electric current is placed over a magnetic compass. The magnetic field lines are found to circle around the current in the wire, as evidenced by the compass needle which always orients itself perpendicular to the wire and parallel to the field lines. If the wire is wrapped into a coil, the magnetic field is bunched up and concentrated inside the coil. If some iron is put in the coil such as to form an electromagnet, the field gets much stronger, but the iron is not essential—it only helps a lot. The magnetic field can exist in empty space if there is an electric current to produce it. The current IS essential.

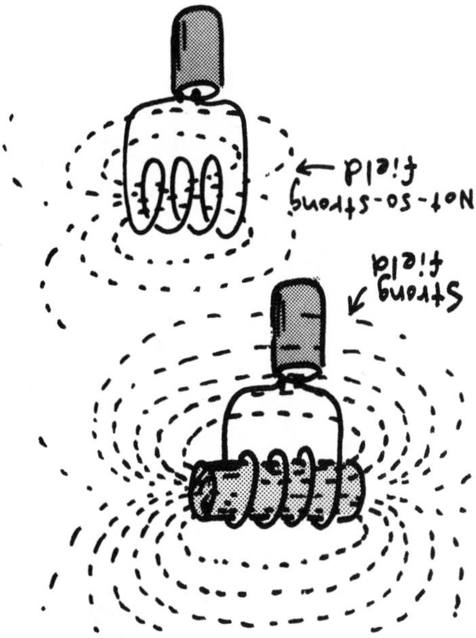

How does the iron help? Iron atoms are themselves tiny electromagnets, for there are electric currents flowing around the iron atoms. Some are flowing clockwise and some anticlockwise so the total effect normally is no effect. If the current in the surrounding coil is clockwise it aligns many of the atomic currents clockwise also. So the current in the wire is now aided by the atomic currents in the iron. Together they make a stronger field than that of the coil.

ON EARTH AS IN HEAVEN

The deflecting force exerted on cosmic rays when they come into the earth's magnetic field is fundamentally the same as the force that turns an electric motor when current flows through its coils.

a) Yes
b) Not so!

ANSWER: ON EARTH AS IN HEAVEN

The answer is: a. The free electron traveling through space experiences a force jointly perpendicular to its flight path and the magnetic field and so it curves around (and around and around and can spend a long time in the field).

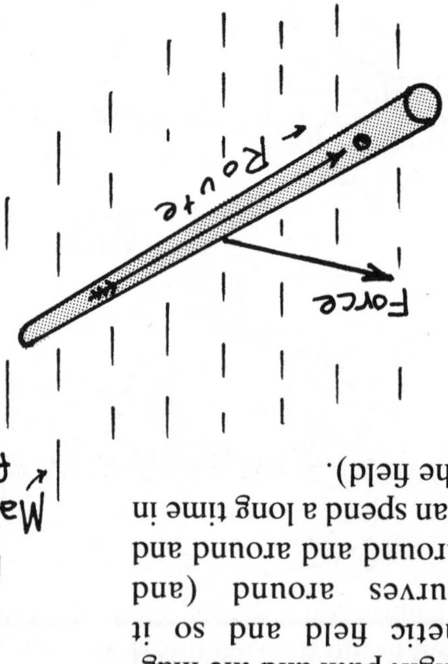

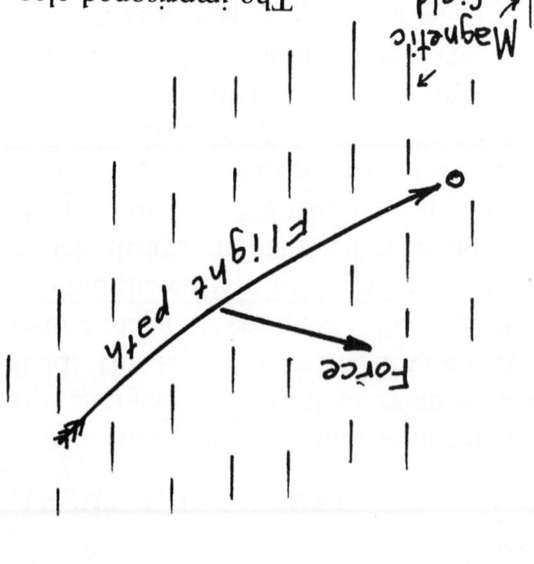

The imprisoned electron traveling through the wire (in the motor coil) also experiences a force jointly perpendicular to its ordered route through the wire and the magnetic field. But the electron cannot get out of the wire. So it pulls the whole wire with it. It is that pull on the wire that makes an electric motor operate.

PINCH

When electric current in two parallel wires is flowing in the same direction, the wires tend to

a) repel each other
b) attract each other
c) exert no force on each other
d) twist at right angles to each other
e) spin

ANSWER: PINCH

The answer is: b. The wires pinch together. What if the currents were flowing in opposite directions? The wires would spread apart. The basic rule of magnetism is usually taken to be that North and South poles attract while South and South or North and North repel. However, the source of magnetism is electric currents, so would it not be simpler to state the rules of magnetic force in terms of the electric currents which make the magnets? Yes. So the "new" rule of magnetism is: currents flowing in the same direction pinch together and currents flowing in opposite directions spread apart. This rule leads immediately to the first rule because when electrons for example flow around an iron cylinder, as shown in the top sketch, one end becomes North and the other end South.

Now when two cylinders are placed next to each other so that electron flows are in the same direction, as shown in the middle sketch, the North pole end of one cylinder faces the South pole end of the other and they pinch together. We can say North and South magnetic poles attract or we can say currents flowing in the same direction pinch.

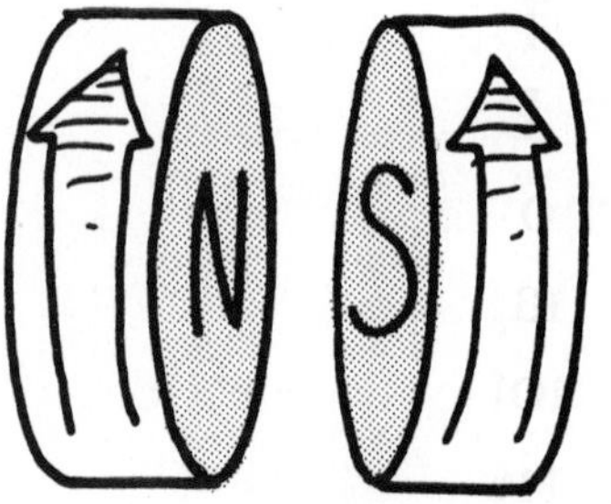

When the cylinders are placed next to each other so that currents are flowing in opposite directions, as shown in the last sketch, the N pole end of one faces the N pole of the other and they repel. We can explain the repulsion by saying like poles repel, or by saying opposite currents spread apart.

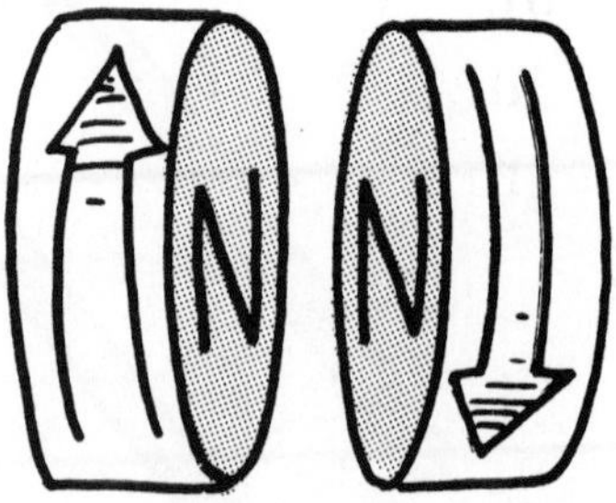

THE MAGNETIC RATS NEST

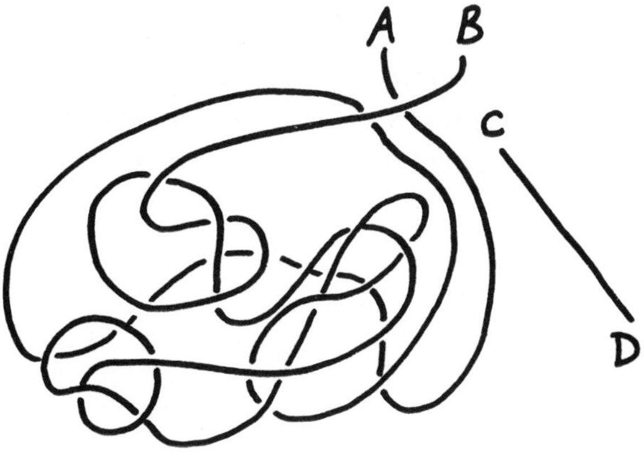

The sketch above shows a long wire running from **A** to **B** which has been tangled into a "rats nest." It also shows a short straight segment of wire **CD**. Now imagine an electric current flowing from **A** to **B** and another one flowing from **C** to **D** (you must have batteries, or something, in your imagination). Because of the current, a certain force is exerted on wire **CD**. Now in your imagination reverse all the electric currents so that the reversed current now flows from **B** to **A** and from **D** to **C**. The force on the short, straight segment of wire will

a) also be reversed
b) be as it was before the current was reversed
c) vanish
d) be in a direction perpendicular to the way it was before the current was reversed
e) be in a direction not mentioned above

ANSWER: THE MAGNETIC RATS NEST

The answer is: b. No doubt the wire **AB** makes a very complicated magnetic field, but whatever that field is, it is exactly reversed by reversing the current.* So that should reverse the force on **CD**? Yes, it should if the current still ran from **C** to **D**. But reversing the current in **CD** reverses the force again. So two reverses puts us back where we started. The force on **CD** will be as it was before the current was reversed.

*We can see that running the current from **B** to **A** must produce a magnetic field which is the EXACT opposite of that produced when the current runs from **A** to **B** by the following reasoning: If we run the current from **A** to **B** and from **B** to **A** at the same time, that is like running no current at all. And zero net current means there must be no magnetic field at all. So, the magnetic field produced by both runs must exactly cancel each other—and to do that they must be exactly equal and opposite to each other.

AC-ING A DC MOTOR

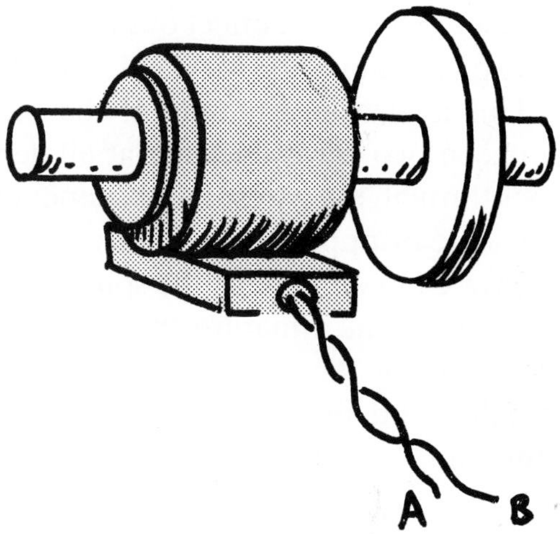

A direct current motor turns clockwise when wire A is connected to the plus (+) and B to the minus (−) side of a battery (the motor contains no permanent magnets). Now if we interchange A and B so that B is plus and A is minus, then

a) the motor will turn anticlockwise
b) the motor will continue to turn clockwise

ANSWER: AC-ING A DC MOTOR

The answer is: b. Inside the motor is a rotating electromagnet (armature) and a stationary electromagnet (stator). Reversing the current through the motor reverses the current through each electromagnet and the net effect of the TWO reverses is no change in force, as explained in THE MAGNETIC RATS NEST.

This is also an example of the principle of relativity. The force between two wires does not at all depend on which way the current goes through the wires. It only depends on whether the currents are in the same or opposite directions.

How can you reverse the spin of the motor? By reversing the current only in the armature or only in the stator.This is most easily done by reversing the position of the "brushes." The brushes serve as switches and in a sense brush the electricity on to and off of the armature.

FARADAY'S PARADOX

This is a coil of wire with a hunk of iron locked in it.

a) If current is made to flow in the wire, the iron becomes a magnet
b) If the iron is a magnet, current is made to flow in the wire
c) Both of the first two statements are true
d) Both of the first two statements are false

ANSWER: FARADAY'S PARADOX

The answer is: a. If current flows in a wire wrapped around some iron (say a nail), it becomes an electromagnet. Making such a magnet is an old standard Cub Scout project. But if a magnet is sitting inside a coil it does not cause a current in the coil or even charge the wires. In the days of Queen Victoria, Michael Faraday* and many of his contemporaries puzzled about this. They thought if current makes magnetism, then by all rights magnetism should make current, but how? While wondering about this, Michael Faraday made his big discovery. A magnet would make a current in the coil, but only if it was moved inside the coil and not locked in one place. After all, it takes energy to make a current and the energy comes from the force that moves the magnet or the coil.

Faraday's discovery was the key to electric generators. A generator just moves a magnet back and forth near a coil (or moves a coil near a magnet) and so makes an electric current flow in the wire. The Prime Minister of England came to Faraday's laboratory to actually see electricity generated in this way. After the demonstration he asked Faraday, "What good is electricity?" Faraday answered that he did not know what good it was, but that he did know some day the Prime Minister would put a tax on it!

---

*At about the same time that Faraday discovered electromagnetic induction, as it is called, an American physicist, Joseph Henry, independently made the same discovery.

METER TO MOTOR

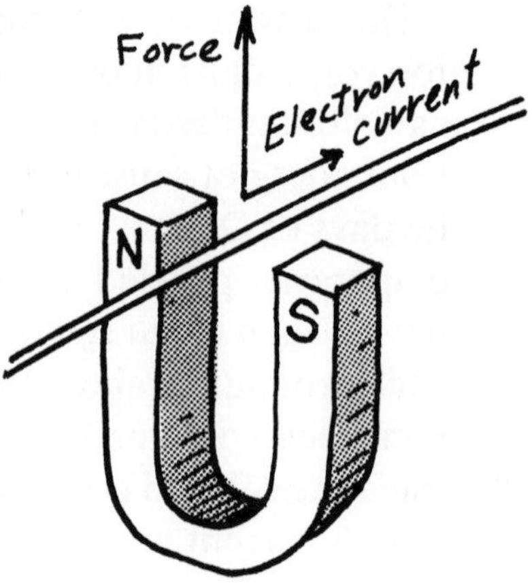

When electrons in a wire flow through a magnetic field in the direction shown, the wire is forced upward. If the current is reversed, the wire is forced downward. If a wire loop is instead placed in the magnetic field and the electrons flow in the direction shown below, the loop will tend to

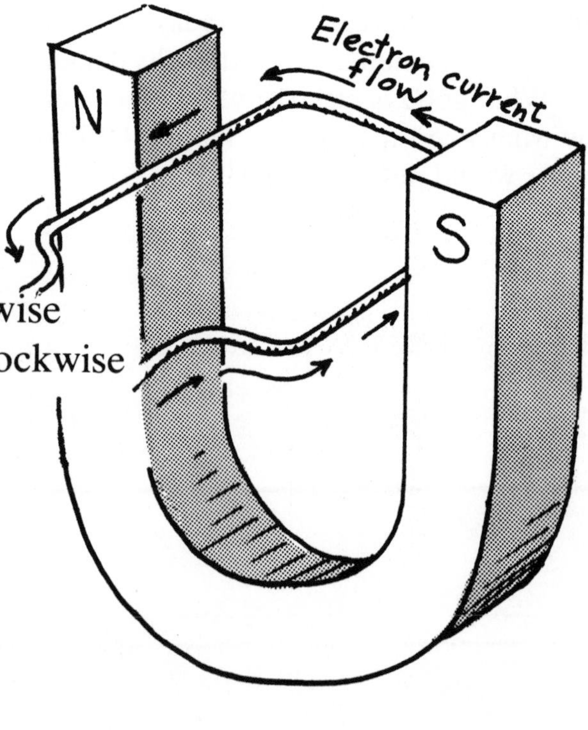

a) rotate clockwise
b) rotate anticlockwise
c) do nothing

ANSWER: METER TO MOTOR

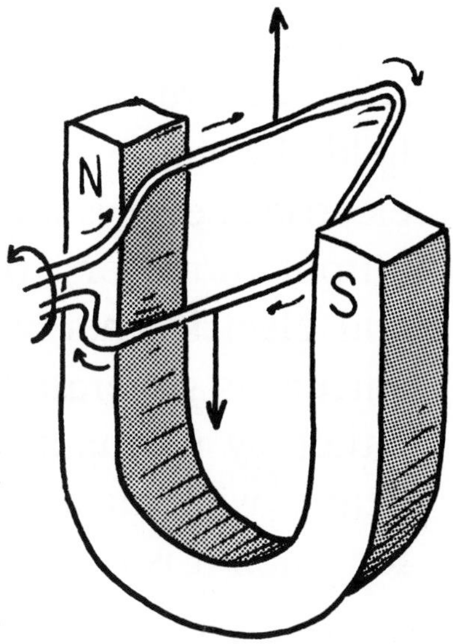

The answer is: b, for the right side is forced up while the left side is forced down as shown. Although this is an easy question to answer, its point is important for this is how electric meters work. Instead of one loop, many loops forming a coil are used, and held by means of a spring. When current is made to flow through the coil the resulting forces twist the coil against the spring—the greater the current, the more the twist, which is indicated by a pointer that gives the reading. It is only one step further to an electric motor, wherein the current is made to change direction with each half turn of the coil so that it turns repeatedly.

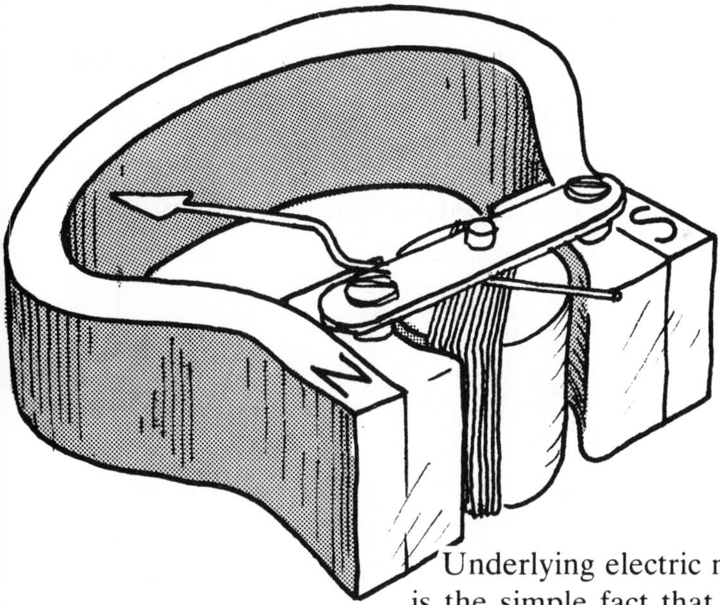

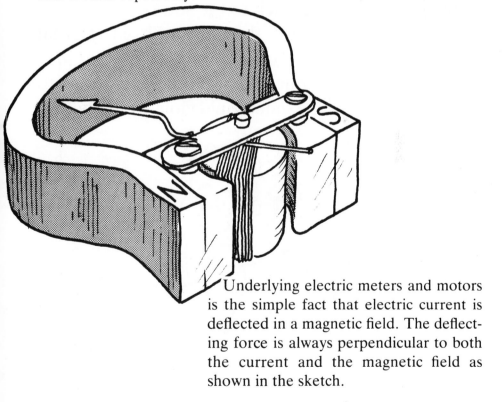

Underlying electric meters and motors is the simple fact that electric current is deflected in a magnetic field. The deflecting force is always perpendicular to both the current and the magnetic field as shown in the sketch.

MOTOR-GENERATOR

Both an electric motor and a generator consist of coils of wire on a rotor that can spin in a magnetic field. The basic difference between the two is whether electric energy is the input and mechanical energy the output (a motor), or mechanical energy is the input and electric energy the output (a generator). Now current is generated when the rotor is made to spin either by mechanical or electric energy—it needn't "care" what makes it spin. So is a motor also a generator when it is running?

a) Yes, it will send an electric energy output through the input lines and back to the source
b) It would if it weren't designed with an internal bypass circuit to prevent this problem
c) No, the device is either a motor or a generator—to be both at the same time would violate energy conservation

ANSWER: MOTOR-GENERATOR

The answer is: a. Every electric motor is also a generator, and in fact, the power company that supplies the input energy in effect gives you a refund for the energy you send back to them. That's because you pay for the *net* current and hence the net energy consumed. If your motor is spinning freely with no external load it will generate almost as much current as it is powered with, so the net current in the motor is very little. Your electric bill is low as a result. The back current, not friction, limits the speed of a free-running motor. When the back current cancels the forward current, the motor can spin no faster. But when your motor is connected to a load and work is done, more current and more energy is drawn from the input lines than is generated back into them. If the load is too great the motor may overheat. If you go to the extreme and put too great a load on a motor such as to prevent it from spinning—like jamming a circular saw in stubborn lumber for example—no back current is generated and the undiminished input current in the motor may be enough to melt the insulation in the motor windings and burn the motor out!

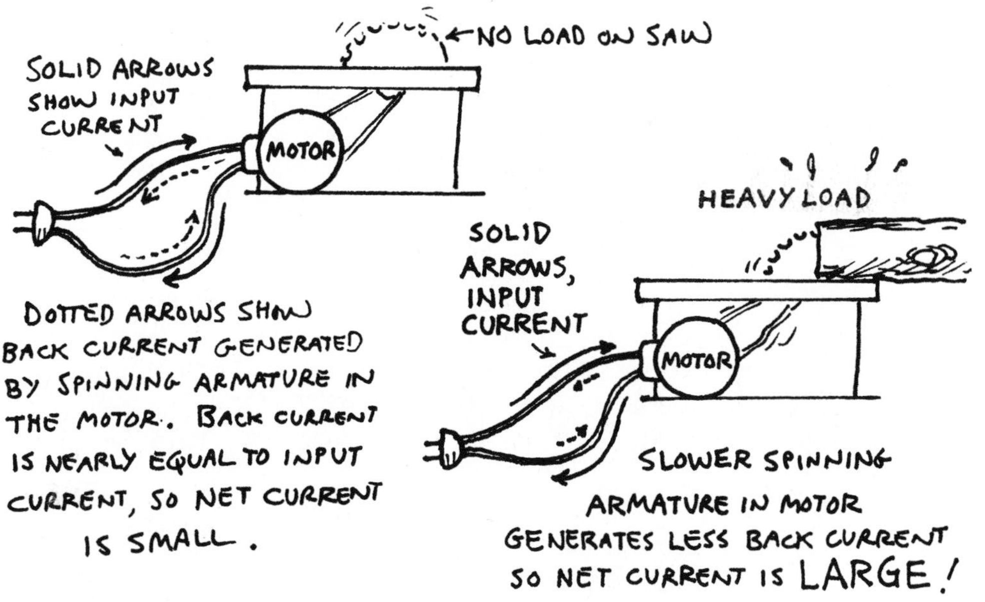
NO LOAD ON SAW
SOLID ARROWS SHOW INPUT CURRENT
MOTOR
DOTTED ARROWS SHOW BACK CURRENT GENERATED BY SPINNING ARMATURE IN THE MOTOR. BACK CURRENT IS NEARLY EQUAL TO INPUT CURRENT, SO NET CURRENT IS SMALL.
HEAVY LOAD
SOLID ARROWS, INPUT CURRENT
MOTOR
SLOWER SPINNING ARMATURE IN MOTOR GENERATES LESS BACK CURRENT SO NET CURRENT IS LARGE!

DYNAMIC BRAKING

At the turn of the century when railroads were first built through the Alps it was proposed that the electric locomotives on long downslopes brake in the following way: the electric motor is disconnected from the power lines and connected instead to a large resistor so that the motor is used as a generator to transform the mechanical energy of the turning wheels to electricity and then into heat. Will this actually make the motor become a brake?

a) Yes
b) No

ANSWER: DYNAMIC BRAKING

The answer is: a. We saw in MOTOR-GENERATOR that a motor is also a generator. The energy of the train rolling downhill turns the generator which converts the energy to electric current and the resistor turns the electric energy to heat. Every unit of heat energy produced is a unit of kinetic energy disposed of, and this brakes the train. In an energy-efficient world the electric current would better be used to power other trains coming uphill.

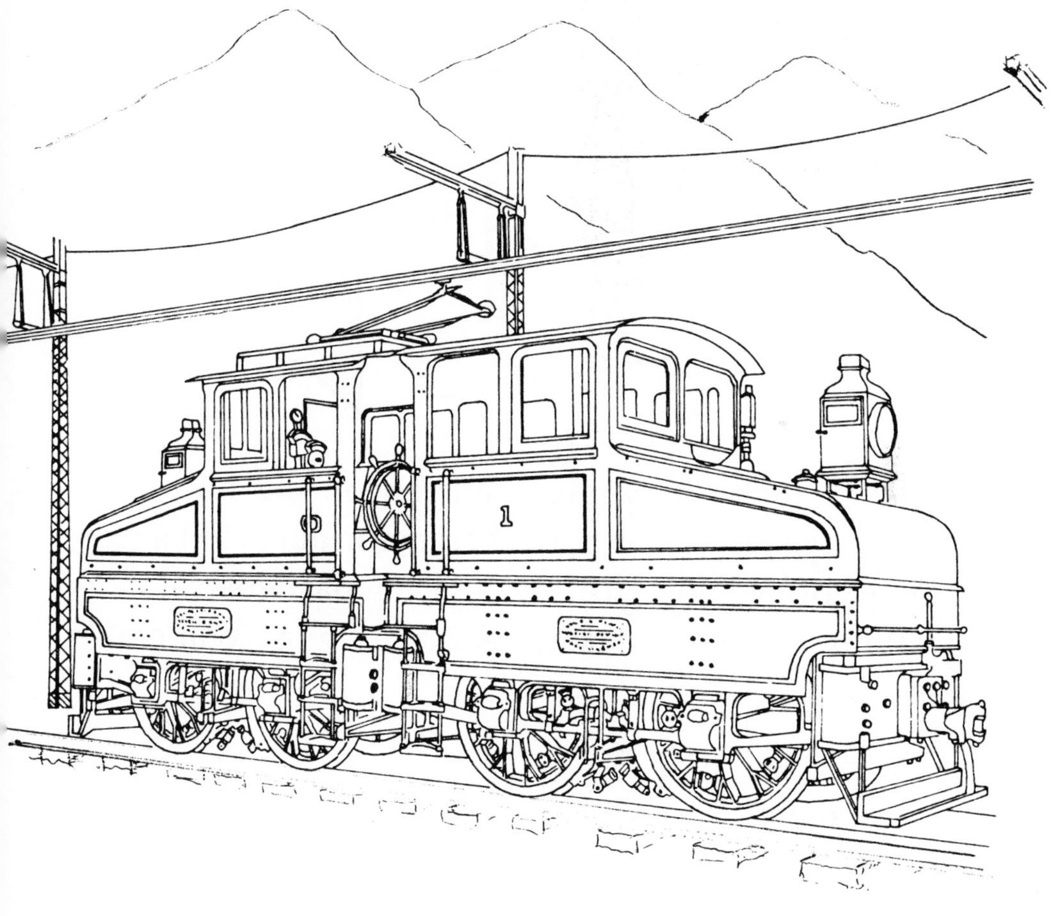

ELECTRIC LEVER

Electromagnetic induction underlies the operation of a transformer, which is simply a hunk of iron with a pair of wire coils around it. Alternating current in the input or primary coil makes the iron an alternating

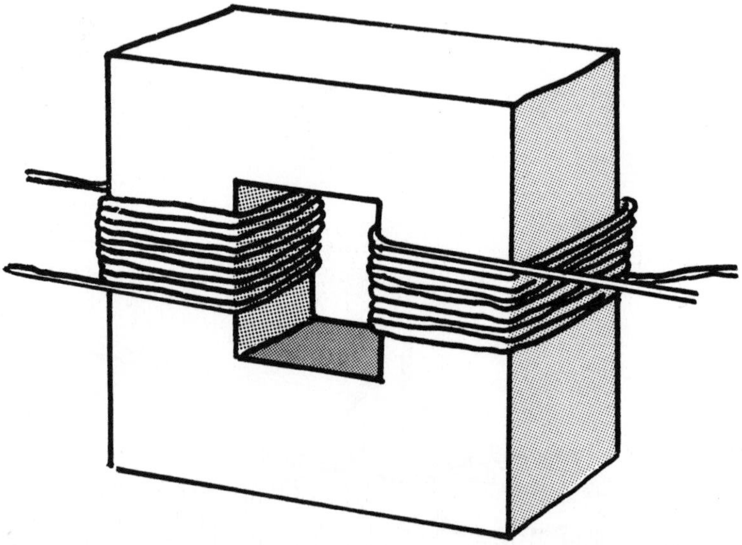

magnet which "generates" current in the output or secondary coil. Into a perfect transformer is fed a particular voltage and current, hence a particular power (wattage). Out of the transformer must come the same

a) amperage
b) voltage
c) wattage
d) amperage, voltage and wattage
e) none of the above

ANSWER: ELECTRIC LEVER

The answer is: c. A transformer is not a source of power, but is a passive device. No more power can come out than goes in. If it is a perfect transformer, all the electric power that goes in comes out—if it is not perfect, some of the power is converted to heat. The unit by which power is measured is the watt. So out of the perfect transformer comes the same wattage that goes in.

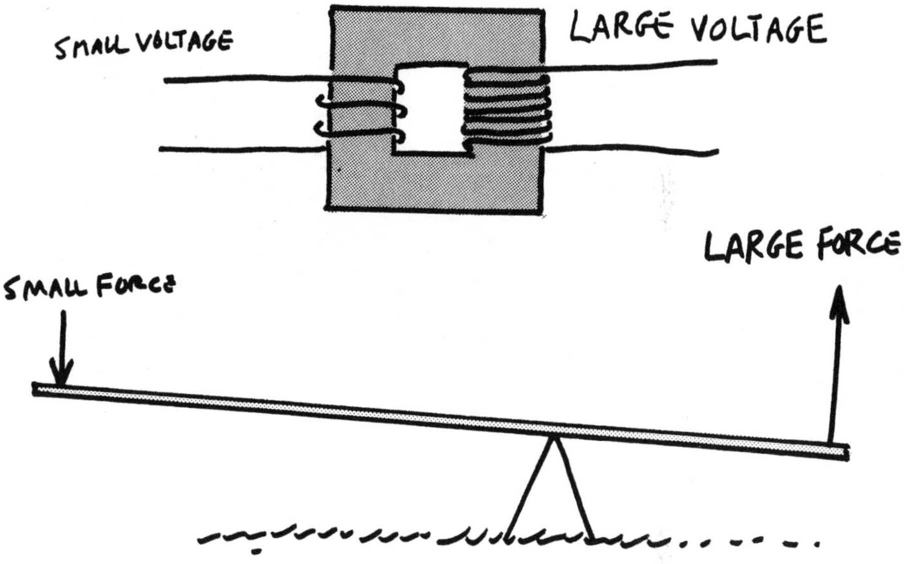

A transformer is very much like a lever. The power that comes out of one end of a lever is equal to the power that goes into the other, because a lever is not a source of power. The lever can convert a small, rapidly moving force to a large, slowly moving force, or vice versa. Similarly, a transformer can convert a small voltage motivating a large electric current into a large voltage motivating a small electric current, or vice versa.

MISUSED TRANSFORMER

A big battery is connected to a lamp by means of a transformer—but no electricity can leave the battery until the switch is pressed down and closed. Only one of the following statements is true. Which is it?

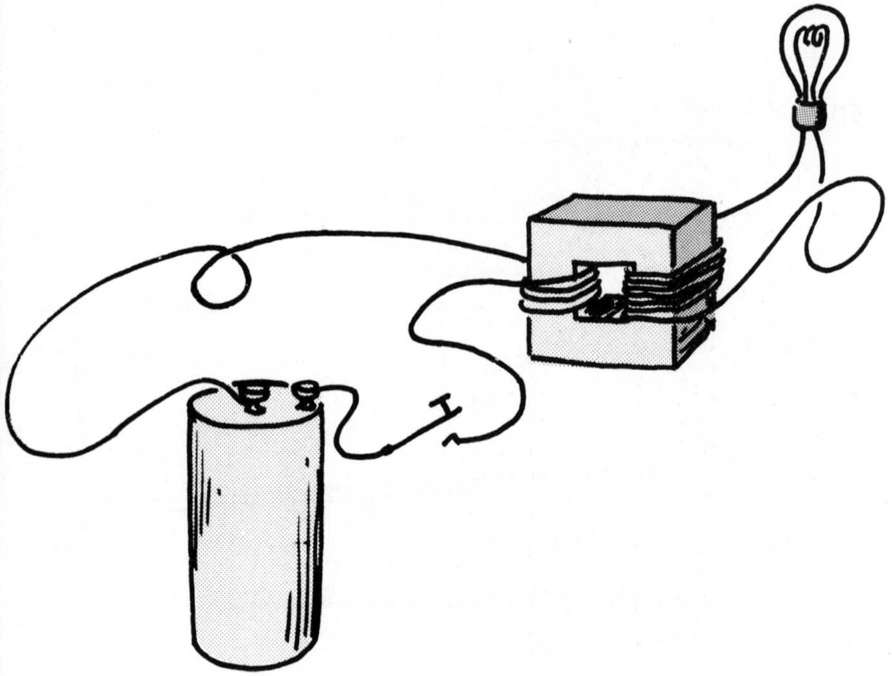

a) As long as the switch is down the lamp is lit
b) The lamp never lights with this arrangement
c) The lamp might light momentarily only when the switch is closed
d) The lamp might light momentarily only when the switch is opened
e) The lamp might light momentarily when the switch is closed and momentarily again when it is opened

ANSWER: MISUSED TRANSFORMER

The answer is: e. The transformer is apparently misused because it is intended for use with alternating current while the battery only puts out direct current. So how can anything happen? Steady current in a transformer produces a steady magnetic field inside the transformer iron. This field threads through the coil connected to the lamp. But if the field doesn't fluctuate or change in some way, no current will be induced in the secondary coil. When the key is first closed, current surges into the primary and the magnetic field in the iron builds up. The growing field induces a current flow in the secondary and through the lamp. When the magnetic field is full grown it no longer is changing. So there is no longer an induction of current through the lamp. Later when the key is opened (disconnecting the battery), the current dies in the primary and likewise the magnetic field must die. But a dying field is a changing field. So again a current flows through the lamp, but this time in the opposite direction. But the lamp does not care which way the current goes; if the current is strong enough, the lamp lights.

BUG

Some people think you can bug a telephone by separating the telephone wires and laying a wire that is connected to a headphone alongside ONE of the telephone wires, as shown in the sketch below. Of course, all of the wires are insulated from each other. Will such a "tap" as this work?

a) Yes b) No

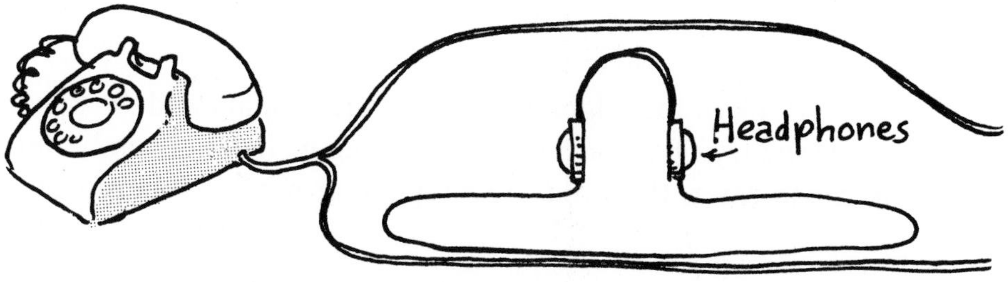

ANSWER: BUG

The answer is: a. This is a very old way to bug a phone line. When current flows in the phone line or in any wire, we know that a magnetic field loops around the current. If we place the bug line nearby, this magnetic field also loops around the bug line. As the current in the phone line alters with the frequency of the spoken voice, the magnetic field alters and extends to the closed loop of the bug line and induces current in it.

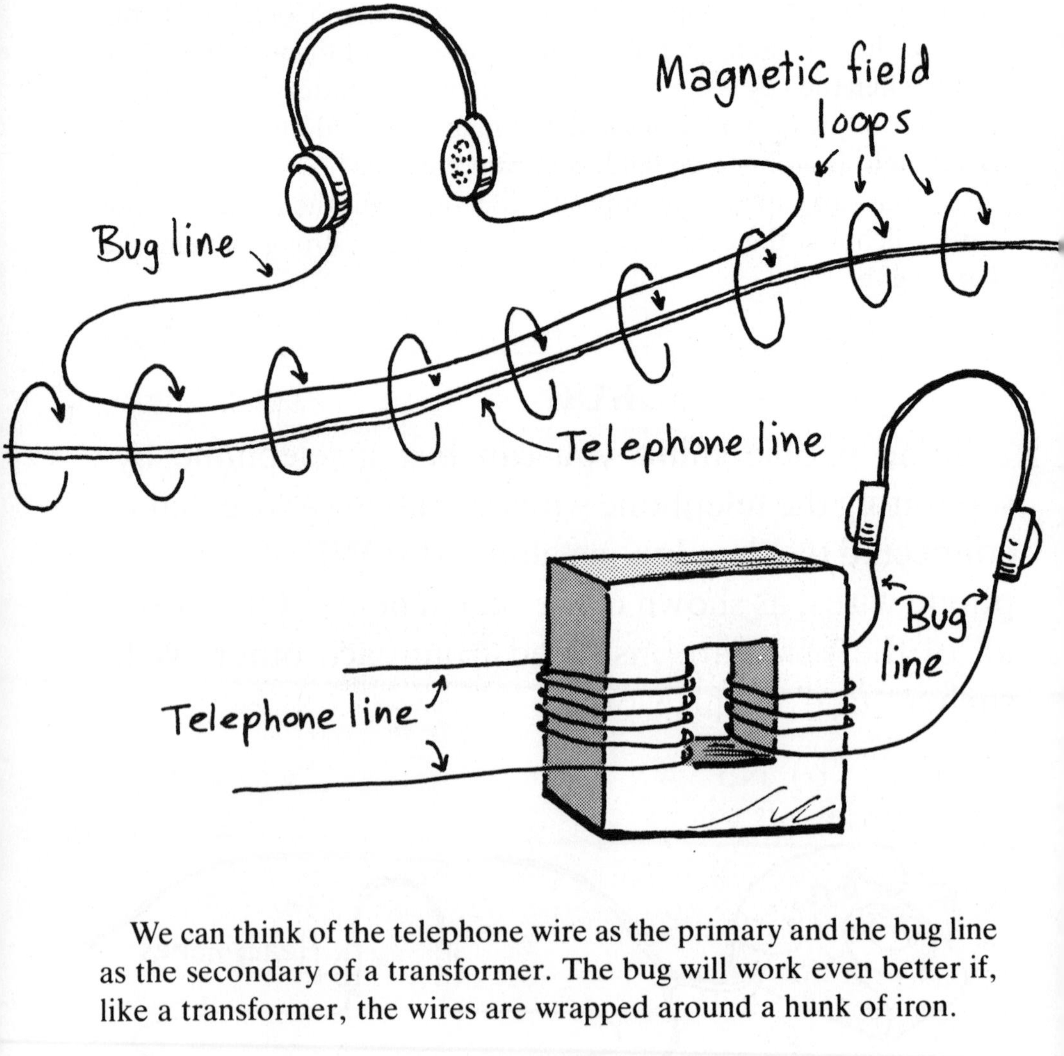

We can think of the telephone wire as the primary and the bug line as the secondary of a transformer. The bug will work even better if, like a transformer, the wires are wrapped around a hunk of iron.

GHOST SIGNALS

Faraday's Law says that if you have a loop of wire and a magnetic field is passing through the loop and the field is changing (getting stronger or weaker), then a current will be made to flow in the wire. Now a telegraph circuit is a loop. The electricity flows from the sender to the receiver through the wire and then back from the receiver to the sender through the earth. Over a century ago, just before the American Civil War, the transatlantic cable was completed, making the biggest electric loop in the world. Oftentimes strange "ghost signals" were heard on the cable, whether or not messages were being sent. After much investigation these signals turned out to be attributed to

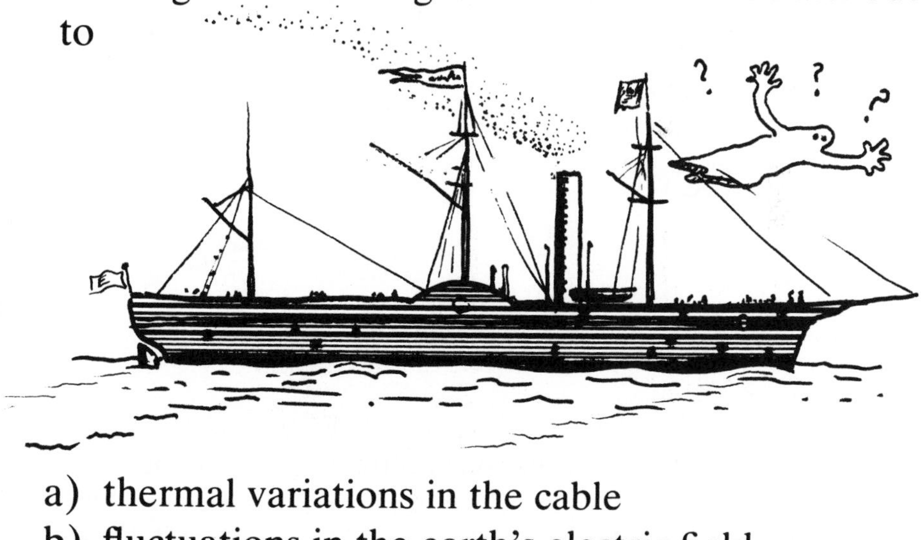

a) thermal variations in the cable
b) fluctuations in the earth's electric field
c) fluctuations in the earth's magnetic field
d) iron ships
e) ghosts

ANSWER: GHOST SIGNALS

The answer is: c. The magnetic field of the earth is not perfectly static, and fluctuations often occur. Sometimes the fluctuations are sufficiently great to be called *magnetic storms* (they occur in the sun as well). These changes in magnetic field intensity in the closed loop made up by the telegraph cable and the earth induce currents that are perceived as the ghost signals. Oddly enough, it might seem that ghost signal currents should not flow in the loop if the telegraph key is open because the loop is closed only when the key is depressed. But the

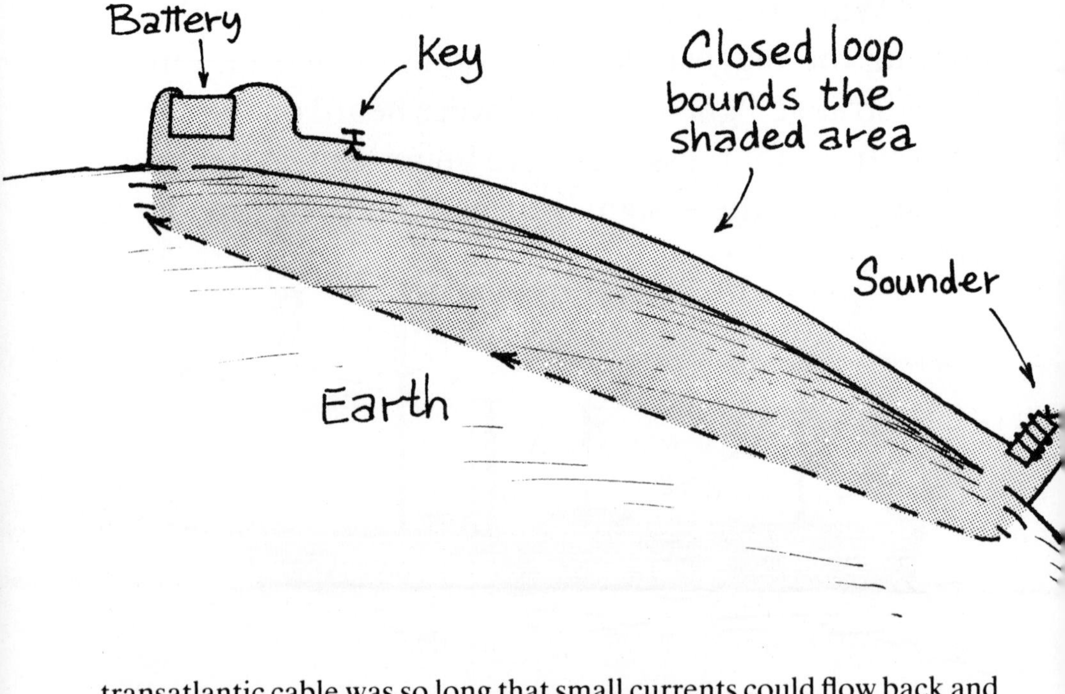

transatlantic cable was so long that small currents could flow back and forth in it even with the key open! Current just swished back and forth around the open loop, piling up at the ends. The length of the cable provided an electric storage capacity, the capacitance, which made the pile-up possible. This was a fascinating discovery at that time. The construction of the transatlantic cable was the "moon project" of its day.

SHOVE IT IN

A light bulb is connected by a thick wire to an AC source as shown in the sketch.

After a piece of iron is shoved into the coil of wire, the light

a) brightens
b) dims
c) is not affected

ANSWER: SHOVE IT IN

The answer is: b. The coil and the electric light bulb are in series. Therefore, the total voltage drop across the series circuit must be the voltage drop in the coil plus the voltage drop in the light—and this total must equal the 110 volts of potential difference supplied by the wall socket. So part of the 110 volt drop happens in the coil and the remaining part of the 110 volt drop is in the light. If there is a big voltage drop in the coil then there can only be a small voltage drop left for the light. A small voltage drop in the light makes it dim. What makes the voltage drop in the coil? Well, part of the coil voltage drop comes from resistance, but that is a small part because the coil is made of thick wire. The main cause of the voltage drop in the coil is the

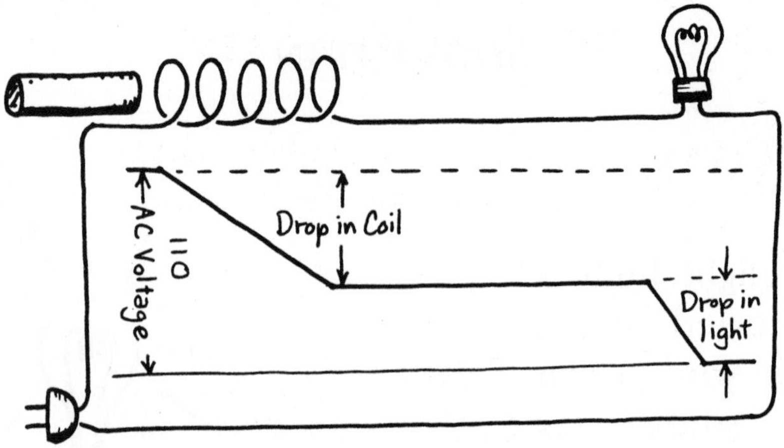

changing magnetic field in the coil. The larger the change per second in the magnetic field, the larger the voltage across the coil. What determines the change in the magnetic field per second? Two things: HOW STRONG the maximum field is and HOW FAST it is changing. We cannot alter how fast the field is changing—that is set by the power company. The voltage provided by the power company fluctuates 60 times per second, so the current in the wire fluctuates 60 times per second and the current makes the magnetic field so the field also fluctuates 60 times per second.

But we can alter the STRENGTH of the magnetic field. How? By putting some iron in the coil. Because of the magnetic "domains" in the iron which align with the magnetic field in the coil, the iron makes a stronger magnetic field. That means there is more field to be changed each 1/60 of a second. That means more voltage in the coil. That means less voltage left for the light bulb, and THAT means a dimmer light. Some old stage light dimmers operated in just that way.

The voltage produced in the coil by the changing magnetic field is always in a direction so as to fight a change in the current. So a changing current fights its own change (called a "self-inductive reactance"). Some people think it is this resistance to change—"electric inertia"—that makes the magnetic field and coil dim the light. Not so. The current is alternating; repeatedly building up and dying down. While the "electric inertia" impedes the current in its build-up stage and so chokes the light, the same "electric inertia" drives current through the light in its dying-down stage and, therefore, force-feeds the light. These two effects, then, exactly cancel each other out.

SHOVE IT IN AGAIN

This is a tricky question. After a piece of iron is shoved into the wire coil, we would find that the light is then

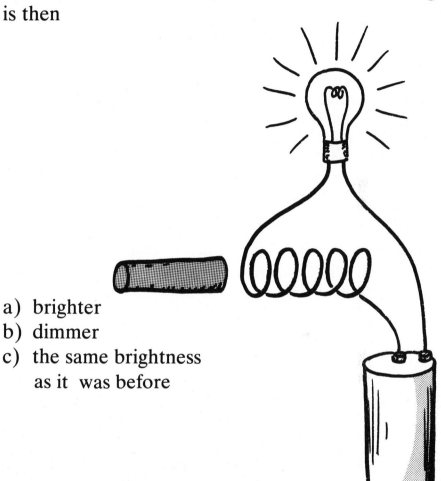

a) brighter
b) dimmer
c) the same brightness as it was before

ANSWER: SHOVE IT IN AGAIN

The answer is: c. We just considered a question like this (SHOVE IT IN), and the answer was that the light gets dimmer. Now you SHOVE IT IN AGAIN and we say the brightness of the light is not affected. What cooks here?

What cooks is that for SHOVE IT IN the power supply was alternating current, 60 cycle AC. The power supply for SHOVE IT IN AGAIN is a battery and a battery puts out direct current, DC. Direct current will not produce a CHANGING magnetic field and the changing magnetic field is essential for the voltage drop in the coil.

So shoving the iron in has absolutely no effect on the light? Not quite. As the iron first goes in it gets magnetized and that takes some energy so the light momentarily dims and when you pull the iron out the light momentarily brightens, but these changes occur ONLY while the iron is moving. The light is not affected *after the iron is in the coil.* A non-moving iron core has no effect on the brightness of the light.

Incidentally, you don't really have to shove it in. The iron is "sucked" in. Why? Loosely speaking, the coil is an electromagnet and magnets "love" iron.

IS EVERYTHING POSSIBLE?

Is it credible that there could be an *unknown universe* in which there were charged bodies, but no electric fields?

a) Yes, such a universe is credible.
b) No, such a universe could never be

ANSWER: IS EVERYTHING POSSIBLE?

The answer is: b. Everything is not possible. In particular, you cannot have charged bodies without also having electric fields. The only way a charged body can be sensed is by its electric field. For example, we can't look and SEE that the uncharged things are "white" and charged ones are "red" or "green." We can only feel, and we feel the electric field surrounding charged things. There would be no way to detect the existence of charged bodies were it not for their fields, because everything they do because of their charge they do by means of their field. Some physicists regard an electric charge as merely a place from where the field lines come out or diverge. This point of view is supported in the expression of the first of Maxwell's four famous field equations.* This equation just says that the divergence of an electric field is equal to the electric charge. In algebra: Div of $E = q$, where Div = divergence, E = electric field and q = charge. But what if q were a minus charge? Then the divergence of the field would be minus? Yes. But what is a negative divergence? A negative divergence is a convergence. It makes a nice mental picture showing the field around a negative charge is the reverse of the field around the positive charge.

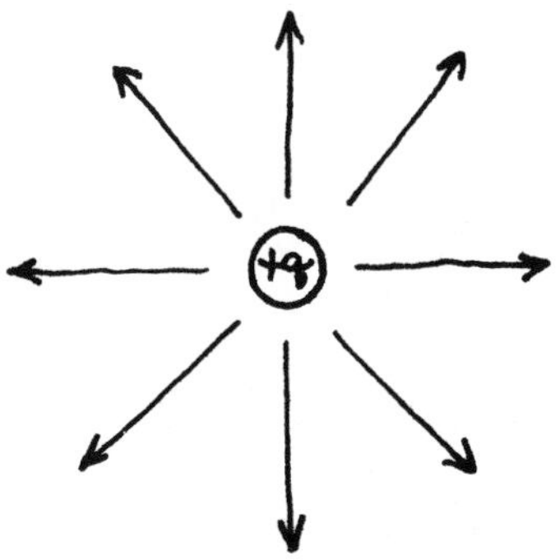

Divergence:

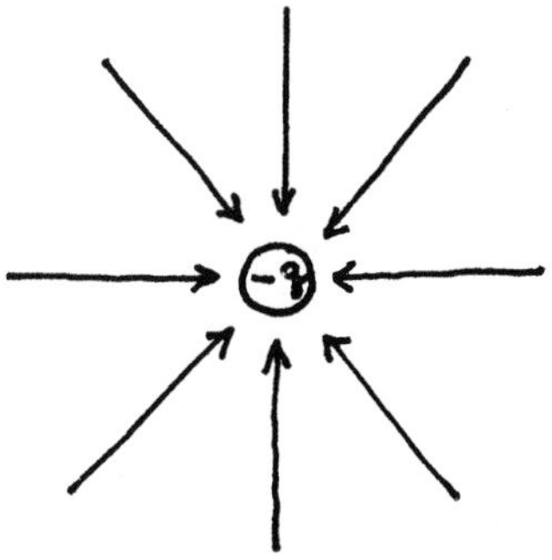

Convergence:

---

*Students of electricity and magnetism are familiar with four fundamental electromagnetic field equations, called *Maxwell's Equations,* after James Clerk Maxwell who in the last century succeeded in putting all the laws and the equations that describe electric and magnetic fields and their interactions with charges and currents into one unified set of equations.

ELECTROMAGNETISM'S HEART

Faraday's law of electromagnetic induction states that a voltage and resulting electric current will be induced in a conducting loop through which a magnetic field is changing with time. Maxwell re-expressed this in terms of fields by stating that a changing magnetic field will induce an electric field. Is the converse found to be the case also—that is, will a changing electric field induce a magnetic field?

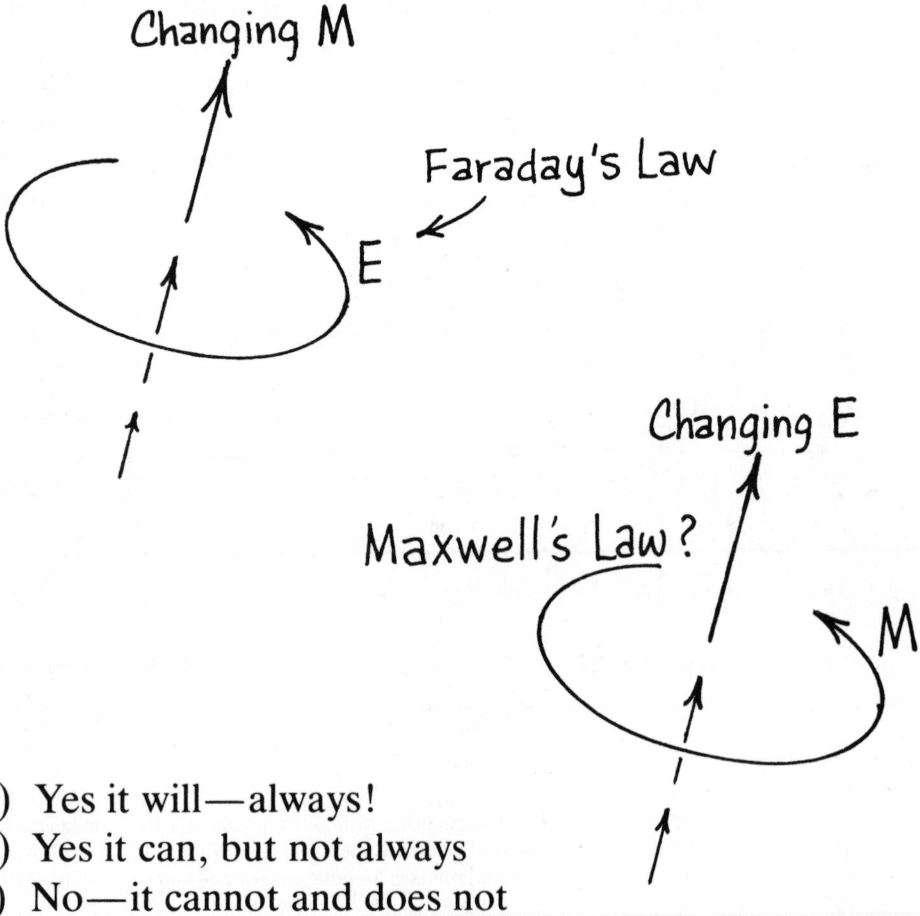

a) Yes it will—always!
b) Yes it can, but not always
c) No—it cannot and does not

ANSWER: ELECTROMAGNETISM'S HEART

The answer is: a. This duality is, in essence, the very heart of electromagnetic theory. It immediately means that once introduced into the universe, an electric field or a magnetic field becomes immortal—it goes on someplace forever. For if you discharge an electrically charged body and so try to kill the associated electric field, or if you destroy a magnet and so try to kill the associated magnetic field, the very death of either of the fields gives rise to a new field of the other kind: electric to magnetic or magnetic to electric. The dying field is a CHANGING FIELD, so it must give rise to a new field of the other kind.

So on and on forever it goes. The electric field collapsing to make a magnetic field, the magnetic field collapsing to make an electric field—an electric field "reincarnated." This perpetual motion is the machinery that propagates radio waves, light waves and even x-ray waves through space.

Travel and travel. Though the broadcasting station is long silent, the candle long out, the radiation lab long closed—the wave goes on. Eternally faithful to its last command: NEVER EVER STOP.

RING AROUND WHAT?

A magnetic field line circles around and joins its own tail, thus forming a ring. What would you expect to find passing through some place in the region enclosed by the ring?

a) an electric field line
b) an electric current
c) a changing electric field line
d) an electric current and/or a changing electric field line

ANSWER: RING AROUND WHAT?

The answer is: d. In the days of Napoleon it was known that a magnetic field circled around a wire that carries an electric current.

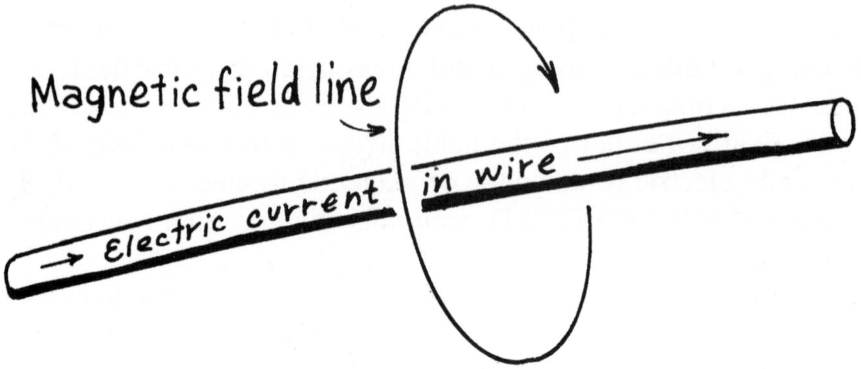

But about the time of the American Civil War it was realized (in England) that a magnetic field could be made in another way. This is the essence of our last question, ELECTROMAGNETISM'S HEART—a *changing* electric field also makes a magnetic field. If the electric field is getting stronger the magnetic field circles it one way. If the electric field is getting weaker the magnetic field circles it the opposite way. If the electric field is not changing—is static—it makes no magnetic field at all.

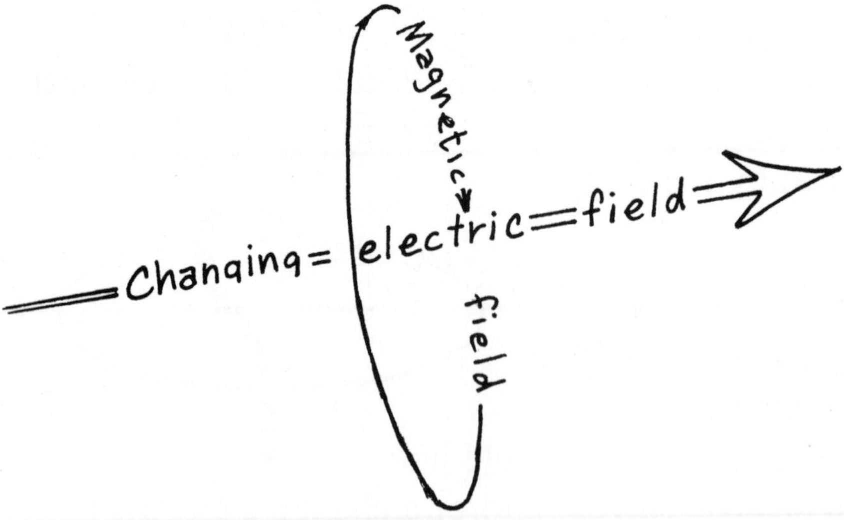

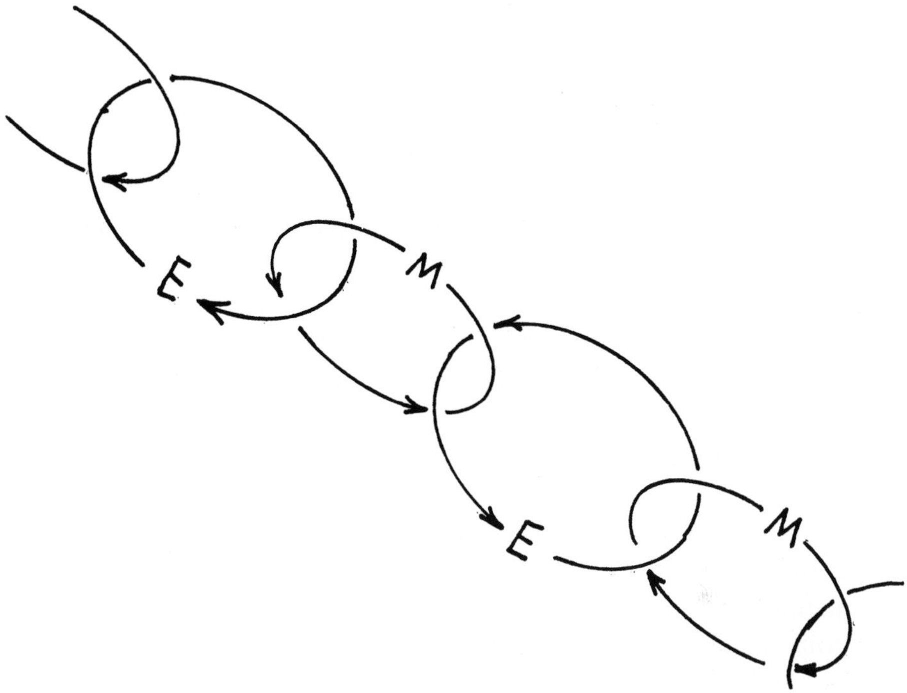

This is quite reminiscent of a transformer. In a transformer a changing magnetic field induces an electric field to circle around it. Now we learn a changing electric field likewise induces a magnetic field to circle around it. The sudden decay of one field results in a cascading of new fields, that form a chain of electric and magnetic fields, each circling the other. The only catch is that they must be CHANGING fields. So the chain could never be a stationary chain. It could never stand still.

THE MIND'S EYE

Is it credible that there could be a universe in which there were electric fields, but no charged bodies?

a) Yes, such a universe is credible
b) No, such a universe could never be

ANSWER: THE MIND'S EYE

The answer is: a. Of course, no one has first-hand knowledge of such a universe, but nevertheless it is possible to reason in your head what would happen—to see it in your mind's eye.

There are several ways to see it. As we just saw in ELECTROMAGNETISM'S HEART, when a magnetic field increases or decreases an electric field is created which curls around the changing magnetic field; when an electric field increases or decreases, a magnetic field is created which curls around the changing electric field. Collapsing or dying fields of one kind induce new fields of the other kind. So on and on it goes—one handing off to the other. The repeated hand-off is called an electromagnetic wave. Once it gets started it is self-sustaining and in no way depends on the original charge that made the first electric field that got it all started. A radio transmitter or star may be destroyed, but once the electric and magnetic fields in the radio signal or starlight are set on their way they will go for millions of years regardless of the fate of their home source.

Yet another way to think about this is to see the charged body's electric field extending through space. Now if the body is neutralized the field must go. But the "news" that the charge has been neutralized cannot travel through all space instantaneously. It travels out from the body at the speed of light. So even though the charge is gone its field remains in distant parts of space where the news has yet to arrive. (The Battle of New Orleans was fought after the War of 1812 was OVER because the "news" had yet to arrive in the Louisiana swamps.)

In the question IS EVERYTHING POSSIBLE? it was concluded that everything is not possible; in particular, you cannot have charged bodies without also having fields. But now it is concluded that you can have fields without charged bodies. Apparently the fields are more basic than the charges! How strange. People used to think the charges were the real things and the fields only abstractions—and now it is the other way around.

DISPLACEMENT CURRENT

A wire connects two oppositely charged plates, as illustrated (such an arrangement is a capacitor). A magnetic compass is located just outside the plates as shown. When the switch is closed, allowing the plates to discharge through the wire, there will be a momentary current in the wire. You might therefore expect that the magnetic compass would be somehow affected. Will it be affected?

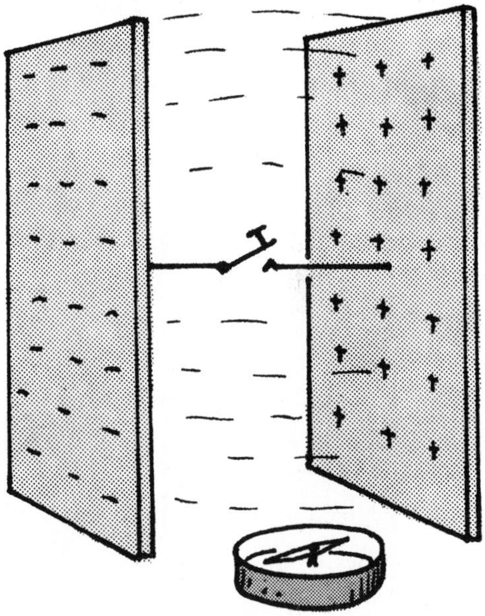

a) Yes, it will be affected
b) No, it will not be affected

ANSWER: DISPLACEMENT CURRENT

The answer is: b. At first you might think that because there will be an electric current in the wire as the plates discharge, that current must produce a magnetic field around the wire, and that magnetic field must affect the compass. But there is also a dying electric field between the plates, and a dying electric field is a changing electric field, and a changing electric field produces a magnetic field around itself just like an electric current. The magnetic effect of the dying electric field is exactly opposite to the magnetic effect of the electric current. So the two cancel each other.

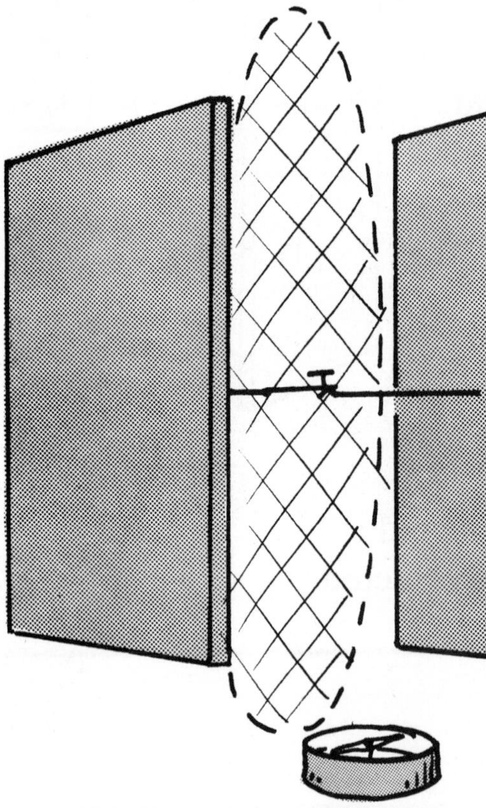

A changing electric field is called a displacement current. The name comes from an old idea that an electric field is really a strain or displacement in the "aether" which some people thought filled empty space.

X-RAYS

When a beam of electrons in a TV picture tube hit the front of the tube and are brought to a stop, they generate some x-rays. Most of these x-rays travel

a) forward in the same direction as the electron beam
b) sideways at right angles to the beam of electrons
c) backwards and opposite to the electron beam
d) equally in all directions

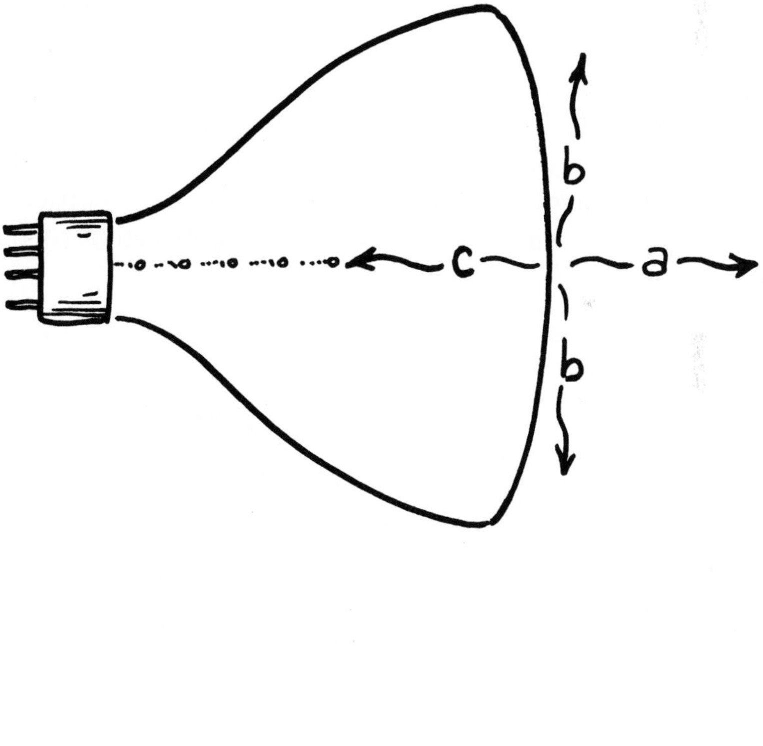

ANSWER: X-RAYS

The answer is: b.

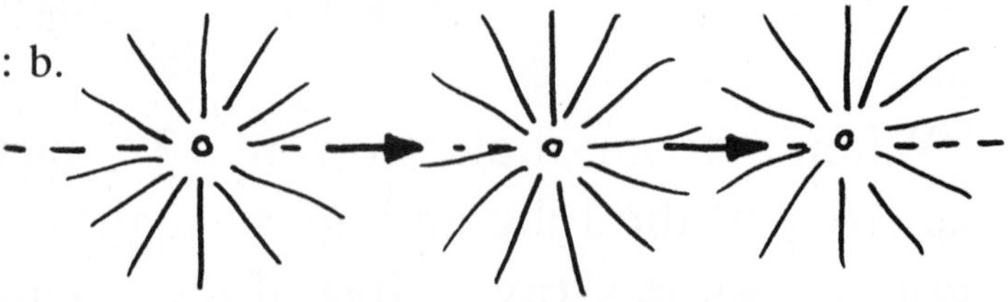

Electrons are surrounded by an electric field. When the electrons move they carry their electric field along with them, as illustrated above. The electric fields extend indefinitely.

If an electron is brought to a sudden stop, the whole FIELD cannot come to a sudden stop. The part of the field near the electron stops first, but the part further away doesn't have the news about the stop so keeps right on moving as though the electron had not stopped. In this way a "kink" develops in the electric field. The kink runs out along the field lines. This electric field kink is the x-ray pulse or wave.

The sketch below shows what happens when an electron hits the front of the TV picture tube at point A and stops. News of the stop has not reached distant parts of the field, which appear to have originated at B. If the news of the stop has reached only as far as the dashed circle, we find a kink in the field at this circle. Geometrically we can see the biggest kinks are sideways to the direction of the electron beam, with no kink directly forward or aft.

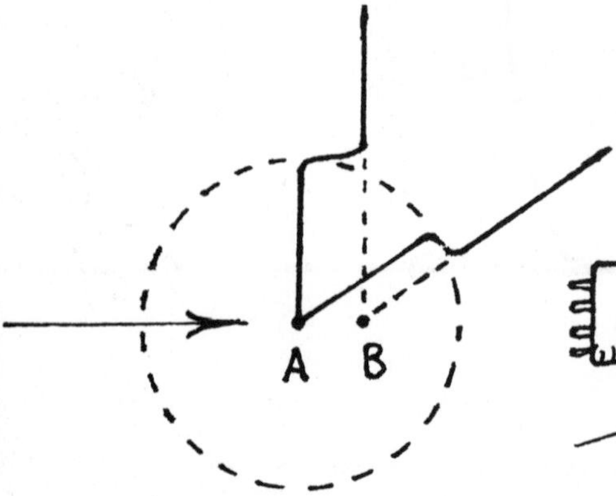

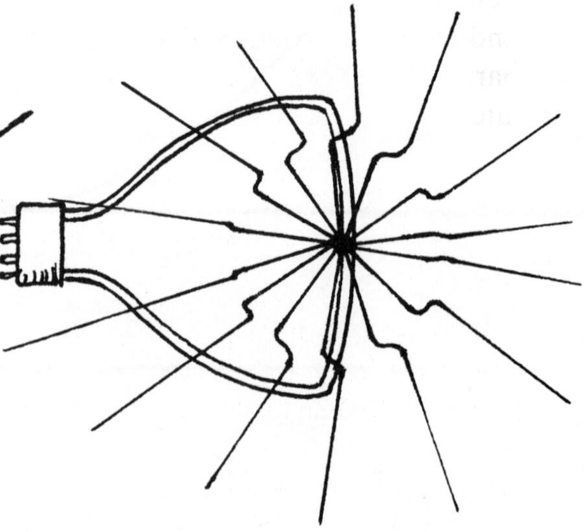

Because the x-rays go off to the sides, x-ray tubes are designed to have x-rays come out of them on the side. The thing that stops the electrons is called the target. The target is slightly tipped to enhance the sending of x-rays off to one side. (The reason the electrons rush to the target in the first place is because the target is positive while the place the electrons come from is negative.)

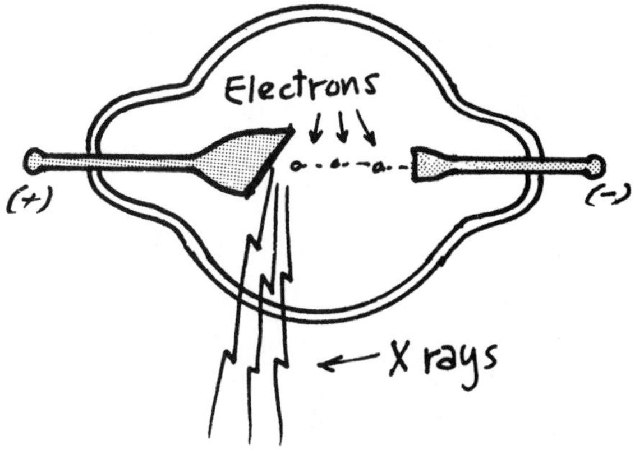

If you can visualize how x-rays are generated you can also visualize how radio waves are made. Radio waves are made by electrons moving back and forth in a wire called the antenna. The bank-and-forth electron motion is smooth and continuous and does not involve any sudden stop as with x-rays.

As the electrons move back and forth in the wire, they move the surrounding electric field along with them and the more distant parts of the field follow later.

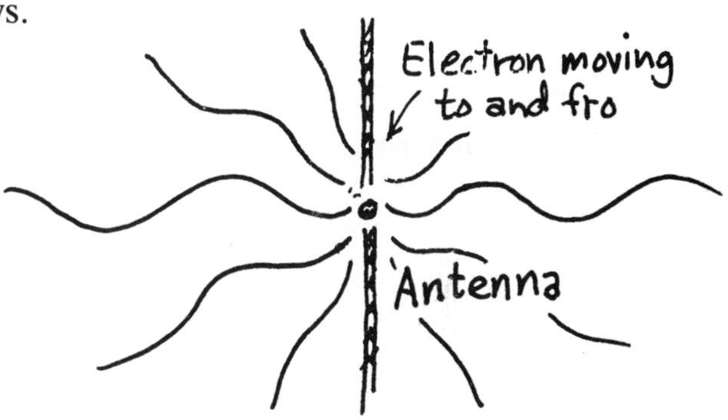

It is like a hand moving a rope back and forth. The hand is the electron and the rope is the electric field. Waves go out along the rope. Electric waves go out along the electric field. The speed of the waves depends on how much tension is in the rope. The frequency of the wave depends on how frequently the hand moves. The speed of the waves in the electric case is the speed of light. It is this speed regardless of whether the waves are light waves, x-rays, or radio waves. The difference between these waves is due completely to the different ways the electron sources move.

SYNCHROTRON RADIATION

If a fast-moving electron is made to move around and around in a circle, it will emit

a) polarized light
b) black body light
c) no radiation at all

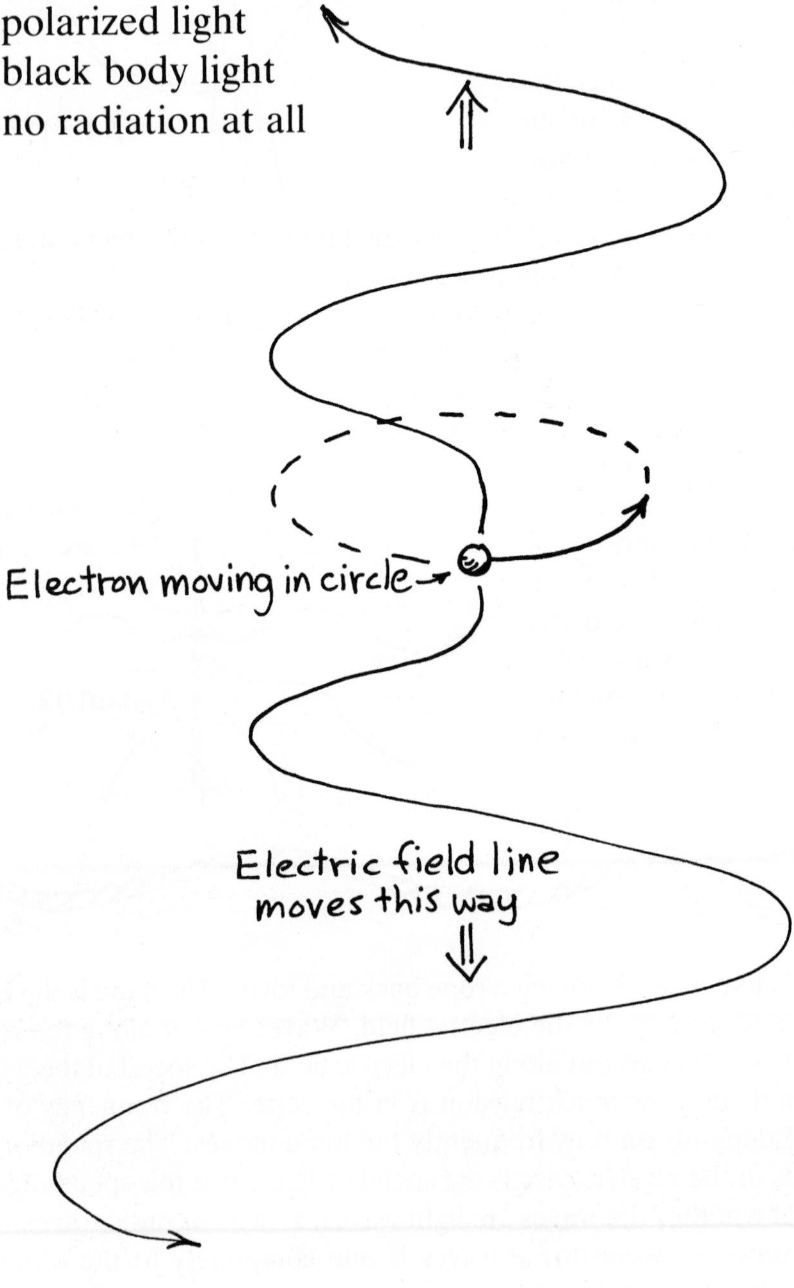

ANSWER: SYNCHROTRON RADIATION

The answer is: a. A very powerful source of x-ray radiation, as well as visible light radiation, is a machine called a SYNCHROTRON which makes electrons move around in a circle (the electrons are trapped in a magnetic field). As the electrons go around, their electric field lines whip back and forth. The whipping field lines are, in fact, the radiation waves. These waves whip back and forth in the plane of the circle around which the electron moves, as in sketch I. They cannot whip up and down, as in sketch II.

A wave that whips back and forth in one direction is said to be polarized in that direction. The wave whips at the same frequency with which the electron goes around the circle (although a relativistic effect messes this up a little), so synchrotron radiation cannot be confused with black body radiation which is a mixture of many frequencies.

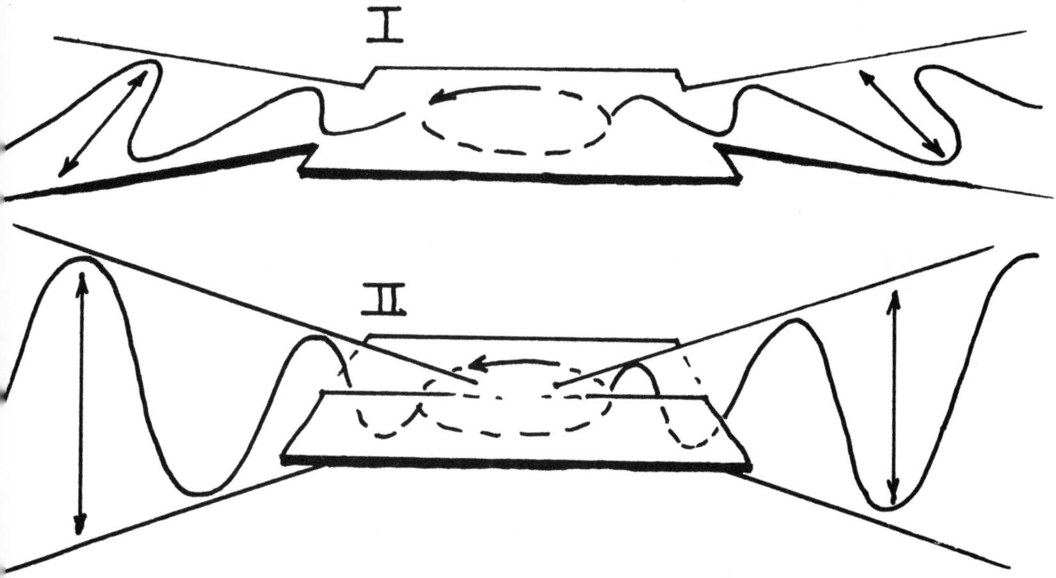

Incidentally, electrons trapped in the magnetic field of the galaxy, in the magnetic fields of the pulsars, and in the magnetic field of the earth (Van Allen Belts), all emit synchrotron radiation at very low (radio) frequencies.

MORE QUESTIONS (WITHOUT EXPLANATIONS)

You're on your own with these questions that parallel those on the preceding pages. Think Physics!

1. Strictly speaking, when a body acquires a positive charge, its mass
a) becomes greater b) becomes less c) doesn't change

2. If two free electrons initially at rest are placed close to each other, the force on each will
a) increase as they move b) decrease as they move
c) remain constant as they move

3. The electric field inside an uncharged copper ball is initially zero. If a negative charge is placed on the ball, the field inside will be
a) less than zero b) zero c) more than zero

4. Two identical capacitors are joined as illustrated to make a larger capacitor, which then has double the
a) voltage b) charge c) both d) neither

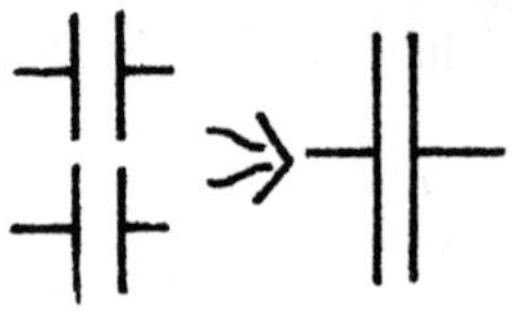

5. A light bulb and battery make up an electric circuit. Current flows
a) out of the battery and into the light bulb
b) through both the battery and the light bulb

6. A pair of light bulbs connected in series to a battery will draw
a) less current than a single bulb would draw
b) the same current that a single bulb would draw
c) more current than a single bulb would draw

7. A pair of light bulbs connected in parallel to a battery will draw
a) less current than a single bulb would draw
b) the same current that a single bulb would draw
c) more current than a single bulb would draw

8. Which has the fatter filament?
a) A 40-watt light bulb b) A 100-watt light bulb

9. The number of amperes that flow through a 60-watt bulb connected to a 120-volt line is
a) ¼ b) ½ c) 2 d) 4

10. The number of electrons delivered to the average American household by power utilities in 1980 was about
a) none b) 110 c) countless billions

11. An electron oscillating back and forth 1000 times each second will generate an electromagnetic wave with a frequency of
a) 0 hertz   b) 1000 hertz)   c) 2000 hertz

12. Which of these statements is always true?
a) Whenever an electric field exists, an electric current exists there also
b) Whenever an electric current exists, an electric field exists there also
c) both statements are true
d) neither statement is true

13. Moving electric charges will interact with a
a) magnetic field   b) electric field   c) both   d) neither

14. When a magnet is plunged into a coil of 10 loops a voltage is induced in the coil. If the magnet is similarly plunged into a coil of 20 loops, then the induced voltage is
a) half   b) the same   c) twice   d) four times

15. A transformer is used to step up
a) voltage   b) energy   c) power   d) all of these   e) none of these

16. A hermit in Vermont once strung a wire from his shack under a power transmission line as shown.

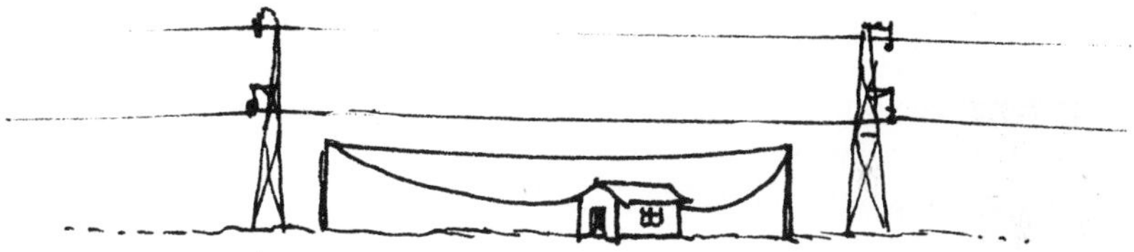

Would this arrangement be successful in bringing power to the hermit's shack?
a) yes   b) no

17. A radio transmitter sends out radio waves by causing electrons to rush to and fro in its aerial wire. Most of the broadcast waves go off in direction
a) I   b) II and III   c) IV and V
d) II, III, IV, and V
e) I, II, III, IV, and V

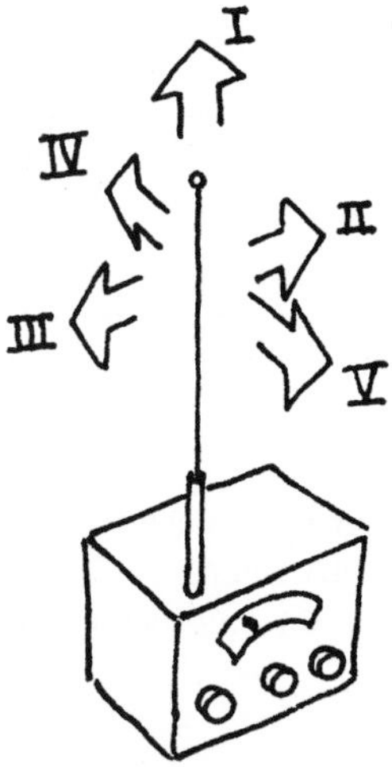

18. As ever more light bulbs are added to this string as indicated, the power it draws

a) increases
b) decreases
c) remains constant

19. As ever more light bulbs are added to this string as indicated, the power it draws

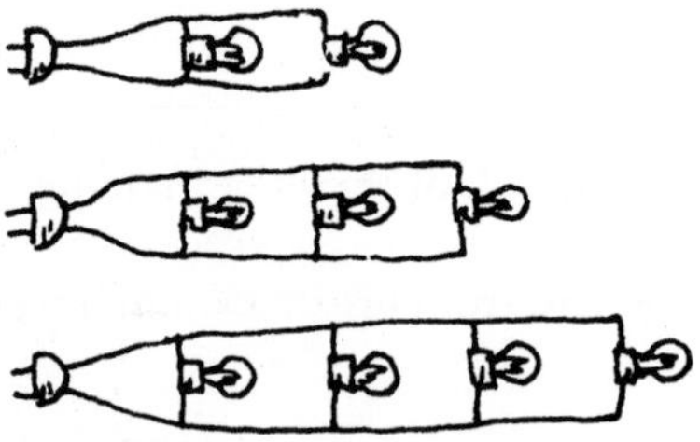

a) increases
b) decreases
c) remains constant

20. As ever more light bulbs are added to this string as indicated, the power it draws

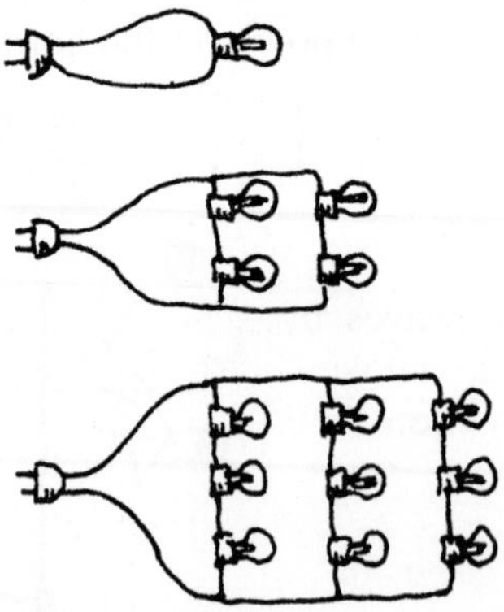

a) increases
b) decreases
c) remains constant

Relativity

Artists know that the apparent shape and size of a thing can change when it is viewed from different angles; it has to do with what they call three-dimensional perspective. The world, as it turns out, really has at least four dimensions, time being the fourth. The important thing is that you can view the same thing from different angles in four dimensions by changing your speed! The same thing, whether it be force, time or the geometry of a box, will be quite different when you view it at a different speed—four-dimensional perspective. That's what Einstein's Relativity is all about.

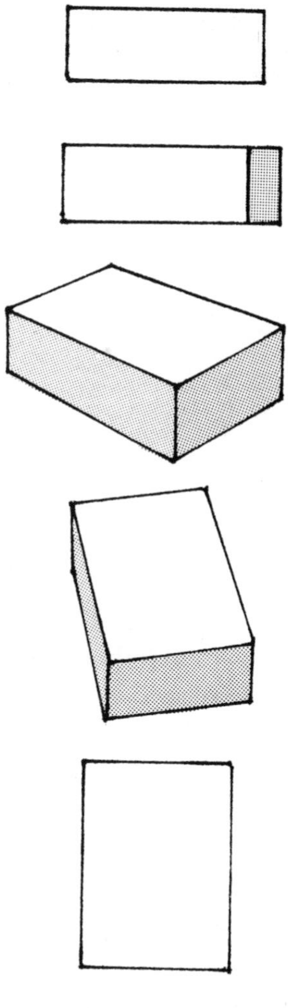

Physics delights in approaching self contradiction, as close as it can, without crossing the line.*

***Or is it just our clumsy perspective that makes it look that way?**

YOUR PERSONAL SPEEDOMETER

If you were traveling with respect to the stars at a speed close to the speed of light* you could detect it because

a) your mass would increase.
b) your heart would slow down.
c) you would shrink.
d) . . . all of the above.
e) You could never tell your speed by changes in you.

ANSWER: YOUR PERSONAL SPEEDOMETER

The answer is: e. The basic idea of Relativity is that if you are in a closed room you have absolutely no way of knowing if that room is moving or not moving. In short, you have no personal speedometer. Now if the room suddenly stops, the person inside can tell. If the room suddenly starts to move the person inside can tell. And if the room is spinning the person inside can tell that it is spinning. But if the room is moving smoothly in a straight line and not accelerating in any way, there is no way for you (inside) to know if the room is moving or not. Even if the room has a window in it and you look out and see something moving toward you, you cannot tell if it is your room moving toward the thing or the thing moving toward your room.

Now if the room's motion affected your mass, heart speed, or size, it would affect all other masses or clocks or sizes in the room in exactly the same way. So there would be nothing in the room that did not change. There would be no fixed reference with which to compare and so there would be no way to detect any change. It is impossible to find a personal speedometer.

*The speed of light is about 300,000 kilometers per second.

DISTANCE IN SPACE, LAPSE IN TIME, SEPARATION IN SPACE-TIME

The personal time lapse for all persons traveling between two events (like lunch today and lunch tomorrow), is the same regardless of which (space-time) path is taken between the two events.

a) True.
b) False.

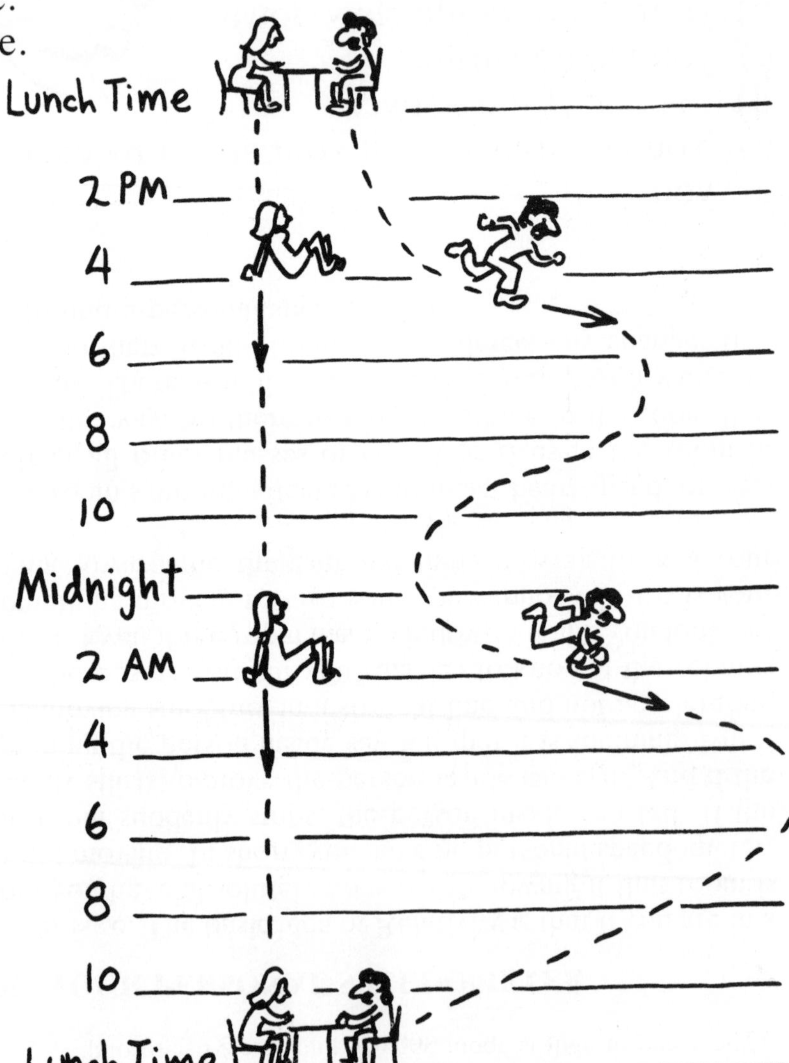

ANSWER: DISTANCE IN SPACE, LAPSE IN TIME, SEPARATION IN SPACE-TIME

The answer is: b. Between your two (lunch) events, you might think there are 24 hours of time. And that is so if you sit here. But if you go for a very fast trip between the two lunches and you carry a watch along with you as you travel you will find that by your watch there is less than 24 hours between the two lunches. The essential idea is that the two lunches are not separated by a certain amount of time, but are separated by a certain amount of space-time. If you sit here and do not travel, there can be no space between the lunches—only time separates them. But if you travel, you put some space between the lunches. Now if the space-time between the lunches stays the same, but the amount of space between them increases, then the amount of time between them must decrease. And decrease it will. If you step on a spacecraft after lunch and buzz around at the speed of light and return for the next lunch, you and your watch will record the passage of zero time—but Mom will have another lunch ready.

The thing of it is just this. You are always traveling. And you are always doing so with a constant speed. Even when you stand still. When you stand still you are traveling through time. If the speed is directed so as to carry you through space then the component remaining to carry you through time is diminished. If the speed is entirely used to carry you through space (at the speed of light), there is nothing left to carry you through time.

COSMIC SPEEDOMETER

If you see a person traveling through space at half the speed of light, you will also see his clocks running

a) at half their normal speed.
b) slower than half their normal speed.
c) slower, but not slowed to half speed.
d) at normal speed.
e) backwards.

ANSWER: COSMIC SPEEDOMETER

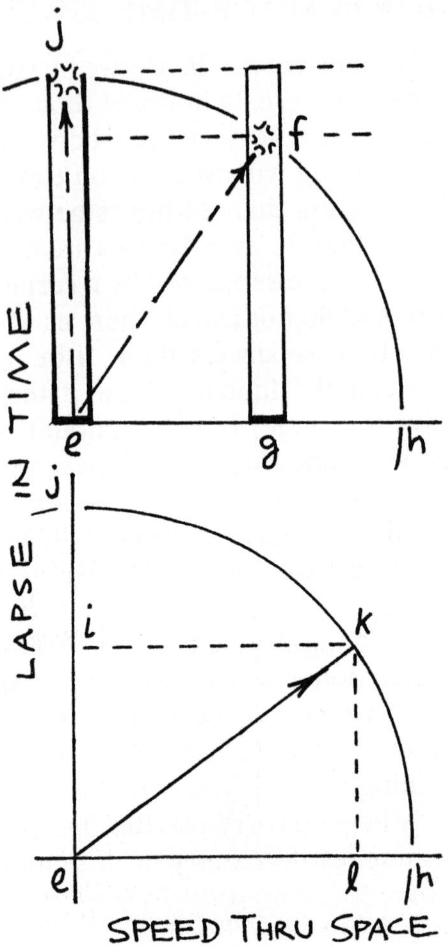

The answer is: c. Picture a light clock. When the clock is stationary a flash goes from the bottom of the clock at **e** to the top at **j** in, say, one second. But if the clock moves at half the speed of light the clock goes from **e** to **g** in one second. (If the clock moved at the speed of light it would go from **e** to **h** in one second.) Now if the clock goes to **g**, the flash can only get to **f** in one second because light can only go so far in one second and that distance is from **e** to **j**, which is the same as the distance from **e** to **f**. (Note the circle.) But **f** is not the top of the clock. The distance between **f** and the top of the clock is about one-seventh of the clock's total height. That means a clock moving through space appears to be moving through time at about 6/7ths of its normal speed.

How fast would it have to go for us to see the clock slow to half speed? Draw a cosmic speedometer (a quarter circle). For a half-speed clock draw the line **i** to **k**, which is halfway up the circle between **j** and **e**. Then draw a line from **k** to **l** which represents the tube of the light clock. The half-speed clock must go from **e** to **l** in one second. The distance from **e** to **l** is approximately 6/7th of the distance from **e** to **h**, so the clock moving through time at half the normal speed moves through space at about 6/7ths the speed of light.

The cosmic speedometer suggests a story which tells you why you see a clock go slow as you see it speed through space. It also tells why you can't see it go faster than the speed of light. The story is that it is always traveling at the same speed. Only the direction of its speed can be changed. If it travels in the direction **e→j** it goes completely through time, and does not go through space at all. But if it goes in the direction **e→h** it goes completely through space, and does not go through time at all. In direction **e→f** it goes mostly through time and a little through space. In direction **e→k** it goes mostly through space and a little through time.

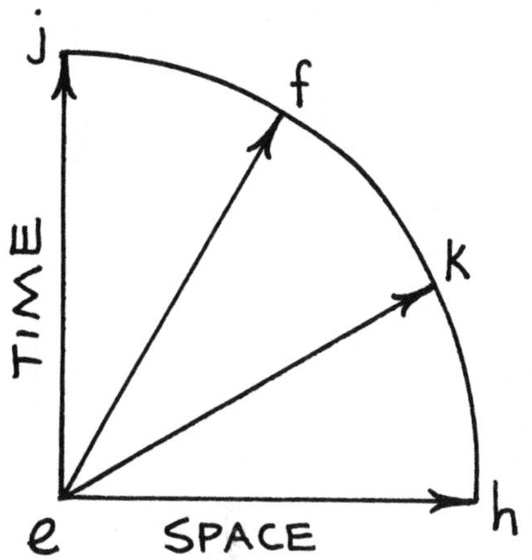

Now you can never see yourself move through space (unless you have out-of-body experiences); therefore you must always see yourself moving completely through time along **e→j**. So with a bit of poetic license it may be said that you are moving with the speed of light through time—and that's as fast as you can go. So, you really know what it feels like to move at the speed of light!

Have you ever asked yourself why time cannot go by faster? Well, when you find the answer to that you will also know why you can't expect to see things go faster than the speed of light.

YOU CAN'T GET THERE FROM HERE

Which of the following facts, if definitely established, would violate the "Theory of Relativity" as we know it today?

a) Things can go faster than the speed of light.
b) Nothing can go faster than the speed of light.
c) If a thing goes faster than light, it quickly slows to a speed less than that of light.

ANSWER: YOU CAN'T GET THERE FROM HERE

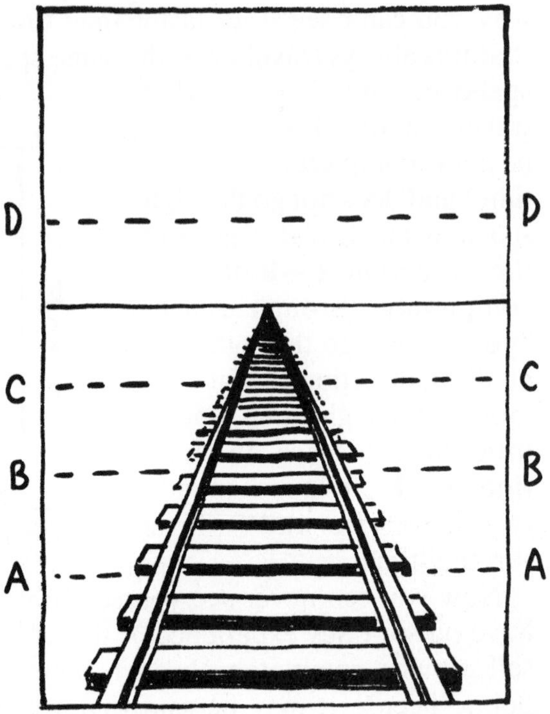

The answer is: c. The "Theory of Relativity" (it should now be called the "Law of Relativity") says if something is going slower than the speed of light, then no matter how much speed you add to it, it will never go faster than the speed of light. Nothing ever stops you (or it) from adding speed, it just happens that the resulting speed is NOT the sum of the added speeds. It is mathematically like walking down a railroad track. You start at line AA, you walk to BB, you can even walk to CC. In fact, nothing ever stops you. You can, if you want, walk forever. But no matter how far you walk, you will never reach DD. That does not mean nothing can be at DD. It just means you can't get there from AA by walking along the track. Similarly, you can't go above the speed of light by adding speed. The speed of light is like the horizon—you can't cross it. That does not mean that nothing can be above the speed of light. It just means that if anything is "up there," it did not get there from here by adding speed. Things might or might not be above it, no one knows; but at this printing nothing has ever been seen above it.

If a thing were found going faster than the speed of light, it would be like finding a second overhead railroad track in the sky. But, and this is the critical point of the story, no matter how far you went along that railroad track in the sky from FF to EE to DD and on forever you could never get to CC below the horizon, that is, below the speed of light.

The theory of relativity would be violated only if something CROSSED the horizon.

Finally, one warning: These horizon pictures are not pictures of the speed of light or the theory of relativity. The horizon picture just has a mathematical characteristic similar to speed addition. It is not unusual to use such mathematical analogies in physics. For example, a force is pictured as an arrow because it has similar mathematical characteristics. But a force is not an arrow —a fact that many physicists almost forget.

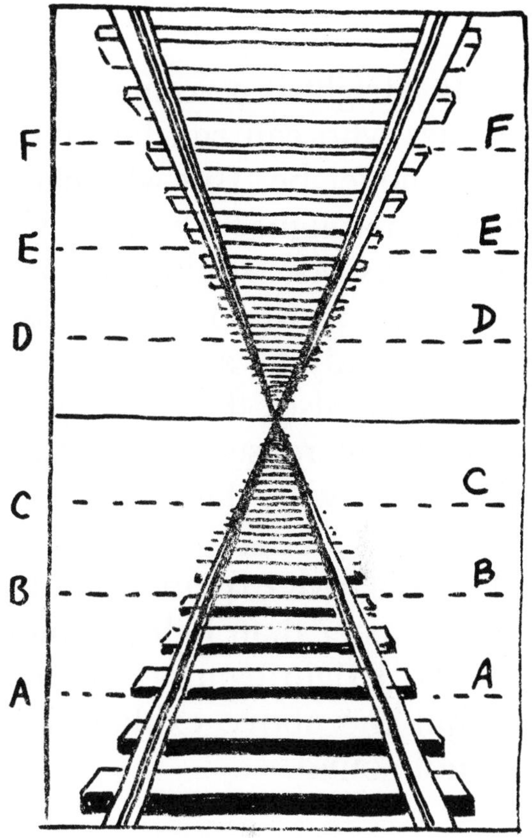

MS. BRIGHT

Ms. Bright is a young lady from the mythology of physics who can run faster than light. Of course, it could never be. But why can she, in fact, never run faster than light? The following reason is sometimes given: As Ms. Bright runs faster and faster her mass increases, so Ms. Bright finds she has become a very massive lady as she approaches the speed of light. She finds, also, that her muscles can no longer cope with the increased mass of her body. And that's it! Try as she will, she cannot go any faster.

a) The above explanation of why Ms. Bright cannot go faster than light is sound.
b) Ms. Bright cannot, in fact, run faster than light, but the above explanation of the reason why is not sound.

ANSWER: MS. BRIGHT

The answer is: b. Nothing ever stops Ms. Bright from running faster. We who are not running see her mass increase, because relative to us she is moving, but relative to herself she is not moving and her mass is her normal mass. Remember there are no "personal speedometers," which means people can't detect their speed by changes in themselves.

If we tried to make Ms. Bright go faster by pushing her with a long stick, her increased mass would resist our push. But if she pushes on herself with her own feet she finds nothing unusual. Why then can she not increase her speed beyond the speed of light? Because when she adds ten kilometers per hour to her speed it does not look like ten kilometers per hour to us. Why not? Because she reckons the hour by her time, but she is in motion, so her hour could be a month for us, and she reckons her kilometer by her space. She is in motion so her kilometer could be a centimeter for us. So, what is ten kilometers per hour for her could be be ten centimeters per month for us. She can increase her speed as much as she wants, but the increases don't add. (Remember YOU CAN'T GET THERE FROM HERE.)

ALMOST INCREDIBLE

You are standing near the middle of a long board which, as you perceive it, falls such that the two ends strike the ground simultaneously. You, therefore, think the board falls FLAT. But Ms. Bright (who is dashing by you near the speed of light) perceives end B striking the ground BEFORE end A and therefore thinks the board was TILTED to the right as it fell.

a) True
b) False

ANSWER: ALMOST INCREDIBLE

The answer, is: a, true. If two things happen at different places, but at the same time, as perceived by you (like end A and B at different places hitting the ground at the same time) then for Ms. Bright, who is in motion relative to you, the same two things can never happen at the same time. Why?

Suppose you are passing by three equally spaced stars which you view through the window of your spaceship. All of a sudden the center star explodes. Should the flash of light from the center star reach the two equally spaced end stars at the same time? You might think so, and it would if you were riding along with the star cluster. But you are overtaking the star cluster and as seen from your window, the flash does not reach the end stars at the same time. It reaches B first.

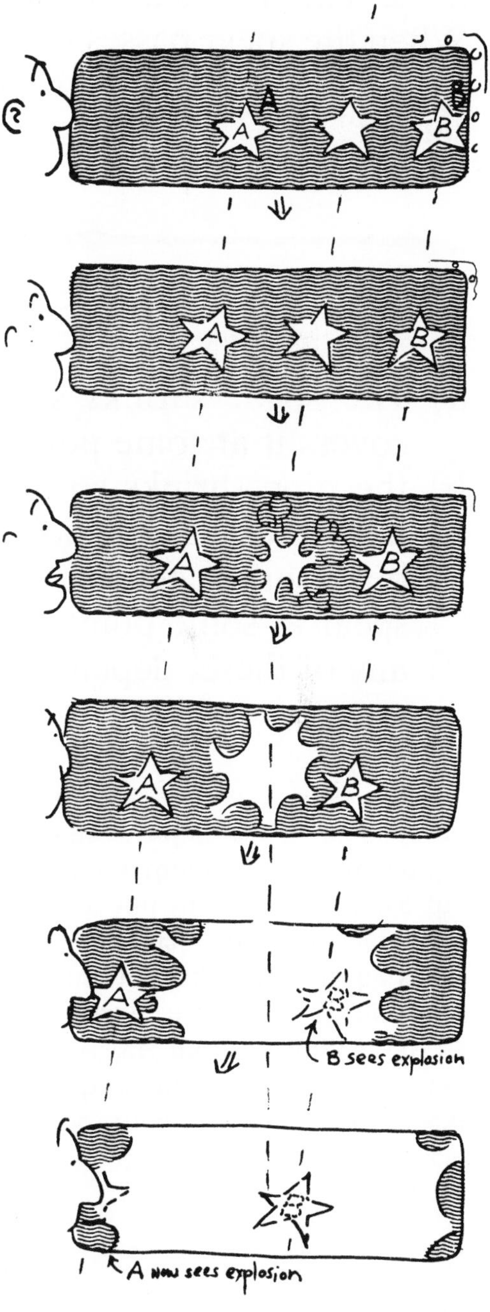

A
B
B sees explosion
A now sees explosion

As Ms. Bright passes the board, it appears to her the board is not just falling, but that the board is falling AND passing her, like the stars passing your spaceship. And so what appears to you to happen at the same time appears to Ms. Bright to happen at B first.

Did it really happen at A and B simultaneously or did it happen at B first? No one can answer that because no one can say who is really moving. It just depends on your reference frame. For you the board fell flat. For Ms. Bright it fell tilted to the right. It is really almost incredible!

HIGH-SPEED SPEAR

A spear 10 meters long is thrown at a relativistic speed through a pipe that is 10 meters long. Both these dimensions are measured when each is at rest. When the spear passes through the pipe, which of the following statements best describes what is observed?

a) The spear shrinks so that the pipe completely covers it at some point
b) the pipe shrinks so that the spear extends from both ends at some point
c) both shrink equally so the pipe just covers the spear at some point
d) any of these, depending on the motion of the observer

ANSWER: HIGH-SPEED SPEAR

The answer is: d. If you are observing the moving spear at a location that is at rest relative to the pipe, then the spear will appear shorter than the pipe and at some point be completely within the pipe. If you're traveling with the spear, however, the pipe appears to be shortened and at some point you'll observe both ends of the spear extending beyond the pipe. Or if you're moving between the spear and the pipe, at half the velocity of the spear in the direction of the moving spear, then both the spear and the pipe will have the same speed relative to you and you'll see them both shortened the same amount. So what really happens is relative—it depends on your point of view, or frame of reference.

MAGNETIC CAUSE

When current flows through a pair of parallel wires they are magnetically attracted if the current in both wires is in the same direction, and repelled if the currents are in opposite directions. This magnetic force is

a) the relativistic result of imbalanced electrostatic forces
b) an outcome of the mass-energy equivalence
c) a fundamental force of nature
d) all of these e) none of these

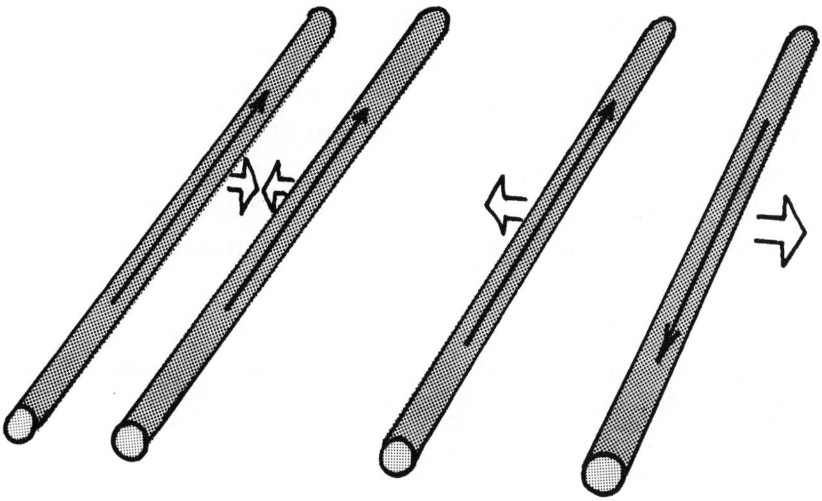

ANSWER: MAGNETIC CAUSE

The answer is: a. Magnetic forces result from a perceived increase in electrostatic charge density by virtue of relativistic length contraction. A meter of wire has as many positive protons as negative electrons in it, and therefore has zero net charge. This is also true when an electron current flows in it, for as many electrons leave one end of the wire as enter the other end.

STATIONARY WIRE

WITH OR WITHOUT CURRENT, ELECTRON IN WIRE SEES AN EQUAL DENSITY OF PLUS AND NEGATIVE CHARGES IN NEIGHBORING PARALLEL WIRE. IT FEELS NO NET EFFECT.

But how does this wire look to an electron moving along parallel to it in the adjacent wire? An electron in either wire sees electrons in the other wire at relative rest, since they move in the same direction at the same average speed. But not so for the protons, which are seen to be moving in a direction opposite the electron current. Due to the perceived relativistic length contraction of the wire, the distance between adjacent protons is reduced. So the moving electron sees a greater density of protons compared to electrons in the neighboring wire. Opposite charges attract in electrostatic fashion

MOVING WIRE

BUT WHEN CURRENT FLOWS IN BOTH WIRES, MOVING ELECTRON SEES AND FEELS A NET PLUS CHARGE DENSITY IN OTHER WIRE – IT FEELS AN ATTRACTIVE FORCE.

and the wires tend to move toward each other. We call this a magnetic attraction, but interestingly enough, it is underlain by simple electrostatics.

Can you account for the repulsion between the wires when the currents flow in opposite directions?

CHASED BY A COMET

A comet is chasing a spacecraft. Let V be the speed of, P the momentum of, and E the energy of the comet as perceived by the astronaut when it hits the spacecraft. In what way would increasing the spacecraft's speed alter the astronaut's perceived values of V, P and E?

a) V, P and E will all be constant and not change at all.
b) V, P and E will all decrease.
c) V and P will get smaller, E will not change.
d) V and E will get smaller, P will not change.
e) E and P will get smaller, V will not change.

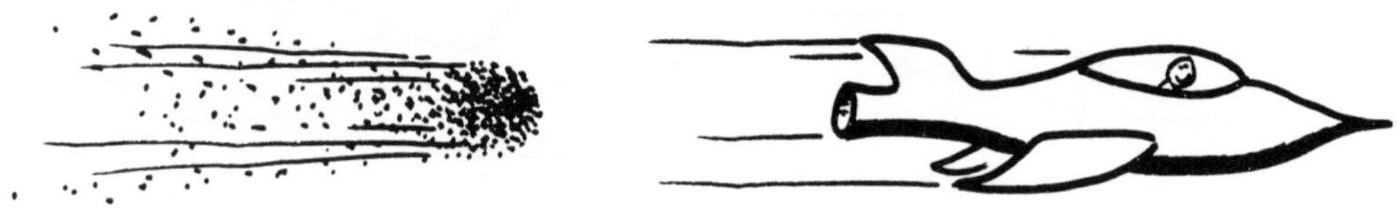

ANSWER: CHASED BY A COMET

The answer is: b. The comet moves fast, but its speed is much, much less than the speed of light. The astronaut might, or might not, be able to outrun the comet. The faster he goes, the slower the comet goes—relative to the astronaut. He perceives the comet going slower as he runs faster (and if he can move fast enough the comet might even be observed to go backwards). As the comet's observed speed is reduced so also is its observed momentum and kinetic energy. The momentum and energy of any punch is reduced by pulling away from the blow—and our astronaut is trying to pull away from the comet.

CHASED BY A PHOTON

This sort of situation troubles many physicists when they begin to think about relativity: A spacecraft is trying to escape a photon (of course, it will never escape). Let V be the speed, P the momentum, and E the energy of the photon as perceived by the astronaut when it hits the spacecraft. In what way will increasing the spacecraft's speed alter the astronaut's perceived values of V, P and E?

a) V, P and E will be constant and not change at all.
b) V, P and E will all decrease.
c) V and P will get smaller, E will not change.
d) V and E will get smaller, P will not change.
e) P and E will get smaller, V will not change.

ANSWER: CHASED BY A PHOTON

The answer is: e. The perceived speed of light will not decrease even if the observer runs away from the light. This strange but experimentally confirmed fact is one of the cornerstones of physics. V will not change. However, as the observer moves faster and faster from approaching light, the Doppler effect makes the light's observed frequency lower and its observed wavelength longer, that is, the light is red-shifted. The redder a photon gets, the less energy and momentum it packs. That's why darkroom safelights are red. So P and E get smaller.

Good old intuition was at least two-thirds right. Intuition would suggest V, P and E would all decrease as the rocket runs from the photon, as indeed happened when the rocket ran from the comet. It might seem strange that the less familiar concepts of momentum and energy more closely obey intuition than the more familiar concept of speed.

WHICH IS MOVING?

The Doppler frequency shift produced by you receding from a source of sound is the same as the shift produced by the source of sound receding from you.

a) True
b) False

The Doppler frequency shift produced by you receding from a source of light is the same as the shift produced by the source of light receding from you

a) True
b) False

ANSWER: WHICH IS MOVING?

The answer to the first question is: b. The answer to the second is: a. There is not one Doppler effect but several variations.

If you rush away from a sound source at the speed of sound, the frequency you receive is lowered to zero. This is because you receive no sound—you outrun it. If instead the sound source rushes away from you at the speed of sound, the frequency received is cut in half because the sound waves are spread out over double the space they would be spread out over if the source were not moving. If you approach a source at the speed of sound its frequency will be doubled, because the speed of sound relative to you is doubled. If the sound source approaches you at the speed of sound its frequency becomes infinite because all the waves are jammed together into one sonic boom.

Which is moving, the source or receiver, makes a big difference in the case of sound. We see that the Doppler effect for sound allows you to distinguish which is moving, the source or the receiver. Applied to light this would allow you to distinguish whether the earth was approaching a star or a star was approaching the earth. This would be a test for determining who was really moving through empty space. But the central idea of relativity is that you can never determine absolute motion—only relative motion. For example, if you are getting closer to a star you cannot distinguish whether you or the star is doing the moving.

So the Doppler effect for light must be different from the Doppler effect for sound. For light the Doppler effect can't betray whether the earth or star moves. It must produce the same shift in both cases.

Now you might ask how does sound get away with violating the principle of relativity? Must not all physics obey the same basic laws? The answer is that in the sound story there is a third party that plays kingmaker. It is the air. It could have been that there was something in space (like ether) that would do for light waves what air does for sound. But the world was not put together that way. Empty space contains nothing that can be used as a kingmaker, that is, as a reference point to tell who is really moving.

LIGHT CLOCK

A rocket ship in space emits brief flashes of light from its beacon at a steady rate of one flash every six minutes, rocket time. These flashes are observed on a distant planet. If the rocket ship approaches the planet at high speed, observers on the planet will see the flashes at intervals

a) less than 6 minutes
b) 6 minutes
c) more than 6 minutes

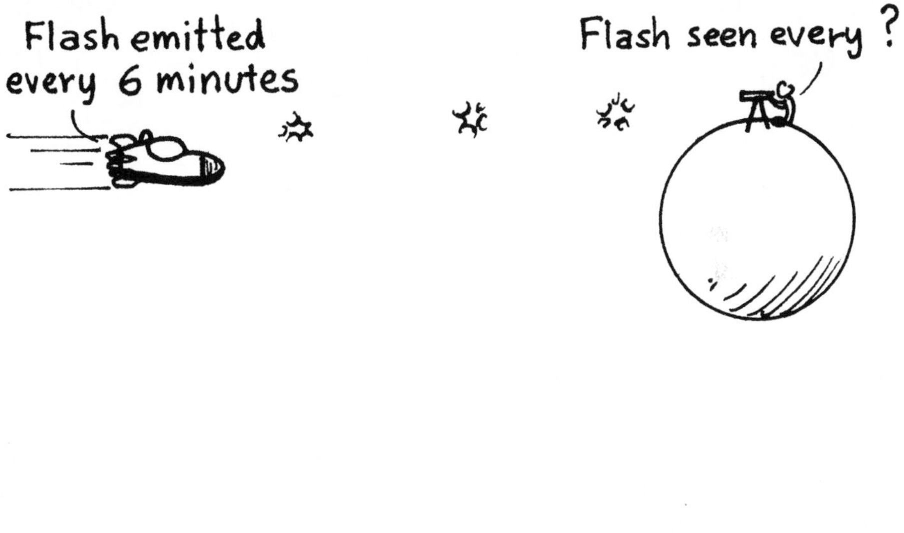

ANSWER: LIGHT CLOCK

The answer is: a, in accord with the Doppler effect. The greater the relative velocity between the sender and the observer, the shorter the perceived time interval. If the ship travels toward the planet at a speed of 0.6*c*, for example, the flashes will be seen at 3-minute intervals.

LIGHT CLOCK AGAIN

A rocket ship that emits flashes of light every 6 minutes travels between two planets, A and B. It travels away from A and toward B. If its flashes are seen at 3-minute intervals on B, then on planet A the flashes are seen at

a) 3-minute intervals
b) 6-minute intervals
c) 9-minute intervals
d) 12-minute intervals

ANSWER: LIGHT CLOCK AGAIN

The answer is: d. This can be shown in the following sequence: Consider Sketch 1 where a sender on earth directs flashes at 3-minute intervals to a distant planet which is at relative rest. An observer on the distant planet receives the flashes at 3-minute intervals. But a rocket moving between the earth and the planet receives them at longer intervals; say the rocket speed is such that they are seen 6 minutes apart. Suppose further that each time the ship receives a flash it emits its own. Einstein's first postulate is that light

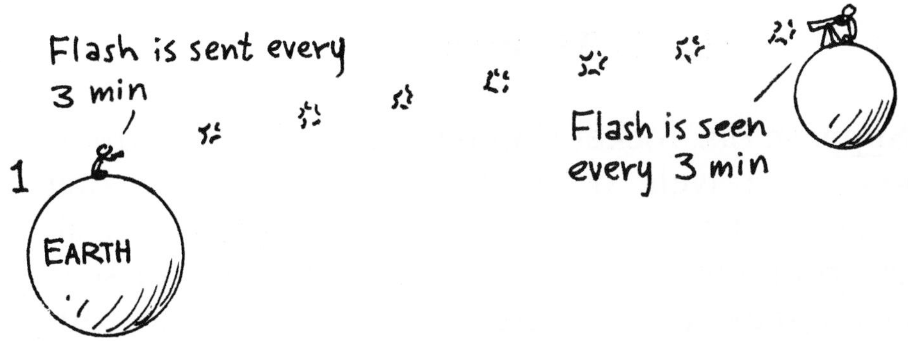
Flash is sent every
3 min
1
EARTH
Flash is seen
every 3 min

has the same speed in all frames of reference, so these flashes travel with those of the earth and are seen on the distant planet at 3-minute intervals, Sketch 2. This is in accord with the previous question, LIGHT CLOCK. But how often would the ship's flashes be seen from the earth? Here we invoke Einstein's second postulate; namely, no observations can discern whether the earth is at rest and the ship is moving or the ship at rest and the earth moving—motion is relative. Since the ship sees the earth's flashes to be twice as long apart (6 minutes rather than 3 minutes) so also will the earth see ths ship's flashes to be twice as long apart—12 minutes rather than 6 minutes.

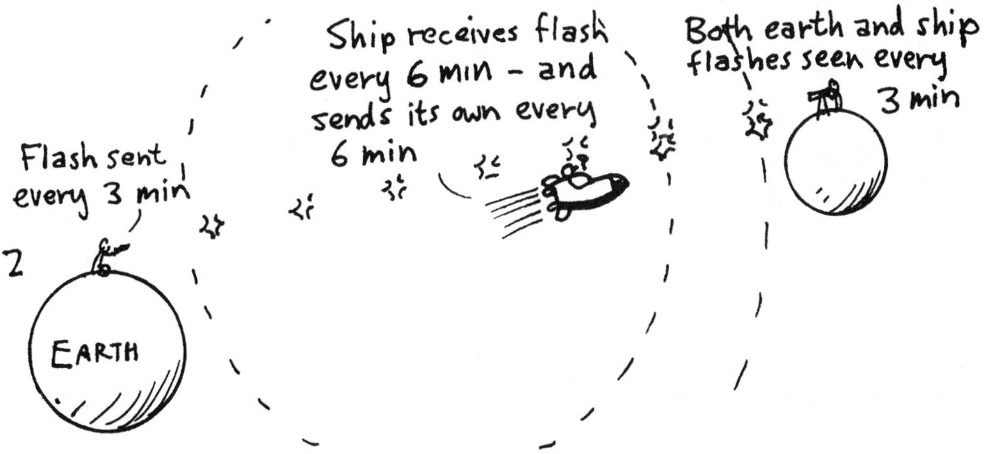
Ship receives flash
every 6 min - and
sends its own every
6 min
Both earth and ship
flashes seen every
3 min
Flash sent
every 3 min
2
EARTH

So the 6-minute flashes emitted by the ship are seen to be 3 minutes apart on the planet it is approaching, but 12 minutes apart on the planet from which it is receding. This reciprocal relationship with light holds for all speeds. If the ship were traveling faster so that the time intervals between flashes were observed at 1/3 or 1/4 their rate for approach, then they would be correspondingly spread out to 3 or 4 times for recession. This simple relationship does not hold for sound waves (recall WHICH IS MOVING?).

TRIP OUT

Suppose our rocket ship departs from the earth at noon and travels at the same high speed for one hour, rocket time. During this hour it emits a flash every 6 minutes, ten in all. An earth observer sees these flashes at 12-minute intervals. When the 10th flash is emitted clocks aboard the rocket ship read 1:00. When the 10th flash is received on earth, clocks on earth read

a) 1:00 o'clock
b) 1:30 o'clock
c) 2:00 o'clock
d) 2:30 o'clock

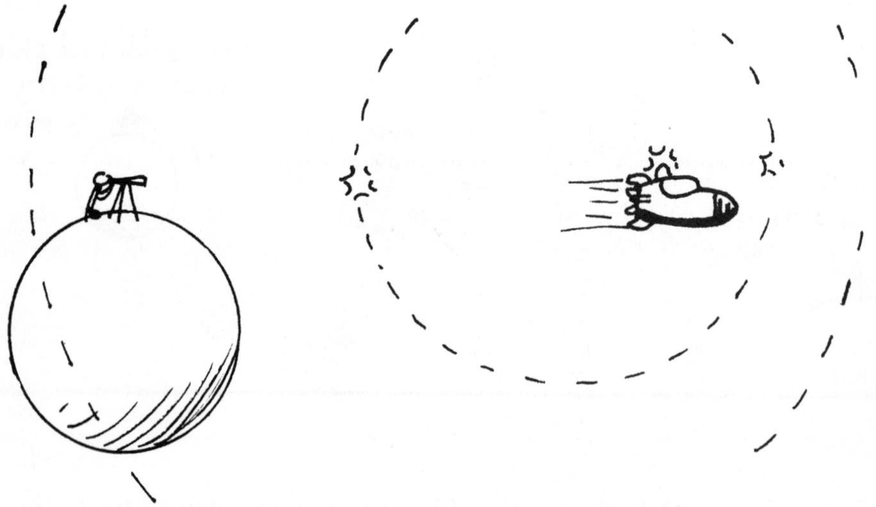

ANSWER: TRIP OUT

The answer is: c, 2:00 o'clock. This may not seem too unusual, for we know that it takes time for the flashes to reach the earth. And the ten flashes are seen at 12-minute intervals on earth, so $10 \times 12 = 120$ minutes $= 2$ hours. Read on.

ROUND TRIP

Suppose our rocket ship is able to abruptly turn around when it emits its 10th flash and then return to earth at the same high speed. It continues sending flashes every 6 minutes and emits 10 in its hour of return. But these flashes are seen at 3-minute intervals on earth. So although a clock aboard the rocket ship will read 2:00 o'clock when the ship gets back to the earth (1 hour out and 1 hour back), clocks on earth will read

a) 2:00 o'clock also
b) 2:30 o'clock
c) none of these

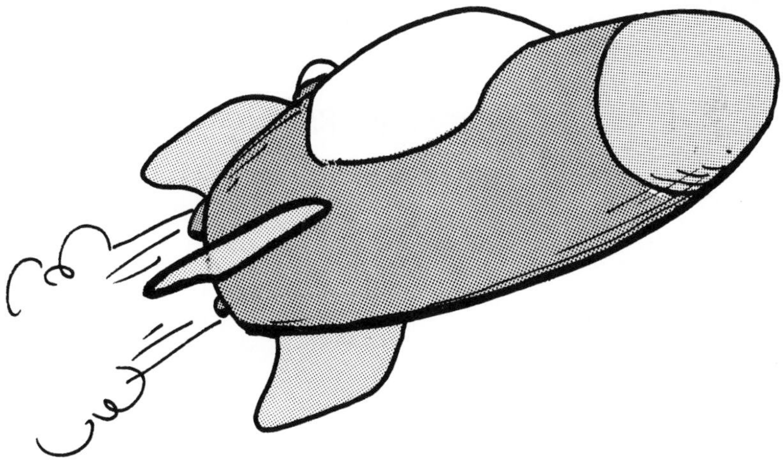

ANSWER: ROUND TRIP

The answer is: b. A person in the high-speed rocket ship will age only 2 hours while those on earth age 2½ hours! If the ship traveled faster, the differences in time would be even more. At $0.87c$, for example, the 2 hours on the ship would appear to take 4 hours from the earth frame of reference; and at $0.995c$, 20 hours. For everyday speeds the time differences are tiny—but there. This is time dilation. One cannot travel through space without altering time. A space traveler is also a time traveler. Two people can be at the same place in space-time, but if one moves away and then returns to the same space, it is at the expense of time.

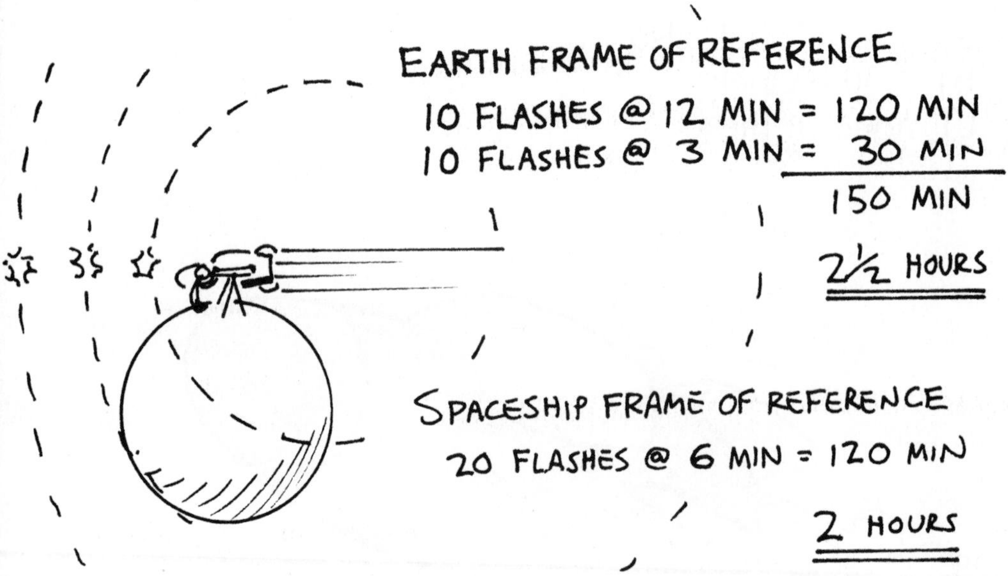

BIOLOGICAL TIME

There are all kinds of clocks: sand clocks, electric clocks, mechanical clocks, light clocks and biological clocks. Now just because motion can be shown to reduce the speed of one kind of clock must it necessarily affect all kinds of clocks equally?

a) Yes, it must affect all clocks equally
b) No, it need not affect all

ANSWER: BIOLOGICAL TIME

Suppose two different kinds of clocks are adjusted to run at the same speed and then sealed in a box. The box is now set in uniform motion. If that motion affects one clock more than the other a person riding inside the sealed box would notice the difference between clocks. The person sealed inside would then have a way to know the box was moving! This would violate a basic tenet of relativity: that there is no way a person in a closed box can distinguish between the states of rest and of uniform motion. So if one clock slows, all clocks, even your body's biological clock, must slow—and slow by exactly the same amount.

STRONG BOX

Suppose an atomic bomb was exploded in a box that was strong enough to contain all the energy released by the bomb. After the explosion the box would weigh

a) more than before the explosion
b) less than before
c) the same as it did before

ANSWER: STRONG BOX

The answer is: c. The atomic bomb turns part of its mass into energy. So the bomb, or what is left of it, weighs less after the explosion. But don't forget about the energy; it too has mass. How much? The energy has exactly as much mass as the bomb lost and all that energy is stuck inside the box. So the energy's mass weighs in with the bomb's mass and that's why the total weight of the box cannot be changed by the explosion.

LORD KELVIN'S VISION

Over a century ago, there lived a physicist named Lord Kelvin, who was Maxwell's teacher and whom we remember when we write Absolute Temperature as 273 K (273 Kelvin).

Lord Kelvin visualized all space filled with an invisible substance called "aether" and by means of this aether he accounted for the existence of both light and matter. At rest, aether was absolutely nothing but the empty vacuum of space. But set into vibration, the vibrations would spread through the aether like sound waves spread through the air. The aether waves were light waves. The atoms of matter were nothing more than tiny vortices in the aether, like a smoke ring is a vortex in air. The aether was frictionless, so once set in rotation the vortices would spin and therefore hold together forever.

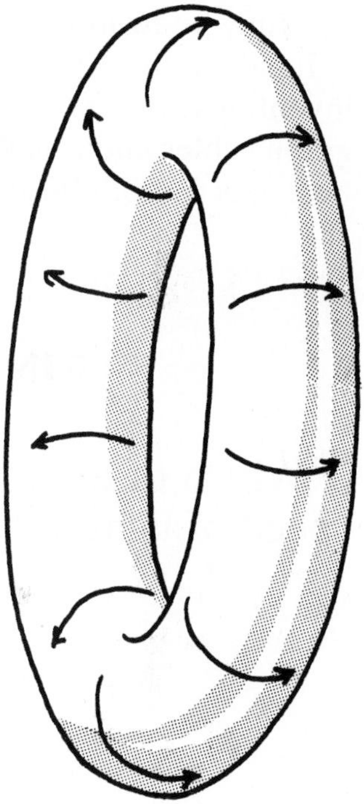

Now pretend Kelvin's vision was correct and atoms were vortices in the aether and you could disrupt and destroy one of the vortex atoms. Would you expect the disruption to result in the release of kinetic energy?

a) Yes b) No

ANSWER: LORD KELVIN'S VISION

The answer is: a. The vortex atom contains kinetic energy just like a flywheel contains kinetic energy. If disrupted, that energy would have to go someplace. So the release of energy from the destruction of matter was, in fact, foreseen over a century before the advent of the atomic bomb.

Lord Kelvin tried to explain all things in the universe as different manifestations of states of one underlying thing—and that is still the goal of physics today. Kelvin's reach exceeded his grasp, but he certainly knew in which direction to reach.

Incidentally, Kelvin was a very practical fellow who made many inventions: devices for ships' compasses, devices for undersea telegraph cables and calculators and on and on. Lord Kelvin was a practical man in both business and science.

EINSTEIN'S DILEMMA

Which of the following statements is correct? The speed of light in free space

a) is always constant
b) is slower in some places than others—the speed of light, therefore, is not always constant.

ANSWER: EINSTEIN'S DILEMMA

It seems almost blasphemous to say it, but the answer is: b. Gravity can bend a light beam and make it curve down. So while the lower side of the beam goes on the inside of the curve from "L" to "l," the high side of the beam goes on the outside of the curve from "H" to "h." The low side has a shorter run than the high side because "L" to "l" is shorter than "H" to "h."

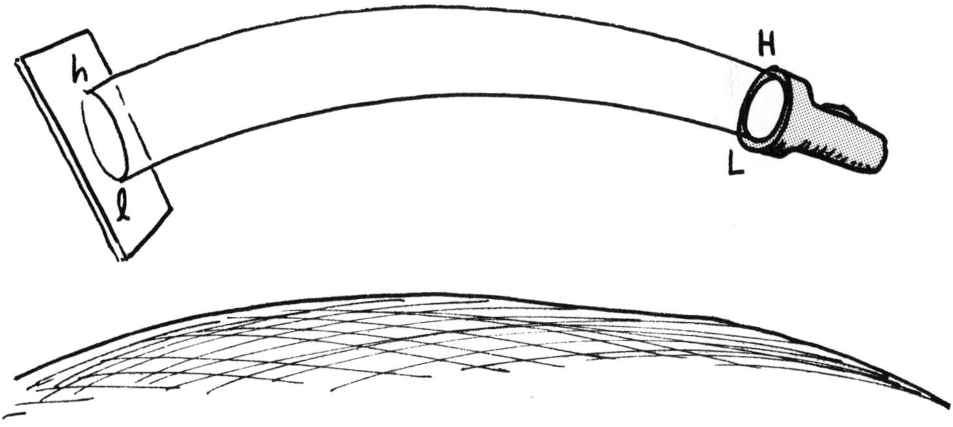

If the beam is to leave the flashlight and arrive at the paper in one piece, the light must move slower on the low side than on the high side.

When Einstein said this his detractors really had a field day. For various reasons there were a lot of people who were hacked off at Einstein, some for political reasons and some because they did not understand his theory even though they knew a lot of physics. Einstein had at first made a big deal out of saying the speed of light was constant, but now he had to say that the speed was not always constant. So he was open for a little disparagement and he caught a bit of flack.

To make things clear for those who would think and listen, Einstein explained it like this: In parts of space where there is no gravity, or so little gravity that you can forget it, you have a special simple case—the SPECIAL theory of relativity. In the SPECIAL theory of relativity, the speed of light is constant. But, in general, there is gravity in space and if you don't want to forget gravity you have a more complicated case—the GENERAL theory of relativity. In the GENERAL theory of relativity, the speed of light is not constant. In the general theory the speed of light is reduced as you get closer to the earth or any other large mass. Therefore, the low side of the beam went slower than the top side because it was closer to the earth.

Now we usually talk about the speed of light in empty space. Of course, in glass or water the speed of light is reduced. That gave some people the idea that the empty space around mass acts as if it had glass or water in it. Some people like to picture it that way, but most people like to picture it another way, as will be explained in TIME WARP.

TIME WARP

Which of the following statements is correct?

a) There are places in space where, even at rest, time is known to run slow.
b) There are no known rest places in space where time runs slow.

ANSWER: TIME WARP

Just like in the science fiction stories, the answer is: a. Einstein said we can think of gravity in the following way. Things don't really fall down—the floor comes up! In fact, the floor accelerates up, like the floor in an accelerating spacecraft.

When a spacecraft in space, away from gravity, fires its rockets, everything in the spacecraft seems to fall to the rear of the ship. The rocket's acceleration makes artificial gravity. Now Einstein says there is no difference between the effects of artificial gravity and real gravity. (Of course, the artificial gravity can last only until the rocket exhausts its fuel.)

Off hand, there seems to be no difference between artificial and real gravity, but think about this: two light flashes are sent from the bottom to the top of the spacecraft. Let's look at the animation of this shown below. The frames in the animation are, say, one second apart.

The distance moved by the rocket between frames is increasing because it is accelerating. Flash "A" starts in frame one and flash "B" starts in frame two exactly one second later. The distance moved by the light is constant. Flash "A" arrives at the top in frame three and flash "B" arrives at the top in frame six.

So the flashes that STARTED one second apart ARRIVED three seconds apart. Suppose there were a long string of flashes starting one second apart. They would arrive as a long string of flashes three seconds apart. The frequency of arrival is lower than the starting frequency.

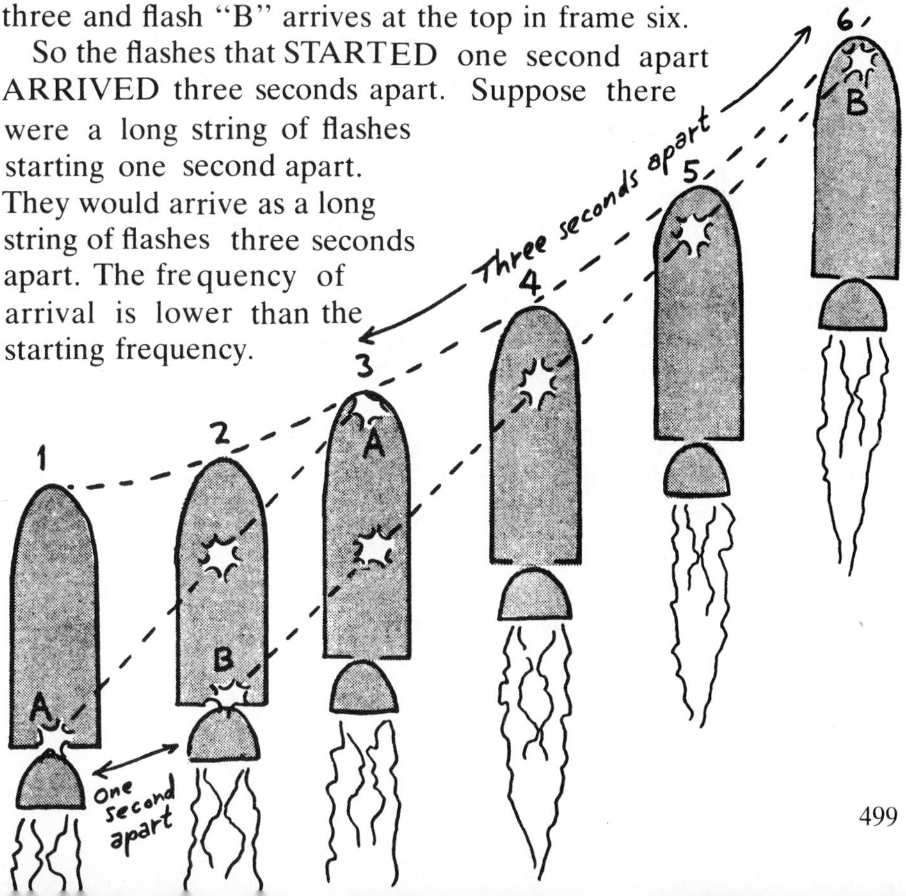

Now according to Einstein, the artificial spacecraft gravity and real gravity are equivalent to each other. If they are equivalent, light flashes starting at the bottom of a tower should arrive at the top of the tower at a lower frequency than they had to start with.

For example, if you send flashes at the rate of 1000 per second from the bottom they should arrive at the top at a lower rate or frequency, say, 999 per second. But that is difficult to believe. Where did the missing flash go? One thousand started at the bottom during any second, but only 999 arrive at the top during any second. Something must be eating one flash per second! But, of course, nothing could be eating flashes.

And now for the stroke of genius. Einstein realized that the only reason the frequency at the top could differ from the frequency at the bottom would be if the clock at the top ran at a different speed than the clock at the bottom. For example, by a half-speed clock there would be two sunsets in twenty-four hours; by a quarter-speed clock there would be four sunsets in twenty-four hours. If the frequency of flashes at the bottom of the tower is larger than the frequency at the top it is because the bottom clock is running slower than the top clock.

Gravity makes time run slow. Mass makes gravity. Mass makes time go slow. So near masses, time runs slower than it does in parts of space away from mass. Your feet age more slowly than your head! How slow can it go? If you have enough mass, you can make time stand still.

If time goes slow, everything goes slow—even light. That explains why light goes slow near mass (remember EINSTEIN'S DILEMMA). So, if you like, you can make light go slow without changing the speed of light. You change the speed of time. If there is enough mass to make time stand still, then light also stands still—paralyzed—in effect, trapped. No light can get out of these mass traps. They are called "black holes."

Black holes exist in theory. Some scientists think they also exist in reality. Einstein himself thought they could not exist in reality. The answer, one way or the other, will likely be found during your lifetime.

But returning to the question: are there places in space where time goes slow? The answer is definitely "yes." You are in such a place right now.

E = MC²

The celebrated equation $E=mc^2$ or $m=E/c^2$ (c is the speed of light) tells us how much mass loss, m, must be suffered by a nuclear reactor in order to generate a given amount of energy, E. Which of the following statements is correct?

a) The same equation, $E=mc^2$ or $m=E/c^2$, also tells us how much mass loss, m, must be suffered by a flashlight battery when the flashlight puts out a given amount of energy, E.

b) The equation $E=mc^2$ applies to nuclear energy in a reactor, but not to chemical energy in a battery.

ANSWER: E = MC²

The answer is: a. If the mass-energy equation, $E=mc^2$, applies to any one form of energy, such as nuclear energy, then it must apply to every kind of energy, including battery energy. It is not hard to see why. Seal a nuclear reactor and a battery in a box. Nothing can enter or leave the box. Now let the reactor put out electric energy and let that energy be put into the battery. As the reactor puts out energy it must lose mass. But no mass can get out of the sealed box. So where could the reactor's lost mass be? The only other place it could be is in the battery. So the battery gains mass as it gains energy, and the battery loses mass as it puts out its energy. Whatever receives the battery's energy also receives some of the battery's mass.

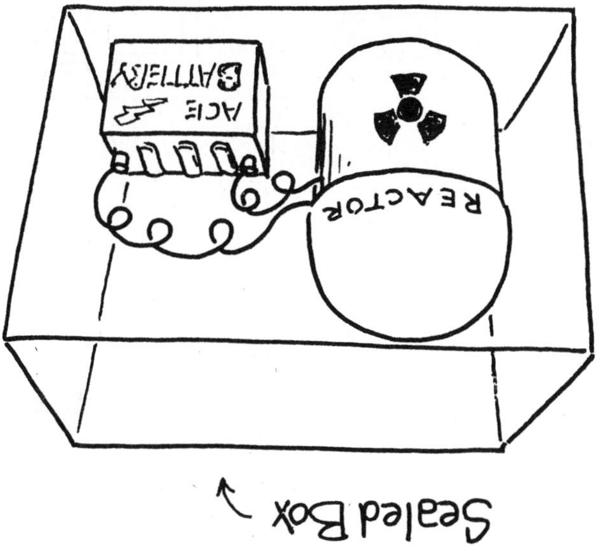

RELATIVISTIC BIKE AND TROLLEY

Consider a motorcycle powered with super-powerful electric batteries, and a common electric streetcar, that are each driven to speeds approaching the speed of light. Measurements of each from our rest frame of reference will indicate an increase in the mass of the

a) motorcycle
b) streetcar
c) both
d) neither

ANSWER: RELATIVISTIC BIKE AND TROLLEY

The answer is: b, despite the widespread misconception among relativity buffs that the mass of a moving thing always increases, going to infinity as the speed of the thing approaches the speed of light. It so happens that the mass of a thing increases not if speed is added to it, but only if *energy* is added to it.

Energy is poured into the streetcar from the powerhouse through the trolley wire. But the motorcycle carries its own energy supply with it. While new energy is added to the streetcar, no new energy is added to the motorcycle. Energy has inertia. So the mass of the streetcar increases with speed while the mass of the motorcycle remains unchanged whatever its speed.

Interestingly enough, all the mass gained by the streetcar is compensated by a like decrease in mass at the power source. If the streetcar gains a thousand kilograms then the mass of fuel and its products at the power plant is minus one thousand kilograms! And with the motorcycle, any gain in mass of the bike and rider is compensated by a like decrease of mass of the battery so there is no net change in mass.

So the mass of all things does not go to infinity simply because their speed goes to the speed of light. After all, light moves at the speed of light and its mass is certainly not infinite.

MORE QUESTIONS (WITHOUT EXPLANATIONS)

You're on your own with these questions that parallel those on the preceding pages. Think Physics!

1. A spaceship travels away from its station at ¾ c. It fires a rocket at ¾ c in a direction directly away from the station. With respect to the space station, the rocket moves at speed
a) less than ¾ c b) ¾ c c) greater than ¾ c but less than c d) 1½ c

2. Nonmaterial things like shadows often exceed the speed of light.
a) True b) False

3. A train measures 110 meters long at rest. A tunnel measures 100 meters long at rest. At low speeds there is no point at which the train is completely inside the tunnel, but at relativistic speeds the train could be seen to be completely within the tunnel from the frame of reference of

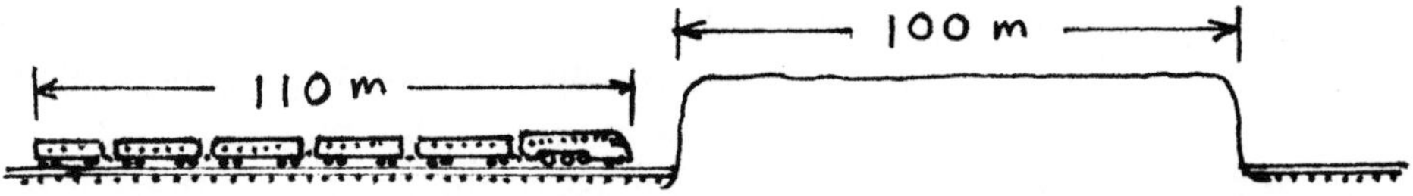

a) the tunnel b) the moving train c) both d) neither

4. When a blinking source of light approaches you at high speed, you measure an increase in the
a) frequency of its light
b) speed of its light
c) both
d) neither

5. According to Einstein's idea of gravity, a remote observer viewing light passing a very massive body would see the speed of light undergo
a) an increase b) a decrease c) no change whatsoever

6. Strictly speaking, from the point of view of a person on the ground floor of a tall skyscraper, a person at the top ages
a) slower b) faster c) the same

Quanta

There is an idea that has been around for thousands of years which has very gradually gained verification. The idea is that this world we live in is the result of another world. A world that lies beneath everything, out of sight because it is too small to see. According to this idea we all are made of very small things called corpuscles or molecules or atoms or nucleons or quarks. It seems that everything, even energy, light, and electricity comes in little packages called quanta.

The dream of physics is that if we could understand how the quanta work—and if the world is just made of quanta—then we could understand how the whole world works.

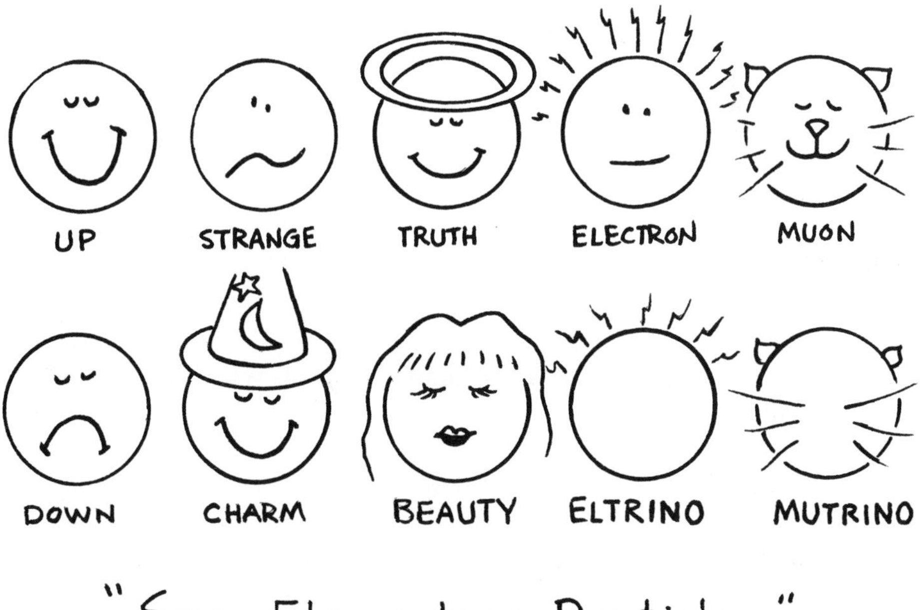

"Some Elementary Particles"

Things which are, were not made of things which do appear.
--- Bible

THE BONES OF DEAD THEORIES

It is said, "Science is built of the bones of dead theories." We say, for example, that it is possible to tell with absolute certainty

a) how an atom looks
b) how an atom does not look
c) both
d) neither

ANSWER: THE BONES OF DEAD THEORIES

The answer is: b. You can't help but get the impression by reading science books that scientists (or science book writers) think they know everything about how the world works. But that is a false impression. All they really know for sure is how the world does NOT work. Why? Because science is not like geometry where proofs are based on logic. In science, the proofs are ultimately based on experiments. The laboratory is the Supreme Court of Science.

Now suppose you think and observe a certain thing to be a square. Does that mean the thing really is a square? No. Under higher magnifying power it might be found it is only approximately a square. So you can never be certain it is a square, but you can be certain it is not a circle.

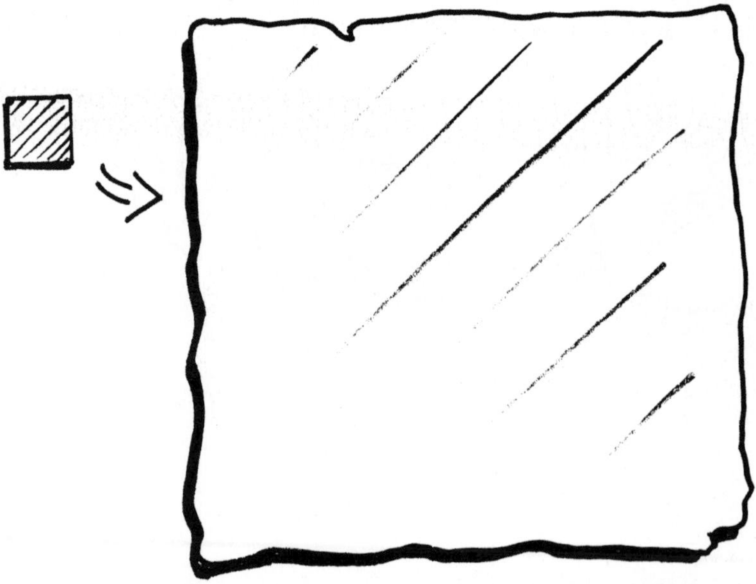

Ideas can be disproved (killed and turned to bones) with certainty, but no idea can live with certainty. No one knows for certain how an atom looks, but everyone knows (for certain) it does not look like a cat.

COSMIC RAYS

Starlight rains down from the night sky. Cosmic rays also rain down from the night sky. The total rain of cosmic ray energy from the sky is

a) much less than
b) about equal to
c) much more than

the total rain of starlight energy from the night sky.

ANSWER: COSMIC RAYS

The answer is: b. So what! So why are poems written about starlight, but seldom about cosmic rays? Because we take the universe to be what we can see. But how much do we really see?

SMALLER AND SMALLER

Which is smaller?

a) an atom
b) a light wave
c) both are about the same size

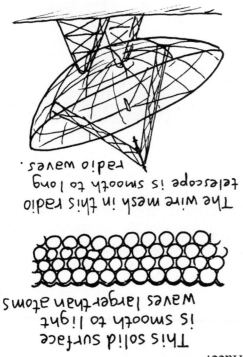

ANSWER: SMALLER AND SMALLER

The answer is: a. How do we know? Because someone made measurements in a laboratory? No! We can tell that atoms must be much smaller than light waves by reflection. If the atoms were not much smaller it would be impossible to make a surface smooth enough to give a good reflection. If a reflection is clear, the wavelength of the reflected wave must be much larger than the size of any bumps on the reflecting surface.

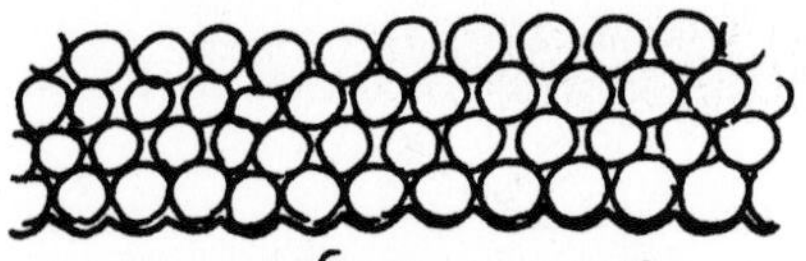

Incidentally, this means that if you are making a telescope mirror for use in the ultraviolet it must be made with a smoother surface than would be required for a visible light telescope. On the other hand, an infrared telescope can get by with a rougher mirror than could be accepted in a regular telescope. When you get to radio waves the mirror surface can be so rough that chicken wire mesh will serve as an acceptable reflector.

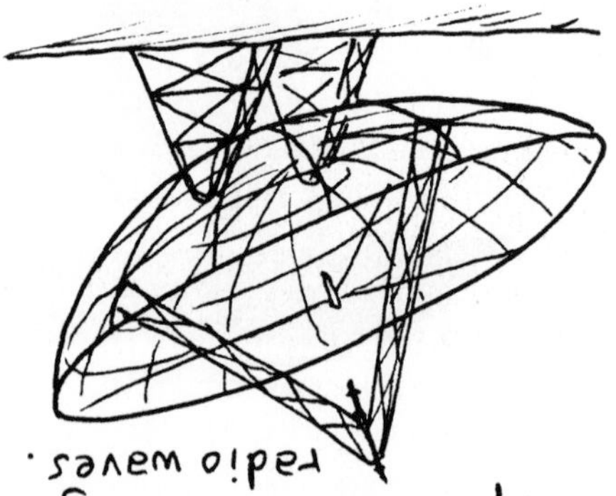

The wire mesh in this radio
telescope is smooth to long
radio waves.

When the paint on an automobile begins to go bad the reflection of the sun from the surface gets a reddish cast. Why? The surface is bad because it has become rough and pitted. This makes it hard for the small blue waves to be effectively reflected while the larger red waves do not suffer as much. This is a nice "sidewalk" demonstration that blue waves are smaller than red waves.

RED HOT

Vega is a blue star. Antares is a red star. Which is hotter?

a) Vega
b) Antares

I see a red neon sign over the corner tavern. Is the neon in the sign as hot as Antares?

a) Yes
b) No.

ANSWER: RED HOT

The answer to the first question is: a. When a solid object is first heated it glows red, then as its temperature continues to go up it glows orange, then yellow, then white. Finally, as its temperature goes still higher it glows blue. Steelmakers use the color of the molten steel to measure its temperature. Glowing gas under very high pressure changes its color with temperature in the same way a glowing molten solid would change color.

The answer to the second question is: b. If the neon sign were as hot as Antares the sign would melt, but in reality you could touch the neon sign tube and it would hardly be warm—so how can the sign glow red? Because what is glowing is not solid or under high pressure. What is glowing is low pressure (neon) gas. The low pressure gas is not putting out as much energy as a solid that glows with the same color. So the low pressure gas can be at a much lower temperature than the solid.

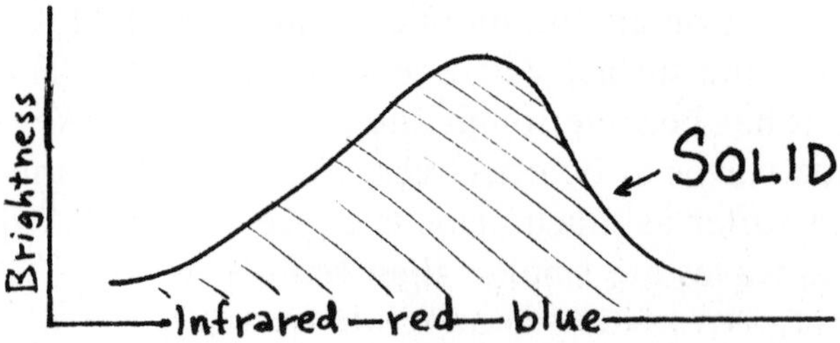

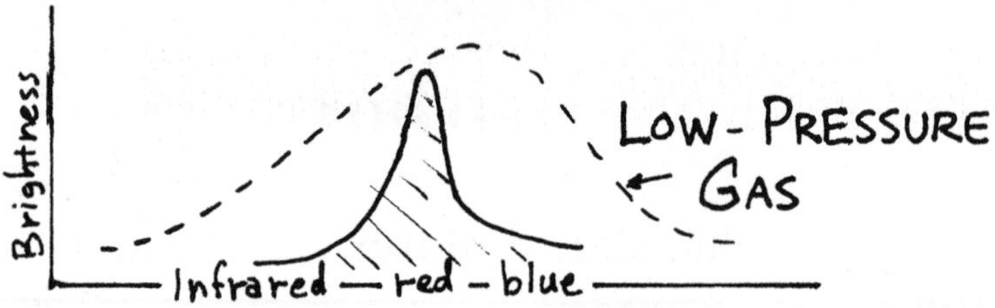

You can see this for yourself with a prism. If you look at a light through a prism it breaks the light up into all the colors it is made of. If you look at a red hot solid or star through a prism you will see ALL the colors. Red will be brightest, but all the other colors will be there along with red. If you look at a red neon sign through a prism you will see ONLY red (and perhaps some other colors very faintly). The low pressure neon gas emits less radiation (fewer colors) than the high pressure gas or solid. So the neon can do its job and be "cool" about it.

LOST PERSONALITY

When light emitted by a glowing gas is passed through a thin slit and then through a prism, a line spectrum is produced. A continuous spectrum will be produced if the gas is

a) a mixture of several kinds of atoms
b) under low pressure
c) under high pressure
d) all of these
e) none of these

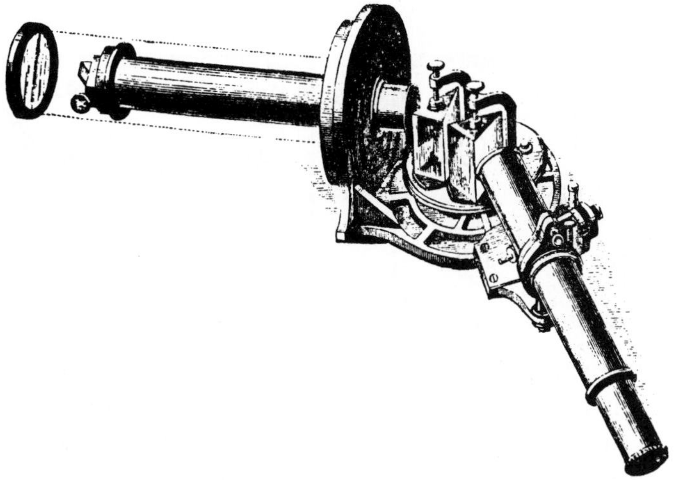

ANSWER: LOST PERSONALITY

The answer is: c. Left to itself, an atom allows its electrons to move in certain orbits. As the atom gains and loses energy its electrons jump back and forth between these allowed orbits. Each jump has a precise energy which means a photon of precisely some color (or frequency or wavelength) is radiated. So the atom (or atoms) when heated do not radiate all colors. They only radiate certain precise colors. This is called a "line spectrum" because you see only a few colored images of the slit through which the light is passed. These images are the lines of the line spectrum. The lines are separated by darkness.

Now suppose the atom is not left to itself. Suppose it is put under high pressure, which means it is crowded against other atoms. The atoms perturb (mess up) each other's orbits. All kinds of new misshapen orbits begin to exist. All kinds of new orbits means all kinds of new jumps. All kinds of new jumps means all kinds of new colors. Soon every color shows up—every color from red to violet—and you have a continuous spectrum. So gas under low pressure shows only a line spectrum, but the same gas in a high pressure star shows a continuous spectrum.

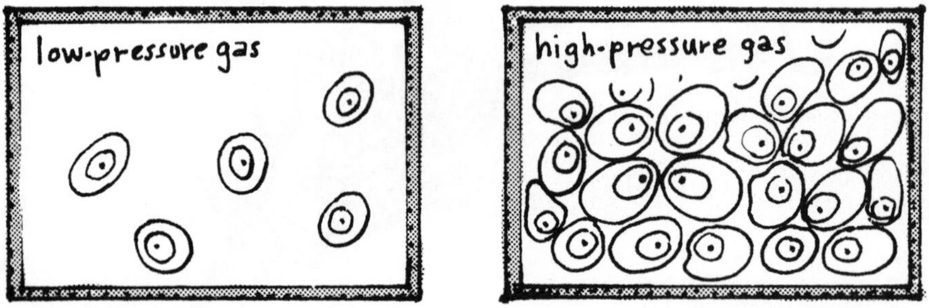

Exactly the same thing happens to bells. Hung separately each bell has its own frequency, color and personality, but crowded together they mess each other up. They lose their individual frequencies, colors and personalities. They don't even sound like bells.

In a low pressure gas, then, the individual personalities of the atoms can be seen. In a high pressure gas or solid the individual personalities of the atoms are lost.

ECONO LIGHT

Both an incandescent light bulb and a fluorescent tube are rated for 40 watts. Which puts out the most light?

a) The light bulb
b) The fluorescent tube
c) Both are equally bright

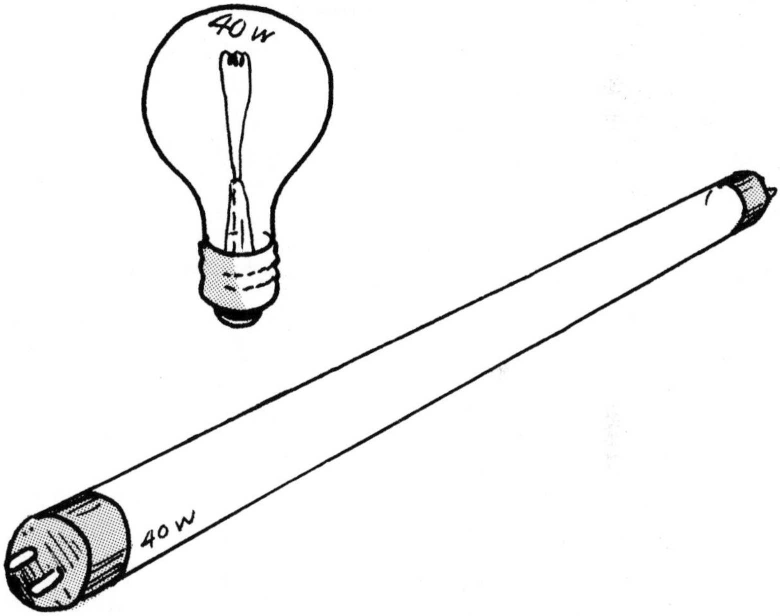

ANSWER: ECONO LIGHT

The answer is: b. This is because the bulb puts out more heat. You can put your hand on the illuminated tube without harm, but the illuminated bulb will burn you very quickly. It takes power to make this heat. The tube puts out roughly four times as much light as the bulb. So most of its energy goes directly into light.

Why this is so has to do with the difference between putting energy into a gas and putting it into a solid. In the gaseous state the atoms are relatively isolated, whereas in a solid they are crowded together. Consider the different effects you'd get in striking an isolated bell and striking a box crammed full of bells. From the isolated bell would come a nice clear ring. Most of the energy put into the bell would come out as a clear sound with a frequency characteristic of the bell. But not so with the box crammed with bells. Sound coming from the box would be indistinct, with many frequencies non-characteristic of any of the single bells. Acoustic people have a name for that kind of sound. They call it *white noise,* because it is a mixture of many frequencies, just as white light is a mixture of many colors.

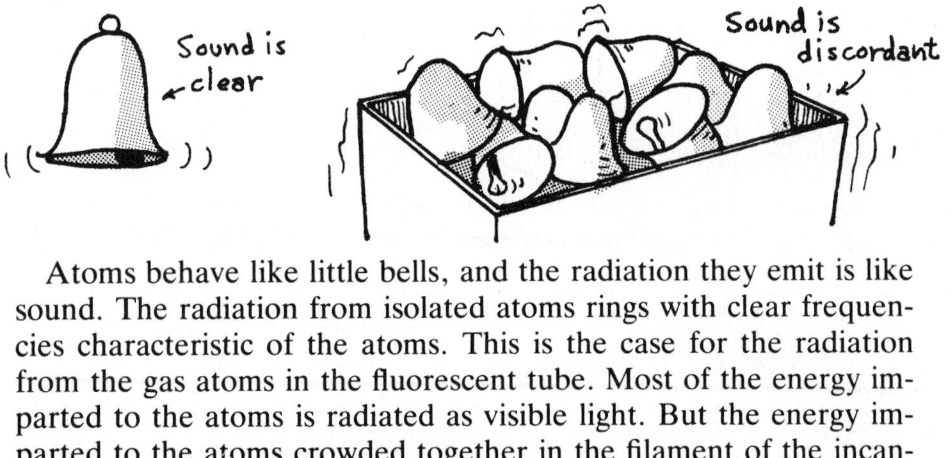

Atoms behave like little bells, and the radiation they emit is like sound. The radiation from isolated atoms rings with clear frequencies characteristic of the atoms. This is the case for the radiation from the gas atoms in the fluorescent tube. Most of the energy imparted to the atoms is radiated as visible light. But the energy imparted to the atoms crowded together in the filament of the incandescent lamp is only partly radiated as visible light. Most of it is emitted as infrared radiation, commonly called "heat radiation." Although it can cook for you, it can't help you to see.

If more electric current is pushed through the lamp filament it runs hotter and more heat and light are emitted. The percentage increase in light exceeds the percentage increase in heat so the bulb becomes more efficient. But it burns out sooner.

SAFE LIGHT

Recall that black-and-white photographic film is more sensitive to blue than red light (which is why the safelight in a photographer's darkroom is red). It then follows that there are

a) more photons in one joule of red light than in one joule of blue light
b) more photons in one joule of blue light than in one joule of red light
c) the same number of photons in one joule of red light and one joule of blue light

ANSWER: SAFE LIGHT

The answer is: a. Photographic film is exposed when one photon of light hits one molecule in the film and causes a chemical reaction.

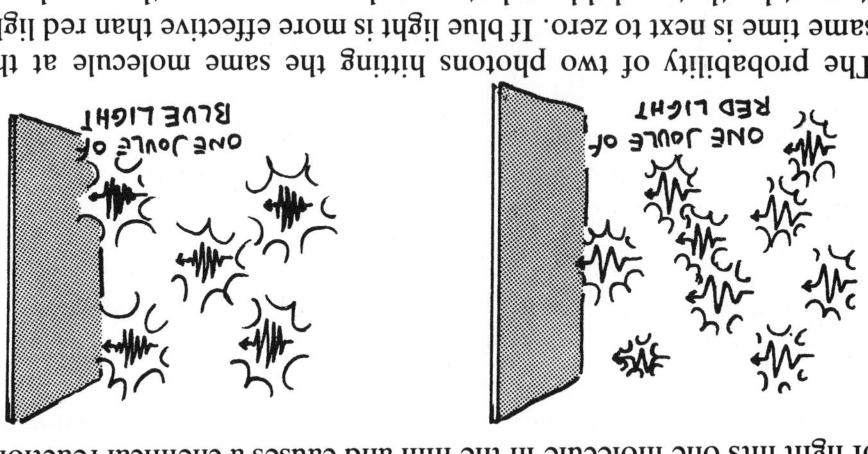

The probability of two photons hitting the same molecule at the same time is next to zero. If blue light is more effective than red light it must be that each blue photon packs more energy than each red photon. When a blue photon hits a film molecule it has what it takes to do the work. A red photon can't handle the job. Of course, the total amount of energy in one joule of red or blue light must be the same. So it must be that there are more photons in the one joule of red light, but each red photon carries less energy.

PHOTONS

Photons are little bundles of light energy. All photons contain the same amount of energy.

a) True
b) False

All yellow photons contain the same amount of energy.

a) True
b) False

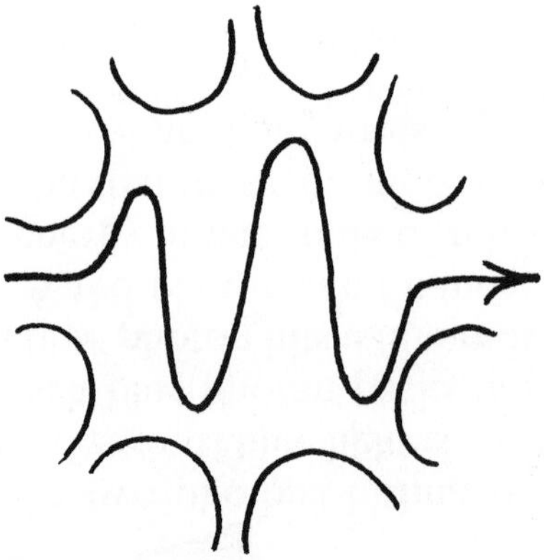

ANSWER: PHOTONS

The answer to the first question is: b. The energy in a photon depends on its color. Red photons have more energy than infrared photons, yellow have more energy than red, blue more than yellow, violet more than blue, ultraviolet more than violet.

The answer to the second question is: a. All yellow photons of the same hue contain some particular amount of energy—no more and no less.

PHOTON CLIP

A beam of yellow light can be cut in half, and each half appears yellow. Can a photon in the yellow beam be "cut in half," and if so, will it still appear yellow?

a) It can be cut in half and yes, it will appear yellow
b) It can be cut in half, but it will not appear yellow
c) It cannot be cut in half, and even if it could it would not appear yellow
d) It cannot be cut in half, but if it could it should appear yellow

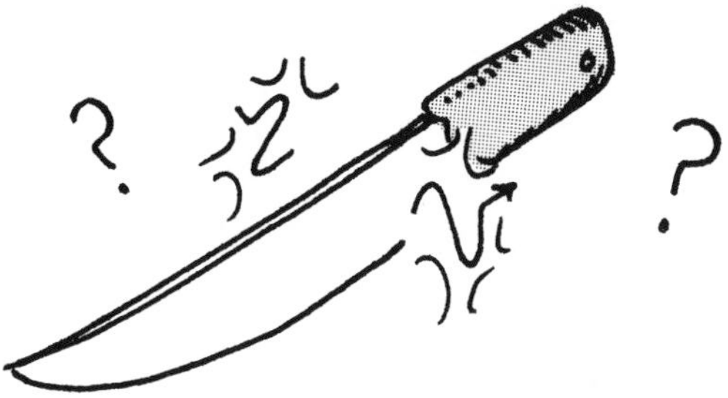

ANSWER: PHOTON CLIP

The answer is: c. If you try to cut a photon you will discover the photon is always on one side of the cut. You can't tell on which side it will be, but it will always be completely on one side of the cut or the other. Although you can't cut a photon, a photon can be absorbed and the energy re-emitted as a pair of photons—but of lower frequencies. This happens in fluorescent materials, where a molecule may absorb a single ultraviolet photon and then emit a pair of red photons. The energy in the two reds adds up to equal the energy in the one ultraviolet.

PHOTON PUNCH

Everyone knows light waves carry energy—that's how solar energy gets here from the sun. But do light waves carry momentum?

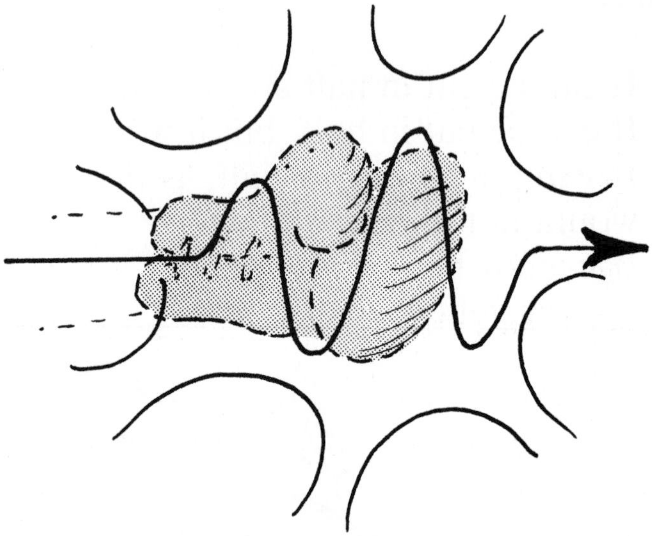

a) All waves which carry energy also carry momentum

b) Light wave energy is pure energy and so carries no momentum

c) Light waves carry momentum as well as energy

ANSWER: PHOTON PUNCH

The answer is: c. Not all waves carry momentum. In fact most waves carry zero net momentum. For example water waves carry energy but they will not push a cork floating on them. The cork bobs

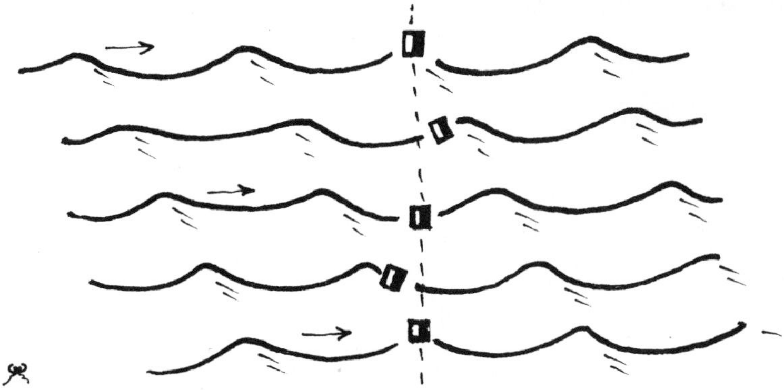

up and down and returns to its starting place. No net momentum gain. The same is true for sound waves.

But light waves are different. Indeed they are unique, for they do carry momentum. It is the momentum of sunlight that pushes the tails of comets away from the sun!

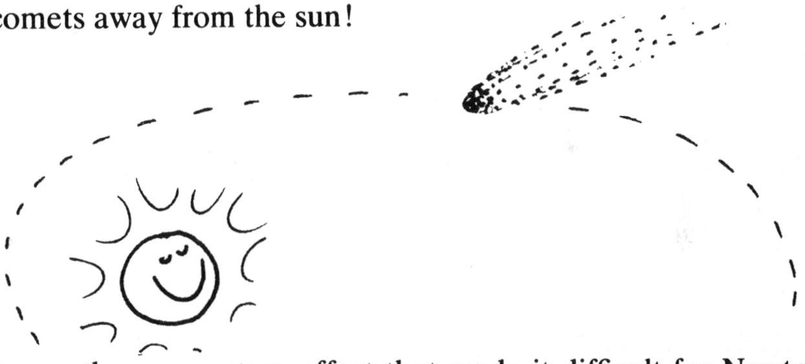

It was the momentum effect that made it difficult for Newton to believe light was waves and it was in part that same effect that made Einstein think of light as particles with mass, which he called photons. Einstein reasoned that if light can push things it must have momentum. Momentum is mass multiplied by velocity. Light therefore not only has velocity, but mass as well.

The force that light exerts on things is called radiation pressure. You might suppose the effect of radiation pressure is always to push things away from the sun. Astonishingly that is not so, as will be shown in the next question.

*The total momentum in a wave is the sum of the momenta in the individual parts of the wave, and disregarding second order effects, this sum is zero, except in the case of light waves. - Lewis Epstein.

SOLAR RADIATION PRESSURE

Can solar radiation pressure blow some things out of the solar system?

a) Yes
b) No

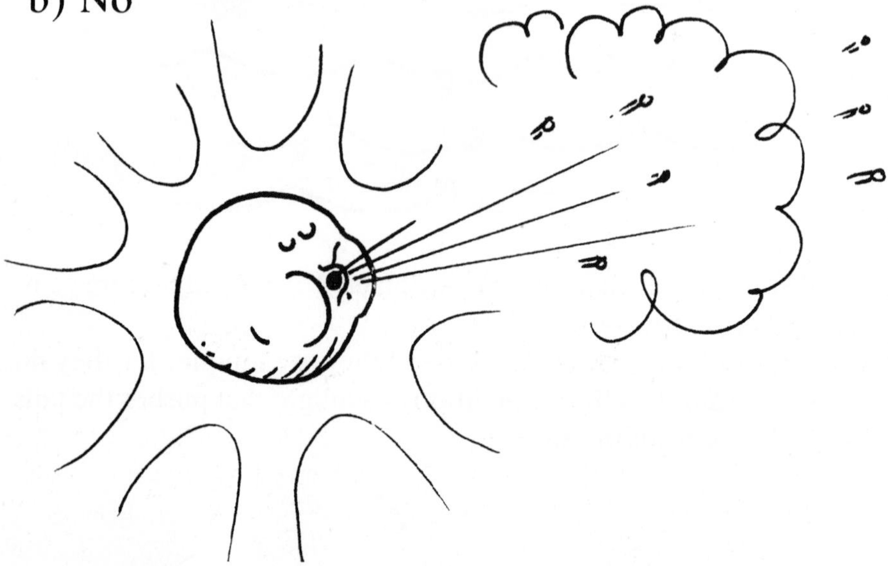

Can solar radiation pressure cause some things to fall into the sun?

a) Yes
b) No

ANSWER: SOLAR RADIATION PRESSURE

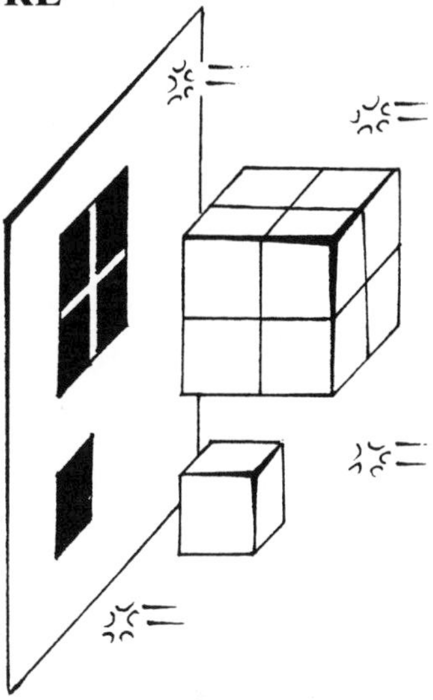

The answer to the first question is: a. Think about a little piece of dust in orbit around the sun. The sunlight pressure on it is proportional to its shadow area or silhouette, but the force of gravity on it is proportional to its mass, which is proportional to its volume. Small particles have more shadow area for EACH cubic inch of volume in them than large particles. That means, if a particle is very small, solar radiation pressure can overwhelm gravity—and that is why comet tails always point away from the sun. That is also why the wind can push small raindrops around (even blow them upward) while gravity rules large raindrops.

The answer to the second question is also a. That might seem to contradict the first part, but there is no contradiction. Suppose a particle is sufficiently large that the force of gravity on it is stronger than the force of radiation pressure—as is the case for all planets and asteroids. Then that particle is bound in the solar system. Now, as we see the particle moving around the sun it seems that sunlight is simply raining down on the particle. (If the orbit is a circle, the sunlight is coming in perpendicular to the particle's direction of motion.) As seen by the particle, however, things are a little different. Rain may fall vertically on a stationary car, but when the car moves, the rain as seen from the moving car comes from the forward direction. So also the sunlight, seen by the moving particle, comes from the forward direction. (Astronomers call this shift "aberration.") The radiation pressure has a component pushing against the orbital motion of the particle. Slowly but inexorably, the particle loses its orbital speed and spirals down into the sun. This is called the Poynting-Robertson Effect—the vacuum cleaner of the solar system!

WHAT'S IN THE OVEN?

Which would most likely be contained in the warm oven in your stove?

a) Two-meter radio waves
b) Two-millimeter radio waves
c) Both
d) Neither

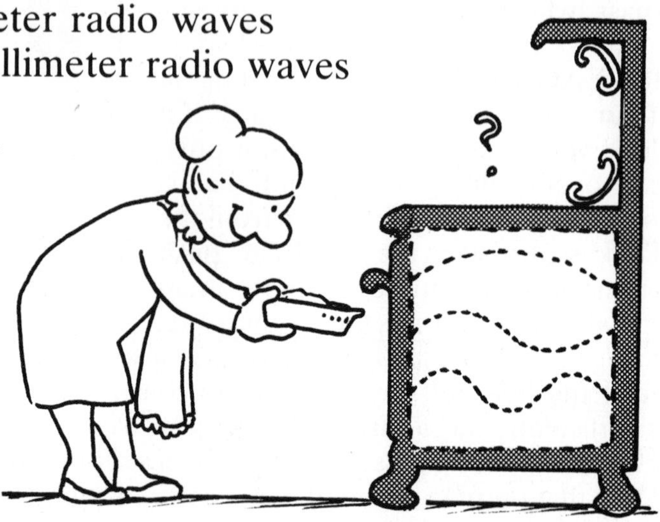

ANSWER: WHAT'S IN THE OVEN?

The answer is: b. The radio wave must literally fit inside the oven and there is no way a two-meter wave could fit into a regular-size home oven.

You could and would find two-millimeter waves inside home ovens—not only in microwave ovens, but in ordinary gas or even wood-fired ovens. The energy in an oven is mostly infrared radiation. Infrared waves are shorter than radio waves, but longer than light waves. Now you must not think there are a lot of two-millimeter waves in the oven, but there are some. The important idea is that there could be some waves of every size in the oven that could fit in there. At first you might think this sounds reasonable, but later you might think: "X-rays are very small waves so they could fit in the oven." Are there any x-rays in a home oven? There are certainly some microwaves in an oven—but asking for x-rays is too much. Why does the logic not hold up here? Why are there no x-rays in a home oven??? We shall see why shortly.

ULTRAVIOLET CATASTROPHE

Suppose you generate a big, long wave by momentarily sloshing a plank in a tank of water, as shown in the sketch. If you don't disturb the water any more, after a while you will find the big, long wave has become

a) a bigger, longer wave
b) many small, shorter waves (ripples)

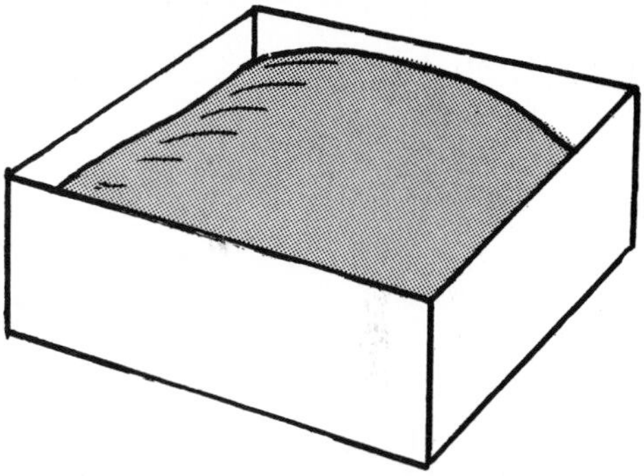

A

If light waves behaved like water waves and you put some yellow light in a tank, after a while you would find the yellow light

a) turned blue
b) turned red
c) remained yellow

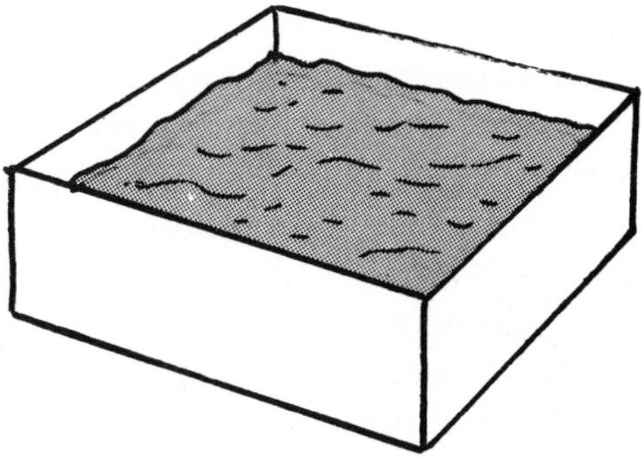

B

Do light waves in a tank in fact behave like water waves in a tank?

a) Yes b) No

ANSWER: ULTRAVIOLET CATASTROPHE

The answer to the first question is: b. Everyone knows the answer to this question from actual experience, but why does the long wave turn into many short ripples? Because the wave energy gets divided up between all the *possible* kinds of waves that can fit in the tank, and there are many, many more short wavelength kinds than long wavelength kinds that can fit in. Theoretically, there are an infinite number of wave kinds that can fit in—most of them exceedingly short.

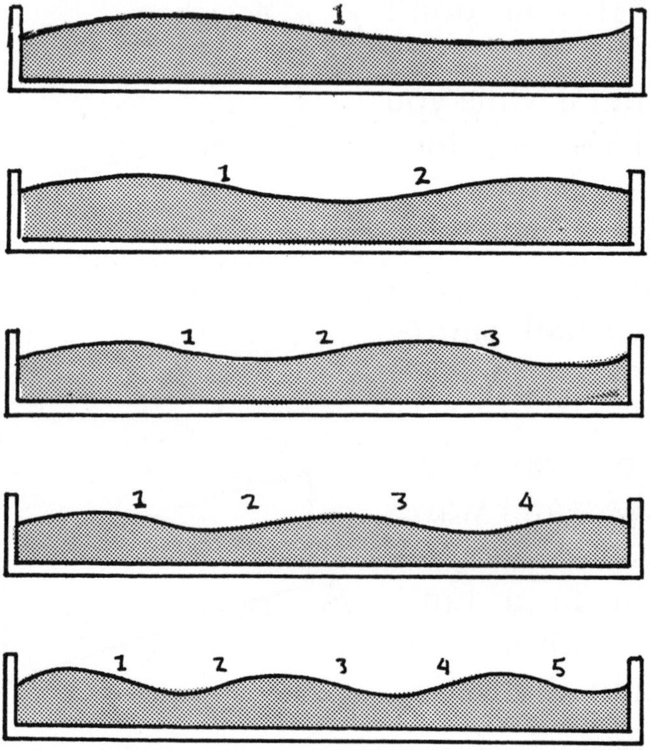

The answer to the second question is: a. Long water waves boxed in a tank turn into short ripple waves. If yellow light waves got shorter they would turn blue. (Red light waves are longer than yellow.) The story would not end there, however, because the blue waves, like the yellow waves, get still shorter and become violet, then ultraviolet, then x-rays.

The answer to the last question must be: b. If you keep on feeding more and more yellow light into a box, allowing none to escape, the inside of the box will become a yellow-hot oven. Now if you shut off the supply of yellow light and simply leave the oven sealed, and *if* the light waves act like water waves, the yellow light would turn blue, then violet, then ultraviolet, etc. All the heat in your home oven would turn to ultraviolet radiation. All the heat and light in the sun would turn to ultraviolet radiation. All the heat and light in the universe would turn to ultraviolet radiation. There would be an ultraviolet catastrophe! (And soon, an x-ray catastrophe.)

Now this ultraviolet catastrophe does not actually happen. Why not? Why does the yellow light wave energy not divide itself equally among all the different sizes of waves like the water wave energy does? First, you must appreciate that the yellow light wave energy does divide itself up into some waves of different sizes. You can see that it has divided itself into some longer waves—red light, and some shorter waves—blue light, by looking at a yellow-hot thing (like the sun) through a prism and seeing red and blue light in addition to the dominant yellow. However, even though some short blue waves creep in, most of the light remains yellow. Why? Because it is difficult to make blue light and more difficult to make ultraviolet. There is a minimum energy required to make a light wave of any particular color or wavelength. The energy in the wave must at least equal the energy in one photon and blue photons require more energy than red. (Remember SAFE LIGHT and PHOTON CLIP.) The dividing up of energy is just like dividing your money up between red, yellow and blue chips at a casino. You can't afford too many blue chips.

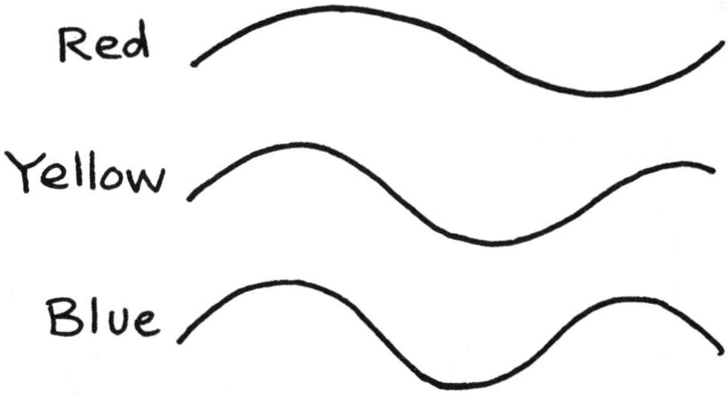

So now you know why a yellow-hot oven (or kiln) remains yellow. The yellow waves cannot become very long radio waves because they will not fit in the oven and the yellow waves cannot become very short ultraviolet or x-ray waves because short waves require so much energy that there would not be enough energy to go around.

DIFFRACTION OR NOT

A beam of photons behaves as a wave, as is evidenced by the phenomenon of diffraction (bending around corners). A beam of particles also behaves as a wave. We call a beam of particles a matter wave. Which of the following statements is correct?

a) All waves suffer diffraction (spread out) on passing through a hole
b) Only matter waves suffer diffraction on passing through a hole
c) Only matter waves do NOT suffer diffraction on passing through a hole

ANSWER: DIFFRACTION OR NOT

The answer is: a. Waves always spread out—diffract—when they go through small holes. If particles did not have wave properties,

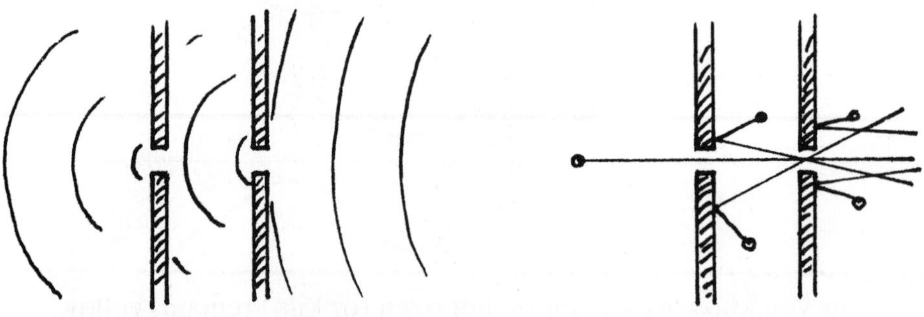

they would go perfectly straight. We can see water waves, but there are many kinds of waves we cannot see: sound waves and light waves, for example. We infer these "invisible" things are waves BY THE WAY THEY ACT. The telltale act we look for is the spreading out.

UNCERTAINTY ABOUT UNCERTAINTY

According to Heisenberg's Uncertainty Principle there must always be an uncertainty h in the

a) momentum of a particle
b) energy of a particle
c) location of a particle in space
d) lifetime of a particle
e) . . . none of the above

ANSWER: UNCERTAINTY ABOUT UNCERTAINTY

The answer is: e. (This builds on the idea introduced in CLIP.)

The Uncertainty Principle comes about because of the wavelike properties of "particles" like electrons and protons. The waves tell the dynamics of the particle—its momentum, its energy and even its angular momentum. The wave runs through space and time. The wavelength of the wave running through space gives the particle's momentum. The frequency of the wave running through time gives the particle's energy. But a wave can't really represent a particle! A particle is located at only one place in space. A wave is not located at only one place. The conflict between a wave and a particle can never be resolved. It can only be compromised and it is compromised by the Uncertainty Principle. The compromise goes like this: If you get a wave that will not run on forever, but will just wave in one location and then kill itself, it will be like a particle. This kind of wave is called a "wave packet."

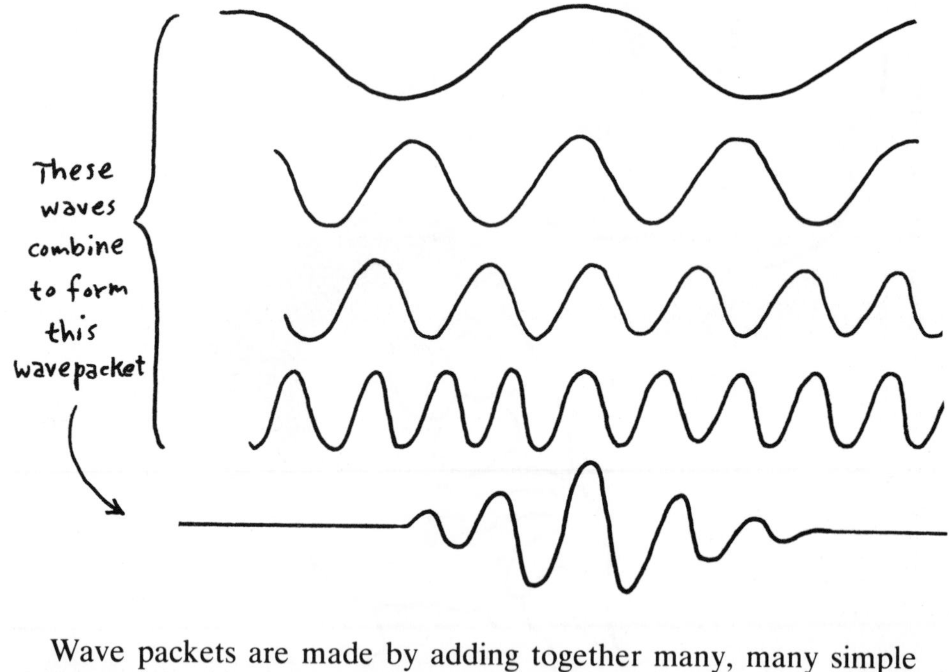
These waves combine to form this wavepacket

Wave packets are made by adding together many, many simple (harmonic) waves that run forever. The effect of adding many waves is that they cancel one another. They do this because the waves have different wavelengths or frequencies so some are in phase while others are not. But they should not cancel each other completely. At one special place, the location where you want the particle to be, all the waves should be in phase. That way the added effect of the waves is constructive in that one special location—and destructive everywhere else.

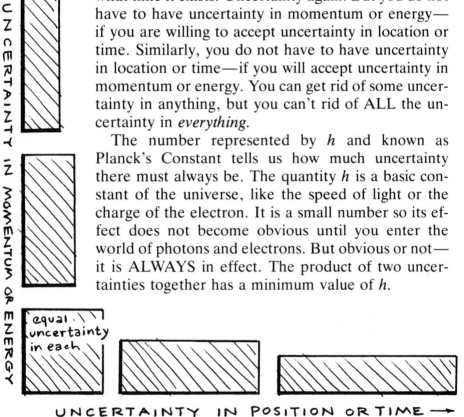

So it takes a MIXTURE of different waves to make a wave packet which we call a particle. Now comes the catch. The wavelength or frequency represents the momentum or energy of the particle. If a mixture of different waves makes a particle, then the particle automatically has a mixture of momenta or energies! That mixture is the "uncertainty." Of course, you can make the particle out of only one wave so that there is no uncertainty in momentum or energy. But one wave will not make a wave packet. One wave runs on forever. So if you make a particle from one wave you cannot tell at what location or what time it exists. Uncertainty again. But you do not have to have uncertainty in momentum or energy—if you are willing to accept uncertainty in location or time. Similarly, you do not have to have uncertainty in location or time—if you will accept uncertainty in momentum or energy. You can get rid of some uncertainty in anything, but you can't rid of ALL the uncertainty in *everything.*

The number represented by *h* and known as Planck's Constant tells us how much uncertainty there must always be. The quantity *h* is a basic constant of the universe, like the speed of light or the charge of the electron. It is a small number so its effect does not become obvious until you enter the world of photons and electrons. But obvious or not—it is ALWAYS in effect. The product of two uncertainties together has a minimum value of *h.*

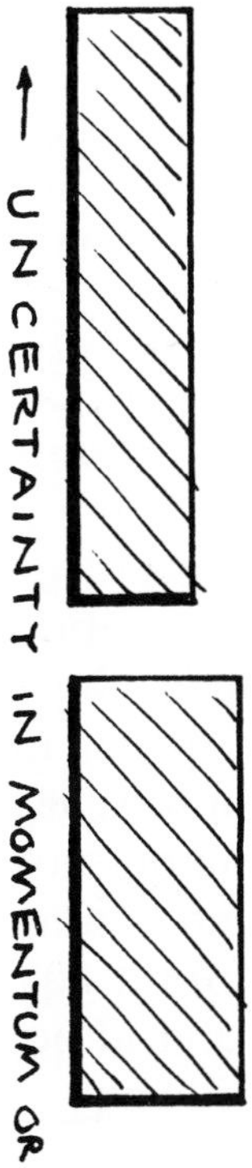

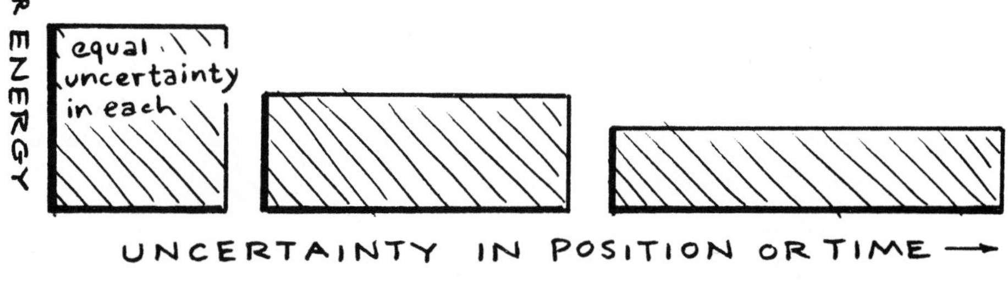

EAT IT

Far out in space a lonely electron meets a lonely proton. They attract each other electrically, and

a) are kept apart by nuclear forces
b) merge together, cancel each other out, and turn to pure energy—which makes the stars shine
c) the proton swallows the electron
d) the electron swallows the proton

ANSWER: EAT IT

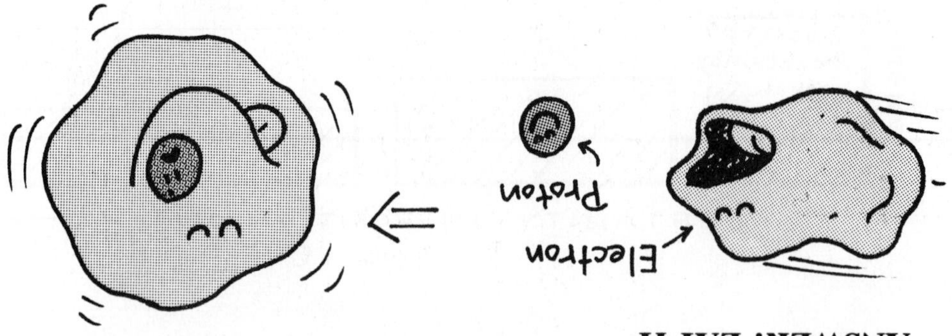

The answer is: d. An electron's wave packet is much larger than the proton, so there is no way the proton can swallow an electron. The proton weighs almost 2000 times as much as the electron, although that is still very little. But just because they have little mass you should *not* suppose they are very small! (After all, empty space has zero mass and yet empty space occupies most of the universe.) In actuality, the electron wave packet swallows the proton! There is a big, light electron wave with a heavy, little proton down at the center of it. The system is as compressed as possible with the available energy. The buzz words are: "It is in its lowest energy state." What is an electron with a proton at its center called? A hydrogen atom.

ORBITING ORBITS

Some people visualize an electron moving around an atomic nucleus as a miniature planet moving around a miniature sun. Is it essential that an atom's electron MOVE AROUND the nucleus of the atom? That is, is it essential that the electron have angular momentum around the nucleus?

a) Yes
b) No

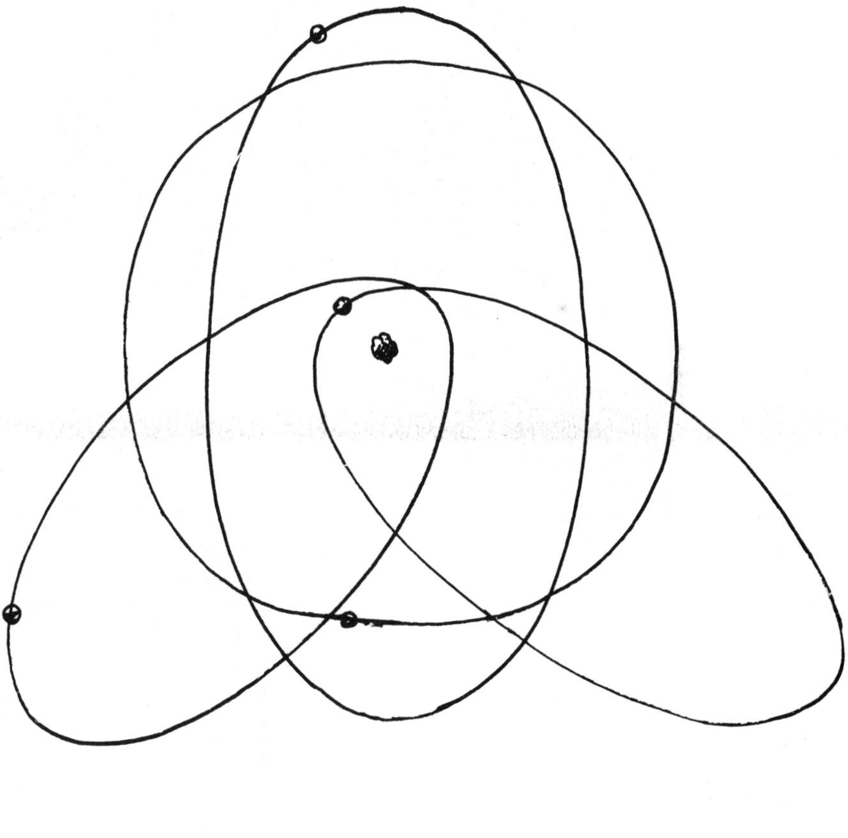

ANSWER: ORBITING ORBITS?

The answer is: b. A satellite needs momentum to orbit around a gravitationally attracting body like the sun, but this situation is very different for electrons in an atom. Electrons may indeed be pulled right into the nucleus. But they fly right out the other side and may be pulled back in again and again. Electrons rarely remain in the nucleus and usually go right through. We can best think of the electron as a wave—a wave too big to fit into the little nucleus. Now the wave can run circles around the atomic nucleus, or it can run back and forth right through the nucleus. In fact, the simplest of all atomic "orbital" electrons in the simplest of all atoms (called the ground state of hydrogen, which is the atomic electron motion first studied by students who major in physics), turns out to be not an orbit at all, but simply a wave running back and forth through the nucleus.

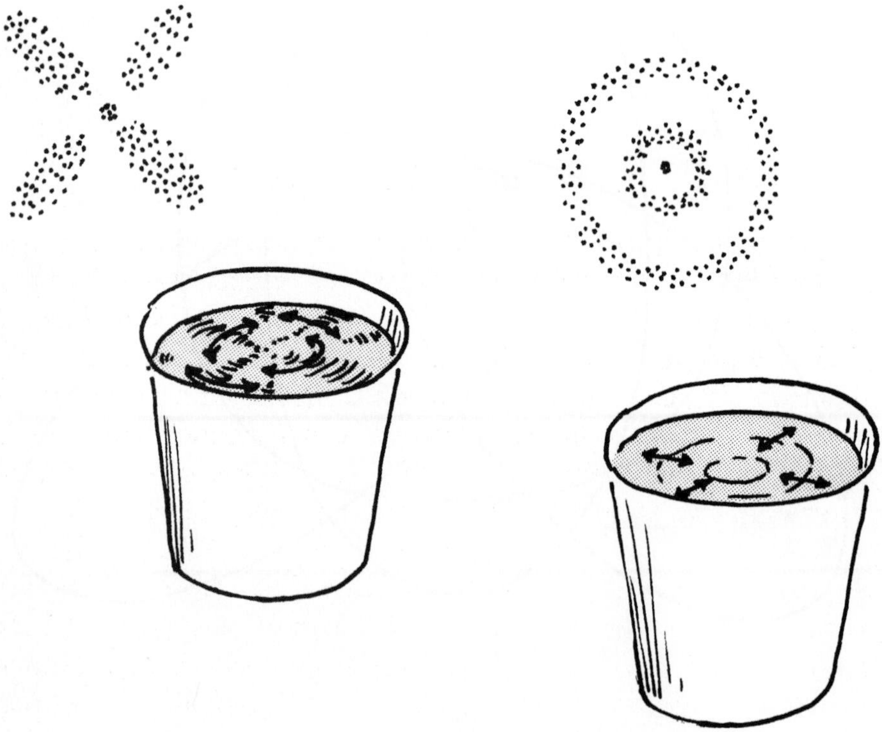

GROUND FLOOR

Atoms can absorb energy from light and/or heat. The absorbed energy lifts the electron wave from a low orbit near the nucleus to a higher orbit. When the atom radiates away the absorbed energy the electron wave drops back down to a lower and smaller orbit. In the smallest orbit, the ground floor orbit, the electron cannot radiate any energy because

a) it has zero kinetic energy
b) the wave will not fit in a lower, smaller orbit
c) both of the above   d) neither of the above

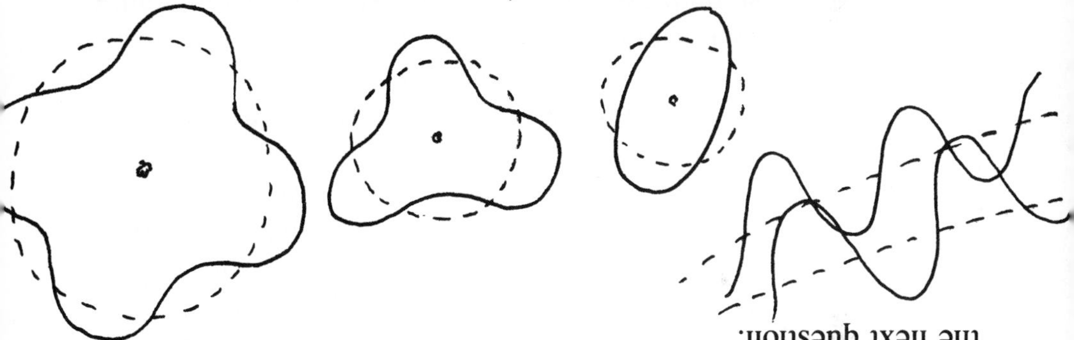

ANSWER: GROUND FLOOR

The answer is: b. The electron wavelength must fit the circumference of the orbit. The minimum-circumference orbit is once around and equals one wavelength. The wave simply cannot fit a smaller and lower orbit. In the ground orbit the electron wave still has kinetic energy (otherwise it could not "wave") but it can't get rid of this energy since it cannot fit into a smaller orbit.

Two important points can be mentioned here. First, the waves in the different orbits are not isolated from each other, though some textbooks erroneously picture them as isolated. They in fact overlap a lot. Second, the electron waves do not fall from high orbits into low orbits by themselves any more than planets fall from higher to lower orbits by themselves. Some outside thing must trigger their fall and consequent release of radiated energy. That is the subject of the next question.

WAVE OR PARTICLE?

Within an atom an electron acts as a

a) wave   b) particle   c) both

An electron wave can interfere with itself.

a) true   b) false

ANSWER: WAVE OR PARTICLE?

The answer to the first question is c because the answer to the second question is a.

When an electron moves in a circle it radiates energy (recall SYNCHROTRON RADIATION) so you might expect an electron orbiting the nucleus of an atom would perpetually put out radiated energy. But atoms don't perpetually radiate energy. Why is this so?

Because the electron is smeared all around the atom's nucleus in a wave. So you can't precisely say where the electron is. All you know is the probability distribution of where it might be. That's where the wave is. The probability of the electron is "melted in a circle around the nucleus" and as long as it stays that way nothing about its known position changes. No change of position means no motion, means no radiation.

How then does an atom ever radiate energy? Eventually the atom is hit by another atom or perhaps by a photon. The impact messes up the nice circular wave by pushing some of the wave into another (lower) orbit. What that means is that there is now probability the electron might be in either orbit. So now you have not one but two waves. These two waves overlap and interfere with each other. At one place the interference may be constructive, at another it is destructive. Where it is constructive there is more probability of finding the electron. Where it is destructive there is less probability. The probability of the electron is no longer "melted in a circle all around the nucleus." The place where the electron is most probable, the wave packet, orbits around the atomic nucleus and puts out syn-

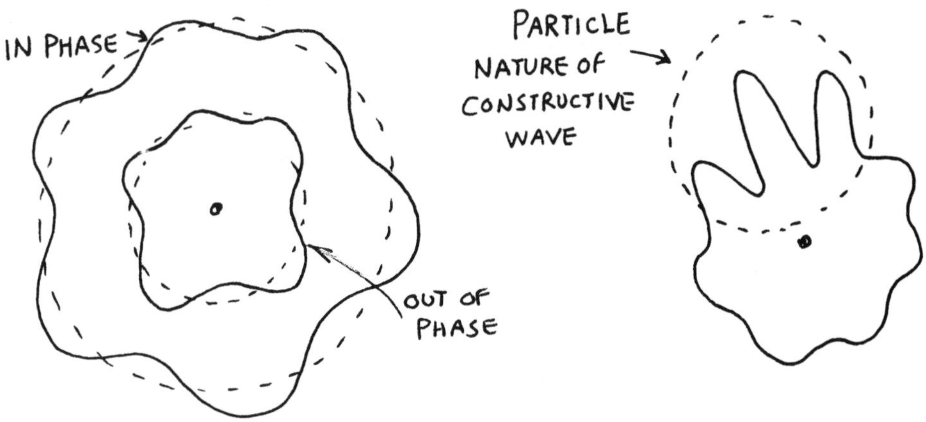

chrotron radiation just as you would expect. So in this case, while the electron's probability is mixed between two orbits, the electron wave interferes with itself and the wave packet produced by that interference acts like a particle.

It is important to note that to get the radiation process started, something had to push part of the wave into a lower orbit.* In a laser that push comes from a passing light wave photon. The passing light wave not only pushes part of the wave into another orbit but given a chance pushes it into exactly the orbit which will make an electron wave packet which will oscillate around the atom at the frequency of the passing wave. It is a resonance effect (recall GETTING IT UP).

Suppose an atom is isolated in space. If it has enough energy it can radiate spontaneously. But what gets the radiation process started? What pushes part of the wave into another orbit? That is a powerful question, and the key to a strange new idea. It seems space, empty space, space which is evenly dark without light waves and photons in it, has phantom photons in it. The phantom photons pop into and out of existence in such a short time that their existence is allowed by the uncertainty principle. (Recall UNCERTAINTY; if the time is very short the energy is very unknown. And if the energy is unknown it might not be zero. So all kinds of things can exist in space for exceedingly short times—weird—yes indeed!) Moreover it is these phantom photons that get spontaneous radiation started. Of course if they do real things they can't be called phantom photons so they are called virtual photons—that's Latin for "phantom."

* This is one of Lewis Epstein's favorite questions.

ELECTRON MASS

Consider the electric charge of a single electron spread throughout infinite space. Work would be required to compress this charge to a tiny volume the size of an electron. The energy required to do this work would be

a) nearly zero
b) equal to the energy equivalent of the electron mass
c) nearly infinite

ANSWER: ELECTRON MASS

The answer is: b. What is matter, but congealed energy! The equivalence of mass and energy is given by Einstein's celebrated equation, $E=mc^2$. The mass equivalent of the potential energy of the electric charge compressed to the size of an electron is simply its mass.

We can look at this another way. Consider the electron to be a wave packet of infinite size. If we compress it we shorten its wavelength. The work required to squeeze an infinite wave packet to the size of an electron yields a wavelength and corresponding frequency and energy that approximates the mass equivalent of the electron. It is interesting to note that these two completely different ways of explaining the mass of the electron differ only by a factor of 137. This pure number is very significant and recurs in physics.

ELECTRON PRESS

The result of continued compression of an electron would be

a) an infinitely dense wave packet
b) more electrons
c) none of the above

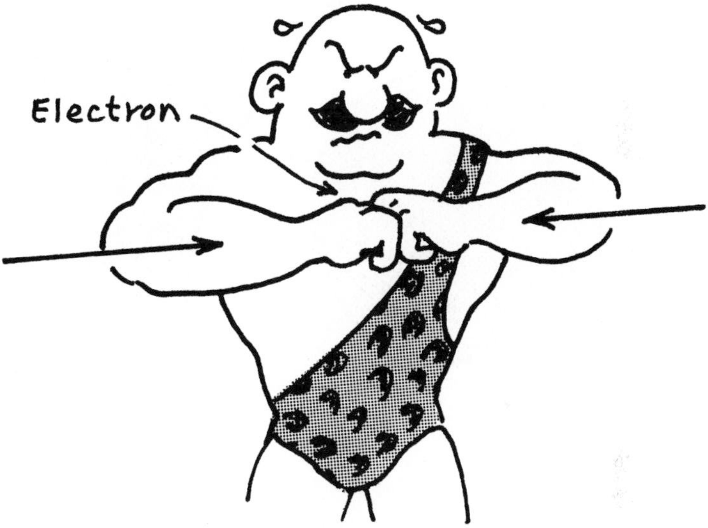

ANSWER: ELECTRON PRESS

The answer is: b. It takes energy to compress a wave packet because shorter waves require higher energy to exist. As you squeeze down on the electron wave packet you put more energy into it. Sooner or later there is enough energy squeezed into it to make an extra electron with a corresponding anti-electron, called a *positron*. Continued pressing makes more electrons rather than a smaller electron.

ANSWER: ANTIMATTER ANTIMASS

The answer is: b. Antielectrons are just like electrons except that they have positive electric charge while electrons have negative electric charge. Could they also have opposite mass? No, they could not, and this is why. When an antielectron meets a regular electron they annihilate each other, releasing a burst of radiant energy, usually powerful X-rays. So energy is left after the mutual annihilation. And energy has mass. If the electron and antielectron had equal but opposite mass, their combined mass would be zero, which would mean there could be no radiation, and that is contrary to the facts. The mass of the X-ray radiation is found to equal the mass of two electrons. So the mass of the electron and antielectron must be equal. Incidentally, not all antimatter is electrically reversed. For example, neutrons and antineutrons both have zero charge but nevertheless put together they destroy each other in a burst of energy.

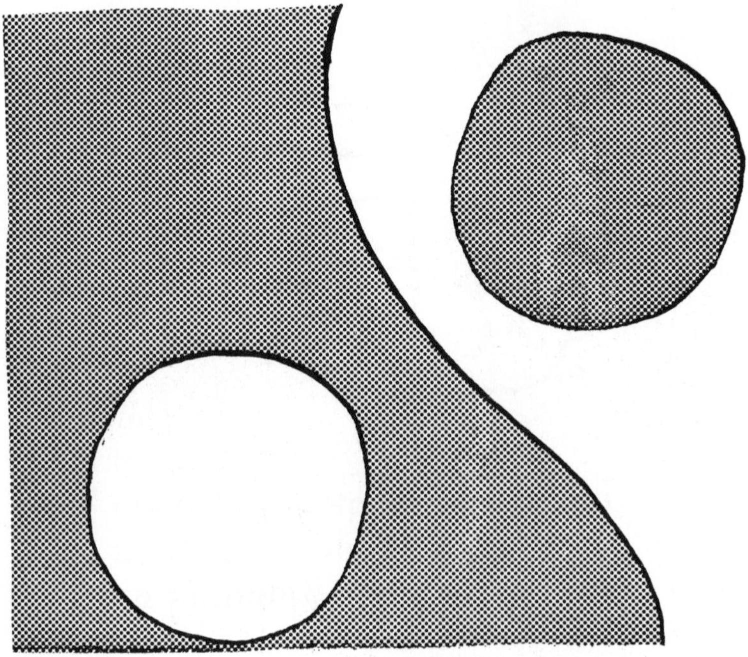

ANTIMATTER ANTIMASS

Does antimatter have antimass?

a) Yes.
b) No.

MAGIC CARPET

An antimatter spacecraft would be pushed up from the earth by the earth's gravity, like a magic carpet.

a) True
b) False

ANSWER: MAGIC CARPET

The answer is: b. The mass of antimatter can be converted to radiation mass. And radiation mass is affected by gravity exactly like all other mass. Remember, the sun's gravitational attraction bends light beams passing it (light is a form of radiation). If a thing has mass it is both affected by gravity and is itself a source of gravity, and antimatter has mass, real mass.

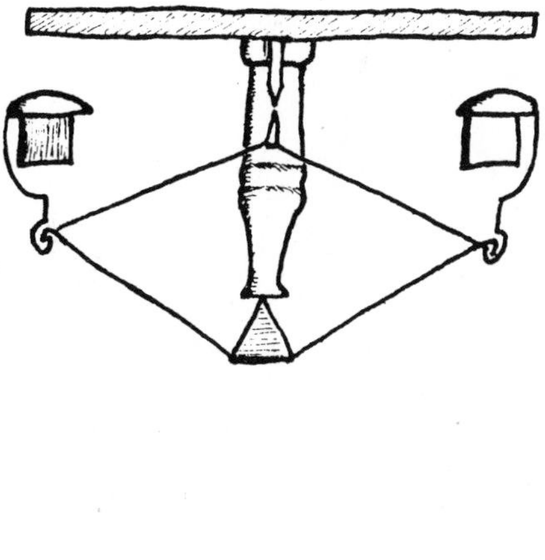

HARD AND SOFT

Galaxies sometimes collide with each other and atomic nuclei sometimes collide with each other.* The sketches illustrate typical collision paths and subsequent deflections. In Sketch I we see that the colliding bodies rebound upon impact, while in Sketch II the bodies continue almost undeviated.

a) Sketch I illustrates a galactic collision.
   Sketch II illustrates a nuclear collision.

b) Sketch I illustrates a nuclear collision.
   Sketch II illustrates a galactic collision.

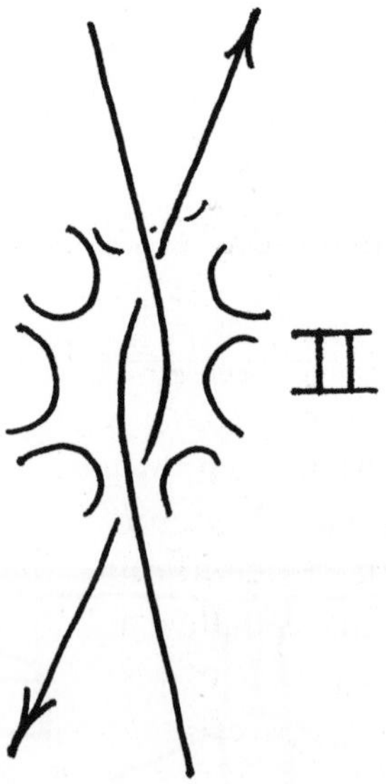

*High speed nuclei come from cosmic rays or shoot out of radioactive atoms and particle accelerators.

ANSWER: HARD AND SOFT

The answer is: b. A galaxy is an aggregate of many billions of stars. We live in a galaxy called the Milky Way. Occasionally distant galaxies are seen to collide. However, this isn't as serious as it seems because the stars in the galaxies are spread over such a huge volume of space that individual stars must seldom hit each other. (You see, the density of stars in space during a collision would only be about twice as great as it is now and the stars are mighty spread out.)

The galaxies are so big that even when they come into contact their centers may be 100,000 light-years apart so the gravitational force between them is not too strong. As they penetrate into each other the gravitational force between them gets weaker, not stronger! (Just as gravitation gets weaker when you burrow into the mass of the earth; recall INNER SPACE.) Since the force is small and the mass of the galaxies is large the galaxies continue on after the collision almost in a straight line. The deflection is small. This is called a soft collision.

Atomic nuclei, on the other hand, are very small, very dense, and very hard. The centers of the colliding nuclei can get quite close—only $\frac{1}{10,000,000,000,000} = 10^{-13}$ centimeters apart—and the force between them is partly electrostatic, which is stronger than gravity, and partly nuclear, which is stronger still. The force between colliding nuclei is large and the mass of the nuclei is small so the paths of the nuclei are violently altered by the collision. The deflection is large. This is called a hard collision.

Now for a bit of history. To begin with no one knew if nuclei were hard or soft—after all, no one has ever seen one. In fact, opinion was that they were soft. Then a man named Rutherford actually shot nuclei from a radioactive material at the nuclei in some thin gold foil and examined how the paths of the nuclei were altered by the collisions.

The deflections were large so Rutherford knew, even though he had never seen them, that the nuclei had to be hard.

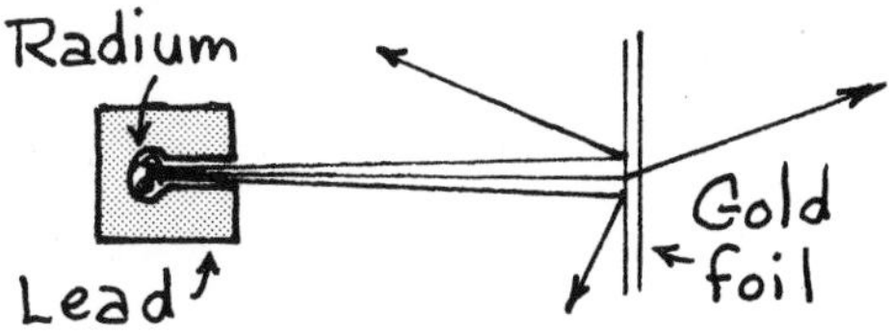

ZEEMAN

We first learned that the sun has a magnetic field

a) by measurements made from spacecraft that went near the sun
b) by the direct effect the sun's magnetic field has on magnetic compasses here on earth
c) because anything that has gravity must also have magnetism
d) by the effect the sun's magnetism has upon the light we get from the sun
e) In fact, we have no knowledge of the sun's magnetic field and no present way to get any.

ANSWER: ZEEMAN

The answer is: d. If you look at the light from a fluorescent lamp through a prism you will see several separate bands of color with darkness between them. These are spectrum lines.

In the later part of the last century the Dutch physicist Zeeman discovered that if a source of light is put in a strong magnetic field each of these lines splits into three components. When the solar spectrum was examined with powerful spectroscopes at the turn of the century it was discovered that the solar lines were split.

FIRST TOUCH

The Zeeman effect fascinated physicists because it was through that effect that they were first (in 1896) able to touch in a controllable way the electrons that were attached to atoms. The touching finger was magnetism, and the effect of the magnetic force was line splitting. The spectrum lines produced by the emission or absorption in a gas located in a magnetic field are split because

a) a magnetic force can slightly shift the color of a photon
b) the magnetic force attracts some gas atoms and repels others
c) the magnetic force alters the way the electrons move in atoms
d) the magnetic force splits photons

ANSWER: FIRST TOUCH

The answer is: c. What causes the spectral lines to split? The split comes from the effect the magnetic field has on the orbits of the electrons in the atoms of the luminous gas. Any changes in the frequency with which the electrons revolve around the nucleus of the gas atoms relates to changes in position of the spectral lines. The magnetic field affects this frequency.

Now an electron moving in a magnetic field experiences a force which depends jointly on the magnetic field direction and the velocity of the electron. If the electron is moving parallel to the magnetic

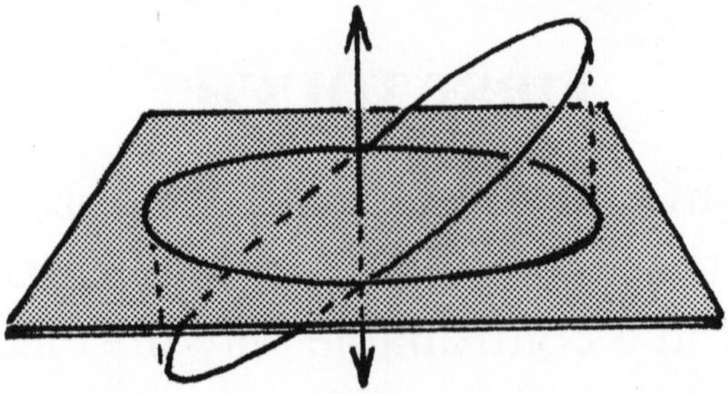

field there is no force but if it is moving perpendicular to the field there is a force which is both perpendicular to the field and to the electron's direction of motion.

Now how does this force affect the electron's orbital motion? The orbital motion of the electron can be pictured as a circular motion perpendicular to the magnetic field combined with an oscillation parallel to the field. The component of motion parallel to the field is unaffected by the field. The orbital motion perpendicular to the field is affected. Depending on which way the electron orbits, the magnetic field produces a force (recall ELECTRON TRAP) which can either increase or decrease the existing centripetal force the nucleus exerts on the electron. So about half the atoms have their electrons shifted to a slightly higher frequency and half are shifted to a lower frequency. But while the frequency of the round-about motion in the plane is increased or decreased depending on the direction of the electron's revolution, the frequency of the perpendicular oscillation is unchanged.

When you look at the radiation from atoms in a direction perpendicular to the magnetic field, you see each spectral line split into three parts. One line due to the parallel motion is unaffected by the field, but there are two lines to either side from electrons orbiting in either clockise or anticlockwise directions. Question: If you look parallel to the field do you see any radiation from the motion parallel to the field? So how many lines do you see when viewed from this angle?

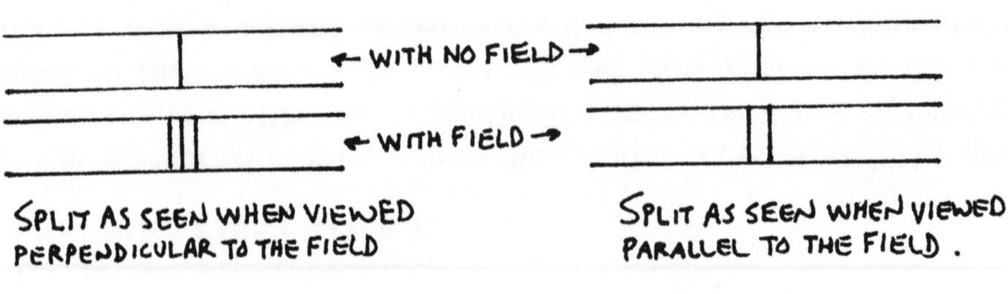

CROWDING

Balls X, Y and Z have equal positive charges on them. The force between which pair of balls is largest?

a) X & Y
b) X & Z
c) Y & Z
d) The force between all pairs is the same.

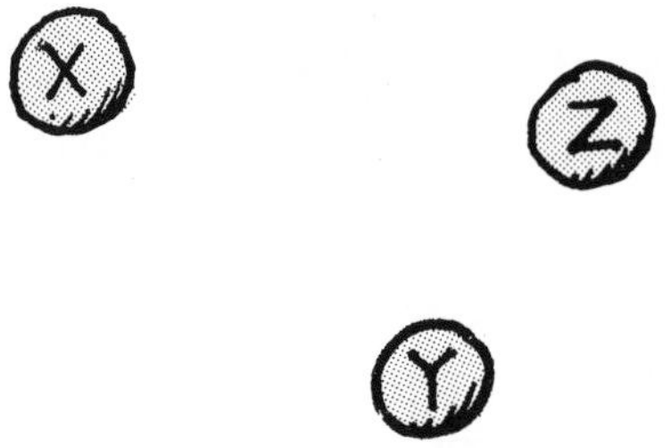

ANSWER: CROWDING

The answer is: c. All balls have the same charge so all repel each other but Y & Z are crowding each other more than Y & X or Z & X so the repulsion between Y & Z is largest. Force decreases as distance increases. This is true for many forces of nature: gravity, magnetism, the strong and weak nuclear forces. But it is not true of all forces between all things; for example, with rubber bands force increases as separation increases.

If you double the distance between balls you might expect the force between them to be reduced to one-half, but it is reduced to one-fourth. Why? We can answer this by stating that electric forces between charged particles obey the inverse-square law, but we can look at this in a completely different way—by the exchange of "virtual photons." The balls exert force on each other by exchanging virtual photons. The virtual photons carry momentum; a photon's momentum is inversely proportional to its wavelength, so when the

distance between balls doubles, the momenta of the photons that fit between them are cut in half. So the force should be cut in half? No. It also takes the photons twice as long to travel between the balls when the distance between them doubles. So the force between balls, which is the rate of exchange of momentum, is again cut in half. Thus the total cut comes to one-fourth.

HALF-LIFE

On a distant planet you are going to leave a base station powered by a radioactive energy supply. But you have a choice between two supplies of equal mass. Which power supply will run the base longest? Supply I uses a radioisotope with a six-month half-life. Supply II uses a different radioisotope which is only half as radioactive (puts out only half as much power) as the first radioisotope, but it has a one-year half-life. The base will run longest with

a) Supply I b) Supply II c) Either

ANSWER: HALF-LIFE

The answer is: b. To see why, let the supply run in your head. Suppose Supply II starts running with power one. After a year it will be down to 1/2 and a year later 1/4 and then 1/8. Supply I is twice as powerful so it starts with power two, but after one year it has gone through two half-lives (each six months long), so it is down to power 1/2 and a year later it is down to 1/8 and then 1/32. So Supply II is certainly to be preferred.

YEAR	0	1	2	3
Supply I	2	$\frac{1}{2}$	$\frac{1}{8}$	$\frac{1}{32}$
Supply II	1	$\frac{1}{2}$	$\frac{1}{4}$	$\frac{1}{8}$

Exactly the same kind of mathematical situation exists in optical communication. Light from a laser is sent down a glass fiber optical light pipe. When light goes through anything which is not perfectly clear half of it is absorbed in a certain distance and then half of the remaining half is absorbed when it goes that certain distance again. A fiber optic light pipe is not perfectly clear.

Now given the choice of doubling laser power or halving the absorption of the glass fiber which would you take? Cutting the transmission loss in the fiber is mathematically equivalent to lengthening the half-life of the radioisotope.

HALVING IT WITH ZENO

Zeno was an old Greek who did a lot of thinking in the days before Aristotle lived. Zeno showed that the logically obvious is often not obvious at all if you really think about it. For example, he said you can never walk completely across a street because before you can cross you must first go half way—then you must go half the remaining half, then half the remaining quarter, then half the remaining eighth, and so on forever and ever. So Zeno argued it should take forever to cross the street. Now if you in fact cut a distance in half, and then the half in half, and so on and on in Zeno-like fashion, would you go on cutting forever?

a) Yes   b) No

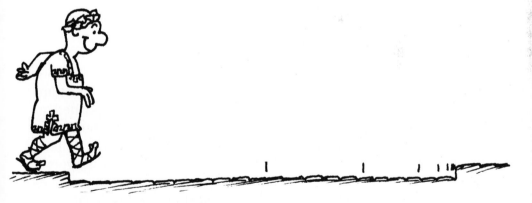

So if you decide to walk across a street, will it take you forever to reach the other side?

a) It certainly will!
b) No way!

ANSWER: HALVING IT WITH ZENO

The answer to the first question is: a. You can always cut a distance in half again and again and again. It may be difficult to cut in actual practice, but in principle you can keep cutting forever. We talk here about cutting or dividing space, not some material object, so we need not worry about cutting atoms.

The second answer is: b. Crossing a street is a common experience so we say it is only common sense. But what about logic? To cross the first half takes a certain amount of time, say 1/2 minute. To go half of the remaining half takes 1/4 minute, and half the remainder takes 1/8 minute, the next interval 1/16 minute, then 1/32 minute and so on. The time to cross is the sum of an infinite number of fractions: $\frac{1}{2}+\frac{1}{4}+\frac{1}{8}+\frac{1}{16}+\frac{1}{32}+\ldots$ forever. But even though the number of fractions is infinite, their sum is not infinite. This is why. Suppose $s=\frac{1}{2}+\frac{1}{4}+\frac{1}{8}+\frac{1}{16}+\ldots$ Now multiply everything by two, so you get

$$2s=\frac{2}{2}+\frac{2}{4}+\frac{2}{8}+\frac{2}{16}+\ldots \text{ or } 2s=1+\frac{1}{2}+\frac{1}{4}+\frac{1}{8}+\ldots$$

Now subtract equations:

$$\begin{aligned} 2s &= 1+\tfrac{1}{2}+\tfrac{1}{4}+\tfrac{1}{8}+\ldots \\ -s &= -\tfrac{1}{2}-\tfrac{1}{4}-\tfrac{1}{8}-\ldots \\ \hline s &= 1 \end{aligned}$$

That means: $1=s=\frac{1}{2}+\frac{1}{4}+\frac{1}{8}+\frac{1}{16}+\ldots$ forever, so you get across the street in one minute!

This analysis also applies to Federal Reserve banking, which controls our money supply. To understand how it works, you must appreciate two things: most large sums of money are deposited in banks, and most large purchases are paid for with money borrowed from banks. Now how much does one dollar buy? One dollar's worth? No! More than that. Here's why. The person who gets a brand new dollar deposits it in the bank which then loans it to someone else who uses it to buy one dollar's worth of goods. That same dollar soon enough is deposited in the bank and is out on loan again. This cycle can be repeated again and again so that in principle that one dollar could be used to pay for an infinite amount of goods! So why doesn't this happen? Because the Federal Reserve Banking Law states that banks can only lend out a certain fraction of each deposited dollar, like say half of it. So you deposit a dollar and the bank lends out \$0.50. When that is redeposited the bank lends out half of the half or \$0.25 and so on. Now how much can the original one dollar buy? Each time it goes around it buys a little less, but if it goes around an unlimited number of times the value in goods it can buy is $\$1 + \$1/2 + \$1/4 + \$1/8 + \$1/16 + \ldots$ which equals \$2.

Some people think that if the Federal Government wants to increase the money supply that the Government has to print more paper—but it does not even have to do that. All it has to do is allow the banks to lend more money, say 3/4 rather than 1/2 of each deposited dollar. Then how much can the one dollar pay for? $\$1 + \$3/4 + \$(3/4)(3/4) + \$(3/4)(3/4)(3/4) + \$(3/4)(3/4)(3/4)(3/4) + \ldots$ And how much is this? Use the old trick. Subtract from this series the same series multiplied by 3/4.

$$\begin{aligned} s &= 1 + 3/4 + (3/4)(3/4) + (3/4)(3/4)(3/4) + \ldots \\ -3/4s &= \quad - 3/4 - (3/4)(3/4) - (3/4)(3/4)(3/4) - \ldots \\ \hline 1/4s &= 1 \end{aligned}$$

Simple algebra then gives $s = 4$; so now one dollar pays for \$4 worth of goods.

Lending out ninety cents on each deposited dollar allows each dollar to pay for how much goods?

The answer is: ten dollars' worth.

So we see that if banks lend out fifty cents on each deposited dollar, each dollar can pay for \$2 worth of goods. If they lend out seventy-five cents on each deposited dollar, each dollar pays for \$4 worth of goods, and lending ninety cents on the dollar pays for ten dollars worth of goods. So it seems like the Government can make as much money as it wants with the greatest of ease by simply allowing banks to lend a greater proportion of deposited money. What's the catch? Inflation!

There are some situtions that fit Zeno's thinking exactly. There are devices that store electricity called capacitors. When they are discharged they lose half their electricity in a certain amount of time. In the following same interval they lose half the remaining fraction and so on and on. So it takes them forever to completely lose all their electricity! Same is true for radioactive materials losing their radioactivity, and also for refrigeration devices that attempt to reach absolute zero by discharging heat.

FUSION VS. FISSION

The natural uranium in the earth was probably formed by the fusion of iron nuclei inside ancient stars. This nuclear fusion

a) cooled the star
b) heated the star
c) could have done either

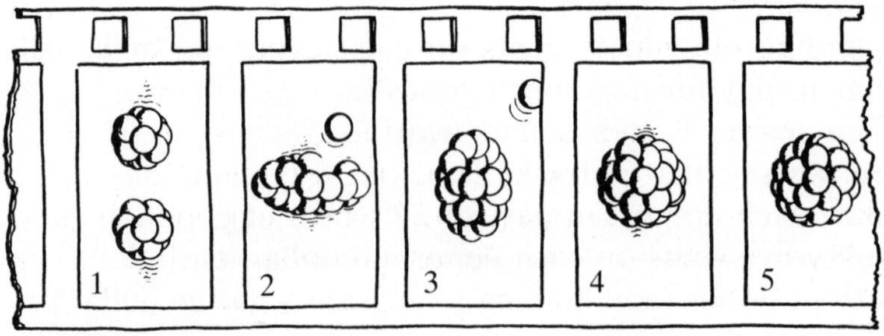

Fusion

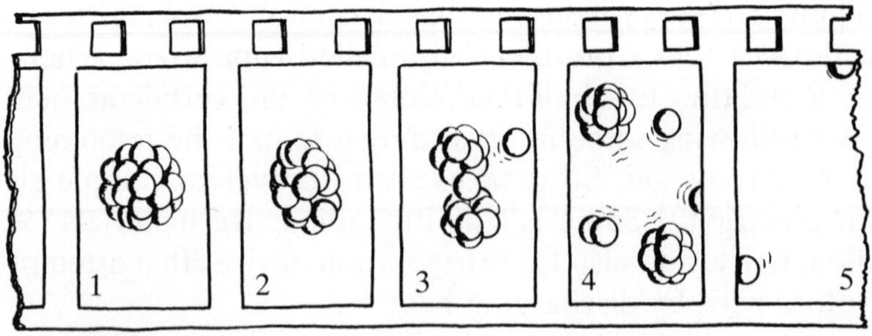

Fission

ANSWER: FUSION VS. FISSION

The answer is: a. You might think nuclear fusion—the pushing together of the nuclei of two light atoms to make the nucleus of a heavy atom—always causes energy to be released, as happens in the sun and in the explosion of a hydrogen bomb. But that is not so. Why? Because nuclear fission—the breaking apart of the nucleus of a heavy atom to form the nuclei of two or more light atoms, can also give off energy, as happens in a nuclear reactor and in a uranium bomb. Now if fusion always released energy and fission always released energy you could repeatedly push together and then break apart an atom and so have a perpetual source of free energy, too good to be true!

If the fission of a heavy uranium nucleus into several iron nuclei gives off energy, as it does in an A-bomb, then the fusion of the iron nuclei to remake the uranium must absorb energy. Likewise if the fusion of two hydrogen nuclei to make helium gives off energy, as it does in an H-bomb, then the fission of helium back into two hydrogen nuclei must absorb energy.

It turns out all nuclei want to be medium weight. Hydrogen is light, iron is roughly medium, and uranium is heavy weight. So hydrogen wants to fuse (though the reaction is not self-starting) and uranium wants to fission. What is meant by saying a nucleus "wants" to be medium weight? The same thing that is meant by saying water "wants" to run downhill—it will give off energy if you let it run down. Of course water will also run uphill, but only if you feed it energy.

Now a lot of people think the universe was mostly all hydrogen when it began. Then inside of stars the hydrogen was fused into ever-heavier elements. As long as the hydrogen was fused into elements no heavier than iron, the fusion gave off energy which made the stars shine. But eventually things must have happened that caused nuclei to be fused into elements heavier than iron—like uranium, because uranium does exist! And that fusion must have soaked up energy from the star and so the star must have been cooled by the formation of uranium inside of it. How much heat did the uranium absorb when it was created? Exactly as much as it gives off in a nuclear reactor or A-bomb.

MORTALITY

Out of one thousand newborn American children, only half, 500, are expected to be alive at age 68 (which gives you only three years to collect Social Security—less if you are a male).

Suppose the radioisotope "Humanitron" has a half-life of 68 years and you start out with 1000 children and 1000 "Humanitron" atoms. You will find that

a) the surviving number of children and atoms will always be approximately equal.
b) during the first 68 years the average number of surviving atoms will be larger than the average number of surviving children, but after 68 years there will always be more surviving children.
c) during the first 68 years the average number of surviving children will be larger than the average number of surviving atoms, but after 68 years there will be more surviving atoms.

ANSWER: MORTALITY

The answer is: c.

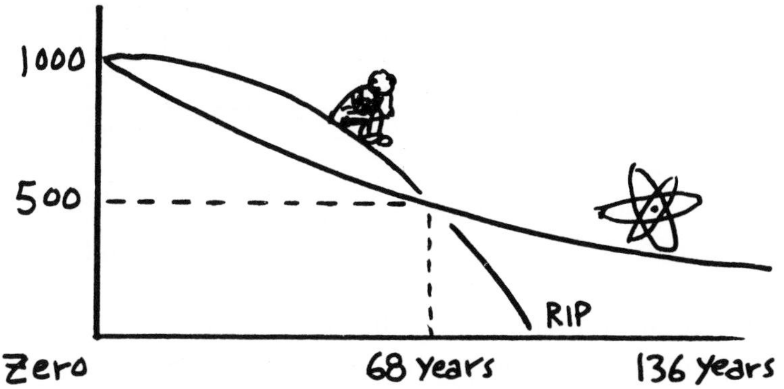

The mortality curves for people and for radioactive atoms have very different shapes. Out of the one thousand babies, 90% are still alive at age 35 and 80% at age 50. But then the death rate rapidly increases. Only 50% reach 68 years, 25% reach 77 years and 10% reach 84 years. One percent reach 92 years, one-tenth of one percent reach 97 years and about one-hundredth of one percent reach 100 years of age. Out of the one thousand "Humanitron" atoms, 50% make it to 68 years, but 25% make it to 136 years.

What makes the curves so different? The probability of a person surviving one more year at age 47 is 99%, but at age 76 the probability of surviving one more year is only 90%. The probability of a "Humanitron" atom surviving one more year is always about 99%. In this sense, it is as if the "Humanitron" atoms were always 47 years old—that is, the percentage of "Humanitron" atoms surviving one more year is always the same as the percentage of 47-year-old people surviving one more year, 99%.

What is the moral of this? Is it something about radioactivity or different kinds of curves? The moral comes closer to home than that. The moral is that it suggests a different way to look at life. We usually look at life as so many years lived (your "age"). A more useful way to look at life might be to consider your expectation of years of life still remaining. The following table makes the conversion from your age to your approximate average remaining life expectation.

Years lived - - - your age

10	15	20	25	30	35	40	45	50	55	60	65	70	75	80	85
55	51	46	42	38	33	29	25	21	18	14	11	9	7	5	4

Years of life still expected

At age ten, your expectation is for 55 more years. Then why, at age 15, after five of those years are spent is your expectation 51 years rather than 50 years?

Your expectation increases for the same reason the expectation of a person running safely across a highway increases after he is part way across. We in life are, in effect, running across a highway with an infinite number of lanes and in each lane there is heavier traffic.

The "Humanitron," too, is running across a highway with an infinite number of lanes, but the traffic in each lane is the same. Think about that!

MORE QUESTIONS (WITHOUT EXPLANATIONS)

You're on your own with these questions that parallel those on the preceding pages. Think Physics!

1. The kind of force that holds electrons in the vicinity of the atomic nucleus is
a) electrostatic b) gravitational c) magnetic
d) none of these

2. Orbital electrons do not spiral into the atomic nucleus principally because of
a) angular momentum
b) electric forces
c) the wave nature of electrons
d) the discreteness of energy states

3 The atomic nucleus consists in part of one or more positively charged particles—the protons. These protons
a) require no force to hold them together
b) are held together by electrostatic forces
c) are held together by gravitational forces
d) are held together by magnetic forces
e) none of the above

4. All the different kinds of force in nature that either attract or repel bodies vary in such a way that the force between the bodies is reduced to ¼ of its initial value if the distance between the bodies is doubled. This statement is
a) true b) false

5. The Uncertainty Principle states that
a) all measurements are essentially incorrect to some degree; no measurement is exact
b) we cannot in principle know both the position and momentum (or energy and time) of a particle with absolute certainty
c) the science of physics is essentially uncertain
d) all of the above
e) none of the above

6. Light intensity from an incandescent source is plotted as a function of frequency as shown in the radiation curve. If this light is first passed through a gas, the resulting radiation curve would most probably look like

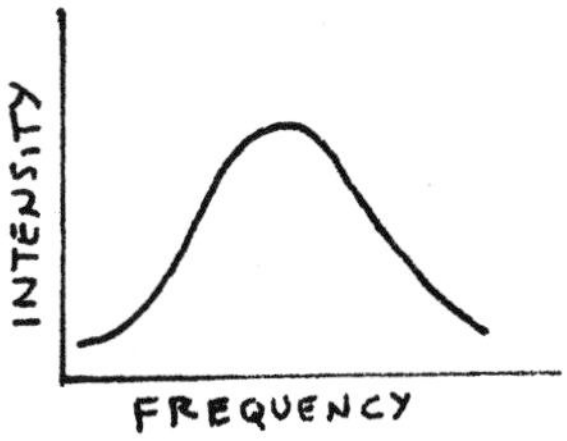

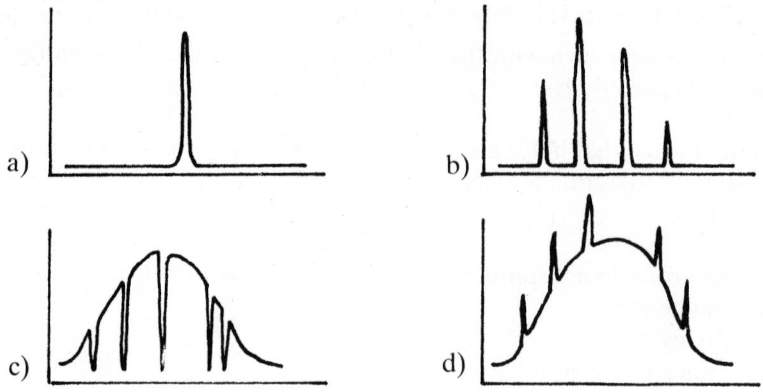

7. The spectrum of light from an incandescent solid is shown in the radiation curve to the right. Light emitted from atoms in the *gaseous* state would more likely produce a curve that looks like

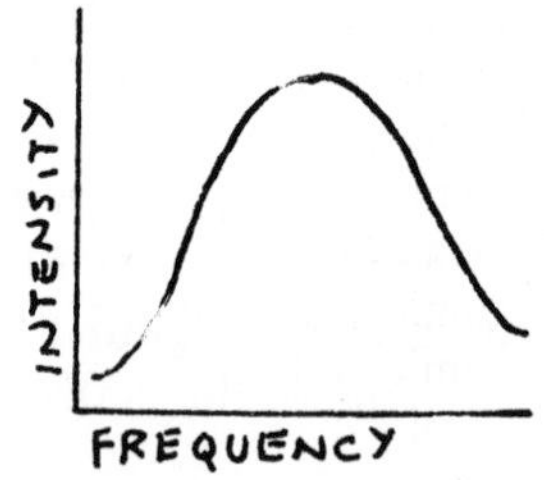

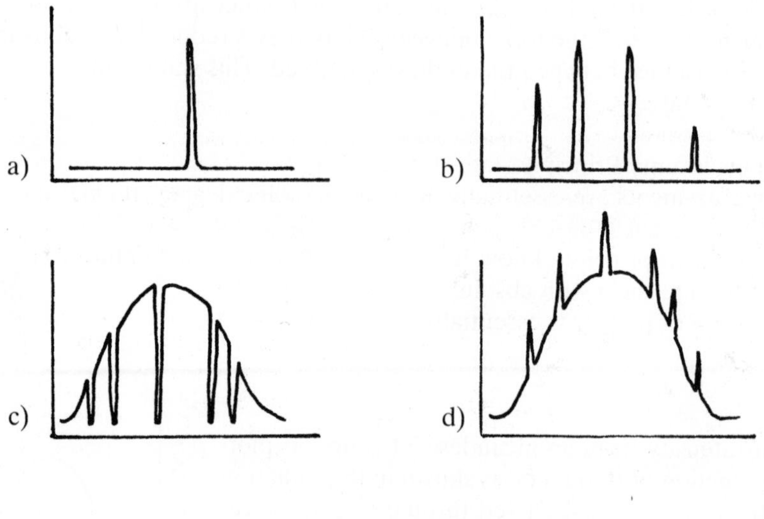

Index

This book, *Thinking Physics,* is available in the **German language.** Read it for your foreign language exam! For details, write to Insight Press, 614 Vermont @ 18th Street, San Francisco, California, 94107 - 2636.

Philosophy, bereft experience, is poison.

Revelations

Well near 5/6 of my living years are past. It is high time to draw some conclusions. I take this opportunity to do it.

Everyone wants to drive the locomotive.
Who wants to puzzle out how the valve gear works?

You must appreciate the question before you can value the answer. Obvious. Very true. No argument about it. But in practice, people almost always demand the solution before grasping the problem; the medicine, before understanding the disease. Here is the problem, the disease.

ALL THE TROUBLE IN THE WORLD

All the trouble in the world begins in high school. In high school, there are two kinds of kid. One kind is interested in how the physical world works. The other kind isn't.[1] The kids interested in how the world works try to make rockets, find a dark place to set up a telescope, see bugs through a magnifier, mess with chemistry in the garage, struggle to trisect an angle, attempt to understand relativity, and in my day, figure how a slide rule worked; in the present day, understand how a computer works inside. The kids disinterested in the outside world look at themselves in the mirror. They study their pimples and dimples, their psychology, sociology, novels, navels, dancing, prancing, and the latest opinions on sex, on race, and on the movies. Eventually, both kinds grow up and most get jobs.

The kids interested in how the world works can end up as scientists, engineers, or medicine men. That work is largely barred to kids who don't take an early interest in the physical world. Nevertheless, many kids disinterested in the physical world go on to college. They study things like psychology and English. What job can an English major get? And English major can become a lawyer, a judge, a senator, an actor, a news commentator, a script writer, a school superintendent, or a rabbi.

---

[1]Disinterested in the physical world does not mean disinterested in grades.

It is these lawyer like people who decide **how** the world **should** work. So it turns out, the kids who were not interested in how the physical world actually works, end up being the adults who decide how the world should work. For example, a judge without a half-day of high school physics decides on auto accident mechanics. For example, the sorority queen who said probability and statistics was for creeps and geeks, dropped the class, and played with politics. Elected to the Senate, she spent six hours in hearings on inspecting four cubic foot suitcases at local airports and then spent sixty minutes in hearings on inspecting 4,000 cubic foot cargo containers stacked 6 x 9 x 12 on the deck of ocean-going freighters entering San Francisco Bay.

For example, the Hollywood movie star, influenced by his high school literature teacher, despised heavy industry and machinery. He persuaded the State of California to make a law requiring Pacific Gas and Electric to mail out thousands of notices declaring the sand, beach and river sand, used by the company for various jobs, contained silicon oxide (silicon oxide is the most abundant compound in the earth's crust; 59% of the crust is silicon oxide) and that silicon oxide might cause cancer. The movie star testified it was better to be safe than sorry. He said a false warning can't hurt you. The movie star was wrong. The harm of false alarms is that serious alarms are not distinguished from the false. When attention was drawn to the Idaho militia, attention was drawn away from Bin Laden's militia.

For example, the high school cheerleader who disliked the dork chem teacher, became a baby doll TV news anchor woman. So anything chemical in the news was bad. Finally someone tricked her into doing a segment on the exposure of children to hydrogen oxide. She ate it up, and didn't realize the casualty numbers were drowning deaths.

The Trouble With College. I. Many, perhaps most, people go to college devoid of questions about how the world works. So why go? They go to avoid hard work, because they are lazy. Yet they hope and **expect** to make big money in consequence of going to college.

II. College is tilted against science students. A physics major must take English and a foreign language. An English or language major need not take physics or any physical science. The physics major must sit in an English and a language class with English and language majors and perform at their level. However, **if** the English or language major takes physics he does not sit in a physics class with physics majors. He sits in a special physics class for poets. Beginning physics at college is taught at three levels: 1) Beginning physics with no math, known as sorority physics or physics for poets. 2) Beginning physics using high school math, for pre-med hopefuls. 3) Beginning physics using calculus for science, math, and engineering nerds. If poets don't have to get their math right in physics class, how come nerds have to get their spelling right in English class?

III. Med school admission is tilted against inquiring minds. Med school hopefuls must stay focused on grades, grades, grades. If a student spends too much time digging into something, a grade will certainly suffer. So the most scientifically curious students are systematically excluded from med school admission. Thus medicine is now advanced largely by chemists and engineers, not M.D.'s. Med school should be as open as military flight school. Students who can't cut it in flight school are washed out after admittance. Likewise, med school students who can't cut it, should be washed out, after admittance, not before. Why? Because pre-admit requirements cannot foretell who will be the best M.D. just as they cannot foretell who will be the best pilot.

For example, the school superintendent, who had no appreciation for manual skills. He discontinued all the machine shop, carpentry shop, etc. classes, in his district. The superintendent never owned or worked on a house, yet was responsible for 23 school houses. Every year he over spent his budget, on academic enrichment and the latest and greatest computer solutions, but never a new roof. In the summer of 1998 he received the N.E.A. superintendent of the year citation. In the fall of 1998, it rained like hell and every school building in his district had interior water damage, totaling one-third of the district's annual budget. In 1999, the voters approved a school rehabilitation bond issue – for the children.

For example, the Berkeley[2] California anti-nuclear activist kid, who became a popular reform rabbi, ate cheese burgers, ate bacon burgers, but declared irradiated meat unkosher.

For example, the most malevolent shyster of all, the lawyer who drove Gilbert Chemistry Sets, the kind with real glassware, real burners and real chemicals, off the market. As long as there were chem sets there would be kids who would become enchanted by the mystery of how the world works. Now Gilbert is gone.[3]

Conclusion: The wrong kids influence the world, the wrong kids run the world. That makes a perpetual mess of the world.

What might be done about it?

---

[2] The first cyclotron, the atom smasher, was invented and built in Berkeley. Berkeley is the only city with an element named after it, Berkelium. The element is very radioactive. Berkeley is a "nuclear free zone."

[3] I laugh at science people who have a fit over any slight to Darwin, but never mention the extinction of Gilbert. Everyone can buy Darwin in almost every bookstore, but no one can buy Gilbert in any store.

As you may know, Air Force Academy cadets are obliged to pass stiff courses in physics and many other sciences in order to graduate. When I was there, to give some talks, one of the officer professors told me this: He said studying physics does not make a pilot a better pilot. In fact, it may well make him a worse pilot. Why? Because in flight he will be tempted to analyze the physics of things or get into the engineering solution of problems, and the temptation will distract the pilot's attention from his job–which is to fly the airplane. That distraction accounts for more than a few accidents. So why teach the Air Force Academy cadets physics? Because the future U.S. Air Force Generals will likely have been former Academy cadets. Generals fly the Air Force. Physics is not required to fly an airplane, but physics is required to fly the Air Force–to plan the future of the Air Force.

Air Force Academy graduates run the Air Force. Law school graduates run the rest of the world. They become president, governor, and mayor. I know, not always, but more often than not.[4] Yet most law school grads don't know which end is up in the physical world.

---

[4] Question: Between 1938 and 1945 Germany occupied 18 major European countries. Stop and think of all the cities, the police departments, tax departments, school departments, sewer departments. Where did Hitler get the **executive** manpower to govern, administrate, and manage 18 different counties? From: a) the Army; b) the Bar (lawyers); c) Scientists. Hint: The Army was up to its neck in a three front war. The scientists that didn't have their heads in the clouds were up to their necks in war production. Military academy students are required to take a class covering the bad things German soldiers did in W.W. II. Are law school students required to study what the German lawyers did?

On a warm autumn afternoon, stand on the corner of Hyde and McAllister Streets in San Francisco, and ask one of the smart set emerging from the University of California law school to give a proof of the simplest, most fundamental and oldest piece of objective information in the world. Something that has been known for 3,000 years: The Pythagorean theorem. Don't bet one in ten can do it. So ask the future lawyers if they can state the theorem, without proof. Go ask them. I dare you. Maybe someone will tell you it has something to do with triangles. What kind of a triangle? Don't expect to hear right triangles. Don't even expect all the law students to know a right angle.

Ok. I see the problem. What is the solution? Oblige law students to pass some stiff science courses? After all, to practice in a U.S. Patent Court, a lawyer must have some serious science or engineering course work. Shouldn't a slip and fall (personal injury) lawyer be qualified to work sophomore friction and gravity questions? How easy it would be. The law school deans, at the stroke of a pen, could make a course requirement so a doctor of law would know a little something about the law of gravity. But it will not happen and would likely have little or no effect if ever it did happen. Why? Because of the unique position and the unique mind set of lawyer like people.

What unique position?

This: If a state governor and/or a lawyer and/or a judge enable a murderer to get out of jail and the murderer murders someone else, the governor, the lawyer or the judge is not responsible for anything. If a medical doctor, an engineer, or a ship's pilot makes an error, there is hell to pay. If the mayor or a councilman understates the cost of a new school house by 200%, or overstates the income from a municipal ballpark by 200%, it is a one day story.

If a dump truck driver underestimates his state withholding tax by 20%, there is the 20% tax, plus a 38% penalty plus the back interest on the tax and the back interest on the penalty to be paid out of the poor schlepper's grocery money.

What about mind set?

This: Behind most legal proceedings is some physical event that occurred in the outside world, for instance, an accident. But the court is 99.9% occupied with procedure, and not with the physical event. Often a lawyer explicitly does not want to know the whole story. A judge is just a lawyer in a black robe. If the judge dislikes something or does not understand something he will rule that thing should not come into evidence.[5] The one thing the lawyer/judge thinks he does understand is legal procedure. So the decision pivots on procedure. The lawyer world thinks the outcome of the case should be decided by the legal expertise of the lawyers. The lawyer world people don't say "facts be damned", but they act it. (Legal expertise includes contacts). How often does the judge bother to visit the incident scene? What do you think of a judge who presides over an 18 month trial but does not once visit the incident scene two miles from his courthouse? How often does the judge let the jury visit the incident scene? How often does the judge instruct the jury **not** to visit the scene, not to do any investigation, research or experiments? There is mandated, printed "jury instruction" to that effect formulated by many "wise" judges in the book of jury instructions. William Rogers, the lawyer in charge of the Space Shuttle Challenger Disaster Investigation Committee, almost kicked the physicist Richard Feynman off the Committee. Why?

---

[5] Trial by jury is a myth. By controlling what comes into evidence, a judge effortlessly controls the jury. Additionally, a judge can give a directed verdict, a summary judgment, a judgment notwithstanding the verdict or dispose of a matter in law and motion court so that the case is taken away from the jury at one stage or the other. What is the jury for? It covers the judge, like the robe; and if the judge wants, it can be used as a tossed coin.

Because Feynman, on his own, went to Cape Canaveral to visit the incident scene. Feynman talked directly to technicians, Feynman did the cold water experiment, and then Feynman publically demonstrated it. If you do what Feynman did you will be **both** fined and jailed for contempt of court.

The kids uninterested in the physical world seldom become interested, even when they grow up and become lawyers and judges. Moreover, they get very pissed off when crap from the physical world interferes with their art.[6] Lawyer world people signal that message to you even before they begin to talk to you. How? By the clothes they wear. An engineer wears overalls and a hard hat, a doctor wears scrubs, a physicist has chalk dust on his pants and diffusion pump oil on his shirt. A lawyer wears a fancy suit, a fancy tie and fancy Italian shoes. What do the fancy rags signal? They signal: "I intend to remain clean. I do not intend to interact with the dirty world. I do not intend to do real work." Work is measured in foot pounds.

The life object of the poet, the lawyer, the rabbi, the actor, the social activist, the governor, is to sell their stuff, to prevail, to win, to con. It is not to understand. The things done by lawyer world people are judged in courts and theaters where other people sit in judgment. The things done by scientists are judged in impersonal experiments where god alone sits in judgment.

---

[6] Their art is word manipulation. For example, lawyers pride themselves on their ability to clip words from a deposition and arrange the clips to say something that was not said. (I have also watched ad people who could clip words from a bad book or movie review to make a good one.) You might wonder why a lawyer, even an excellent lawyer, is represented by another, often inferior lawyer, when the expert lawyer is personally involved in a legal proceeding. The reason is this: What a client says in a proceeding is evidence and is under oath. What the lawyer says is not evidence and is not under oath. A lawyer has license to take liberty with truth. That is why the lawyer speaks for the client.

Ok, ok, ok, but what serious harm have the lawyer world people done? They misguide us. They consume our output and have squandered our assets. (They have also demoralized us.) Stop and reflect.

I recall San Francisco in 1939, before World War II. Not many people went to college so life was not flooded with lawyer type people.[7] Typically one man supported himself, his wife and three kids. So each person lived off 1/5 or 20% of a 1939 man's output. Now in 1999, both man and wife work to support themselves and one kid. Assume one man = one woman. So now each person has available for support 2/3 or 67% of one 1999 man/woman's output. Does the output of a 1939 man equal the output of a 1999 man/woman? The 1939 man used a hammer, the 1999 man uses a nail gun. The 1939 sailor used a sextant and chronometer, the 1999 sailor uses G.P.S. The 1939 manual typewriter has become a word processor, the Western Union messenger boy has become a fax phone.

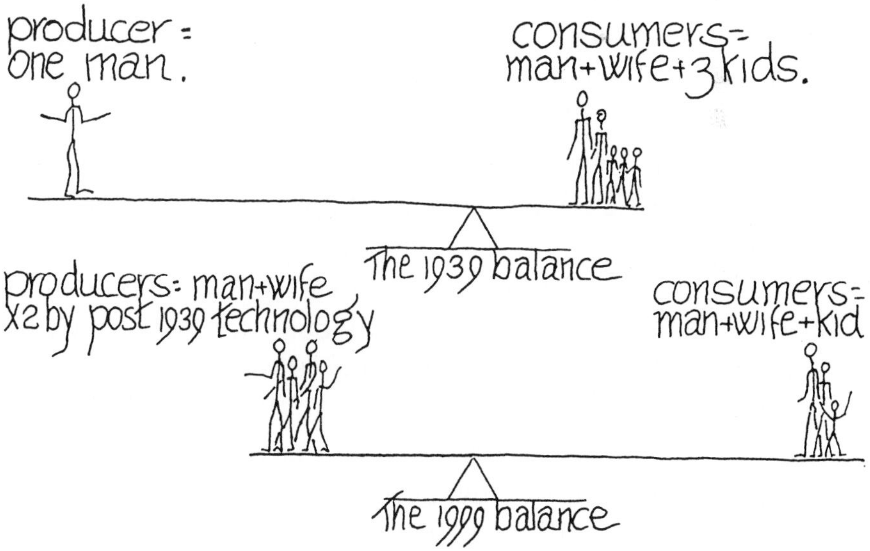

[7] By 1999, San Francisco had four major law schools, one major med school and zero major engineering schools.

Nuclear magnetic resonance has almost replaced the time, skill, and manpower 1939 M.D.'s spent on exploratory surgery and divining x-ray film. Yet, because of concocted med school entrance restrictions, med costs out run the C.P.I. by a 2:1 ratio. In 1939, a dish washer was a person; now it's a machine. Microwaves have traveled from the rad lab to the kitchen. Image stabilizers are pushing well-made camera tripods into junk stores.

The output of a 1999 man/woman is at **least** double the output of a 1939 man. So 67% of a year 1999 man/woman's output is equivalent to twice 67% of a 1939 man's output. Twice 67% = 134%. Remember in 1939, each person lived off 20% of a 1939 man's output. In 1999 each person should have available for support the equivalent of 134% of one 1939 man's output. Our 1999 standard of living should be six times higher than the 1939 standard, 134% / 20% = 6. Is it?

In 1939 a man worked an eight-hour day and retired at 65. In 1999 the man still works eight hours but the federal retirement age is about to be increased. I live in a 1939 house and I can eat no more than the 1939 man did– except I can't eat red meat, he could. In 1939 a young man could have sex without fear of death. A middle aged man could take his pay home without having six different hands in his pocket and an old man could walk the city streets without dread of mugging.

Our standard of living has not increased sixfold since 1939. At best, it has doubled, this mainly due to penicillin. Before 1939, death from infection overshadowed life. To double the 1939 standard of living, double the output available to a 1939 individual. In 1939 the output available to an individual was 20% of a 1939 man's output. Double 20% is 40% of a 1939 man's output. In 1999 each person lives off the equivalent of 40% of a 1939 man's output. **But** in 1999 each person should have the equivalent of 134% of one 1939 man's output available for consumption. That leaves 134% - 40% = 94% of a 1939 man's output to be accounted for. Where in hell does the 94% go?

The unaccounted for 94% of a 1939 man's output represents 70% of what should be available to each 1999 person for individual consumption, 94%/134% = 70%. The unaccounted for 94% of a 1939 man's output is more than enough to support four 1939 people. That is, if you could get your hands on the unaccounted for output that should be available to you in 1999, and ship it back to 1939 in a time machine, it could support four 1939 people by 1939 standards. Where is all that output going? What happens to it?

It is squandered.

It is squandered on layer on layer on layer of government. All these layers must be penetrated before the 1999 man can use his nail gun to build one house. It is squandered in dumpsters full of two-year-old computers. It is squandered on skyscrapers full of offices, full of swivel chairs. It is squandered on meetings and committees and memos and e-mails and depositions and reports and presentations and hearings and staff this and staff that and staff up your. . . .

It is squandered on food stamps for overweight people, on needles for drugees, on pay for farmers not to farm, on gifts of federal real estate (army bases, shipyards, airfields, dockyards, railyards, Presidios, mansion houses and hotels) to city politicians and their friends. It is squandered on one government agency suing another government agency and collecting judgments against each other. It is squandered on all kinds of loans never repaid. [8]

---

[8] Your insurance, social security, retirement investment, and pension fund money is lent out by kids fresh out of the college where they went to get knowledge. Surprise. The loan can't be repaid! So the loan is "forgiven," or exchanged for equity. That sounds good. I wish Citibank would forgive the mortgage on my house, and take equity in my sour investments. To make a loan which is repaid requires someone who knows how the world works – not how it should work.

The Trouble With Administrators. Periodically, every well-fed organization, from Pharaoh's government to the local public school system, chokes to death, or near death, on administrators. How come? It is just arithmetic. One boss/administrator is enough for one worker, two workers, three workers . . . yet at some number, say ten workers, a second administrator appears.

In theory, double the number of workers will do double the work. Will double the number of administrators, administrate double the number of workers? How do administrators spend their day? Answer: Directing workers and meeting with other administrators. As the number of workers grows, so does the number of administrators. In this example, each administrator administrates ten workers, so each administrator's time directing workers is fixed. However, the time each administrator spends with other administrators grows in proportion to the number of administrators. Eventually, the administrators don't have time enough for meetings with other administrators. So the administrators require deputy administrators to help them. And then assistant deputy administrators to help the deputy administrators.

The administrators are elated, rather than alarmed, as they are elevated on the pyramid of executives growing under them. The pyramid is taken as proof of success. And it insulates ever more administrators from contact with dirty work at the bottom.

A tad of high school algebra illuminates how this part of the world works:

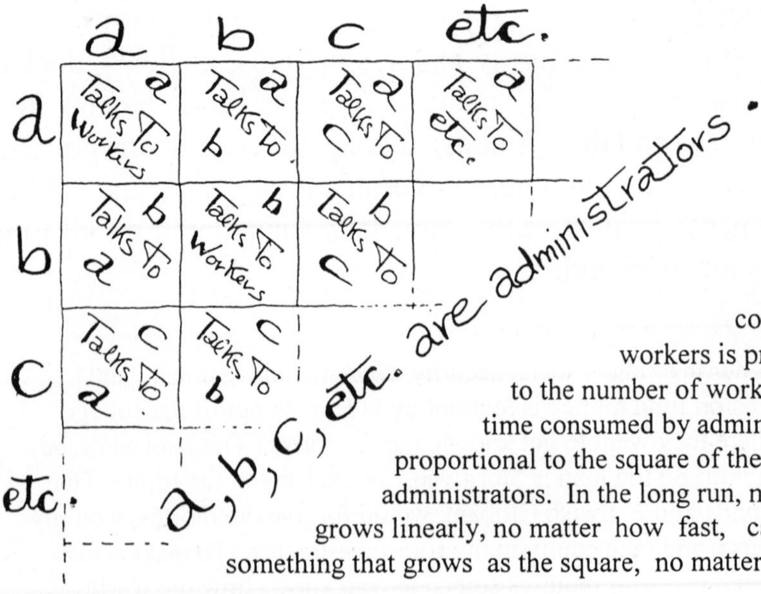

Total time consumed by workers is proportional to the number of workers. Total time consumed by administrators is proportional to the square of the number of administrators. In the long run, nothing that grows linearly, no matter how fast, can support something that grows as the square, no matter how slow.

It is squandered on welfare buying Viagra for convicted sex maniacs. It is squandered on redoing steep hill (35% grade) San Francisco sidewalks with cub cuts for wheel chair traffic.

It is squandered when railroad right of ways of astronomical value are cut up. It is squandered when a high tech bomber is flown from the central U.S.A. to central Asia and back, to blow up a mudhut that could be blown up with a World War I bomb dropped from a World War I biplane. It is squandered when cities go into the apartment house business, the sports business, or the gas and electric business, and lose their pants.

It is squandered on schools and teachers. (California public schools spent $396 per student per week in 2004). The students express their appreciation by vandalizing the schools and "dissing" the teachers. How much better the outcome could easily be if the school tax money was redirected to science museums and public libraries which are in fact appreciated. Money is squandered hiring teachers to reduce the number of students per teacher. But competent teachers are scarce, so schools are filled with incompetent teachers, who do vastly more harm than good. How much more reasonable it would be to triple the few competent teachers' pay and let them teach big classes and if necessary employ night club bouncers to eject trouble makers. It is squandered on mindless destruction and reconstruction of public buildings. It is squandered on giant social engineering programs which have ruined the inner part of almost every big U.S.A. city. It is squandered on social engineering which has encouraged the least responsible individuals to produce six children and the most responsible individuals to produce zero children. That is devastating.

It is squandered by spending money for the sake of spending. I would like a diode light for my bicycle and I will get one when my old light's bulb burns out or wears out.

But all over San Francisco the old stop and go street lights were replaced by new diode lights in eleven months. Someone had money to blow. It is squandered by replacing old toilets with new toilets which use 30% less water but must be flushed twice to flush once. It is squandered by putting stuff in auto gasoline which reduces air pollution 10%, increases fuel consumption 10%, and contaminates ground water when it leaks.
It is squandered by putting stuff in diesel fuel which ruined half the diesel truck engines in California– the state paid for repairs.

It is squandered by so burdening the Southern Pacific Railroad with regulation and taxation that the company abandoned its almost sea level line from San Francisco Bay through Niles Canyon to Livermore.[9] Then the tax money was spent to build a new railroad from the Bay to Livermore running up 900 feet over the Hayward hills and down the other side. Now forever each train will eat more energy. But that is okay because the taxpayers own the new railroad, and pay for the power to get the trains up the hill and the brakes to get them back down the other side.

It is squandered by closing the railroad tunnel between Santa Cruz and San Jose and replacing it with a freeway which runs up 1,800 feet from San Francisco Bay over the Santa Cruz mountains and back down the other side to the Pacific Ocean.
And this between Silicon Valley where the smart people work and the University of California, Santa Cruz, where the smart people teach! It is squandered by building a freeway to run fuel-guzzling freight trucks from Sacramento, California up 7,200 feet over the top of the Sierra Nevada and then back down to Reno, Nevada, rather than completing the planned railroad tunnel under the mountains; the mountains which the Sierra conservation clubs' name extol. The top is Donner Pass.

---

[9] At Livermore is the Lawrence Rad Lab's big hydrogen fusion laser.

Through it goes every big rig truck moving between central California and the Midwestern or Eastern states.

It is squandered by having a hundred cops drive around town in a hundred cop cars looking to catch the bad guys red-handed. Most crime is committed by a small number of bad guys who do it again and again. To catch the bad guys, why not set up decoy automobiles to steal and decoy houses to burglarize?

It is squandered on hundreds of fireman sleeping in firehouses all over town. When there is no fire or a small fire there are too many firemen. When there is a big fire there are never enough firemen. What is the answer? Do what the United States Army does. It has the regular standing Army backed up by the National Guard. So also the standing fire department should be backed by a Municipal Guard.

It is squandered on prisoners in jail. A prisoner should not be an expense. A prisoner should be, and historically was, a source of income. After all, the word "prisoner" comes from the words "to seize a prize".[10] It is squandered by closing the efficient Sacramento California Nuclear power reactor. Why? Because "no nukes is good nukes."[11] But no nukes = no electricity = rolling blackouts. Emergency power had to be imported from Canada and electricity prices shot to the moon. The State of California will be paying the monster bill with its taxpayers' money for a decade.

All this squandering is a consequence of the authority of grown kids who dictate how the world should work before bothering to find out how the world does work.

What is to be done about this situation? I don't know, and at this late phase of my game, I almost don't care.

---

[10] Prisoners are yet a prize to lawyer world people.

[11] Once a month, ABC, BBC, NBC, CBS, or NPR runs a radiation hazard scare-- but never, never, never any risk numbers. Sometimes even a power line or cell phone radiation scare, but never a half-word on TV transmitter tower radiation or picture tube x-rays.

I set out to describe the disease, not the cure. (That means I get the disease named after me).

I have only this thought. Some of the foolish kids who want to change the world before they know how the world works are, at heart, idealists. They desire to make the world better. Pre-high school kids maybe influenced by the following: When I was a child, I recall my father, Harry, talking to his sister, Lena. They stood in the dining room of her austere San Francisco Victorian, I lay on the floor listening. My father said: "To be a good Jew you must first be a scientist." According to my father, a "good Jew," helps the world. According to my father, a scientist studies how the world works. My father was saying you must know how the world works before you attempt to help it. Otherwise, you are like a layman attempting abdominal surgery.

A few days later, I related what my father said to a religious in-law, Sammy. Sammy said my father was wrong. Sammy believed you could be good by fervently wishing to be good. Sammy thought a person could be good without understanding how the world worked. Then I understood why Lena's husband Louie, called Sammy "Sammy the Goo Goo".[12]

---

[12] Goo Goo implied childlike. It was Louie's homily on the axiom: Bereft history, forever a child = without experience always naive, always in danger. Whence the corollary: Philosophy, bereft experience, is poison. Now Sammy is sixty years gone. During those years, steam engines and vacuum tubes and polio vanished, hearts were transplanted, hydrogen fused in a lab, mammals cloned, blackholes sensed, the far side of the moon seen, the near side walked on, a spacecraft shot out of the solar system, the big bang was heard, continents demonstrated to drift on the earth's surface, the four-color and Fermat's last theorem both proven, and Maxwell's electric theory enlarged and extended down into the subatomic world. Not a lick of this thanks to lawyer world people. And still, still Sammy's childish view of life is mighty yet.

If you liked

THINKING Physics

we have something else you will like:

"What an extraordinary, unique, and magical book! I wonder if Einstein saw his own theory so directly and simply."

R. Pisani, GALOIS-GUITON

"Relativity for right-brain thinkers, the best I have seen."

L. Sessions, ASTRONOMY MAGAZINE

"Epstein intrigues the beginner with his clarity and the expert with unexpected relationships ignored by others."

T. Page, CHOICE

Recall when you last asked:

"Why can't you go faster than light

or what makes space shrink and time go slow?" Did you ever get a good answer, something you could visualize? Like this:

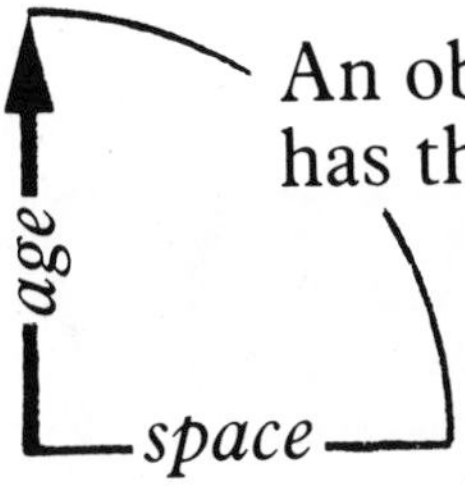

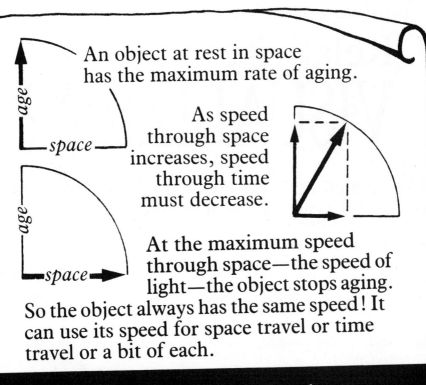

An object at rest in space has the maximum rate of aging.

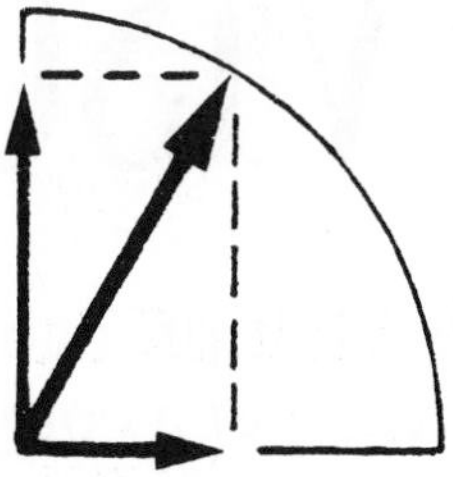

As speed through space increases, speed through time must decrease.

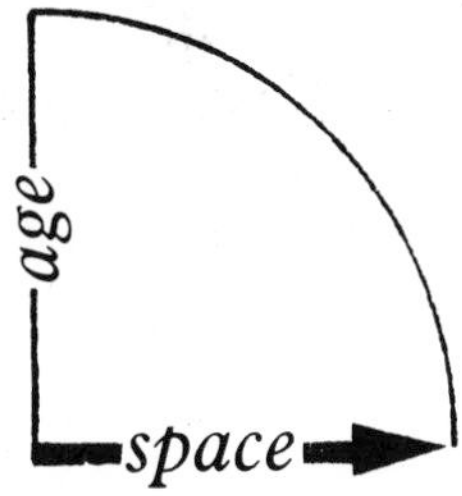

At the maximum speed through space—the speed of light—the object stops aging.

So the object always has the same speed! It can use its speed for space travel or time travel or a bit of each.

This book shows how to

master relativity via diagrams and intuition rather than algebra. If you have tried to learn relativity before, try again with ***Relativity Visualized.***

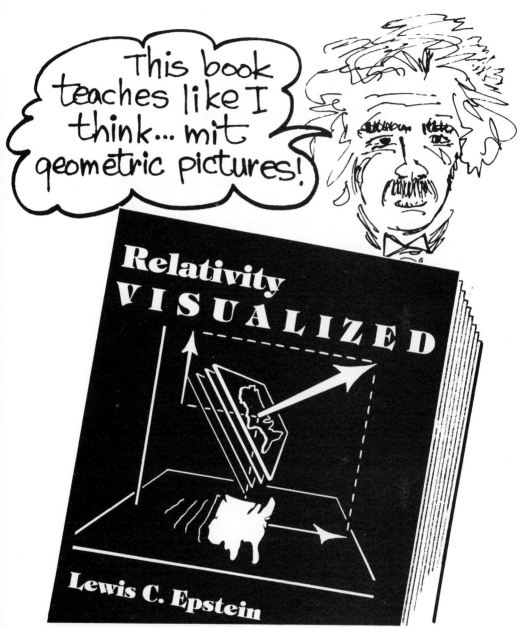

1 print	$ 6
2 prints	11
3 prints	15
4 prints	18
5 or more prints	4 each

Tax and postage are included.
All orders must be prepaid.

A. The Medieval Astronomer
11″ × 13″

B. The Great Telescope of the Lick Observatory
12″ × 18″

A.A. Lower base casting bolted firmly to stone pier.
B.B. Second base casting bolted to AA, but capable of azimuthal adjustment on same by screw *b*.
C.C. Upper base casting bolted to B, but capable of vertical adjustment on same by means of levelling screws inside frame, not visible in woodcut.
P. Polar pillar, inside which revolves the polar axis.
S. Stay or strut under polar pillar, giving direct support under the principal bearing of polar axis.
D.D. Cross head, a hollow casting bolted to polar axis, and in which the declination axis revolves. At one end of this cross head is
d.d. the declination circle 18 inches diameter, and divided on gold alloy, and
e.e. the declination clamp, into which gears
f.f. a slow motion screw, worked from eye-end by
g'.g'. a pair of bevel wheels and handle which give the slow motion in declination.
h.h. is the clamping handle in declination.
i.i. are the cords and lever for right ascension clamping.
K.K. The cords for the slow motion in right ascension.

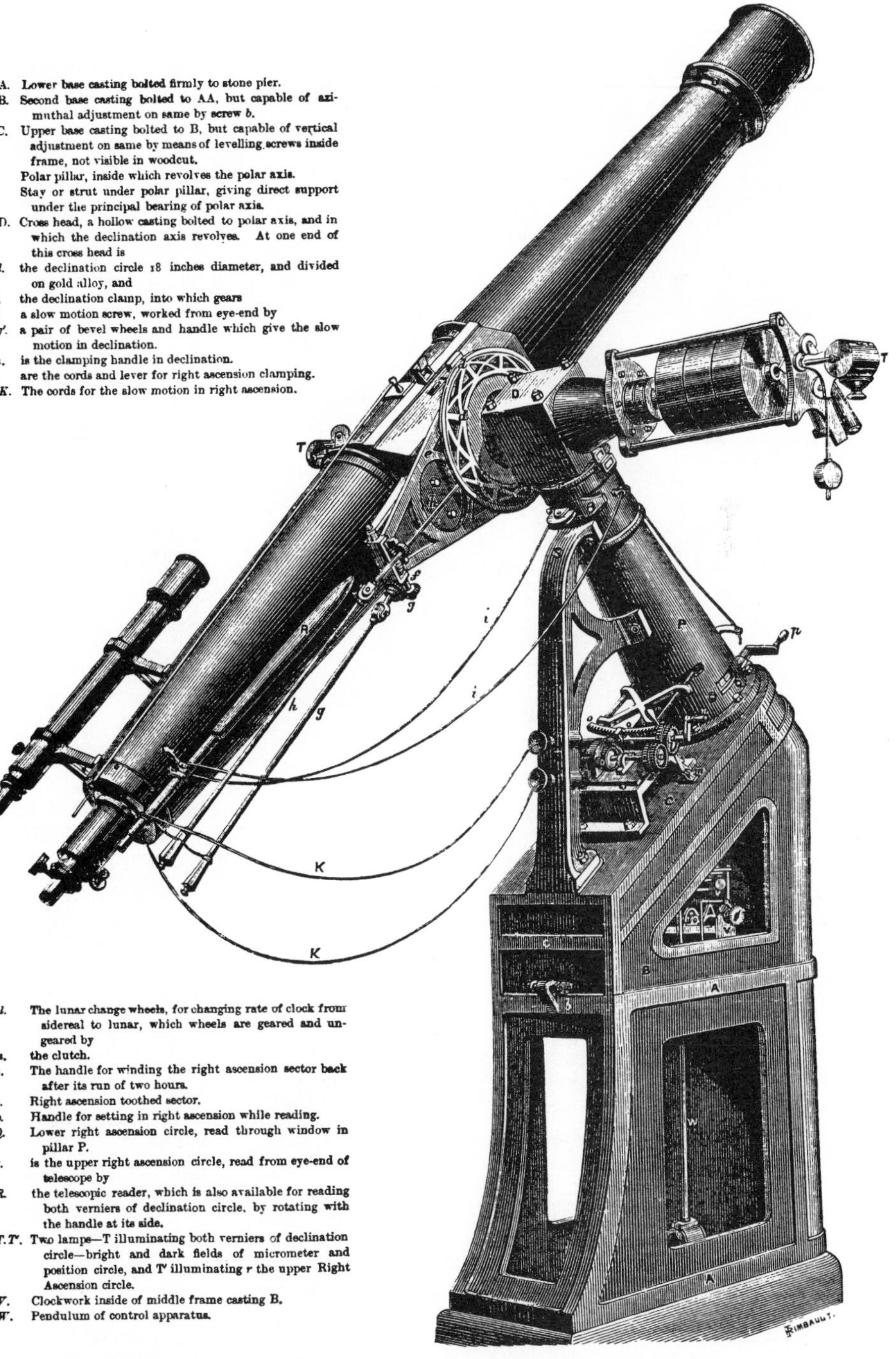

l.l. The lunar change wheels, for changing rate of clock from sidereal to lunar, which wheels are geared and ungeared by
m. the clutch.
n. The handle for winding the right ascension sector back after its run of two hours.
o. Right ascension toothed sector.
p. Handle for setting in right ascension while reading.
Q. Lower right ascension circle, read through window in pillar P.
r. is the upper right ascension circle, read from eye-end of telescope by
R. the telescopic reader, which is also available for reading both verniers of declination circle, by rotating with the handle at its side.
T.T'. Two lamps—T illuminating both verniers of declination circle—bright and dark fields of micrometer and position circle, and T' illuminating *r* the upper Right Ascension circle.
V. Clockwork inside of middle frame casting B.
W. Pendulum of control apparatus.

C. The Equatorial of the Cork Observatory

13″ × 20″

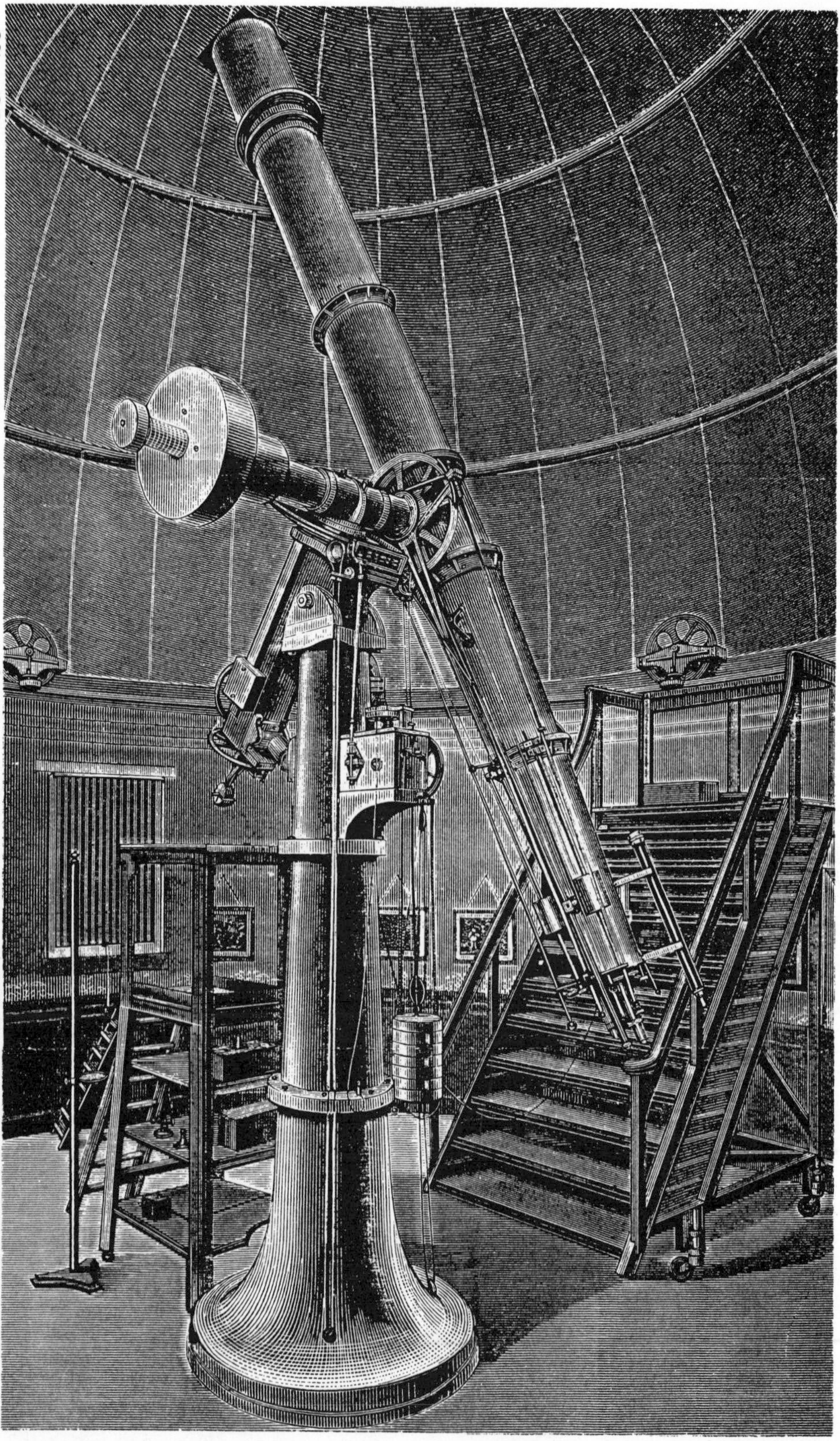

D. 15-Inch Refractor of Mr. Wigglesworth's Observatory 13″ × 20″

From time to time, the author of this book
Lewis Carroll Epstein
is available to give presentations at bookstores, schools, museums, science clubs, teachers meetings and for radio programs.

To discuss arrangements contact:

INSIGHT PRESS
614 Vermont Street
San Francisco, California
94107-2636
Tel: 415-826 - 3488
Fax: 415-821 - 2811
email: thinking@prodigy.net

38

Congresswoman Nancy Pelosi
U.S. House of Representatives
The Capitol Building
Washington D.C.

Summer 2005

Dear Congresswoman Nancy Pelosi:

During the first Clinton administration, I wrote the letter on the facing page. It was never answered. Now, to protect our old South Korean friends, we face a nuclear show down with North Korea. Yet, our S. Korean friends make zero effort to protect our intellectual property.

Look below and see pages from *Thinking Physics* presently being pirated by our S. Korean friends; "crime in progress," as the police radio dispatcher says. You may not read Korean, but you will recognize the illustrations.

Why post U.S. Army soldiers in S. Korea to protect people who rob us? You are my Congresswoman. Please represent me.

Lewis Epstein
A poor old physics teacher

욕조에 떠 있는 전함

욕조에 전함이 뜰 수 있을까?* 물론 욕조가 엄청나게 큰 것이든지 전함이 매우 작은 것이라고 상상해야 한다. 두 경우 모두 배의 아래쪽에는 물이 둘러싸고 있어야 한다. 예컨대 전함의 무게는 100톤(매우 작은 배)이고 욕조에 담긴 물의 무게는 100 kg이라고 하자. 배는 뜰까 아니면 바닥에 닿을까?

a) 배 주변에 충분한 물이 있다면 배는 뜬다
b) 배의 무게가 물의 무게보다 무거우므로 바닥에 닿는다

* 이것은 우리 아버지가 좋아하는 물리학 문제였다. —L. 엡스타인

165

정지 거리

매시 10 km의 속도로 달리는 차가 있다. 운전자가 브레이크를 밟은 후 차는 3 m를 더 앞으로 나갔다. 똑같은 차가 매시 20 km의 속도로 달리다가 브레이크를 밟으면 얼마나 더 앞으로 나갈까?

a) 3 m
b) 6 m
c) 9 m
d) 12 m
e) 15 m

70